Chagas Disease: Celebrating the 115th Anniversary of the Discovery of *Trypanosoma cruzi*

Chagas Disease: Celebrating the 115th Anniversary of the Discovery of *Trypanosoma cruzi*

Guest Editor

Michel Tibayrenc

MDPI

Basel • Beijing • Wuhan • Barcelona • Belgrade • Novi Sad • Cluj • Manchester

Guest Editor
Michel Tibayrenc
MIVEGEC (IRD 224-CNRS
5290-UM1-UM2
Institute of Research for the
Development
Montpellier
France

Editorial Office
MDPI AG
Grosspeteranlage 5
4052 Basel, Switzerland

This is a reprint of the Special Issue, published open access by the journal *Pathogens* (ISSN 2076-0817), freely accessible at: https://www.mdpi.com/journal/pathogens/special_issues/5Q5CV68TOS.

For citation purposes, cite each article independently as indicated on the article page online and as indicated below:

Lastname, A.A.; Lastname, B.B. Article Title. *Journal Name* **Year**, *Volume Number*, Page Range.

ISBN 978-3-7258-5883-5 (Hbk)
ISBN 978-3-7258-5884-2 (PDF)
https://doi.org/10.3390/books978-3-7258-5884-2

Contents

About the Editor

Michel Tibayrenc

Michel Tibayrenc, Institut de recherche pour le développement (France), has worked on the genetics and evolution of infectious diseases for 45 years. He is a director of research emeritus at the French Institut de Recherche pour le Développement (IRD), founder and editor-in-chief emeritus of *Infection, Genetics and Evolution* (Elsevier), editor-in-chief of the new journal *Disease Biology, Genetics and Socioecology* (DBGS; Melbourne, Australia) and founder and principal organizer from 1996 until 2021 of the international congresses MEEGID (molecular epidemiology and evolutionary genetics of infectious diseases), supported and managed by Elsevier. He is the author of more than 200 international papers (H index = 50). He has published 6 scientific books: (1) American trypanosomiasis: Chagas disease. 100 years of discovery (Elsevier; first editor: Jenny Telleria; 2010; reedited 2017); (2) Genetics and evolution of infectious diseases (Elsevier; 2010; reedited 2017 and 2024); (3) On Human Nature (Elsevier/Academic press; 2016; coeditor: Francisco J. Ayala; 2nd edition planned 2026); (4) What makes us humans (2020; Nova science publishing; coeditor: Francisco J. Ayala).; (5) Spanish edition: Lo que nos hace humanos (SAL TERRAE; coauthor: Francisco J. Ayala); (6) French edition: Notre humaine nature (éditions rue de Seine, Paris; coauthor: Francisco J. Ayala). With his collaborator Jenny Telleria, he is the founder, scientific advisor and honorary member of the Bolivian Society of Human Genetics.

pathogens

MDPI

Article

Genetic Identification of Brazilian Mammalian Hosts of *Trypanosoma cruzi*: Improving Blood Meal Source Discrimination in Vector-Borne Transmission

Quezia Moura Oliveira [1,2,†], Thaíla Santos Pessanha [1,†] and Alena Mayo Iñiguez [1,*]

1 Laboratório de Parasitologia Integrativa e Paleoparasitologia—LPIP, Instituto Oswaldo Cruz, Fiocruz, Avenida Brasil 4365, Rio de Janeiro 21040-360, RJ, Brazil; queziamouraeea@gmail.com (Q.M.O.); thailasantospessanha@gmail.com (T.S.P.)

2 Programa de Pós-Graduação em Biologia Parasitária, Instituto Oswaldo Cruz, Fiocruz, Avenida Brasil 4365, Rio de Janeiro 21040-360, RJ, Brazil

* Correspondence: alena@ioc.fiocruz.br or alenainiguez@gmail.com

† These authors contributed equally to this work.

Abstract: The detection of food sources of blood-sucking vectors is essential for a better understanding of the hosts, reservoirs, and other fauna that participate in the transmission web of hemoparasites. The molecular identification of triatomine blood meal sources (BMSs) has been shown to be highly sensitive and taxonomically specific when compared to the immunological method. The application of molecular cloning makes it possible to identify multiple BMS species and/or different individuals/haplotypes of the same vertebrate species in a single triatomine specimen. In Brazil, the molecular detection of BMSs is incipient, with insufficient genetic information on the species of animals involved in the transmission of *Trypanosoma cruzi*. In this work, we evaluated the sensitivity and specificity of a molecular approach using molecular cloning for the detection of multiple Brazilian mammalian species. The DNA was extracted from blood clots of 13 species of canids, bats, xenarthral, marsupials, and rodents. Serial proportions were used to formulate mixtures combining taxonomically close (belonging to the same family or order) and taxonomically distant (different families) species. The results showed that GenBank lacks reference sequences for some native species tested, such as the sylvatic rodent, *Necromys lasiurus*, and the wild canid, *Lycalopex gymnocercus*, for *cyt*b and 12S rDNA, and the rodent *Oecomys cleberi* for 12S rDNA. The study also demonstrated that it is possible to detect multiple different species, even for those that are taxonomically close. This approach was proven to be efficient for the detection of species in equal and even in disparate unequal proportions, which could represent complementary information about the diversity of potential hosts of *T. cruzi*. The detection of multiple BMS species in mixed samples provides a more comprehensive and accurate landscape of *T. cruzi* transmission in nature.

Keywords: blood clot; Brazilian fauna; molecular cloning

1. Introduction

American trypanosomiasis (AT), an infection caused by the *protozoan Trypanosoma cruzi* Chagas, 1909 (Kinetoplastida, Trypanosomatidae), is primarily a wild enzootic that is widely distributed in the mammalian fauna of Brazilian biomes [1]. In humans, this infection can cause Chagas disease (CD), which was discovered in 1909 by the Brazilian physician Carlos Chagas and is still considered today a neglected disease that can result in a debilitating and often fatal chronic condition [2]. It is estimated that about eight million

https://doi.org/10.3390/pathogens14060579

people are infected with CD in Latin America, and more than seventy million are at risk of contracting the infection [3]. In Brazil, the prevalence of CD varies from 1.0 to 2.4%, equivalent to 1.9 to 4.6 million people infected with *T. cruzi* [4].

Trypanosoma cruzi can potentially infect all mammalian species, which play significant roles in the transmission cycle [5]. In Brazil, although more than 700 mammalian species have been described based on morphological data, genetic information is only available for a limited number [6]. This limitation hinders genetic studies for species identification, as many of them lack curated reference data. Certain mammals are frequently infected by *T. cruzi*, have been recognized as reservoirs of the parasite [5], and serve as food sources for the triatomine vector.

Brazilian marsupial opossums (Didelphimorphia) are classic reservoirs of *T. cruzi* and other trypanosomatids in the wild [7]. *Trypanosoma cruzi* was isolated and detected by PCR in the species *Didelphis albiventris* (Lund, 1840) and *Monodelphis domestica* (Wagner, 1842) found in the Caatinga region [8,9]. Infections in opossums are well-documented, and their parasite–host interactions have been extensively studied [7,8,10]. Bats (Chiroptera) have a long evolutionary history with trypanosomatids, and it has been postulated that they are the ancestral hosts responsible for introducing trypanosomatid species into the New World [11]. *Trypanosoma cruzi* infection has been reported in more than 20 species across all Brazilian biomes [8,12–15]. Rodents (Rodentia) are important hosts of *T. cruzi*, and some species represent the main protagonists in the transmission cycle in Northeast Brazil [16]. Synanthropic and wild rodents have been found to be infected in all Brazilian biomes of the country [8,9,17]. Wild and domestic canids (Carnivora) are commonly found infected, despite domestic dogs maintaining low parasitemia in Brazil [18]. Armadillos (Cingulata) are classic and ancient hosts, and their role as a wild reservoir was first defined in the discovery of CD by Dr. Carlos Chagas [2]. All these orders of mammals have been indicated as blood meal sources (BMSs) for *T. cruzi* vectors [16,19–23].

Siqueira (1960), using the immunological technique called precipitin, was the first researcher to demonstrate that it is possible to determine the BMSs of triatomines [24]. The precipitin test and Enzyme-Linked Immunosorbent Assay (ELISA) are based on an antigen–antibody reaction using sera from known mammals, birds, and reptiles, resulting in the formation of a precipitate when combined [24]. However, it is important to note that both tests have taxonomic limitations, providing results only up to the order or family level [25,26]. This leads to low sensitivity and specificity, failing to reflect the actual diversity of BMSs in hematophagous vectors. On the other hand, the molecular identification of BMSs proved to be highly sensitive and taxonomically specific compared to immunological methods. In addition, molecular cloning allows the identification of multiple BMS species and/or different individuals/haplotypes of the same species in a single triatomine specimen [6–8].

Studies of the BMSs of hematophagous insects have been applied in epidemiological surveys of infectious diseases in order to better understand the ecology of vectors and hosts, reservoirs, and other non-host environmental fauna that participate in the transmission of hemoparasites.

Recent molecular studies have improved our understanding of host diversity in the *T. cruzi* transmission network [16,22,27–29]. Using this approach, species of mammals, birds, and reptiles have been suggested to play different roles in the transmission networks of *T. cruzi* in populations of the *Triatoma brasiliensis* complex from the Brazilian semi-arid region, State of Rio Grande do Norte [16,30]. Almeida and collaborators (2016) [28] suggest that rodents are the main BMSs, with distinct roles, such as *Kerodon rupestris* (Wied-Neuwied, 1820) in nature and *Galea spixii* (Wagler, 1831), potentially connecting *T. cruzi* networks in the wild and peridomestic ecotopes. It is known that different profiles of BMSs are consistent with local wild and domestic fauna. Furthermore, variations in blood meal preferences in some vector species, even in sympatry, were suggested [28].

Despite recent studies on the BMSs of triatomines, the deficiency of genetic information available in GenBank for wild mammals involved in *T. cruzi* transmission has been reported [28,31]. This absence of genetic data was observed in the study by Almeida and collaborators (2016) [28], in which the identification of some BMSs was taxonomically limited to the genera *Proceratophrys* and *Mabuya*. The difficulty in the taxonomic identification of species was also observed by Valença-Barbosa and collaborators (2022), showing the family Mustelidae as BMSs [31]. The cut-off for detecting BMSs at the species level was only considered when sequences presented a coverage higher than 97% and genetic identity above 97% and in the 75–97% range for genus/family identification [28].

Considering that BMS studies provide information on a wide diversity of species based on PCR and Sanger sequencing methodologies, the discrimination of multiple BMSs is exclusively obtained by methods such as molecular cloning or next-generation sequencing. From this, the following questions arise: (1) Is there bias in identifying the BMSs with higher proportions? (2) Are the techniques based on clone sequencing efficient at revealing secondary sources with equal or minor proportions? (3) Is the fauna of endemic mammals in Brazil recognized as hosts of *T. cruzi* represented in GenBank? The current work seeks to answer basic methodological questions to enrich the genetic identification of BMSs in the country.

In the current study, we evaluated the sensitivity and specificity of DNA barcoding and molecular cloning for detecting multiple Brazilian mammalian species that are close and distant taxonomically in individual samples. We hypothesized that molecular cloning is effective at discriminating mammal species, even at equal proportions or with a very low secondary representation in the sample.

2. Materials and Methods

2.1. Blood Clot Samples to Evaluate the Specificity and Sensitivity of the BMS Molecular Approach

To evaluate the specificity and sensitivity of the BMS molecular targets, blood clots from 14 species of mammals involved in the transmission of *T. cruzi* DNA were selected (Table 1). The dataset includes species from five orders—*Didelphis albiventris*, *Didelphis aurita* (Wied-Neuwied, 1826), *Monodelphis domestica*, *Marmosa demerarae* (O. Thomas, 1905) (Didelphimorphia), *Canis lupus familiaris* (Linnaeus, 1758), *Cerdocyon thous* (Linnaeus, 1766), *Lycalopex gymnocercus* (G. Fischer, 1814) (Carnivora), *Oecomys cleberi* (Locks, 1981), *Necromys lasiurus* (Lund, 1841), *Rattus rattus* (Linnaeus, 1758 (Rodentia), *Desmondus rotundus* (E. Geoffroy, 1810), *Sturnira lilium* (É. Geoffroy, 1810), *Carollia perspicillata* (Linnaeus, 1758) (Chiroptera), and *Dasypus novemcinctus* (Linnaeus, 1758) (Cingulata)—from 3 Brazilian biomes, including the Atlantic Forest, Pampa, and Cerrado. Mammal identification and blood clot collection procedures were conducted by veterinary professionals as part of previously performed fieldwork [32]. Blood clot collection was authorized by the Brazilian Institute of the Environment and Renewable Natural Resources (IBAMA) license no. 19037–1 and no. 10070–2.

Table 1. Data regarding the samples collected from the Brazilian mammalian species for the present study.

Mammalian Species	Common Name in Brazil	Collection Site [1]	Brazilian Biome
Canis lupus familiaris	Domestic dog	Mangaratiba-RJ	Atlantic Forest
Cerdocyon thous	Bush dog	Mamamguape-PB	Atlantic Forest
Lycalopex gymnocercus	Field fox	Alegrete-RS	Pampa
Oecomys cleberi	Tree mouse	Cumari-GO	Cerrado
Necromys lasiurus	Pixuna	Mamamguape-PB	Atlantic Forest
Rattus rattus	Rat	Paraty-RJ	Atlantic Forest
Desmodus rotundus	Vampire bat	Mamamguape-PB	Atlantic Forest
Sturnira lilium	Fruit bat	Mamamguape-PB	Atlantic Forest
Carollia perspicillata	Short-tail bat	Mamamguape-PB	Atlantic Forest
Didelphis albiventris	White-ear possum	Mamamguape-PB	Atlantic Forest
Monodelphis domestica	Short-tailed cuica	Mamamguape-PB	Atlantic Forest
Marmosa (*Micoureus*) *demerarae*	Cuica	Mamamguape-PB	Atlantic Forest
Didelphis aurita	Black-ear possum	Mamamguape-PB	Atlantic Forest
Dasypus novemcinctus	Armadillo	Mamamguape-PB	Atlantic Forest

[1] Brazilian States: RJ—Rio de Janeiro, PB—Paraíba, RS—Rio Grande do Sul, GO—Goiás.

2.2. DNA Extraction from Blood Clots

The flow of the methodological procedures is described in Figure 1. Blood clots were collected and stored in ethanol. DNA was extracted based on the ammonium acetate precipitation protocol [14]. The absolute ethanol was removed, and then the sample was centrifuged at 17,900× *g* for 10 min in buffer (38 mM NaCl, 10 mM EDTA, 5 mM Tris-Cl) to discard any residual ethanol. The supernatant was removed, and the pellet was resuspended in 200 µL of Digsol buffer (120 mM NaCl, 20 mM EDTA, 50 mM Tris-Cl, 1% SDS) and 20 µL of proteinase K at 20 mg/mL. The reaction was incubated in a thermo-shaker at 55 °C for 3 h. After the incubation, 400 µL of 4 M ammonium acetate was added. The DNA was resuspended in 25 µL of buffer (10 mM Tris-HCl pH 7.4; 1 mM EDTA, pH 8.0) and stored at −20 °C until molecular characterization.

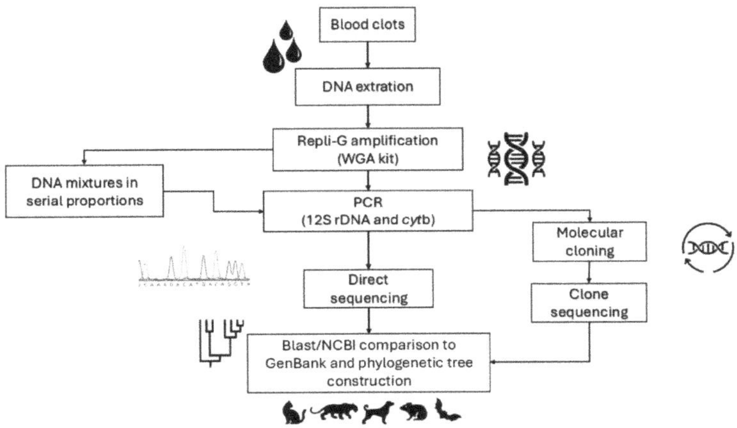

Figure 1. Diagram illustrating the flow of the methodological procedures applied in this study for the use of blood clots from different mammals through vertebrate BMS barcoding using both direct sequencing and clone sequencing approaches.

2.3. Molecular Identification of Mammalian Species and *T. cruzi* Infection

The detection of mammalian species was based on the DNA barcoding approach with the application of the mitochondrial DNA targets *cyt*b (358 bp) and 12S rDNA (215 bp) [33,34]. The PCR method was applied, with a final volume of 25 µL, including High Fidelity Buffer [600 mM Tris-SO_4 (pH 8.9), 180 mM (NH_4) $2SO_4$], 2.5 mM $MgSO_4$, 2 mM each dNTP, 2.0 U of Platinum® *Taq* DNA Polymerase High Fidelity (Invitrogen, Paisley, Scotland), and 40 ng of DNA. To check if these blood clots were molecular positive for *T. cruzi* infection and their DTUs, PCR assays were performed based on the 18S rDNA and *cyt*b molecular targets (200 pb) [35,36]. Positive (PCRs using *Mus musculus* DNA) [37] and negative (PCRs without DNA) PCR controls were always included.

PCR products were subjected to agarose (2%) gel electrophoresis, stained with GelRed (Biotium, Inc., Fremont, CA, USA), and visualized under UV light. Amplicons were purified using the Gel Band Purification Kit (GE Healthcare, Chicago, IL, USA), according to the manufacturer's instructions. DNA direct sequencing was conducted using the Big Dye Terminator v 3.1 Cycle Sequencing Kit (Applied Biosystems, Foster City, CA, USA) and the ABI 3730 sequencer (Applied Biosystems, Waltham, MA, USA) at the RPT01A/FIOCRUZ sequencing facility. Sequences were deposited in GenBank (accession numbers: PV023314-PV023326).

Molecular cloning was performed using the pGEM®-T Easy Vector Systems protocol (Promega, Madison, WI, USA). Cloning products were sequenced as described above (Figure 1). All sequences obtained were submitted for editing, alignment, and visualization using Lasergene SeqMan^{TM} v.7.0 (DNASTAR, Madison, WI, USA) and Bioedit v.7.0.5 (Department of Microbiology, North Carolina State University, Raleigh, NC, USA) [38,39]. BLAS TNsearches were performed at NCBI (http://blast.ncbi.nlm.nih.gov/Blast.cgi (accessed on 31 May 2025)) to identify mammalian species using the cut-off values of 97% query coverage and 98% genetic identity. Phylogenetic trees were constructed using the MEGA v.11 software package, utilizing the Neighbor-Joining (NJ) and Maximum Likelihood (ML) methods. Statistical support for the branches was generated by 500 bootstrap replicates.

3. Results

3.1. Evaluation of the Specificity of Mammalian Species Identification Using BMS Molecular Targets and T. cruzi Infection

All blood clots were negative for *T. cruzi* infection and positive for the two BMS molecular targets *cyt*b and 12S rDNA (Figure S1). Regarding the molecular identification of the mammalian species, it was observed that there are no sequences available in GenBank for the wild canid field fox *Lycalopex gymnocercus* or the wild mouse *Oecomys cleberi* for the two molecular targets *cyt*b and 12S rDNA (Table 2). In addition, *Necromys lasiurus* (Pixuna) did not have 12S rDNA sequences available in GenBank (Table 2). The 12S rDNA and *cyt*b markers demonstrated specificity in species discrimination in all the orders except for didelphids (Table 2). Marsupials of the genus *Didelphis* showed high intra-genus genetic proximity, showed no 12S rDNA genetic distance between *D. aurita* and *D. marsupialis*, and 1.84% genetic distance between *D. aurita* and *D. albiventris*, the species selected for this study (see below). Therefore, we decided not to perform *Didelphis* intra-genus mixing.

The 12S rDNA *D. aurita* sequence revealed a maximum genetic identity of 99.08% with the only two *D. aurita* sequences available in the database and, importantly, with the *D. marsupialis* sequence (NC057518). In addition, the *D. aurita* sequence obtained showed high genetic identity with *D. albiventris* at 98.16%. For the *cyt*b marker, the *D. aurita* sequence obtained displayed a maximum genetic identity of 99.40% for the available *D. aurita* sequences. There are fourteen *cyt*b sequences in the database for this species. *Didelphis aurita* showed 98.50% genetic identity with *D. marsupialis* and 96.40% with *D. albiventris*. Species that are not distributed in Brazil but showed a high genetic similarity with *D. aurita* in this study were *D. pernigra* (NC057519) at 97% and *D. virginiana* (NC001610) at 95.09%.

The 12S rDNA sequences generated from *D. albiventris* in this study revealed a maximum genetic identity of 99.36% with *D. albiventris* from GenBank (Table 2). There are only two sequences deposited in the database for this marker. *Didelphis albiventris* also showed high genetic identity with three other species of the same genus: *D. imperfecta* (NC057517) at 98.73% and both *D. marsupialis* and *D. aurita* at 98.09%. For the *cyt*b target, the *D. albiventris* sequence had 99.85% maximum identity with the available *D. albiventris* data. *Didelphis albiventris* also exhibited a high genetic identity with three other species of the same genus: *D. perniga* with 96.26% genetic identity and *D. marsupialis* and *D. aurita* both with 95% genetic identity.

The *Marmosa demerarae* 12S rDNA sequence obtained presented a genetic identity of 99.05% with *M. demerarae*. It also showed 97.14% identity with *M. murina* (KX381772). For the *cyt*b target, the *M. demerarae* sequence had a higher genetic identity of 98.80% with *M. demerarae*. Other species of the same genus were also identified: *M. murina* with 97.90%, *M. limae* (MN978639) with 97.31%, and *M. constantiae* (MK496216) with 96.47%. The 12S rDNA sequences generated by the *Monodelphis domestica* pointed to a maximum genetic identity of 97.36% with the available *M. domestica* sequences. For the *cyt*b target, the *M. domestica* sequence had a higher genetic identity of 100% with *M. domestica* (Table 2).

Table 2. Results of the genetic analysis of mammalian species detection using the 12S rDNA and *cyt*b rDNA BMS molecular markers.

Mammal Species	Blast/NCBI Genetic Identity Result	Maximum Genetic Identity	GenBank Accession Number	Blast/NCBI Genetic Identity Result	Maximum Genetic Identity	GenBank Accession Number
	Molecular Target 12S rDNA			**Molecular Target *cytb***		
Canis lupus familiaris	*Canis lupus familiaris*	100%	KU291093	*Canis lupus familiaris*	99%	DQ309764
Lycalopex gymnocercus	*Lycalopex sechurae*	98.63%	KT228284	*Lycalopex sechurae*	99.06%	KT447686
Cerdocyon thous	*Lycalopex sechurae*	98.00%	KT448284	ND	ND	ND
Oecomys cleberi	*Oecomys bicolor*	98.92%	JF693852.1	*Oecomys cleberi*	98.88%	KR190443
Necromys lasiurus	*Akodon montensis*	97.70%	KF769456	*Bolomys lasiurus*	98.57%	KR190443
Rattus rattus	*Rattus rattus*	100%	OM574970.1	*Rattus rattus*	99.00%	KT232247
Desmodus rotundus	*Desmodus rotundus*	97.00%	HG003310	*Desmodus rotundus*	99.00%	HG003310
Sturnira lilium	*Sturnira lilium*	100%	KC747657	*Sturnira lilium*	100%	KC747657
Carollia perspicillata	*Carollia perspicillata*	100%	HG003309	*Carollia perspicillata*	100%	FJ589715.1
Didelphis albiventris	*Didelphis albiventris*	99.36%	NC057515.1	*Didelphis albiventris*	99.85%	AF089802
Monodelphis domestica	*Monodelphis domestica*	99.53%	AJ508398	*Monodelphis domestica*	100%	KM071418
Marmosa (Micoureus) demerarae	*Marmosa (Micoureus) demerarae*	99.05%	AJ628384.1	*Marmosa (Micoureus) demerarae*	98.80%	AJ487006
Didelphis aurita	*Didelphis aurita*	99.08%	NC057518.1	*Didelphis aurita*	99.40%	GU112886
Dasypus novemcinctus	*Dasypus yesepi*	100%	NC028570	*Dasypus novemcinctus*	99.00%	KU253494

ND: The analysis was not performed.

The 12S rDNA sequence of the *Lycalopex gymnocercus* canid species showed a maximum genetic identity of 98.63% with *Lycalopex sechurae* and with other canid genera: 97.26% genetic identity with *Speothos venaticus* (NC053974), 96.80% with *Chrysocyon brachyurus* (KJ508409), and 95.89% with *Canis lupus familiaris*. The *cyt*b sequence of *L. gymnocercus* also displayed a maximum genetic identity of 99.06% with *L. sechurae*, and with canids *Speothos venaticus* at 97.64%, and *C. l. familiaris* at 95.17%. One wild canid sample selected from a cloth bank as *Cerdoncyon thous* was identified as *L. sechura* based on 12S rDNA (Table 2). Therefore, a misidentification of the blood clot, probably with *L. gymnocercus*, was presumed. The sample was omitted from the *cyt*b analysis and molecular cloning procedure. The *C. l. familiaris* sequence of the 12S rDNA target demonstrated the highest genetic identity of 100% with at least one thousand sequences of the species.

The *Necromys lasiurus* rodent sequence of the 12S rDNA marker showed a maximum genetic identity of 97.70% with *Akodon montensis*. There are no sequences available in the database for the *Necromys* genus. *Neotoma floridana* (AF294340) species, from a different genus, was also identified with 95.79% genetic identity. For the *N. lasiurus cyt*b target, the maximum genetic identity was 98.57% with the species *Bolomys lasiurus* and 95.14% with *B. temchuki* (AY273913) and *Oligoryzomys fornesi* (EU192158). The 12S rDNA *Rattus rattus* sequence presented 100% genetic identity with *R. rattus* sequences. The species *Sundamys infraluteus* (NC036728) was also identified with 98.17% genetic identity. The *cyt*b sequence had the highest genetic identity of 99% with the first one hundred *R. rattus* sequences deposited in the database. The *Oecomys cleberi* rodent 12S rDNA sequence obtained a maximum genetic identity of 98.92% with *O. bicolor*. There are no sequences deposited for this marker in the database for *O. cleberi*. Other species were also identified: *Oecomys auyantepui* (KX381574) with 97.85% genetic identity and *Hapalomys delacouri* (MZ159976), *Euryoryzomys macconnelli* (KX381426), and *Acomys cahirinus* (JN571145) all with 96.76% genetic identity. For the *cyt*b marker, *O. cebleri* had the highest genetic identity of 99.16% with *O. cleberi* and 98.88% with *O. bicolor*.

The 12S rDNA sequence of the chiropteran *D. rotundus* pointed to a maximum genetic identity of 97% with *D. rotundus* and 90.60% with another bat species, *Choeronycteris mexicana*. For the *cyt*b target, a sequence obtained from *D. rotundus* revealed the highest genetic identity of 99% with the available *D. rotundus* sequences, and the second highest identification with *Diaemus youngii* (FJ155475), with 89.05% genetic identity. The 12S rDNA sequence of the chiropteran *Sturnira lilium* showed the maximum identity of 100% with *S. lilium*, 99.17% with *S. magna* (AY395845), and 98.60% with *S. tildae* (NC022427). For *cyt*b, the *S. lilium* sequence presented 100% genetic identity with *S. lilium* and 99.3% with *S. tildae* (NC022427). The 12S rDNA sequence of *Carollia perspicillata* showed maximum identity with *C. perspicillata*, 99.07% with *C. bravicauda* (NC066073), 98.14% with *C. costanea* (NC065677), and 96.74% with *Micronycteris nicefori* (AY395830). The *C. perspicillata cyt*b sequence had 100% genetic identity with *C. perspicillata* for the first one hundred sequences available for this genetic marker.

The 12S rDNA sequence of the xenarthral *Dasypus novemcinctus* had the highest genetic identity of 100% with *D. yepesi*; 99.53% with the three species *D. novemcinctus*, *D. sabanicola* (NC028568), and *D. bellus* (MW963147); and 99.07% with the three species *D. septemcinctus* (KT818546), *D. hybridus* (NC028565), and *D. guianensis* (PP916004). There are only nine sequences available in GenBank for this marker. For the *cyt*b target, the sequence of *D. novemcinctus* had a maximum identity of 99% with *D. novemcinctus*, 98.73% with *D. yesepi* (NC028570), and 92.72% with *D. pilosus* (NC028567).

NJ and ML phylogenetic trees for each order were constructed using the sequences obtained from the 14 species and their best-scoring sequences in GenBank. Both reconstructions showed the same topology. The ML phylogeny that includes the Didelphimorphia species from the present study showed three genus-specific clusters: *Didelphis* spp., *Monodelphis* spp., and *Marmosa* spp. (Figure 2A). The sequences obtained from *Monodelphis domestica* (CO12) and *Marmosa demerarea* (CO13) were grouped into each genus cluster and were closely related to their corresponding GenBank species reference with high bootstrap values (Figure 2A). However, the *Didelphis* spp. cluster was moderately supported, and the sequences from *D. aurita* (CO01) and *Didelphis albiventris* (CO11) were poorly defined (Figure 2A), confirming our option of not conducting serial mixtures between these species.

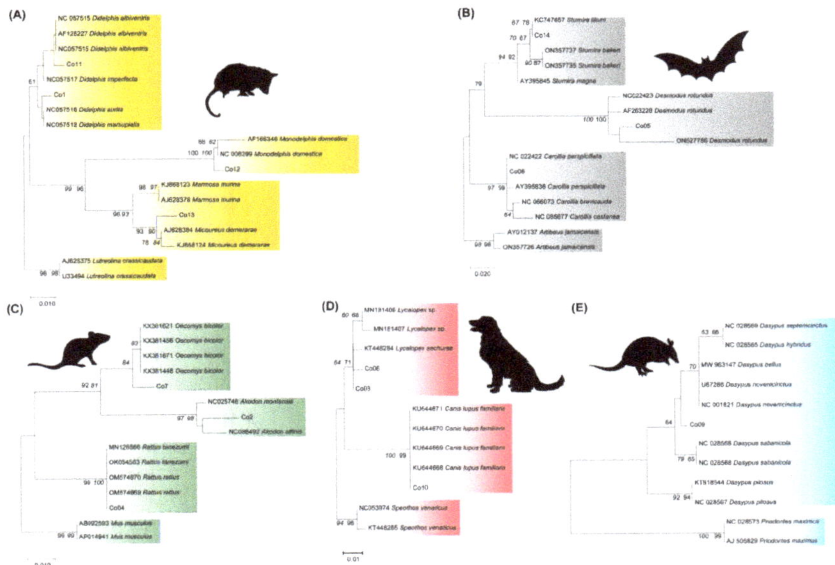

Figure 2. ML phylogenetic tree for the 12S rDNA target from the blood clots recovered from Brazilian mammals. (**A**) Opossums: CO01 = *Didelphis aurita;* CO11 = *Didelphis albiventris*, CO12 = *Monodelphis domestica*, CO13 = *Micoureus demerarea.* (**B**) Chiropterans: CO05 = *Desmodus rotundus*, CO06 = *Carollia perspicillata*, CO14 = *Sturnira lilium* (**C**) Canids: CO 03 and 08 = *Lycalopex schurae,* CO10 = *Canis lupus familiaris.* (**D**) Armadillo: CO09 = *Dasypus novemcinctus.* (**E**) Rodents: CO02 = *Akodon montensis*, CO04 = *Rattus rattus*, CO07 = *Oecomys bicolor.* The tree was constructed in MEGA v.11 software using ML and NJ methods and the Kimura 2-parameter model plus the gamma distribution with 500 bootstrap replicates. Reference sequences are indicated by species name and the GenBank numbers. The species used as outgroups were (**A**) the opossum *Lutreolina crassicaudata* (Desmarest, 1804); (**B**) chiropteran *Artibeus jamaicensis* Leach, 1821; (**C**) rodent *Mus musculus* (Linnaeus, 1758); (**D**) canids *Speothos venaticus* (Lund, 1842); and (**E**) armadillo *Priodontes maximus* (Kerr, 1792). Bootstrap values are shown on the internal nodes as ML = regular font and NJ = italic font.

The Chiropteran phylogenetic tree demonstrated that the *D. rotundus* (CO05), *C. perspicillata* (CO06), and *S. lilium* (CO14) sequences were located in each corresponding genus clade and were closely related to the GenBank species reference with high bootstrap values (Figure 2B).

The Rodentia phylogenetic reconstructions showed *R. rattus* (CO04) grouping in a species cluster (Figure 2C). However, *Oecomys cleberi* (CO07) was located in an *O. bicolor* species cluster, and *Necromys lasiurus* (CO02) grouped into an *Akodon* spp. clade (Figure 2C), reflecting the absence of these native species in the GenBank dataset, which limits accurate species-level mammalian identification and could consequently produce an incomplete and mistaken interpretation of the scenario of host–parasite transmission.

The ML 12S rDNA tree from canids (Figure 2D) showed two main well-defined clusters, one from *Canis lupus familiaris*, which includes the CO10 sequence, and a second one from the *Lycalopex* genus (Figure 2C). With the absence of a GenBank deposit of *L. gymnocercus* (CO08), the species sample (CO8) grouped into the *Lycalopex* genus clustered with the *L. sechura* sequence. The canid cloth sample named *Cerdoncyon thous* (CO3) clustered with *L. sechura*. Therefore, the misidentification of the blood clot sample with *L. gymnocercus* was confirmed. Finally, the Armadillo 12S rDNA phylogenetic analysis revealed the *Dasypus novemcinctus* (CO09) sequence near a cluster that included *D novemcinctus* reference sequences (Figure 2E).

3.2. Sensitivity Assessment of Detecting Multiple Blood Clot DNAs

Clone sequencing throughout molecular cloning demonstrated the detection of all 13 species used in the 13 multiple mixtures in equal proportions (5/5), including those that are close and those that are unrelated taxonomically (Figure 3). It was possible to detect the species in a lower concentration in 17/26 multiple mixtures of mammalian species in disparate proportions (Figure 3B,C). In the proportions of 1/9 and 9/1, the species in the lower proportion were detected in 9/13 mixtures and 8/13 mixtures, respectively. All 13 species with a higher proportion were detected in the mixed samples, as expected. A total of 9/26 species with lower proportions in multiple mixtures were not detected (Figure 3B,C).

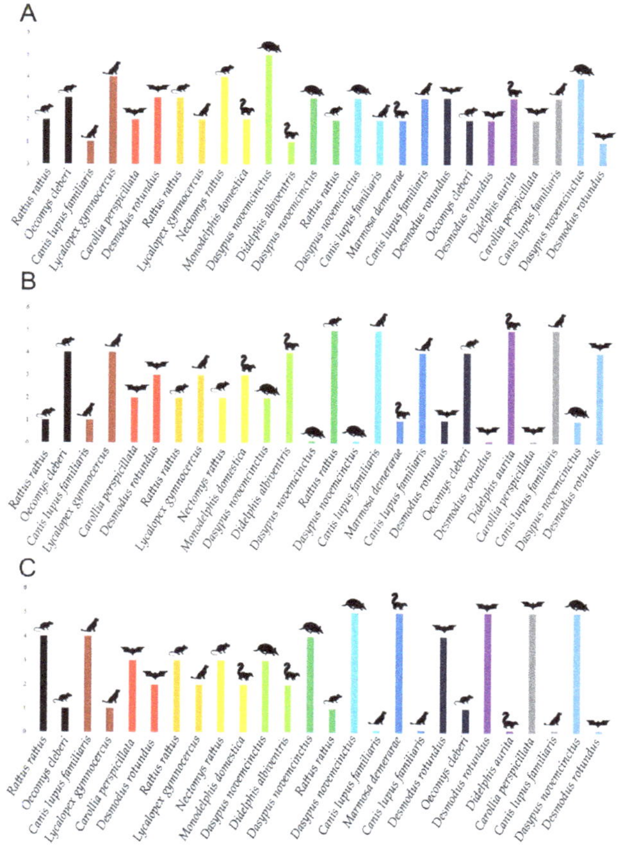

Figure 3. Number of clones detected from 39 multiple mixtures of 13 mammalian species' DNA. (**A**) Number of clones detected in the proportion of 5/5. (**B**) Number of clones detected in the proportion

of 1/9 (C). Number of clones detected in the proportion of 9/1. Bars in the same color show the 13 mixtures performed. The animal symbols represent the orders, and the names of species included in each intrafamilial and interfamilial mixture are shown.

4. Discussion

4.1. Specificity of Molecular Species Identification

Identifying triatomine BMSs increases our understanding of the hosts involved in the transmission cycle of *T. cruzi* in nature. However, this analysis is usually performed using the precipitin technique, which has low sensitivity and specificity and does not permit an observation of the feeding eclecticism of these insect vectors. Moreover, knowledge of potential BMSs is essential, as this technique detects specific antibodies in the blood that respond to the animal protein origin.

On the other hand, with the application of molecular biology techniques, such as PCR and Sanger sequencing using DNA barcoding, it is possible to identify the BMSs at the species level. However, this method relies on a robust and complete dataset of the fauna from the area of the *T. cruzi* transmission network studied. There is a lack of available sequences for species in GenBank, particularly for mammals native to Brazil [16,22,28].

In the current work, we first evaluated the DNA barcoding approach in discriminating species of mammals involved in the transmission of *T. cruzi* in Brazil. It was observed that the BMS molecular markers 12S rDNA and *cyt*b demonstrated specificity in the discrimination of the species from the orders selected, except for didelphids. Opossum species *Didelphis aurita* and *D. albiventris* showed a close genetic distance (less than 1%). This genetic identification has already been observed previously by Steiner et al. (2015), who identified the phylogenetic relationships of 19 species of *Didelphis* using two nuclear markers, the non-coding transthyretin intron 1 (TTR) and the inter-photoreceptor coding retinoid-binding protein exon 1 (IRBP), and the same two mitochondrial genes that were used in the present study, *cyt*b and 12S rDNA [40].

For the marsupial species *M. demerae,* the subgenus *Micoureus* has already been clarified in the literature by Voss and Jansa (2009), even though the authors mentioned the possibility of keeping *Micoureus* as a genus due to the monophyly of the group [41]. It was possible to observe an inter-genus genetic distance between *M. demerarae* and *M. domestica,* and between the *Monodelphis* and *Didelphis* genera [41]. The high genetic distance among these genera has already been reported by Agrizzi and collaborators (2005), who constructed a phylogenetic tree using *cyt*b gene sequences [42].

Regarding the canid species from the present study, the genetic distance considering the reference sequences of GenBank does not show genetic proximity among them. This was previously shown by Fuentes-González and Muñoz-Durán (2012), who studied the phylogenetic relationships of 35 canid species [43]. *Canis lupus familiaris* and *C. lupus* have a low genetic distance for both molecular markers used in this study. Studies using mitochondrial DNA markers demonstrated genetic proximity between these species, and it was suggested that *C. lupus* is the only ancestor of *C. l. familiaris* [44,45]. Even though *C. lupus* and *C. l. familiaris* are genetically close species, the question of whether they are two distant species has been discussed [45]. The molecular markers showed specificity in discriminating the canid species. The blood clot labeled as *C. thous* was identified as *L. gymnocercus* through the sequence analysis. Thus, this species was excluded from the mixture analysis in the present work due to the incorrect name of the blood clot sample.

It is known that a lack of reference sequences deposited in GenBank can lead to the incorrect identification of species. In BMS studies, this lack can cause the erroneous association of species that contribute to the *T. cruzi* life cycle. In the present work, *L. gymnocercus*

GenBank deposits for both markers were absent, generating a maximum identity of the Brazilian field fox with *L. sechurea*, which is a species that does not occur in Brazil. This canid species inhabits the coastal deserts of northern Peru and southern Ecuador [44]. According to the Taxonomic Catalog of the Fauna of Brazil (Catálogo Taxonômico da Fauna do Brasil—http://fauna.jbrj.gov.br/fauna/listaBrasil (accessed on 31 May 2025), two *Lycalopex* species occurred in the country, *L. gymnocercus* and *L. vetulus* (Lund, 1842), both absent from GenBank. Therefore, we strongly recommend targeted sequencing efforts for absent and underrepresented species in the GenBank genetic database, some of which have been repeatedly highlighted in the present study.

The same situation was observed for the lack of an *O. cleberi* 12S rDNA reference sequence in the database, which generated genetic identification with other species of the genus, such as *O. bicolor*, *O. auyantepui*, and other rodent genera like *Hapalomys delacouri*, *E. macconnelli*, and *A. cahirinus*. *Oecomys auyantepui* is only found in the Brazilian Amazon biome [46], *O. catherinae* occurs in the Brazilian Atlantic Forest and Cerrado biomes [47], and *O. bicolor* occurs in the Pantanal, Cerrado, and Amazon biomes [48]. According to Musser and Carleton (2005), the genus *Oecomys* presents an unclear taxonomy [49]. This was confirmed by Suárez-Villota and collaborators (2017), who studied phylogenetic relationships and the delineation of *Oecomys* species using multiple cytogenetic, morphological, and molecular (mtDNA) analyses. The authors concluded that species of the genus *Oecomys* form a species complex and highlighted the importance of combining cytogenetics, morphology, and geographic information with molecular analyses in the delimitation of species [42]. The taxonomic revision of this group was also suggested by Rocha and collaborators (2015) [50]. *Euryoryzomys macconelli* occurs in the Brazilian Amazon biome, and the genetic distance from the *Oecomys* genus has already been demonstrated [51].

The 12S rDNA *Necromys lasiurus* sequence obtained in the current study was identified by BLAST/NCBI as *Akodon montensis*. No sequences of the *Necromys* genus are deposited for this target in GenBank. *Necromys lasiurus* and *A. montensis* belong to the same subfamily Sigmodontinae [52,53], which could be a reason why *A. montensis* appears as a BLAST/NCBI result. Using the *cyt*b target, a sequence of *Bolomys lasiurus* presented 99% genetic identity with the *N. lasiurus* from this study. According to the Integrated Taxonomic Information System, *B. lasiurus* is currently an invalid species, a synonymous species of *N. lasiurus*. In this sense, we highlight the importance of taxonomically updating the GenBank deposits to avoid incorrect nomenclature and missing synonymy information. This failure could raise unrealistic ecological premises about the number and role of species serving as BMS, and whether the blood-feeding behavior of vectors is broad or restricted, with critical implications for understanding vector–host relationships.

It is important to mention that the three rodent species studied in this work do not present genetic proximity, as can be seen in the rodent phylogenetic study carried out by Vilela and collaborators (2014) [54].

The three chiropteran species selected for this work, *C. perspicillata*, *D. rotundus*, and *S. lilium*, showed high genetic distances among them, as can be seen in the work by Baker and Hoofer (2003), who performed a phylogenetic analysis of different genera of bats using mtDNA [55].

The bat species *Sturnira lilium* and *S. tildae* showed genetic distances of 1.4% and 0.7% for 12Sr DNA and *cyt*b, respectively. The genus *Sturnira* is characterized as a group that contains complex phylogenetic relationships [56,57]. Due to their genetic similarity, *S. tildae* and *S. lilum* were suggested to be congeneric species [57] but not sister species [58].

In the present study, *C. perspicillata* and *C. brevicauda* showed high genetic proximity, but this can be explained because *C. perspicillata* and *C. brevicauda* are sister species and closely related to *C. castanea* [59]. The difficulty in identifying these species based only on morphological characters was reported [55].

The Xenarthra species *D. novemcinctus* showed high genetic proximity with *D. yepesi* for the 12S rDNA marker and with *D. novemcinctus* for the *cyt*b marker. The high genetic proximity of these species was shown by a systematic analysis using the *cyt*b marker [60]. The species *D. yepesi*, which showed the highest genetic identity with *D. novemcinctus*, is not considered a valid species by the Integrated Taxonomic Information System (ITIS). We found evidence of a synonymy issue with *D. mazzai* (Yepes, 1933), a valid species that occurs in northern Argentina, but has no genetic sequences deposited in GenBank for any gene [61]. These genes are widely used for the identification of BMSs, and their absence directly affects the accurate identification of the species acting as a BMS, thereby impairing the understanding of transmission dynamics.

It is important to mention that birds and reptiles, considered refractory to *T. cruzi* infection, also served as vector BMSs. However, the current evidence shows that these animals can be infected by *T. cruzi*, and, specifically, reptiles were proposed as silent hosts of the parasite [62,63]. This type of research should be stimulated to provide a larger and more diverse dataset of the Brazilian fauna, not only for BMS studies but also for other enzootic diseases, biodiversity studies, biogeography, taxonomy, and molecular systematics.

4.2. Sensitivity in Detecting Multiple DNA Samples from Blood Clots

DNA barcoding studies for the identification of triatomine BMSs are restricted to the detection of a single BMS, which limits the spectrum of potential species that participate in the parasite transmission network or in the maintenance of vector colonies. The application of this approach and subsequent molecular cloning allows the discrimination of multiple BMS species and/or different individuals of the same species by a haplotype analysis [64,65].

In the present study, the molecular cloning analysis demonstrated superior detection of the species with higher proportions, as expected. However, it was possible to detect species with lower concentrations in the different mixtures performed. For mixtures of the same proportion, both species were always detected. Relevant species present in smaller proportions were detected in 54% of the mixtures. From these results, it is possible to state that gene cloning using the 12S rDNA target is effective at detecting multiple BMSs and individuals of the same taxon, even in extremely unequal proportions. The specificity of the 12S rDNA target in the detection of DNA from vertebrate animals has previously been commented [64], and it has even been mentioned that this target is superior for identifying degraded samples compared to the *cyt*b target. The sensitivity of the cloning technique for detecting multiple BMSs has already been mentioned in previous works [64,65]. However, this technique was only applied once previously in triatomine BMS studies in Brazil [36]. The next-generation sequencing method is also efficient at retrieving multiple BMSs in a single vector [66]. However, in comparison, molecular cloning is low-cost, relatively short, and less time-consuming, and does not require very specialized operators to perform the technique and to process the results.

4.3. The Importance of the Specificity and Sensitivity in Recovered Mammalian DNA for a Better Understanding of the Life Cycle of T. cruzi

The role of marsupials in maintaining the life cycle of *T. cruzi* is already well supported because these mammals share habitats with the triatomine vectors and feed on these insects and on reservoir species, such as small rodents, which can potentially be infected [10]. Legey and collaborators (1999) highlighted that some didelphids are important reservoirs of *T. cruzi* due to their ability to control the infection and not present clinical changes [67]. Some marsupial species, such as *D. aurita* and *D. marsupialis*, circulate in different ecotopes and can assist in the dispersion of the agent in sylvatic, domestic, and peridomestic environments [10,67]. In addition to moving between different ecotopes, marsupials can use all parts of the forest, being in contact with the different transmission cycles of *T. cruzi* [5]. The omnivorous diet of marsupials favors oral transmission by the predation of infected triatomines or other small mammals. Another marsupial species that was also identified as the main reservoir of *T. cruzi* in an outbreak of Chagas disease in Brazil was *M. domestica* [68]. *Marmosa demerarae* was found to be infected by the parasite [69]. The xenarthral species *D. novemcinctus* was identified as the second most important reservoir of *T. cruzi* after *D. marsupialis* [70].

In the domestic environment, *C. l. familiaris*, the domestic dog, is also reported as a major reservoir host of *T. cruzi* and as a risk factor for human infection [71–73]. Dogs may live inside houses and wander through wild ecotopes where they can acquire *T. cruzi* infection from sylvatic triatomines and/or infected prey. These mammals that circulate between domestic and wild ecotopes play very important roles in connecting the transmission network of the parasite [74]. In general, canids can maintain the *T. cruzi* infection throughout their lives [75].

Wild canids are part of a poorly studied taxon in terms of their possible roles in the transmission cycle of *T. cruzi* in nature [76]. These mammals are known to explore large areas and different habitats and can be considered parasite bioaccumulators, especially those that can be transmitted orally, such as *T. cruzi* [76]. This characteristic of wild canids to explore different ecotopes is an extremely critical aspect of the dispersion of the parasite [77].

Rodents are considered the main prey animals that circulate in peridomestic ecotopes [68]. These small mammals connect sylvatic, peridomestic, and domestic ecotopes, transporting the parasite to human dwellings, where they are transmitted to humans and domestic animals. For these reasons, these hosts are important in the study of *T. cruzi* transmission [78].

Almeida and coauthors (2024) [21] proposed that for the transmission cycle of the parasite to occur and lead to CD, five interactions are necessary: parasite, vectors, hosts, environments, and human action. The study highlights the role of humans in producing artificial domestic and peridomestic environments, such as the accumulation of wood piles, which favor the colonization of vectors and small mammals [21].

Rattus rattus, a synanthropic animal, has been suggested as a *T. cruzi* reservoir that connects different ecotopes [79]. The authors also mentioned that the proximity of these species to the domestic environment favors the contact of the parasite with the human population. The wild species of rodent, *O. cleberi*, was previously reported to be infected by *T. cruzi* with a high diversity of trypanosomatid species [80]. On the other hand, *N. lasiurus* is one of the most abundant species in the Brazilian Cerrado and is considered a significant host of *T. cruzi* in the biome [8].

For bats, *D. rotundus* and *C. perspicillata* were recently identified as reservoirs of *T. cruzi* in the Atlantic Forest biome [13]. This was not the first report in which bats were considered reservoirs of *T. cruzi*. In Colombia, one of the main food sources detected in triatomines infected by *T. cruzi* was bats [81]; in Ecuador, chiropter was identified as being responsible for the connection between the wild and household ecotopes [82]; in Argentina, *D. rotundus* was found to be infected with *T. cruzi* [83]; and in Venezuela, bats were identified as dispersers of the parasite [84]. The application of the NGS technique reinforced the roles of bats as hosts/reservoirs and as bioaccumulators of *T. cruzi* [13].

5. Conclusions

The present study aimed to respond to some basic questions regarding the BMS molecular methodology. The first question concerned the possibility of a bias in identifying BMSs with higher proportions. This was answered since we showed that species with very low proportions of 1/9 and 9/1 were detected in 17/26 of mixed samples, although species with higher proportions were identified in all 26 mixtures.

The second question concerned whether cloning sequencing is efficient at revealing secondary sources or equal proportions of BMSs. Our results demonstrated the detection of two mammalian species in equal proportions in all 26 mixed samples, and, as mentioned above, the secondary or low representative species were also recovered in 17/26 mixed samples. We are aware that the low concentration detection in mixed samples was partial, with a number of false-negative results (9/26). Future efforts to enhance assay sensitivity are necessary to address the false-negative results.

Finally, we questioned if the fauna of endemic mammals in Brazil recognized as hosts of *T. cruzi* are represented in GenBank. Despite the limited number of mammalian species evaluated, 3/13 species did not have GenBank sequences available for the two BMS molecular markers applied.

Thus, the study provides more detailed information on multiple BMS detection protocols that could reflect a succession of vector contact events, the diversity of mammalian hosts, and the complexity of the parasite transmission web.

Therefore, the data confirmed our hypothesis that molecular cloning is effective at distinguishing mammalian species, even with equal proportions or with a very low secondary representation in the sample. The integration of these methodologies in the detection of multiple BMSs could provide more detailed information that could reflect a series of vector contact events, the diversity of mammalian hosts, and the complexity of the parasite transmission network. Future research should further amplify the analysis of more Brazilian mammalian species and explore these dynamics in the wild to obtain effective information about the ecology of *T. cruzi* transmission in nature.

Supplementary Materials: The following supporting information can be downloaded at: https://www.mdpi.com/article/10.3390/pathogens14060579/s1, Figure S1. Amplification of mammalian DNA in blood clots and electrophoresis on an agarose gel. (A) Amplification of the molecular target 12S rDNA. Lane 1—Promega DNA Ladder 100 bp, Lane 2—CO01 = *Didelphis aurita*, Lane 3—CO02 = *Akodon montensis*, Lane 4—CO03 = *Lycalopex schurae*, Lane 5—CO04 = *Rattus rattus*, Lane 6—CO05 = *Desmodus rotundus*, Lane 7—CO06 = *Carollia perspicillata*, Lane 8—CO07 = *Oecomys bicolor*, Lane 9—CO08 = *Lycalopex schurae*, Lane 10—CO09 = *Dasypus novemcinctus*, Lane 11—CO10 = *Canis lupus familiaris*, Lane 12—CO11 = *Didelphis albiventris*, Lane 13—CO12 = *Monodelphis domestica*, Lane 14—CO13 = *Miccureus demerarea*, Lane 15—CO14 = *Sturnira lilium*, Lane 16—negative control. (B) Amplification of the molecular target *cyt*b. Lane 1—Promega DNA Ladder 100 bp, Lane 2—CO01 = *Didelphis aurita*, Lane 3—CO02 = *Akodon montensis*, Lane 4—CO03 = *Lycalopex schurae*, Lane 5—CO04 = *Rattus rattus*, Lane 6—CO05 = *Desmodus rotundus*, Lane 7—CO06 = *Carollia perspicillata*, Lane 8—CO07 = *Oecomys bicolor*, Lane 9—CO08 = *Lycalopex schurae*, Lane 10—CO09 = *Dasypus novemcinctus*, Lane 11—CO10 = *Canis lupus familiaris*, Lane 12—CO11 = *Didelphis albiventris*, Lane 13—CO12 = *Monodelphis domestica*, Lane 14—CO13 = *Micoureus demerarea*, Lane 15—CO14 = *Sturnira lilium*, Lane 16—negative control.

Author Contributions: Conceptualization, A.M.I.; methodology, T.S.P. and Q.M.O.; software, Q.M.O.; validation, A.M.I.; investigation, T.S.P., Q.M.O. and A.M.I.; resources, A.M.I.; data curation, A.M.I.; writing—original draft preparation, T.S.P. and Q.M.O.; writing—review and editing, T.S.P., Q.M.O. and A.M.I.; supervision, A.M.I.; funding acquisition, A.M.I. All authors have read and agreed to the published version of the manuscript.

Funding: This research was funded by Fundação de Amparo a Pesquisa do Rio de Janeiro (http://www.faperj.br/) (A.M.I., Grant Numbers E-26/210.131/2023 and E-26/201.069/2021; Q.M.O., E-26/200.067/2021, 260339); Conselho Nacional de Desenvolvimento Científico e Tecnológico (http://www.cnpq.br/) (A.M.I., 315634/2021-9); and Coordenação de Aperfeiçoamento de Pessoal de Nível Superior (https://www.gov.br/capes/pt-br) (T.S.H., 88882.332462/2019-01; Q.M.O., 88887.644154/2021-00).

Institutional Review Board Statement: Not applicable.

Informed Consent Statement: Not applicable.

Data Availability Statement: All sequences generated in this study have been submitted to the GenBank database under the accession numbers: PV023314-PV023326.

Acknowledgments: We thank the PDTIS/FIOCRUZ genomic sequencing platform, Laboratório de Biologia de Tripanossomatideos IOC/FIOCRUZ, for donating the blood clots for this study, and the Laboratório de Parasitologia Integrativa e Paleoparasitologia for their valuable technical assistance during the study.

Conflicts of Interest: The authors declare no conflicts of interest.

References

1. Jansen, A.M.; Xavier, S.C.D.C.; Roque, A.L.R. Landmarks of the Knowledge and *Trypanosoma cruzi* Biology in the Wild Environment. *Front. Cell. Infect. Microbiol.* **2020**, *6*, 10. [CrossRef]
2. Chagas, C. Descoberta do *Tripanozoma cruzi* e verificação da Tripanozomiase Americana: Retrospecto histórico. *Mem. Inst. Oswaldo Cruz* **1922**, *15*, 67–76. [CrossRef]
3. WHO. Chagas Disease (American Trypanosomiasis). Available online: https://www.who.int/health-topics/chagas-disease (accessed on 23 July 2024).
4. Ministério da Saúde. Situação Epidemiológica. Available online: https://www.gov.br/saude/pt-br/assuntos/saude-de-a-a-z/d/doenca-de-chagas/situacao-epidemiologica (accessed on 24 July 2024).
5. Jansen, A.M.; Xavier, S.C.C.; Roque, A.L.R. The multiple and complex and changeable scenarios of the *Trypanosoma cruzi* transmission cycle in the sylvatic environment. *Acta Trop.* **2015**, *151*, 1–15. [CrossRef]
6. Abreu, E.F.; Casali, D.; Costa-Araújo, R.; Garbino, G.S.T.; Libardi, G.S.; Loretto, D.; Loss, A.C.; Marmontel, M.; Moras, L.M.; Nascimento, M.C.; et al. Lista de Mamíferos do Brasil. Zenodo. 2022. Available online: https://zenodo.org/record/7469767 (accessed on 10 January 2025).
7. Deane, M.P.; Lenzi, H.L.; Jansen, A. *Trypanosoma cruzi*: Vertebrate and invertebrate cycles in the same mammal host, the opossum *Didelphis marsupialis*. *Mem. Inst. Oswaldo Cruz* **1984**, *79*, 513–515. [CrossRef]
8. Jansen, A.M.; Xavier, S.C.d.C.; Roque, A.L.R. *Trypanosoma cruzi* transmission in the wild and its most important reservoir hosts in Brazil. *Parasites Vectors* **2018**, *11*, 502. [CrossRef]
9. Dario, M.A.; Furtado, C.; Lisboa, C.V.; De Oliveira, F.; Santos, F.M.; D'Andrea, P.S.; Roque, A.L.R.; Das Chagas Xavier, S.C.; Jansen, A.M. Trypanosomatid Richness Among Rats, Opossums, and Dogs in the Caatinga Biome, Northeast Brazil, a Former Endemic Area of Chagas Disease. *Front. Cell. Infect. Microbiol.* **2022**, *12*, 851903. [CrossRef]
10. Carreira, J.C.A.; Jansen, A.M.; De Nazareth Meirelles, M.; Costa ESilva, F.; Lenzi, H.L. *Trypanosoma cruzi* in the Scent Glands of *Didelphis marsupialis*: The Kinetics of Colonization. *Exp. Parasitol.* **2001**, *97*, 129–140. [CrossRef]
11. Hamilton, P.B.; Cruickshank, C.; Stevens, J.R.; Teixeira, M.M.G.; Mathews, F. Parasites reveal movement of bats between the New and Old Worlds. *Mol. Phylogenetics Evol.* **2012**, *63*, 521–526. [CrossRef]
12. Alves, F.M.; Rangel, D.A.; Vilar, E.M.; Pavan, M.G.; Moratelli, R.; Roque, A.L.R.; Jansen, A.M. *Trypanosoma* spp. Neobats: Insights about those poorly known trypanosomatids. *Int. J. Parasitol. Parasites Wildl.* **2021**, *16*, 145–152. [CrossRef]
13. Dario, M.A.; Moratelli, R.; Schwabl, P.; Jansen, A.M.; Llewellyn, M.S. Small subunit ribosomal metabarcoding reveals extraordinary trypanosomatid diversity in Brazilian bats. *PLoS Negl. Trop. Dis.* **2017**, *11*, e0005790. [CrossRef]
14. Torres, J.M.; De Oliveira, C.E.; Santos, F.M.; Sano, N.Y.; Martinez, É.V.; Alves, F.M.; Tavares, L.E.R.; Roque, A.L.R.; Jansen, A.M.; Herrera, H.M. Trypanosomatid diversity in a bat community of an urban area in Campo Grande, Mato Grosso do Sul, Brazil. *Infect. Genet. Evol.* **2024**, *118*, 105563. [CrossRef] [PubMed]
15. Austen, J.M.; Barbosa, A.D. Diversity and Epidemiology of Bat Trypanosomes: A One Health Perspective. *Pathogens* **2021**, *10*, 1148. [CrossRef] [PubMed]

16. Lilioso, M.; Reigada, C.; Pires-Silva, D.; Fontes Fvon, H.M.; Limeira, C.; Monsalve-Lara, J.; Folly-Ramos, E.; Harry, M.; Costa, J.; Almeida, C.E. Dynamics of food sources, ecotypic distribution and *Trypanosoma cruzi* infection in *Triatoma brasiliensis* from the northeast of Brazil. *PLoS Negl. Trop. Dis.* **2020**, *14*, e0008735. [CrossRef]
17. Santos, F.M.; Sano, N.Y.; Liberal, S.C.; Dario, M.A.; Nantes, W.A.G.; Alves, F.M.; da Silva, A.R.; De Oliveira, C.E.; Roque, A.L.R.; Herrera, H.M.; et al. Kinetoplastid Species Maintained by a Small Mammal Community in the Pantanal Biome. *Pathogens* **2022**, *11*, 1205. [CrossRef]
18. Jansen, A.M.; Roque, A.L.R.; Xavier, S.C.C. Trypanosoma cruzi enzootic cycle. In *American Trypanosomiasis Chagas Disease*; Elsevier: Amsterdam, The Netherlands, 2017; pp. 265–282. Available online: https://linkinghub.elsevier.com/retrieve/pii/B9780128010297000125 (accessed on 30 January 2025).
19. Urbano, P.; Hernández, C.; Ballesteros, N.; Vega, L.; Alvarado, M.; Velásquez-Ortiz, N.; Martínez, D.; Barragán, K.; Ramírez, A.; Páez-Triana, L.; et al. Exploring dietary differences among developmental stages of triatomines infected with *Trypanosoma cruzi* in different habitats. *Int. J. Parasitol.* **2024**, *54*, 559–568. [CrossRef]
20. Urbano, P.; Hernández, C.; Velásquez-Ortiz, N.; Ballesteros, N.; Páez-Triana, L.; Vega, L.; Urrea, V.; Ramírez, A.; Muñoz, M.; Ibarra-Cerdeña, C.N.; et al. Transmission ecology of *Trypanosoma cruzi* by *Rhodnius prolixus* (Reduviidae: Triatominae) infesting palm-tree species in the Colombian Orinoco, indicates risks to human populations. *PLoS Negl. Trop. Dis.* **2024**, *18*, e0011981. [CrossRef]
21. Almeida, C.E.; Máximo, M.M.; Pires-Silva, D.; Takiya, D.M.; Valença-Barbosa, C.; Viana, M.C.; Reigada, C.; Iñiguez, A.M.; Harry, M.; Folly-Ramos, E. From molecules to ecosystems: Insights into a network of interactions for a Chagas disease outbreak using *Triatoma brasiliensis* as natural samplers. *Acta Trop.* **2024**, *251*, 107107. [CrossRef]
22. Da Silva Leal, A.R.; Da Silva Ferreira, A.L.; De Araujo-Pereira, T.; Torres De Sousa, R.L.; Furtado Campos, J.H.; Tavares Dos Reis, R.; Finamore-Araujo, P.; Diotaiuti, L.; Ferreira, F.C.; Mendonça, V.J.; et al. Eco-epidemiological aspects and risk factors associated with human Chagas disease in rural areas of the state of Piauí, Brazil. *BMC Infect. Dis.* **2024**, *24*, 1335. [CrossRef]
23. Daflon-Teixeira, N.F.; Coutinho, C.; Gomes, T.F.; Toma, H.K.; Duarte, R.; Bóia, M.N.; Carvalho-Costa, F.A.; Almeida, C.E.; Lima, M.M. Multiple Approaches to Address Potential Risk Factors of Chagas Disease Transmission in Northeastern Brazil. *Am. J. Trop. Med. Hyg.* **2019**, *100*, 296–302. [CrossRef]
24. De Siqueira, A.F. Estudos Sôbre a Reação Identificação de Sangue da Precipitina Aplicada à Ingerido por Triatomíneos. *Rev. Inst. Med. Trop. São Paulo* **1960**, *2*, 41–53.
25. Silva, M.B.A.; de Menezes, K.R.; de Farias, M.C.G.; Andrade, M.S.; Victor, C.C.A.; Lorosa, E.S.; Jurberg, J. Description of the feeding preferences of triatominae in the Chagas disease surveillance study for the State of Pernambuco, Brazil (Hemiptera: Reduviidae). *Rev. Soc. Bras. Med. Trop.* **2017**, *50*, 543–546. [CrossRef] [PubMed]
26. Alencar, J.; De Mello, C.F.; Gil-Santana, H.R.; de Leão Giupponi, A.P.; Araújo, A.N.; Lorosa, E.S.; Guimarães, A.E.; dos Santos Silva, J. Feeding Patterns of Mosquitoes (Diptera: Culicidae) in the Atlantic Forest, Rio de Janeiro, Brazil. *J. Med. Èntomol.* **2015**, *52*, 783–788. [CrossRef] [PubMed]
27. Verly, T.; Costa, S.; Lima, N.; Mallet, J.; Odêncio, F.; Pereira, M.; Moreira, C.J.D.C.; Britto, C.; Pavan, M.G. Vector competence and feeding-excretion behavior of *Triatoma rubrovaria* (Blanchard, 1843) (Hemiptera: Reduviidae) infected with *Trypanosoma cruzi* TcVI. *PLoS Negl. Trop. Dis.* **2020**, *14*, e0008712. [CrossRef]
28. Almeida, C.E.; Faucher, L.; Lavina, M.; Costa, J.; Harry, M. Molecular Individual-Based Approach on Triatoma brasiliensis: Inferences on Triatomine Foci, *Trypanosoma cruzi* Natural Infection Prevalence, Parasite Diversity and Feeding Sources. *PLoS Negl. Trop. Dis.* **2016**, *10*, e0004447. Available online: https://repositorio.unesp.br/handle/11449/172623 (accessed on 6 April 2019). [CrossRef]
29. Valença-Barbosa, C.; Fernandes, F.A.; Santos, H.L.C.; Sarquis, O.; Harry, M.; Almeida, C.E.; Lima, M.M. Molecular Identification of Food Sources in Triatomines in the Brazilian Northeast: Roles of Goats and Rodents in Chagas Disease Epidemiology. *Am. J. Trop. Med. Hyg.* **2015**, *93*, 994–997. [CrossRef]
30. Lilioso, M.; Pires-Silva, D.; Fontes, F.V.H.M.; Oliveira, J.; da Rosa, J.A.; Vilela, R.V.; Folly-Ramos, E.; Almeida, C.E. *Triatoma petrocchiae* (Hemiptera, Reduviidae, Triatominae): A Chagas disease vector of *T. brasiliensis* species complex associated to reptiles. *Infect. Genet. Evol.* **2020**, *82*, 104307. [CrossRef]
31. Valença-Barbosa, C.; Finamore-Araujo, P.; Moreira, O.C.; Alvarez, M.V.N.; Borges-Veloso, A.; Barbosa, S.E.; Diotaiuti, L.; de Souza, R.d.C.M. High Parasitic Loads Quantified in Sylvatic Triatoma melanica, a Chagas Disease Vector. *Pathogens* **2022**, *11*, 1498. [CrossRef]
32. Rodrigues, M.S.; Lima, L.; Xavier, S.C.D.C.; Herrera, H.M.; Rocha, F.L.; Roque, A.L.R.; Teixeira, M.M.G.; Jansen, A.M. Uncovering *Trypanosoma* spp. diversity of wild mammals by the use of DNA from blood clots. *Int. J. Parasitol. Parasites Wildl.* **2019**, *8*, 171–181. [CrossRef]
33. Kocher, T.D.; Thomas, W.K.; Meyer, A.; Edwards, S.V.; Paabo, S.; Villablanca, F.X.; Wilson, A.C. Dynamics of mitochondrial DNA evolution in animals: Amplification and sequencing with conserved primers. *Proc. Natl. Acad. Sci. USA* **1989**, *86*, 6196–6200. [CrossRef]

34. Kitano, T.; Umetsu, K.; Tian, W.; Osawa, M. Two universal primer sets for species identification among vertebrates. *Int. J. Leg. Med.* **2006**, *121*, 423–427. [CrossRef]
35. Borghesan, T.C.; Ferreira, R.C.; Takata, C.S.; Campaner, M.; Borda, C.C.; Paiva, F.; Milder, R.V.; Teixeira, M.M.; Camargo, E.P. Molecular Phylogenetic Redefinition of *Herpetomonas* (Kinetoplastea, Trypanosomatidae), a Genus of Insect Parasites Associated with Flies. *Protist* **2013**, *164*, 129–152. [CrossRef] [PubMed]
36. Pessanha, T.S.; Herrera, H.M.; Jansen, A.M.; Iñiguez, A.M. "Mi Casa, Tu Casa": The coati nest as a hub of *Trypanosoma cruzi* transmission in the southern Pantanal biome revealed by molecular blood meal source identification in triatomines. *Parasites Vectors* **2023**, *16*, 26. [CrossRef] [PubMed]
37. Pessanha, T.S.; Pires, M.; Iñiguez, A.M. Molecular Detection of Blood Meal Source up to Three Months since the Last Meal: Experimental Starvation Resistance in Triatomines. *Acta Trop.* **2022**, *232*, 106507. [CrossRef]
38. Edgar, R.C. MUSCLE: A multiple sequence alignment method with reduced time and space complexity. *BMC Bioinform.* **2004**, *5*, 113. [CrossRef]
39. Nicholas, K.B.; Nicholas, H.B. GeneDoc: A Tool for Editing and Annotating Multiple Sequence Alignments. 1997. Available online: https://api.semanticscholar.org/ (accessed on 31 May 2025).
40. Steiner, C.; Tilak, M.-K.; Douzery, E.J.; Catzeflis, F.M. New DNA data from a transthyretin nuclear intron suggest an Oligocene to Miocene diversification of living South America opossums (Marsupialia: Didelphidae). *Mol. Phylogenetics Evol.* **2005**, *35*, 363–379. [CrossRef]
41. Voss, R.S.; Jansa, S.A. Phylogenetic Relationships and Classification of Didelphid Marsupials, an Extant Radiation of New World Metatherian Mammals. *Bull. Am. Mus. Nat. Hist.* **2009**, *322*, 1–177. [CrossRef]
42. Agrizzi, J. Molecular Diagnosis of Atlantic Forest Mammals Using Mitochondrial DNA Sequences: Didelphid Marsupials. *TOZJ* **2012**, *5*, 2–9. [CrossRef]
43. Fuentes-González, J.A.; Muñoz-Durán, J. Filogenia de los cánidos actuales (Carnivora: Canidae) mediante análisis de congruencia de caracteres bajo parsimonia. *Actual. Biol.* **2017**, *34*, 85–102. [CrossRef]
44. Savolainen, P.; Zhang, Y.-P.; Luo, J.; Lundeberg, J.; Leitner, T. Genetic Evidence for an East Asian Origin of Domestic Dogs. *Science* **2002**, *298*, 1610–1613. [CrossRef]
45. Vilà, C.; Savolainen, P.; Maldonado, J.E.; Amorim, I.R.; Rice, J.E.; Honeycutt, R.L.; Crandall, K.A.; Lundeberg, J.; Wayne, R.K. Multiple and Ancient Origins of the Domestic Dog. *Science* **1997**, *276*, 1687–1689. [CrossRef]
46. Gentry, A. *Mammal Species of the World*, 3rd ed. In *A Taxonomic and Geographic Reference*; Don, E.W., DeeAnn, M.R., Eds.; Johns Hopkins University Press: Baltimore, ML, USA, 2005; Volume 2, 2142p, ISBN 0-8018-8221-4. A nomenclatural review. *Bull. Zool. Nomencl.* **2006**, *63*, 215–219.
47. Asfora, P.H.; Palma, A.R.T.; Astúa, D.; Geise, L. Distribution of *Oecomys catherinae* Thomas, 1909 (Rodentia: Cricetidae) in northeastern Brazil with karyotypical and morphometrical notes. *Biota Neotrop.* **2011**, *11*, 415–424. [CrossRef]
48. Mendes-Oliveira, A.C. *Pequenos Mamíferos Não-Voadores da Amazônia Brasileira*; Sociedade Brasileira de Mastozoologia: Rio de Janeiro, RJ, Brazil, 2014; Volume 1, 333p, ISBN 8563705024.
49. Suárez-Villota, E.Y.; Carmignotto, A.P.; Brandão, M.V.; Percequillo, A.R.; Silva, M.J.D.J. Systematics of the genus *Oecomys* (Sigmodontinae: Oryzomyini): Molecular phylogenetic, cytogenetic and morphological approaches reveal cryptic species. *Zool. J. Linn. Soc.* **2017**, *184*, 182–210. [CrossRef]
50. Rocha, R.G.; Justino, J.; Leite, Y.L.R.; Costa, L.P. DNA from owl pellet bones uncovers hidden biodiversity. *Syst. Biodivers.* **2015**, *13*, 403–412. [CrossRef]
51. Kocher, A.; de Thoisy, B.; Catzeflis, F.; Huguin, M.; Valière, S.; Zinger, L.; Bañuls, A.; Murienne, J. Evaluation of short mitochondrial metabarcodes for the identification of Amazonian mammals. *Methods Ecol. Evol.* **2017**, *8*, 1276–1283. [CrossRef]
52. Müller, L.; Gonçalves, G.L.; Cordeiro-Estrela, P.; Marinho, J.R.; Althoff, S.L.; Testoni, A.F. DNA Barcoding of Sigmodontine Rodents: Identifying Wildlife Reservoirs of Zoonoses. *PLoS ONE* **2013**, *8*, e80282. [CrossRef]
53. FUNASA; Ministério da Saude. Manual de Controle de Roedores. In *Assessoria de Comunicação e Educação em Saude*; FUNASA: Tokyo, Japan, 2002.
54. Vilela, J.F.; Mello, B.; Voloch, C.M.; Schrago, C.G. Sigmodontine rodents diversified in South America prior to the complete rise of the Panamanian Isthmus. *J. Zool. Syst. Evol. Res.* **2013**, *52*, 249–256. [CrossRef]
55. Baker, G.C.; Smith, J.J.; Cowan, D.A. Review and re-analysis of domain-specific 16S primers. *J. Microbiol. Methods* **2003**, *55*, 541–555. [CrossRef]
56. Pacheco, V.; Patterson, B.D. Phylogenetic Relationships of the New World Bat Genus *Sturnira* (Chiroptera: Phyllostomidae). *Bull. Am. Mus. Nat. Hist.* **1991**, *206*, 101–121.
57. Dinelli, L.L. Três Espécies Crípticas em *Sturnira lilium* (Chiroptera: Phyllostomidae): Evidências Baseadas em Genes Mitocondriais. Master's Thesis, Universidade Federal do Espírito Santo, Centro de Ciências Humanas e Naturais, Vitória, Brazil, March 2014. Available online: http://repositorio.ufes.br/handle/10/3844 (accessed on 31 May 2025).

58. Villalobos, F.; Valerio, A.A. The phylogenetic relationships of the bat genus *Sturnira* Gray, 1842 (Chiroptera: Phyllostomidae). *Mamm. Biol.* **2002**, *67*, 268–275. [CrossRef]
59. Lemos, T.H.; Da Cunha Tavares, V.; Moras, L.M. Character variation and taxonomy of short-tailed fruit bats from *Carollia* in Brazil. *Zoologia* **2020**, *37*, e34587. [CrossRef]
60. De Machado, J.C.A. Biogeografia Cladística e Sistemática Molecular de Tatus do Gênero *Dasypus* e Uma Abordagem Filogeográfica da Espécie *D. novemcinctus* (Mammalia, Cingulata, Dasypodidae). Master's Thesis, Universidade do Estado do Rio de Janeiro, Rio de Janeiro, Brazil, 2017. Available online: http://www.bdtd.uerj.br/handle/1/5874 (accessed on 31 May 2025).
61. Feijó, A.; Cordeiro-Estrela, P. The Correct Name of the Endemic *Dasypus* (Cingulata: Dasypodidae) from North western Argentina. *Zootaxa* **2014**, *3887*, 88–94. [CrossRef] [PubMed]
62. Martínez-Hernández, F.; Oria-Martínez, B.; Rendón-Franco, E.; Villalobos, G.; Muñoz-García, C.I. *Trypanosoma cruzi*, beyond the dogma of non-infection in birds. *Infect. Genet. Evol.* **2022**, *99*, 105239. [CrossRef] [PubMed]
63. Botto-Mahan, C.; Correa, J.P.; Araya-Donoso, R.; Farías, F.; San Juan, E.; Quiroga, N.; Campos-Soto, R.; Reyes-Olivares, C.; González-Acuña, D. Lizards as Silent Hosts of *Trypanosoma cruzi*. *Emerg. Infect. Dis.* **2022**, *28*, 1250–1253. Available online: https://wwwnc.cdc.gov/eid/article/28/6/22-0079_article.htm (accessed on 1 August 2023). [CrossRef]
64. Stevens, L.; Dorn, P.L.; Hobson, J.; de la Rua, N.M.; Lucero, D.E.; Klotz, J.H.; Schmidt, J.O.; Klotz, S.A. Vector Blood Meals and Chagas Disease Transmission Potential, United States. *Emerg. Infect. Dis.* **2012**, *18*, 646–649. [CrossRef]
65. Stevens, L.; Monroy, M.C.; Rodas, A.G.; Dorn, P.L. Hunting, Swimming, and Worshiping: Human Cultural Practices Illuminate the Blood Meal Sources of Cave Dwelling Chagas Vectors (*Triatoma dimidiata*) in Guatemala and Belize. *PLoS Negl. Trop. Dis.* **2014**, *8*, e3047. [CrossRef]
66. Arias-Giraldo, L.M.; Muñoz, M.; Hernández, C.; Herrera, G.; Velásquez-Ortiz, N.; Cantillo-Barraza, O.; Urbano, P.; Cuervo, A.; Ramírez, J.D. Identification of blood-feeding sources in *Panstrongylus, Psammolestes, Rhodnius* and *Triatoma* using amplicon-based next-generation sequencing. *Parasites Vectors* **2020**, *13*, 434. [CrossRef]
67. Legey, A.P.; Pinho, A.P.; Xavier, S.C.; Marchevsky, R.; Carreira, J.C.; Leon, L.L.; Jansen, A.M. *Trypanosoma cruzi* in marsupial didelphids (*Philander frenata* and *Didelphis marsupialis*): Differences in the humoral immune response in natural and experimental infections. *Rev. Soc. Bras. Med. Trop.* **2003**, *36*, 241–248. [CrossRef]
68. Roque, A.L.R.; Xavier, S.C.C.; da Rocha, M.G.; Duarte, A.C.M.; D'andrea, P.S.; Jansen, A.M. *Trypanosoma cruzi* Transmission Cycle Among Wild and Domestic Mammals in Three Areas of Orally Transmitted Chagas Disease Outbreaks. *Am. J. Trop. Med. Hyg.* **2008**, *79*, 742–749. [CrossRef]
69. Herrera, L.; D'Andrea, P.S.; Xavier, S.C.C.; Mangia, R.H.; Fernandes, O.; Jansen, A.M. *Trypanosoma cruzi* infection in wild mammals of the National Park 'Serra da Capivara' and its surroundings (Piauí, Brazil), an area endemic for Chagas disease. *Trans. R. Soc. Trop. Med. Hyg.* **2005**, *99*, 379–388. [CrossRef]
70. Raccurt, C.P. *Trypanosoma cruzi* in French Guinea: Review of accumulated data since 1940. *Med. Trop.* **1996**, *56*, 79–87.
71. Beard, C.B.; Pye, G.; Steurer, F.J.; Rodriguez, R.; Campman, R.; Peterson, A.T.; Ramsey, J.; Wirtz, R.A.; Robinson, L.E. Chagas Disease in a Domestic Transmission Cycle in Southern Texas, USA. *Emerg. Infect. Dis.* **2003**, *9*, 103–105. [CrossRef] [PubMed]
72. Cardinal, M.V.; Lauricella, M.A.; Ceballos, L.A.; Lanati, L.; Marcet, P.L.; Levin, M.J.; Kitron, U.; Gürtler, R.E.; Schijman, A.G Molecular epidemiology of domestic and sylvatic *Trypanosoma cruzi* infection in rural northwestern Argentina. *Int. J. Parasitol.* **2008**, *38*, 1533–1543. [CrossRef] [PubMed]
73. Morales, M.E.; Cimino, R.; Mackern-Oberti, J.P.; Martín, C.M.-S.; Cattan, P.E.; Superina, M. Eco-epidemiological Survey of *Trypanosoma cruzi* in Dogs from Mendoza, Argentina. *Ecohealth* **2025**, *22*, 79–90. [CrossRef]
74. Ramírez, J.D.; Turriago, B.; Tapia-Calle, G.; Guhl, F. Understanding the role of dogs (*Canis lupus familiaris*) in the transmission dynamics of *Trypanosoma cruzi* genotypes in Colombia. *Veter. Parasitol.* **2013**, *196*, 216–219. [CrossRef]
75. Rocha, F.L.; Roque, A.L.R.; De Lima, J.S.; Cheida, C.C.; Lemos, F.G.; de Azevedo, F.C.; Arrais, R.C.; Bilac, D.; Herrera, H.M.; Mourão, G.; et al. *Trypanosoma cruzi* Infection in Neotropical Wild Carnivores (Mammalia: Carnivora): At the Top of the *T. cruzi* Transmission Chain. *PLoS ONE* **2013**, *8*, e67463. [CrossRef]
76. Sillero-Zubiri, C. Family Canidae (dogs). In *Handbook Mammals of the World*; Lynx Ediciones: Barcelona, Spain, 2009; pp. 352–447.
77. Butler, J.; du Toit, J.; Bingham, J. Free-ranging domestic dogs (*Canis familiaris*) as predators and prey in rural Zimbabwe: Threats of competition and disease to large wild carnivores. *Biol. Conserv.* **2004**, *115*, 369–378. [CrossRef]
78. Pinto, C.M.; Ocaña-Mayorga, S.; Lascano, M.S.; Grijalva, M.J. Infection by trypanosomes in marsupials and rodents associated with human dwellings in Ecuador. *J. Parasitol.* **2006**, *92*, 1251–1255. [CrossRef]
79. Cab-Romero, M.; da Costa, A.P.; Costa, J.O.J.; de Oliveira Manhães, I.B.; da Silva, R.E.; Ruiz-Piña, H.A.; Reyes-Novelo, E.A.; Escobedo-Ortegona, J.; Tonhosolo, R.; Marcili, A. Synanthropic mammals in transmission cycle of *Trypanosoma cruzi* in Yucatán, Mexico. *Braz. J. Glob. Health* **2020**, *1*, 21–28. [CrossRef]
80. Brandão, E.M.V.; Xavier, S.C.C.; Carvalhaes, J.G.; D'Andrea, P.S.; Lemos, F.G.; Azevedo, F.C.; Cássia-Pires, R.; Jansen, A.M.; Roque, A.L.R. Trypanosomatids in Small Mammals of an Agroecosystem in Central Brazil: Another Piece in the Puzzle of Parasite Transmission in an Anthropogenic Landscape. *Pathogens* **2019**, *8*, 190. [CrossRef]

81. Hernández, C.; Salazar, C.; Brochero, H.; Teherán, A.; Buitrago, L.S.; Vera, M.; Soto, H.; Florez-Rivadeneira, Z.; Ardila, S.; Parra-Henao, G.; et al. Untangling the transmission dynamics of primary and secondary vectors of *Trypanosoma cruzi* in Colombia: Parasite infection, feeding sources and discrete typing units. *Parasites Vectors* **2016**, *9*, 620. [CrossRef] [PubMed]
82. Pinto, C.M.; Ocaña-Mayorga, S.; Tapia, E.E.; Lobos, S.E.; Zurita, A.P.; Aguirre-Villacís, F.; MacDonald, A.; Villacís, A.G.; Lima, L.; Teixeira, M.M.G.; et al. Bats, Trypanosomes, and Triatomines in Ecuador: New Insights into the Diversity, Transmission, and Origins of *Trypanosoma cruzi* and Chagas Disease. *PLoS ONE* **2015**, *10*, e0139999. [CrossRef] [PubMed]
83. Argibay, H.D.; Orozco, M.M.; Cardinal, M.V.; Rinas, M.A.; Arnaiz, M.; Segura, C.M.; Gürtler, R.E. First finding of *Trypanosoma cruzi* II in vampire bats from a district free of domestic vector-borne transmission in Northeastern Argentina. *Parasitology* **2016**, *143*, 1358–1368. [CrossRef] [PubMed]
84. Añez, N.; Crisante, G.; Soriano, P.J. *Trypanosoma cruzi* congenital transmission in wild bats. *Acta Trop.* **2009**, *109*, 78–80. [CrossRef]

pathogens

MDPI

Review

Adipose Tissue in Chagas Disease: A Neglected Component of Pathogenesis

Vitória França dos Santos Pessoa [1], Mariana Hecht [1], Nadjar Nitz [1] and Luciana Hagström [1,2,*]

1 Interdisciplinary Laboratory of Biosciences, Faculty of Medicine, University of Brasilia, Brasília 70910-900, Brazil; vitoria.pessoa@aluno.unb.br (V.F.d.S.P.); marianahecht@gmail.com (M.H.); nadjarnitz@gmail.com (N.N.)

2 Faculty of Physical Education, University of Brasília, Brasília 70910-900, Brazil

* Correspondence: loubex@hotmail.com; Tel.: +55-61996687772

Abstract: Chagas disease (CD), caused by the protozoan *T. cruzi*, is a serious public health issue with high morbidity and mortality rates. Approximately 7 million people are infected, mostly in Latin America. The pathogenesis is multifactorial and poorly elucidated, particularly regarding the role of adipose tissue (AT). This review aims to explore the complex relationship between *T. cruzi* and AT, focusing on the possible role of this tissue in CD, as well as to explore the impact of diet on the progression of the disease. *T. cruzi* infects adipocytes, affecting their function. Chronic infection alters adipose physiology, contributing to systemic inflammation and metabolic disturbances. Adipokines are dysregulated, while markers of inflammation and oxidative stress increase within AT during CD. Additionally, the immune response and clinical aspects of CD may be influenced by the host's diet. High-fat diets (HFDs) impact parasite burden and cardiac pathology in murine models. The complex interaction among *T. cruzi* infection, AT dysfunction, and dietary factors underscore the complexity of CD pathogenesis. Despite accumulating evidence suggesting the role of AT in CD, further research is needed to elucidate its clinical implications. Understanding the bidirectional relationship between AT and *T. cruzi* infection may offer insights into disease progression and potential therapeutic targets, highlighting the importance of considering adipose physiology in CD management strategies.

Keywords: *T. cruzi*; adipocytes; high-fat diet; pathophysiology

1. Introduction

Although Chagas disease (CD) is endemic in Latin America, it affects about 6 to 7 million individuals worldwide. Due to the migration of infected people, this neglected tropical disease is found, for example, in the United States of America, as well as in several countries in Europe and Africa [1]. CD is caused by the flagellated protozoan *T. cruzi*, an obligate intracellular parasite known for its ability to persist in mammalian hosts indefinitely after the initial infection [2,3]. Thus, CD is a life-threatening illness with distinct acute and chronic phases. Most acute infections are asymptomatic or present nonspecific symptoms. During this phase, which lasts 4 to 8 weeks post-infection, parasitemia is elevated [2,4].

Conversely, in the chronic phase, parasitemia is absent or very low. Most patients will remain asymptomatic for life. However, approximately 30% of infected people will develop cardiac and/or gastrointestinal symptoms around 30 years after the initial infection [2,4]. The drugs currently available for the treatment of CD are effective in the acute phase but are not satisfactory in the chronic phase [5].

https://doi.org/10.3390/pathogens14040339

This gap in effective chronic-phase treatments underscores the need for a deeper understanding of the disease's progression. Despite decades of research, the pathogenesis of CD is still debated and questioned, leading to the development of theories aimed at elucidating the mechanisms underlying the characteristic symptoms of its late phase. In fact, CD pathogenesis is multifactorial and very complex, influenced by various factors such as parasite strain, the persistence of *T. cruzi* in different host tissues, and the host's metabolic and immunological status (for review, see [6]). Thus, two primary theories have emerged: parasite persistence and autoimmunity mechanisms [4] (Figure 1).

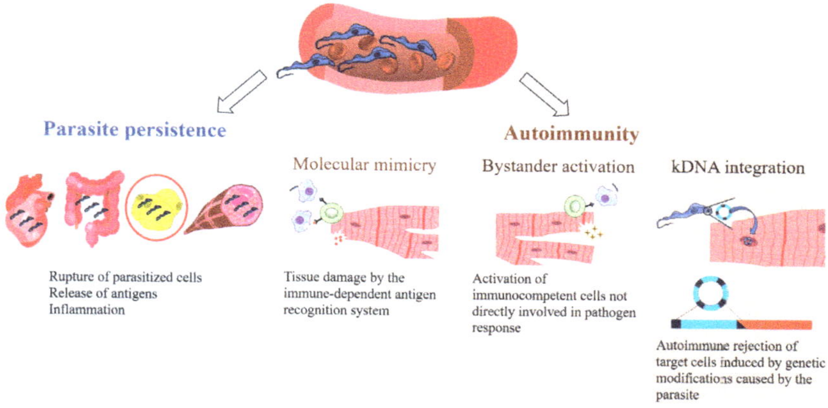

Figure 1. Main theories on the pathogenesis of Chagas disease. According to the parasite persistence theory, *T. cruzi* replicates within host cells, causing the rupture of infected cells and releasing antigens that trigger the host's immune response, leading to inflammation. The autoimmune theory suggests that the immune system targets the host's own tissues, causing chronic inflammation through the following mechanisms: molecular mimicry, bystander activation, and potentially *T. cruzi* kinetoplast DNA (kDNA) integration. Erythrocytes (https://beta.scidraw.io/drawing/634), heart (https://beta.scidraw.io/drawing/277), intestine (https://beta.scidraw.io/drawing/296), adipocyte (https://doi.org/10.5281/zenodo.3926133), myocyte (https://doi.org/10.5281/zenodo.3926129), and macrophage (https://beta.scidraw.io/drawing/221) were obtained from https://scidraw.io/. Accessed on 19 February 2025.

The first theory proposes that tissue damage initiates with parasite infection. *T. cruzi* demonstrates wide infective versatility, invading diverse tissues and cell types [2,7], including macrophages, skeletal and cardiac muscle cells, epithelial cells, and adipocytes [8]. Actively proliferating within cells, the parasite causes their mechanical rupture, leading to microvascular spacing, myocytolysis, and alterations in enteric and cardiac nerves [7] (Figure 1). Importantly, *T. cruzi* can evade the immune system, enabling its persistence throughout the host's lifespan [9–11]. Recent studies corroborate the role of parasite persistence in the pathophysiology of CD [12–14].

Numerous findings on the tissue distribution of *T. cruzi* have further supported the theory of parasite persistence [9,13,15,16]. Quantitative PCR (qPCR) and bioluminescence imaging systems have demonstrated that *T. cruzi* circulation is highly dynamic, showing a tropism for the gastrointestinal tract and adipose tissue (AT), among other tissues (for review, see [10]). These tissues may act as reservoirs for the parasite in the chronic phase of the disease and facilitate its eventual migration to the myocardium, resulting from years of myocarditis and cardiac fibrosis [17,18]. Amastigotes of *T. cruzi* were observed in cultures of adipocytes differentiated from human AT-derived stem cells, reinforcing the idea that AT also serves as a reservoir for the parasite in humans [19]. However, the association between the presence of *T. cruzi* in AT and CD symptoms remains inconclusive. The need for ultrasensitive techniques to detect the parasite in injured areas suggests that *T. cruzi* alone is not the single cause of tissue injury, indicating a potential damage amplification mechanism [20].

In this context, even though it remains controversial, autoimmunity has been proposed as a mechanism triggered by *T. cruzi* in the host. This self-reactive hypothesis proposes that, regardless of the initial cause, a powerful immunological stimulus disrupts the body's surveillance mechanism, playing a critical role in cardiac tissue damage [6,21]. This hypothesis was initially demonstrated in vitro by Santos-Buch and Teixeira [22], who observed the destruction of cardiac fibers in rabbit fetuses by lymphocytes derived from chronic chagasic rabbits, whereas control rabbit lymphocytes did not attack cardiac cells [22]. Thus, it was found that immunocompetent lymphocytes from chagasic rabbits could destroy myofibers and neurons of the parasympathetic nervous system in healthy, uninfected rabbits [23].

Several mechanisms have been proposed to explain autoimmunity, including the cross-recognition of similar antigens between the parasite and host tissues (molecular mimicry), activation of immunocompetent cells not directly involved in pathogen response (bystander activation), and potentially the genetic material exchange between the parasite and host (integration of minicircles of *T. cruzi* kinetoplast DNA [kDNA]) [6,21,23] (Figure 1).

Although distinct, these theories (parasite persistence and autoimmunity) are not mutually exclusive. It is highly probable that CD results from complex interactions between the parasite, the host's immune system, and other genetic factors [8,24].

In this context, AT emerges as an additional element in this intricate chain, contributing to the persistence of the parasite by acting as a reservoir [25–27] and providing an environment rich in inflammatory mediators [27], which may aid in the development of chronic Chagas cardiomyopathy (CCC). Thus, this review will explore the known mechanisms by which AT influences CD pathogenesis, including its role in immune modulation, metabolic alterations, and parasite–host interactions. Additionally, this review will investigate the impact of diet on the physiology of AT and CD. We will primarily focus on white adipose tissue (WAT), as the involvement of brown or beige AT in the context of CD remains underexplored.

Of note, due to *T. cruzi*'s ability to invade a wide range of cells and tissues, Añez and Crisante [28] propose that the parasite has no preference, suggesting that specific tropism does not exist [28]. Nevertheless, most studies discuss the parasite's preference for infecting certain tissues over others (due to different factors, e.g., the strain) and, therefore, use the concept of tissue tropism.

It is also important to highlight that only a few research groups study the role of AT in the context of *T. cruzi* infection, and most experiments have been conducted in animal models. Although the results from murine models cannot be directly translated to human CD, they remain crucial for understanding the disease. Actually, studies in humans are scarce. However, research on differentiated human adipose cells showed that *T. cruzi* parasites these cells [19,29,30]. Similarly, in chronically infected patients, *T. cruzi* DNA was detected in AT [26].

2. Complexity of Adipose Tissue: Structure and Function

AT is an intricate multicellular organ exhibiting significant biological variability depending on the depot level, anatomical location, and metabolic state. It exerts extensive physiological actions influencing processes such as immunity and inflammation (for review, see [31]). Actually, beyond the adipocytes, many cell types are present, interacting spatially and functionally [32], including vascular components, nerve tissue, immune cells, endothelial cells, fibroblasts, and other adipocyte precursors [33,34]. Immune cells (e.g., macrophages, neutrophils, eosinophils, dendritic cells, and lymphocytes) are an important part of AT that secrete various pro-inflammatory cytokines that can modulate the physiological state of the organism (for review, see [35]). Indeed, due to the abundance of immune cells within AT, adipogenic cells, and their precursors, this tissue may play a role in combating infections [33].

WAT and brown adipose tissue (BAT) are the main types of AT, currently recognized as an endocrine organ with distinct functions [34]. WAT is the predominant type in human adults [36] and is mainly involved in lipid storage and release to maintain energy homeostasis. Adipocytes, the principal cell type of WAT, are characterized by a large, unilocular lipid vacuole in the cytoplasm, which facilitates substantial lipid storage (for review, see [31,37]), while containing few mitochondria [38] (Figure 2). Beyond their role in energy storage, white adipocytes in WAT secrete various bioactive substances, including peptides (e.g., adipokines) [38], proteins [39], and microRNAs (miRNAs) [40], which influence systemic metabolism and modulate the immune response. WAT plays a critical role in energy balance, insulin sensitivity, and inflammation. In particular, it secretes pro-inflammatory adipokines, such as tumor necrosis factor-alpha (TNF-α) and interleukin-6 (IL-6), which influence the immune system and contribute to inflammatory processes [41]. These cells also serve as a reservoir for immune cell infiltration, as almost all types of innate and adaptive immune cells are found in WAT, further underscoring their involvement in both metabolic and immune regulation [41].

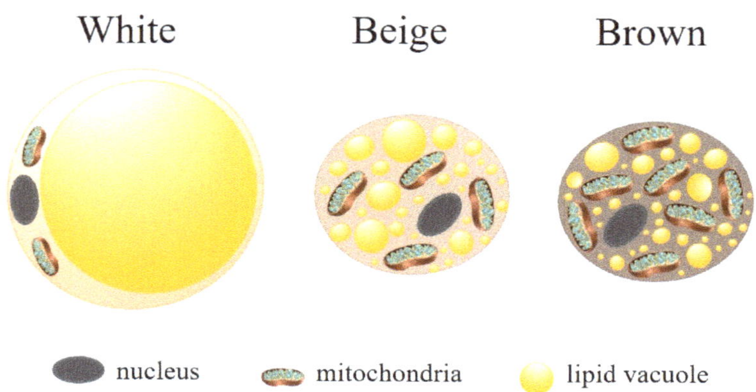

Figure 2. Main types of adipocytes. There are three main types of adipocytes, each differing functionally and morphologically. White adipocytes are large and contain a single lipid droplet and few mitochondria. Their primary function is energy storage. Brown adipocytes have numerous small lipid droplets and a high number of mitochondria. Their main function is thermogenesis. Beige (or brite) adipocytes display intermediary characteristics, highlighting the remarkable plasticity of adipose tissue. Adipocyte (https://doi.org/10.5281/zenodo.3926133) and mitochondria (https://doi.org/10.5281/zenodo.7590755) were obtained from https://scidraw.io/. Accessed on 19 February 2025.

Although BAT is found in human adults in small depots localized in specific areas (neck, supraclavicular, axillary, paravertebral, mediastinal, and epigastric regions), it is mainly present in small mammals (e.g., mice) and human newborns (for review, see [37]). BAT is more metabolically active than WAT, playing a role in thermogenesis (non-shivering heat generation and dissipation of energy), weight loss, and the secretion of adipokines [31,32,42]. Unlike WAT, brown adipocytes contain several cytoplasmic lipid droplets of different sizes and numerous mitochondria (Figure 2). The high iron and cytochrome content resulting from the dense mitochondrial population, along with extensive vascularization, imparts the characteristic brown coloration to BAT [43].

Another type of AT is the beige or brite (brown-like in white) adipocytes. These cells emerge in WAT depots in response to external stimuli, such as prolonged cold exposure in response to sympathetic activation [31,43]. Beige adipocytes contain multilocular lipid vacuoles and abundant mitochondria (Figure 2). Functionally, beige adipocytes resemble BAT expressing brown adipocyte genes (for review, see [44]). Their role in thermogenesis and the regulation of insulin resistance and obesity has been studied. Beige adipocytes also possess a secretory capacity that exerts endocrine and paracrine effects (for review, see [31]).

During gestation and lactation, subcutaneous WAT transdifferentiates into pink AT (PAT), a mammary glandular tissue that acquires a pink color and is capable of producing and secreting milk. PAT contains multiple cytoplasmic lipid droplets, an apical surface with microvilli, a large and round nucleus, a well-developed endoplasmatic reticulum, a Golgi complex, and granules containing milk (for review, see [44]). The conversion of WAT to BAT or beige AT, as well as the WAT to PAT, is reversible, reinforcing the enormous plasticity of AT [45].

This review will primarily focus on WAT, as the involvement of BAT or beige AT in the context of CD remains underexplored.

3. Pathogens in White Adipose Tissue

Various infectious agents accumulate at different sites within the AT, a tissue with great longevity that is rich in nutrients. These characteristics facilitate the persistence of pathogens within the host during chronic infections [18,41]. Noteworthily, the immune response to the presence of foreign microorganisms varies depending on the site within the AT where they are localized, such as the cytoplasm of adipocytes, interstitial spaces, inside vessels, immune cells, and stromal vascular fraction [18]. Additionally, AT interacts with cells from other tissues through the release of adipokines and cytokines, which probably influence the progression of diseases [37,41].

The relationship between microorganisms, host AT, variations in host fat accumulation, and pathogenesis of infections is complex and not yet fully elucidated. Indeed, microbes can disrupt the normal functioning of AT [18], and the accumulation or decrease of fat in different organs can impair their normal functions. The dysregulation of AT can lead to systemic impairments, affecting other physiological systems [46].

Various pathogens persist in AT during infection, such as *Mycobacterium tuberculosis* [47,48], *Plasmodium berghei* [49,50], SARS-CoV-2 [41,51], and *Trypanosoma brucei* [52–54]. *Leishmania infantum* was also observed by immunolabeling at 40 weeks post-infection in WAT, as well as in BAT. Furthermore, it has been demonstrated that *L. infantum*, present in AT, maintains its infectivity [55]. Barthelemy et al. [41] provide a list of pathogens (bacteria, viruses, and parasites) documented to infect AT and eventually persist within them [41].

T. brucei and *T. congolense* were found in different AT deposits in naturally infected livestock animals. Subcutaneous WAT and gonadal WAT exhibited the highest rates of trypanosome detection [55]. This finding reinforces the role of AT as reservoirs for trypanosomes not only in mice [56] but also in natural hosts (cows, goats, and sheep) [52]. Moreover, it has been proposed that *T. brucei* within AT exhibits resistance to pharmacological treatment [54], underscoring the necessity to further elucidate the interactions among parasites, AT, and the host.

In the case of *T. cruzi*, AT serves as an early target for the parasite and acts as a reservoir during chronic infection [7,25,57]. The bidirectional interaction between the AT and *T. cruzi* may imply two distinct consequences. First, since *T. cruzi* resides in the AT, it benefits from the long half-life of adipocytes in both humans and mice, which provides the parasites with easy access to nutrients, including lipids and fatty acids [18]. Both are essential for the growth and survival of *T. cruzi* [58]. This may allow the parasites to remain in a quiescent state [18], evading the host's immune response [59]. Indirectly, the specific location of *T. cruzi* in AT might protect the heart. On the other hand, even though the dormant forms of *T. cruzi* in AT may have a minor role in disease progression, they sustain the parasite persistence in the host [10,60], leading to an inflammatory response that would contribute to tissue damage, potentially reactivating the disease [25].

Amastigotes, the intracellular replicative form of *T. cruzi*, are found in the cytoplasm [3]. Thus, in WAT, they are close to lipid droplets [18]. Given that trypanosomes exhibit limited fatty acid oxidation during their life cycle, it was suggested that they may rely on local lipolysis to satisfy their energy requirements [58] and sustain these replicative stages.

4. Consequences of *T. cruzi* Presence in Adipose Tissue

Numerous studies underscore the importance of AT as a critical target for *T. cruzi* infection [25,61,62]. Indeed, the presence of the parasite in fat cells alters the function of AT, leading to local and systemic inflammation, metabolic abnormalities, and changes in cardiac physiology, thereby aggravating CD [17,60,63–65]. The involvement of epicardial and pericardial AT in CCC has also been suggested [65].

Several mechanisms may be involved, including adipocyte apoptosis or enhanced adipogenesis [64], mitochondrial dysfunction [63,65], oxidative stress [65–68], and endoplasmic reticulum stress in the myocardium [68]. Additionally, infection-specific extracellular vesicles (adiposomes) released by AT have been shown to modulate immune and metabolic gene expression in various cell types, such as cardiomyocytes and macrophages [65]. Undeniably, various pathways are altered during acute infection and continue to exhibit changes throughout the chronic phase [60], facilitating the parasite's persistence within the host [65].

The presence of *T. cruzi* within AT has been demonstrated both in cell culture [19,25,30,69] and in murine models [70–75]. Bioluminescence imaging, which enables the visualization of parasite-retaining regions, confirms the attraction of *T. cruzi* to AT [9,76] with implications for the progression of CD. In mice, it was observed that cardiac parasitism was lower compared to the WAT and BAT at 15 days post-infection (dpi) [25,57]. Likewise, Zaki et al. [75] reported a higher parasite burden in epididymal fat than in the heart during the early stages of infection [75]. This preferential tropism of *T. cruzi* for AT in the acute phase may explain the relatively reduced cardiac damage observed in this period [17].

In the acute phase of *T. cruzi* infection, significant alterations are observed in the transcriptional program of adipocytes [25]. Consequently, this leads to pro/anti-inflammatory imbalance in AT and impaired lipolysis [77].

The immune function of AT can be modulated by *T. cruzi* infection [27]. Fifteen/seventeen days post-infection with *T. cruzi*, mice exhibit macrophage infiltration into AT [27,57,78]. The interaction between macrophages and AT results in persistent inflammation [79]. This phenomenon remains in the chronic phase, concomitant with the persistence of parasites contributing to the pathophysiology of CD [27]. Additionally, infected animals showed a decrease in body fat, which may also contribute to the pathogenesis of CCC [25,78]. Many factors can be responsible for the atrophy and loss of adipocytes, such as the inability of adipocytes to store fatty acids and the lysis caused by parasites [78]. The reduction in adipocytes allows the translocation of parasites to other tissues and may trigger pro-inflammatory responses with immune cell infiltration, potentially leading to the reactivation of the infection [25].

It was also suggested that inflammatory factors released into the bloodstream during infection may participate [30,78]. In this context, the supernatant of the indirect cultivation between AT differentiated from human AT-derived stem cells subsequently infected by *T. cruzi* and human peripheral blood mononuclear cells (PBMCs) indicated enhanced levels of pro-inflammatory cytokine interleukin-6 (IL-6) and a decrease in tumor necrosis factor-alpha (TNF-α) [30].

In mice, an upregulation of inflammatory markers in AT during acute *T. cruzi* infection was observed [25,53,78,80], including IL-6 and TNF-α [25,78]. In chronic chagasic patients (n = 190), IL-6 levels were elevated compared to the control group, while TNF-α levels showed only an upward trend [81].

Through the secretion of multiple adipokines, the AT controls the metabolism of various organs, such as the heart. Thus, AT dysfunction in disease states may contribute to adverse clinical outcomes (for review, see [82]). Recognizing that adipose tissue modulates the metabolism of various organs, including the heart, through the secretion of multiple adipokines, numerous researchers have investigated its role in the pathogenesis of CD. The inflammatory phenotype in mice in the early phase of *T. cruzi* infection was associated with a decrease in peroxisome proliferator-activated receptor gamma (PPAR-γ), adiponectin (AD), and leptin [78]. The transcription factor PPAR-γ is a regulator of adipogenesis [77]. AD, an adipocyte hormone, has anti-inflammatory and cardioprotective actions [82] and plays critical roles in energy regulation, glucose, and lipid metabolism (for review, see [83]). Leptin, primarily produced by adipocytes and released in proportion to body fat deposits, is a pro-inflammatory adipokine and a key mediator in the pathophysiology of cardiovascular disease, being involved in vascular remodeling, oxidative stress, and hypertrophy in the heart (for review, see [84]).

PPAR-γ has been observed to be suppressed during *T. cruzi* infection [77]. Conversely, Nagajyothi et al. [57] demonstrated differential responses of BAT and WAT to acute *T. cruzi* infection, with a reduction in PPAR-γ expression and an upregulation of TNF-α in BAT, whereas PPAR-γ was elevated in WAT. These contrasting responses may be attributed to variations in parasite load between the two types of adipocytes [57]. Additionally, the activating transcription factor-7 (ATF-7), a protein required for adipocyte differentiation, showed no changes during *T. cruzi* infection [77].

Despite the anti-inflammatory and cardioprotective properties [82], the role of AD in cardiovascular disease remains controversial, as well as in Chagas cardiomyopathy. Due to its anti-inflammatory and cardioprotective properties [82], AD has been widely investigated in Chagas cardiomyopathy. However, its role in CD and other cardiovascular disorders remains controversial. Some authors have reported that AD levels are reduced during acute *T. cruzi* infection in both in vitro [69] and in vivo [25,27,57,77] studies. Indeed, Santamaría et al. [77] observed lower AD concentrations in murine plasma and a lack of AD protein expression in WAT in the acute phase of CD [77]. Combs et al. [25] showed in mice that AD plasma levels were reduced at 30 dpi compared to controls; however, at 60 and 90 dpi, the values were similar [25].

On the other hand, other studies stated that AD overexpression can occur in chronic *T. cruzi* infection [64,65], potentially resulting in adverse effects. In two different studies with 15 patients with CCC, the amount of serum AD was significantly higher [85] or tended to increase compared to controls [86]. The reduced fat mass in the group with CCC could explain these findings [85,86]. In fact, in the context of fat loss, adipocytes may fail to express adequate levels of AD [65].

Leptin plays a role in innate and adaptative immunity [41]. Fernandes et al. [84] demonstrated that serum leptin levels were reduced in patients with CD (indeterminate chronic phase, with electrocardiogram abnormalities and heart failure) compared to healthy individuals [84]. More recently, Barbosa-Ferreira et al. [85] and Dabarian et al. [86] found no differences in leptin and AD levels between individuals with CCC and the control group.

In these studies, body mass index (BMI) was similar between groups [85,86]. In contrast, when the BMI was higher in patients chronically infected than in controls, leptin levels were increased, and AD tended to decrease [81].

In a study involving the ablation of fat cells in mice during the indeterminate form of the chronic phase (90 dpi), Lizardo et al. [64] demonstrated an increase in cardiac pathology, reinforcing the idea that adipocytes contribute to the progression of CCC. In the hearts of these mice, there was a higher immune cell infiltration, more pronounced histological changes, and an increased number of parasites compared to infected mice without fat cell ablation. Several signaling pathways might be implicated [64]. Furthermore, the parasite persistence in AT in chronically infected mice was shown, suggesting that the presence of *T. cruzi* and the associated inflammatory infiltrates alter AT physiology (i.e., imbalance between adipogenesis/lipolysis) [64].

Markers of oxidative stress that have been associated with heart disease show increased levels of both WAT and BAT in CD. Simultaneously, the antioxidant response in these tissues is impaired in *T. cruzi*-infected mice. Thus, CCC has been associated with a high degree of inflammation and oxidative stress within AT [67].

While several studies have documented the presence of *T. cruzi* in WAT and/or BAT [25,26,87,88], the distinct consequences of parasitism in each tissue type, mainly in BAT, have received limited attention. Given their distinct cellular, molecular, and physiological characteristics, including antagonistic functions [41], the adipocyte response infection on WAT and BAT is probably different [89].

In view of the above, the presence of *T. cruzi* in AT plays a critical role in the progression of CD, influencing various pathological processes such as inflammation, metabolic disturbances, and immune dysregulation. The parasite's preference for AT, particularly in the acute phase, leads to alterations in adipocyte function, including apoptosis, impaired lipolysis, and changes in adipokine secretion. These disturbances contribute to the persistence of *T. cruzi* and exacerbate cardiac pathology during the chronic phase of the disease.

5. Influence of a High-Fat Diet on the Physiology of Adipose Tissue and Chagas Disease

Human health is closely linked to dietary habits. Most of the global population follows Western diets, characterized by high calorie and fat intake. This dietary pattern is becoming prevalent in Latin America, an endemic region for CD [90], often leading to overweight and obesity, which may have effects on the interaction between *T. cruzi* and the human host [17]. Indeed, more than half of the female population in Latin America is either overweight or obese [91].

A study involving 66 patients with chronic CD in Brazil found that 83% exhibited a sedentary lifestyle and 94% were overweight or obese [92]. In Argentina, 42% of 190 *T. cruzi*-chronically infected individuals had BMI values compatible with obesity [81]. Despite these findings, there is still no consensus that *T. cruzi* infection causes obesity, that obese individuals are more susceptible to poor prognosis of CD, or that the parasite load is higher in overweight individuals [7,25,26]. However, it is established that the high-fat diet (HFD) has metabolic consequences and induces a pro-inflammatory state in AT [75]. The question that remains unresolved is whether an HFD improves or exacerbates CD.

Of note, Janoschek et al. [93] emphasize that a wide variety of diets are used in murine studies. The nutritional composition is quite variable, underscoring the lack of standardization. The most commonly used HFD in studies over the past decade provides 60% of its calories from fat. The studies selected for this review, which allow an analysis of the influence of HFD on AT and CD, have employed this diet, except for the study by Zaki et al. [75], where the HFD contained 20% fat, 20% protein, and 48% carbohydrate [94].

Without *T. cruzi* infection, mice fed with HFDs for different periods (16 to 24 weeks) had increased body weight, body fat percentage, and fasting blood glucose level, confirming the metabolic effects of this type of diet. In addition, changes in immune patterns were found with the enhanced accumulation of immune cells in AT, increased serum cytokines levels, and an enlargement of adipocyte size [95]. The influx of macrophages and effector T cells in AT was also observed during human AT inflammation [96].

Likewise, in the context of CD, the infiltration of macrophages in AT was presented in both early and late infection in a murine model of diet-induced obesity. In this model, a medium-fat diet (MFD, 14%) was administered in association with fructose in water (5%) and a single dose of streptozotocin [27]. In infected mice fed with HFD, the number of inflammatory cells in AT and heart at 30 dpi is greater than that observed in *T. cruzi*-infected mice fed with a standard diet (SD, 10% fat) [17,73].

The consumption of MFD resulted in both AT and systemic inflammatory response. Circulating levels of IL-6 were higher in acute and chronic phases compared to a group submitted to the MFD but not infected, while TNF-α showed a significant increase exclusively during the early infection [27]. In a model in which mice started HFD 8 weeks before infection, plasma TNF-α levels [73], as well as its concentrations in the heart [70,73], were elevated in the acute phase of infection, while the measurements in AT were similar between acutely infected and uninfected mice on HFD [73]. In the same way, the production of interferon-gamma (IFN-γ) in AT tissue was equivalent in groups submitted to HFD infected or not [73]. When comparing acutely *T. cruzi*-infected animals fed an SD to the HFD, divergent results concerning the levels of IFN-γ in cardiac and AT could be observed in different studies (Figure 3).

The expression of leptin, AD, and PPAR-γ, markers associated with lipogenesis and adipogenesis, were lower in the WAT of mice fed an SD compared to those fed an HFD during acute infection, whereas these adipokines were elevated in the heart, particularly in mice on the HFD [17]. In the chronic phase, serum leptin levels were reduced in uninfected MFD mice compared to their infected counterparts at 12- and 24 weeks post-infection. Conversely, AD levels were decreased in MFD-infected mice relative to uninfected ones in the same period [27].

There was a discrepancy regarding mortality and parasitemia during the acute phase. When comparing mice on HFD to those on SD, the survival rate was reported as higher [17,70], lower [71], or similar between the groups [74]. These differences may perhaps be attributed to the *T. cruzi* strain used in the studies (Brazil strain for [17,70]; VL-10 strain for [71]; Colombian strain for [74]).

Similarly, *T. cruzi*-infected mice on HFD exhibited varying levels of parasitemia, which were decreased [17,75], enhanced [70,71,73], or similar to those on SD [74]. Regarding parasite load, heart tissue showed a lower parasite burden with HFD [17], while an increase was observed in WAT [17,74]. Although Souza et al. [74] did not observe an elevation in parasitemia in infected mice fed with HFD compared to SD, the parasitic load in both heart and AT increased at 30 dpi with this dietary pattern [74] (Figure 3). Once again, some of these differences may be attributed to the experimental design, the key features of which are outlined in Table S1 (Supplementary Materials).

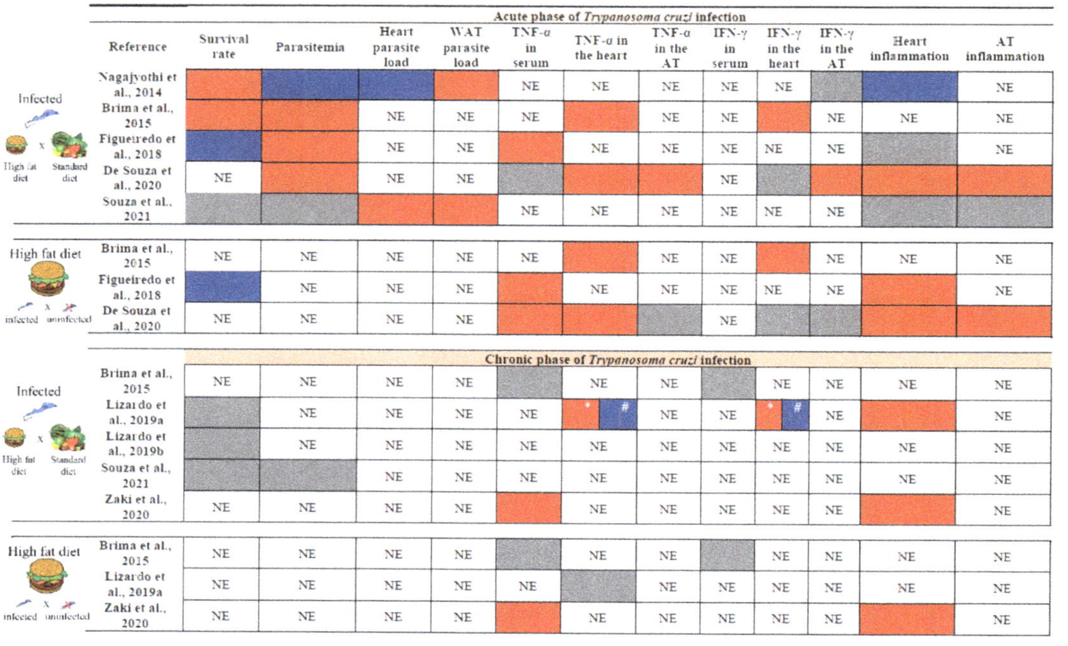

Figure 3. Impact of a high-fat diet on some variables analyzed in the acute and/or chronic phases of *T. cruzi* infection according to key studies. The upper section presents the results of experiments conducted during the acute phase of *T. cruzi* infection, while the bottom section corresponds to the chronic phase. In each section, the top panel compares infected animals fed with a high-fat diet (HFD) to those fed with a standard diet (SD), and the bottom panel compares infected and uninfected mice fed an HFD. The red box indicates an increase in the variable, the blue box indicates a decrease, and the gray box indicates no change. It is important to note that differences in results may be attributed to variations in experimental protocols. All studies were performed in murine models [17,63,70–75]. WAT: white adipose tissue. TNF-α: tumor necrosis factor-alpha. IFN-γ: interferon-gamma. AT: adipose tissue. NE: Not evaluated. *: Increase in 120 dpi. #: Decrease in 160 dpi.

When associated with *T. cruzi* infection, HFD promotes adipogenesis [17] and cardiac lipidopathy, which may change the course of the acute phase. Indeed, Nagajyothi et al. [17] suggest that HFD has a protective effect against myocardial damage during the acute phase of *T. cruzi* infection, likely due to these mechanisms. Histology of the heart from infected HFD-fed mice displayed a reduction in fibrosis and inflammation compared to infected SD-fed mice. Additionally, infected mice on HFD had only a 20% reduction in fat compared to uninfected HFD-fed mice, whereas infected SD-fed mice showed significant lipolysis [17].

In the chronic phase of infection, a link was suggested among the consumption of HFD, inflammatory status in the AT, and the progression and severity of CD. During this stage of the disease, the presence of inflammatory infiltrate was higher in the heart [63,75], as well as serum levels of TNF-α in mice fed with HFD and infected with *T. cruzi* [75] (Figure 3).

On the other hand, the ablation of WAT in mice did not increase the number of parasites but led to a redistribution of *T. cruzi* in different tissues, resulting in a decreased parasitic load in the epididymal WAT. These mice also showed higher cardiac inflammatory infiltrate compared to infected wild-type mice. In other words, fat ablation may exacerbate the inflammatory response during *T. cruzi* infection, potentially intensifying cardiac damage and Chagas cardiomyopathy [75].

According to Lizardo et al. [63], mice exposed to an HFD and infected with *T. cruzi* are more susceptible to cardiac alterations in the late chronic phase (160 dpi). These alterations affected left and right ventricular function, accompanied by tissue damage characterized by heightened inflammation and hypertrophy. Furthermore, HFD exacerbates mitochondrial stress, which the authors suggest may contribute to the pathogenesis of CCC [63].

Altogether, these findings indicate that dietary lipid content influences the progression of CD by impacting immune responses, inflammation, and lipid metabolism. In fact, *T. cruzi* demonstrates a strong affinity for host lipoproteins and cholesterol, utilizing lipids and the low-density lipoprotein receptor (LDLr) for cell invasion [97].

It is essential to note that HFD consumption exhibits mixed effects on CD, depending on the disease stage, *T. cruzi*'s genetic background, the host's immune response, and the overall metabolic status, including AT metabolism (Figure 3). This underscores the intricate nature of CD's pathogenic process. It is important to emphasize that the lack of agreement regarding assay design may explain divergent results and, at times, lead to inconsistencies when comparing studies.

The literature review pointed out that a greater number of analyses were performed in the acute phase of disease, yielding both positive and negative results (Figure 3). In the chronic phase of infection, although a certain number of variables presented in Figure 3 do not change with the consumption of HFD, the results of the different cardiac analyses indicate that this diet is detrimental in the long term, contributing to the progression of CCC [63,75].

Thus, the studies conducted to date suggest that the consumption of an HFD can modulate the progression of CD by influencing immune responses, inflammatory processes, and lipid metabolism. The HFD induces notable alterations in AT, characterized by increased infiltration of inflammatory cells and modifications in adipokine expression. Additionally, this dietary pattern impacts parasite load and cardiac responses. While the effects of HFD during the acute phase of infection exhibit variability, in the chronic phase, it appears to exacerbate cardiac inflammation and mitochondrial dysfunction, thereby amplifying myocardial damage and contributing to the progression of CCC (Figure 3). Together, these results highlight the necessity for further investigation into the impact of diet on the course of *T. cruzi* infection.

6. Conclusions

Despite the evidence, limited attention has been given to the clinical importance of AT in the pathological progression of CD. This review underscores the potential role of AT in the pathogenesis of CD, as it is a complex organ composed of several cell types with immunological and hormonal functions through the release of numerous adipokines and cytokines involved, among others, in inflammatory pathways. Moreover, *T. cruzi* persists within adipocytes, contributing to systemic inflammation and metabolic abnormalities that can exacerbate CD severity (Figure 4). The interplay between cardiac tissue and AT warrants further investigation, as both are affected by *T. cruzi* infection, and cardiac fat appears to play a role in the regulation of metabolic processes and myocardial inflammation. Furthermore, the influence of HFD on CD progression also reveals complexities. While HFD may offer protection against acute cardiac damage, it could potentially exacerbate inflammation and myocardial alterations in the chronic phase, thereby worsening CCC. The influence of normal and dysfunctional AT on the severity of CD remains unclear and should be further investigated, especially the role of myocardial fat. Elucidating the role of AT in CD pathogenesis may allow advances in treatment strategies for the disease.

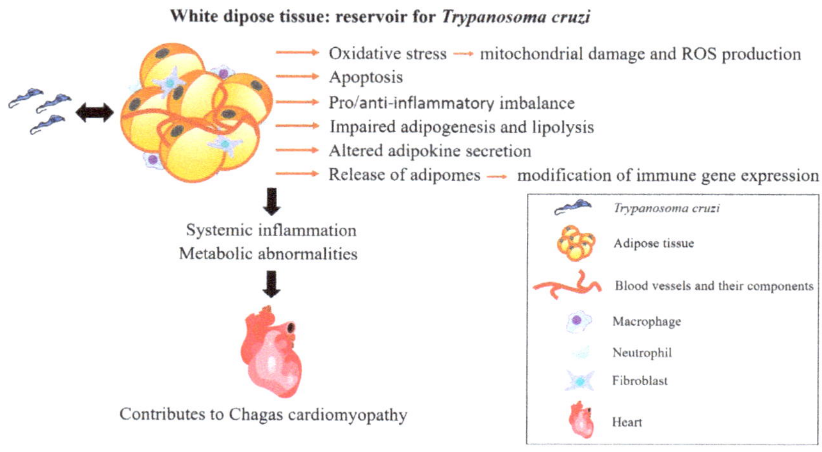

Figure 4. Consequences of the interaction between *T. cruzi* and white adipose tissue. In the acute phase of infection, adipose tissue is a target for *T. cruzi*. In the chronic phase, it serves as a reservoir for the parasite, contributing to the persistence of the infection. The inflammation triggered by *T. cruzi* persists in adipose tissue, leading to several effects, including oxidative stress, adipocyte apoptosis, an imbalance in pro- and anti-inflammatory factors, as well as impaired adipogenesis and lipolysis. These effects are influenced by the parasite's genetics and the host's immune response. Together, these factors contribute to the development of chronic Chagas cardiomyopathy. ROS: reactive oxygen species. Heart (https://beta.scidraw.io/drawing/277) and macrophage (https://beta.scidraw.io/drawing/221) were obtained from https://scidraw.io/. Accessed on 19 February 2025.

Supplementary Materials: The following supporting information can be downloaded at https://www.mdpi.com/article/10.3390/pathogens14040339/s1, Table S1: Key characteristics of the main articles on the impact of high-fat diet on Chagas disease.

Author Contributions: L.H. conceived the work and prepared the figures. L.H., V.F.d.S.P. and M.H. wrote the manuscript. N.N. helped shape the final report. Funding acquisition, L.H. All authors have read and agreed to the published version of the manuscript.

Funding: This research was funded by Fundação de Apoio à Pesquisa do Distrito Federal (FAPDF), grant number 00193.00000194/2019-13, and Coordenação de Aperfeiçoamento de Pessoal de Nível Superior (CAPES) (23106.111911/2023-83).

Acknowledgments: We thank the financial support provided by DPI/BCE/UnB, under the Public Notice No. 001/2025 DPI/BCE/UnB, which covered the publication fee for this article.

Conflicts of Interest: The authors declare no conflicts of interest.

Abbreviations

The following abbreviations are used in this manuscript:

CD	Chagas disease
AT	adipose tissue
HFD	high-fat diet
qPCR	quantitative PCR
kDNA	kinetoplast DNA
CCC	chronic Chagas cardiomyopathy
WAT	white adipose tissue
BAT	brown adipose tissue
miRNAs	microRNAs

PAT	pink adipose tissue
PBMCs	peripheral blood mononuclear cells
IL-6	interleukin-6
TNF-α	tumor necrosis factor-alpha
PPAR-γ	peroxisome proliferator-activated receptor gamma
AD	adiponectin
ATF-7	activating transcription factor-7
BMI	body mass index
dpi	days post-infection
MFD	medium-fat diet
IFN-γ	interferon-gamma
SD	standard diet
LDLr	low-density lipoprotein receptor

References

1. WHO Chagas Disease (Also Known as American Trypanosomiasis). 2024. Available online: https://www.who.int/news-room/fact-sheets/detail/chagas-disease-(american-trypanosomiasis) (accessed on 29 August 2024).
2. Fernandes, M.C.; Andrews, N.W. Host Cell Invation by *T. cruzi*: A Unique Strategy That Promotes Persistence. *FEMS Microbiol.* **2013**, *36*, 734–747. [CrossRef] [PubMed]
3. Martín-Escolano, J.; Marín, C.; Rosales, M.J.; Tsaousis, A.D.; Medina-Carmona, E.; Martín-Escolano, R. An Updated View of the *T. cruzi* Life Cycle: Intervention Points for an Effective Treatment. *ACS Infect. Dis.* **2022**, *8*, 1107–1115. [CrossRef] [PubMed]
4. Guarner, J. Chagas Disease as Example of a Reemerging Parasite. *Semin. Diagn. Pathol.* **2019**, *36*, 164–169. [CrossRef] [PubMed]
5. Ribeiro, V.; Dias, N.; Paiva, T.; Hagström-Bex, L.; Nitz, N.; Pratesi, R.; Hecht, M. Current Trends in the Pharmacological Management of Chagas Disease. *Int. J. Parasitol. Drugs Drug Resist.* **2020**, *12*, 7–17. [CrossRef]
6. De Bona, E.; Lidani, K.C.F.; Bavia, L.; Omidian, Z.; Gremski, L.H.; Sandri, T.L.; de Messias Reason, I.J. Autoimmunity in Chronic Chagas Disease: A Road of Multiple Pathways to Cardiomyopathy? *Front. Immunol.* **2018**, *9*, 1842. [CrossRef]
7. Nagajyothi, F.; Machado, F.S.; Burleigh, B.A.; Jelicks, L.A.; Scherer, E.; Mukherjee, S.; Lisanti, M.P.; Weiss, L.M.; Garg, N.J.; Tanowitz, H.B. Mechanisms of *T. cruzi* Persistence in Chagas Disease. *Cell Microbiol* **2012**, *14*, 634–643. [CrossRef]
8. Pérez-Mazliah, D.; Ward, A.I.; Lewis, M.D. Host-Parasite Dynamics in Chagas Disease from Systemic to Hyper-Local Scales. *Parasite Immunol.* **2021**, *43*, e12786. [CrossRef]
9. Lewis, M.D.; Fortes Francisco, A.; Taylor, M.C.; Burrell-Saward, H.; Mclatchie, A.P.; Miles, M.A.; Kelly, J.M. Bioluminescence Imaging of Chronic *T. cruzi* Infections Reveals Tissue-Specific Parasite Dynamics and Heart Disease in the Absence of Locally Persistent Infection. *Cell. Microbiol.* **2014**, *16*, 1285–1300. [CrossRef] [PubMed]
10. Lewis, M.D.; Kelly, J.M. Putting Infection Dynamics at the Heart of Chagas Disease. *Trends Parasitol.* **2016**, *32*, 899–911. [CrossRef]
11. Ramírez-Toloza, G.; Ferreira, A. *T. cruzi* Evades the Complement System as an Efficient Strategy to Survive in the Mammalian Host: The Specific Roles of Host/Parasite Molecules and *T. cruzi* Calreticulin. *Front. Microbiol.* **2017**, *8*, 1667. [CrossRef]
12. Echeverría, L.E.; Rojas, L.Z.; Rueda-Ochoa, O.L.; Gómez-Ochoa, S.A.; González Rugeles, C.I.; Díaz, M.L.; Marcus, R.; Morillo, C.A. Circulating *T. cruzi* Load and Major Cardiovascular Outcomes in Patients with Chronic Chagas Cardiomyopathy: A Prospective Cohort Study. *Trop. Med. Int. Health* **2020**, *25*, 1534–1541. [CrossRef] [PubMed]
13. Khan, A.A.; Langston, H.C.; Costa, F.C.; Olmo, F.; Taylor, M.C.; McCann, C.J.; Kelly, J.M.; Lewis, M.D. Local Association of *T. cruzi* Chronic Infection Foci and Enteric Neuropathic Lesions at the Tissue Micro-Domain Scale. *PLoS Pathog.* **2021**, *17*, e1009864. [CrossRef]
14. de Oliveira, M.T.; Schmidt, A.; da Silva, M.C.; Donadi, E.A.; da Silva, J.S.; Marin-Neto, J.A. Parasitic Load Correlates With Left Ventricular Dysfunction in Patients With Chronic Chagas Cardiomyopathy. *Front. Cardiovasc. Med.* **2021**, *8*, 741347. [CrossRef] [PubMed]
15. Marcon, G.E.B.; de Albuquerque, D.M.; Batista, A.M.; Andrade, P.D.; Almeida, E.A.; Guariento, M.E.; Teixeira, M.A.B.; Costa, S.C.B. *T. cruzi*: Parasite Persistence in Tissues in Chronic Chagasic Brazilian Patients. *Mem. Inst. Oswaldo Cruz* **2011**, *106*, 85–91. [CrossRef]
16. Silberstein, E.; Serna, C.; Fragoso, S.P.; Nagarkatti, R.; Debrabant, A. A Novel Nanoluciferase-Based System to Monitor *T. cruzi* Infection in Mice by Bioluminescence Imaging. *PLoS ONE* **2018**, *13*, e0195879. [CrossRef] [PubMed]
17. Nagajyothi, F.; Weiss, L.M.; Zhao, D.; Koba, W.; Jelicks, L.A.; Cui, M.H.; Factor, S.M.; Scherer, P.E.; Tanowitz, H.B. High Fat Diet Modulates *T. cruzi* Infection Associated Myocarditis. *PLOS Negl. Trop. Dis.* **2014**, *8*, e3118. [CrossRef] [PubMed]
18. Tanowitz, H.B.; Scherer, P.E.; Mota, M.M.; Figueiredo, L.M. Adipose Tissue: A Safe Haven for Parasites? *Trends Parasitol.* **2017**, *33*, 276–284. [CrossRef] [PubMed]

19. Moreira, L.R.; Silva, A.C.; Oliveira, C.N.d.C.; Júnior, C.D.d.S.; Nascimento, A.V.; Oliveira, K.K.D.S.; Soares, A.K.d.A.; Saraiva, K.L.A.; Cavalcanti, M.d.P.; de Lorena, V.M.B. Benznidazole Treatment Decreases IL-6 Levels in *T. cruzi*-Infected Human Adipocytes Differentiated from Adipose Tissue-Derived Stem Cells. *Mem. Inst. Oswaldo Cruz* **2023**, *118*, e220295. [CrossRef]
20. Chao, C.; Leone, J.L.; Vigliano, C.A. Chagas Disease: Historic Perspective. *Biochim. Biophys. Acta-Mol. Basis Dis.* **2020**, *1866*, 165689. [CrossRef] [PubMed]
21. Bonney, K.M.; Engman, D.M. Autoimmune Pathogenesis of Chagas Heart Disease: Looking Back, Looking Ahead. *Am. J. Pathol.* **2015**, *185*, 1537–1547. [CrossRef]
22. Santos-Buch, C.A.; Teixeira, A.R.L. The Immunology of Experimental Chagas' Disease. *J. Exp. Med.* **1974**, *140*, 38–53. [CrossRef]
23. Teixeira, A.R.L.; Hecht, M.M.; Guimaro, M.C.; Sousa, A.O.; Nitz, N. Pathogenesis of Chagas' Disease: Parasite Persistence and Autoimmunity. *Clin. Microbiol. Rev.* **2011**, *24*, 592–630. [CrossRef]
24. Wesley, M.; Moraes, A.; Rosa, A.d.C.; Carvalho, J.L.; Shiroma, T.; Vital, T.; Dias, N.; de Carvalho, B.; Rabello, D.D.A.; Borges, T.K.D.S.; et al. Correlation of Parasite Burden, Kdna Integration, Autoreactive Antibodies, and Cytokine Pattern in the Pathophysiology of Chagas Disease. *Front. Microbiol.* **2019**, *10*, 1856. [CrossRef] [PubMed]
25. Combs, T.P.; Nagajyothi; Mukherjee, S.; De Almeida, C.J.; Jelicks, L.A.; Schubert, W.; Lin, Y.; Jayabalan, D.S.; Zhao, D.; Braunstein, V.L.; et al. The Adipocyte as an Important Target Cell for *T. cruzi* Infection. *J. Biol. Chem.* **2005**, *280*, 24085–24094. [CrossRef] [PubMed]
26. Ferreira, A.V.M.; Segatto, M.; Menezesb, Z.; Macedo, A.M.; Gelaped, C.; de Oliveira Andrade, L.; Nagajyothif, F.; Schererg, P.E.; Teixeira, M.M.; Tanowitzf, H.B. Evidence for *T. cruzi* in Adipose Tissue in Human Chronic Chagas Disease. *Microbes Infect.* **2011**, *13*, 1002–1005. [CrossRef] [PubMed]
27. Cabalén, M.E.; Cabral, M.F.; Sanmarco, L.M.; Andrada, M.C.; Onofrio, L.I.; Ponce, N.E.; Aoki, M.P.; Gea, S.; Cano, R.C. Chronic *T. cruzi* Infection Potentiates Adipose Tissue Macrophage Polarization toward an Anti-Inflammatory M2 Phenotype and Contributes to Diabetes Progression in a Diet-Induced Obesity Model. *Oncotarget* **2016**, *7*, 13400–13415. [CrossRef] [PubMed]
28. Añez, N.; Crisante, G. The Tissue Specific Tropism in *T. cruzi*. Is It True? *Acta Trop.* **2021**, *213*, 105736. [CrossRef]
29. da Silva, A.C.; Moreira, L.R.; Oliveira, C.N.d.C.; Júnior, C.D.d.S.; do Ó, K.P.; Oliveira, K.K.D.S.; Melo, M.G.N.D.; Soares, A.K.d.A.; Cavalcanti, M.d.P.; Vasconcelos, L.R.S.; et al. Dynamics of the *T. cruzi* Infection in Adipose Tissue: Assessing Gene Expression of PNPLA2, FASN, and ACAT1 under Benzonidazole Treatment and Indirect Mononuclear Immune Cells Interaction. *Mol. Biochem. Parasitol.* **2024**, *258*, 111618. [CrossRef]
30. Moreira, L.R.; Silva, A.C.; Costa-Oliveira, C.N.d.; Silva-Júnior, C.D.d.; Oliveira, K.K.d.S.; Torres, D.J.L.; Barros, M.D.; Rabello, M.C.d.S.; de Lorena, V.M.B. Interaction between Peripheral Blood Mononuclear Cells and *T. cruzi*-Infected Adipocytes: Implications for Treatment Failure and Induction of Immunomodulatory Mechanisms in Adipose Tissue. *Front. Immunol.* **2024**, *15*, 1280877. [CrossRef]
31. Lizcano, F.; Arroyave, F. Control of Adipose Cell Browning and Its Therapeutic Potential. *Metabolites* **2020**, *10*, 471. [CrossRef]
32. Michailidou, Z.; Gomez-Salazar, M.; Alexaki, V.I. Innate Immune Cells in the Adipose Tissue in Health and Metabolic Disease. *J. Innate Immun.* **2022**, *14*, 4–30. [CrossRef] [PubMed]
33. Hegde, V.; Dhurandhar, N.V. Microbes and Obesity-Interrelationship between Infection, Adipose Tissue and the Immune System. *Clin. Microbiol. Infect.* **2013**, *19*, 314–320. [CrossRef]
34. Rutkowski, J.M.; Stern, J.H.; Scherer, P.E. The Cell Biology of Fat Expansion. *J. Cell Biol.* **2015**, *208*, 501–512. [CrossRef]
35. Ray, A.; Bonorden, M.J.L.; Pandit, R.; Nkhata, K.J.; Bishayee, A. Infections and Immunity: Associations with Obesity and Related Metabolic Disorders. *J. Pathol. Transl. Med.* **2023**, *57*, 28–42. [CrossRef]
36. McGown, C.; Birerdinc, A.; Younossi, Z.M. Adipose Tissue as an Endocrine Organ. *Clin. Liver Dis.* **2014**, *18*, 41–58. [CrossRef] [PubMed]
37. Wang, S.; Liu, Y.; Chen, J.; He, Y.; Ma, W.; Liu, X.; Sun, X. Effects of Multi-Organ Crosstalk on the Physiology and Pathology of Adipose Tissue. *Front. Endocrinol.* **2023**, *14*, 1198984. [CrossRef]
38. Recinella, L.; Orlando, G.; Ferrante, C.; Chiavaroli, A.; Brunetti, L.; Leone, S. Adipokines: New Potential Therapeutic Target for Obesity and Metabolic, Rheumatic, and Cardiovascular Diseases. *Front. Physiol.* **2020**, *11*, 578966. [CrossRef] [PubMed]
39. Scherer, P.E. The Multifaceted Roles of Adipose Tissue—Therapeutic Targets for Diabetes and beyond: The 2015 Banting Lecture. *Diabetes* **2016**, *65*, 1452–1461. [CrossRef]
40. Thomou, T.; Mori, M.A.; Dreyfuss, J.M.; Konishi, M.; Wolfrum, C.; Rao, T.N.; Winnay, J.N.; Grinspoon, S.K.; Gorden, P.; Kahn, C.R. Adipose-Derived Circulating MiRNAs Regulate Gene Expression in Other Tissues. *Nature* **2017**, *542*, 450–455. [CrossRef]
41. Barthelemy, J.; Bogard, G.; Wolowczuk, I. Beyond Energy Balance Regulation: The Underestimated Role of Adipose Tissues in Host Defense against Pathogens. *Front. Immunol.* **2023**, *14*, 1083191. [CrossRef]
42. Kajimura, S.; Spiegelman, B.M.; Seale, P. Brown Adipose Tissue and Its Therapeutic Potential. *Cell Metab.* **2015**, *22*, 546–559. [CrossRef] [PubMed]
43. Oelkrug, R.; Polymeropoulos, E.T.; Jastroch, M. Brown Adipose Tissue: Physiological Function and Evolutionary Significance. *J. Comp. Physiol. B Biochem. Syst. Environ. Physiol.* **2015**, *185*, 587–606. [CrossRef] [PubMed]

44. Corrêa, L.H.; Heyn, G.S.; Magalhaes, K.G. The Impact of the Adipose Organ Plasticity on Inflammation and Cancer Progression. *Cells* **2019**, *8*, 662. [CrossRef] [PubMed]
45. Giordano, A.; Smorlesi, A.; Frontini, A.; Barbatelli, G.; Cint, S. White, Brown and Pink Adipocytes: The Extraordinary Plasticity of the Adipose Organ. *Eur. J. Endocrinol.* **2014**, *170*, R159–R171. [CrossRef]
46. Rajala, M.W.; Scherer, P.E. Minireview: The Adipocyte—At the Crossroads of Energy Homeostasis, Inflammation, and Atherosclerosis. *Endocrinology* **2003**, *144*, 3765–3773. [CrossRef]
47. Ayyappan, J.P.; Ganapathi, U.; Lizardo, K.; Vinnard, C.; Nagajyothi, J.F.; Subbian, S.; Perlin, D.S. Adipose Tissue Regulates Pulmonary Pathology during TB Infection. *MBio* **2019**, *10*, 1–16. [CrossRef]
48. Beigier-Bompadre, M.; Montagna, G.N.; Kühl, A.A.; Lozza, L.; Weiner, J.; Kupz, A.; Vogelzang, A.; Mollenkopf, H.J.; Löwe, D.; Bandermann, S.; et al. Mycobacterium Tuberculosis Infection Modulates Adipose Tissue Biology. *PLoS Pathog.* **2017**, *13*, e1006676. [CrossRef]
49. Franke-Fayard, B.; Janse, C.J.; Cunha-Rodrigues, M.; Ramesar, J.; Büscher, P.; Que, I.; Löwik, C.; Voshol, P.J.; Den Boer, M.A.M.; Van Duinen, S.G.; et al. Murine Malaria Parasite Sequestration: CD36 Is the Major Receptor, but Cerebral Pathology Is Unlinked to Sequestration. *Proc. Natl. Acad. Sci. USA* **2005**, *102*, 11468–11473. [CrossRef]
50. Mejia, P.; Humberto Treviño-Villarreal, J.; De Niz, M.; Meibalan, E.; Longchamp, A.; Reynolds, J.S.; Turnbull, L.B.; Opoka, R.O.; Roussilhon, C.; Spielmann, T.; et al. Adipose Tissue Parasite Sequestration Drives Leptin Production in Mice and Correlates with Human Cerebral Malaria. *Sci. Adv.* **2021**, *7*, eabe2484. [CrossRef]
51. Moser, J.; Emous, M.; Heeringa, P.; Rodenhuis-Zybert, I.A. Mechanisms and Pathophysiology of SARS-CoV-2 Infection of the Adipose Tissue. *Trends Endocrinol. Metab.* **2023**, *34*, 735–748. [CrossRef]
52. Amisigo, C.M.; Amegatcher, G.; Sunter, J.D.; Gwira, T.M. Adipose and Skin Distribution of African Trypanosomes in Natural Animal Infections. *Parasites Vectors* **2024**, *17*, 215. [CrossRef] [PubMed]
53. Machado, H.; Bizarra-Rebelo, T.; Costa-Sequeira, M.; Trindade, S.; Carvalho, T.; Rijo-Ferreira, F.; Pacheco, B.R.; Serre, K.; Figueiredo, L.M. Trypanosoma Brucei Triggers a Broad Immune Response in the Adipose Tissue. *PLoS Pathog.* **2021**, *17*, e1009933. [CrossRef]
54. Trindade, S.; De Niz, M.; Costa-Sequeira, M.; Bizarra-Rebelo, T.; Bento, F.; Dejung, M.; Narciso, M.V.; López-Escobar, L.; Ferreira, J.; Butter, F.; et al. Slow Growing Behavior in African Trypanosomes during Adipose Tissue Colonization. *Nat. Commun.* **2022**, *13*, 7548. [CrossRef] [PubMed]
55. Schwing, A.; Pisani, D.F.; Pomares, C.; Majoor, A.; Lacas-Gervais, S.; Jager, J.; Lemichez, E.; Marty, P.; Boyer, L.; Michel, G. Identification of Adipocytes as Target Cells for Leishmania Infantum Parasites. *Sci. Rep.* **2021**, *11*, 21275. [CrossRef] [PubMed]
56. Trindade, S.; Rijo-Ferreira, F.; Carvalho, T.; Pinto-Neves, D.; Guegan, F.; Aresta-Branco, F.; Bento, F.; Young, S.A.; Pinto, A.; Van Den Abbeele, J.; et al. Trypanosoma Brucei Parasites Occupy and Functionally Adapt to the Adipose Tissue in Mice. *Cell Host Microbe* **2016**, *19*, 837–848. [CrossRef]
57. Nagajyothi, F.; Desruisseaux, M.S.; MacHado, F.S.; Upadhya, R.; Zhao, D.; Schwartz, G.J.; Teixeira, M.M.; Albanese, C.; Lisanti, M.P.; Chua, S.C.; et al. Response of Adipose Tissue to Early Infection with *T. cruzi* (Brazil Strain). *J. Infect. Dis.* **2012**, *205*, 830–840. [CrossRef] [PubMed]
58. de Aquino, G.P.; Gomes, M.A.M.; Salinas, R.K.; Laranjeira-Silva, M.F. Lipid and Fatty Acid Metabolism in Trypanosomatids. *Microb. Cell* **2021**, *8*, 262–275. [CrossRef]
59. Zuzarte-Luís, V.; Mota, M.M. Parasite Sensing of Host Nutrients and Environmental Cues. *Cell Host Microbe* **2018**, *23*, 749–758. [CrossRef]
60. Tanowitz, H.B.; Jelicks, L.A.; Machado, F.S.; Esper, L.; Qi, X.; Desruisseaux, M.S.; Chua, S.C.; Scherer, P.E.; Nagajyothi, F. Adipose Tissue, Diabetes and Chagas Disease. *Adv. Parasitol.* **2011**, *76*, 235–250. [CrossRef]
61. Chevillard, C.; Nunes, J.P.S.; Frade, A.F.; Almeida, R.R.; Pandey, R.P.; Nascimento, M.S.; Kalil, J.; Cunha-Neto, E. Disease Tolerance and Pathogen Resistance Genes May Underlie *T. cruzi* Persistence and Differential Progression to Chagas Disease Cardiomyopathy. *Front. Immunol.* **2018**, *9*, 2791. [CrossRef]
62. Nagajyothi, F.; Weiss, L.M.; Silver, D.L.; Desruisseaux, M.S.; Scherer, P.E.; Herz, J.; Tanowitz, H.B. *T. cruzi* Utilizes the Host Low Density Lipoprotein Receptor in Invasion. *PLoS Negl. Trop. Dis.* **2011**, *5*, e953. [CrossRef] [PubMed]
63. Lizardo, K.; Ayyappan, J.P.; Cui, M.H.; Balasubramanya, R.; Jelicks, L.A.; Nagajyothi, J.F. High Fat Diet Aggravates Cardiomyopathy in Murine Chronic Chagas Disease. *Microbes Infect.* **2019**, *21*, 63–71. [CrossRef]
64. Lizardo, K.; Ayyappan, J.P.; Oswal, N.; Weiss, L.M.; Scherer, P.E.; Nagajyothi, J.F. Fat Tissue Regulates the Pathogenesis and Severity of Cardiomyopathy in Murine Chagas Disease. *PLoS Negl. Trop. Dis.* **2021**, *15*, e0008964. [CrossRef] [PubMed]
65. Thangavel, H.; Dhanyalayam, D.; Kim, M.; Lizardo, K.; Sidrat, T.; Lopez, J.G.; Wang, X.; Bansal, S.; Nagajyothi, J.F. Adipocyte-Released Adipomes in Chagas Cardiomyopathy: Impact on Cardiac Metabolic and Immune Regulation. *iScience* **2024**, *27*, 109672. [CrossRef]
66. Nagajyothi, J.F.; Weiss, L.M. Advances in Understanding the Role of Adipose Tissue and Mitochondrial Oxidative Stress in *T. cruzi* Infection. *F1000Research* **2019**, *8*, 1–17. [CrossRef]

67. Wen, J.-J.; Nagajyothi, F.; Machado, F.S.; Weiss, L.M.; Scherer, P.E.; Tanowitz, H.B.; Garg, N.J. Markers of Oxidative Stress in Adipose Tissue during *T. cruzi* Infection. *Parasitol. Res.* **2014**, *113*, 3159–3165. [CrossRef] [PubMed]
68. Ayyappan, J.P.; Lizardo, K.; Wang, S.; Yurkow, E.; Nagajyothi, J.F. Inhibition of SREBP Improves Cardiac Lipidopathy, Improves Endoplasmic Reticulum Stress, and Modulates Chronic Chagas Cardiomyopathy. *J. Am. Heart Assoc.* **2020**, *9*, e014255. [CrossRef]
69. Nagajyothi, F.; Desruisseaux, M.S.; Thiruvur, N.; Weiss, L.M.; Braunstein, V.L.; Albanese, C.; Teixeira, M.M.; De Almeida, C.J.; Lisanti, M.P.; Scherer, P.E.; et al. *T. cruzi* Infection of Cultured Adipocytes Results in an Inflammatory Phenotype. *Obesity* **2008**, *16*, 1992–1997. [CrossRef] [PubMed]
70. Brima, W.; Eden, D.J.; Mehdi, S.F.; Bravo, M.; Wiese, M.M.; Stein, J.; Almonte, V.; Zhao, D.; Kurland, I.; Pessin, J.E.; et al. The Brighter (and Evolutionarily Older) Face of the Metabolic Syndrome: Evidence from *T. cruzi* Infection in CD-1 Mice. *Diabetes/Metab. Res. Rev.* **2015**, *31*, 346–359. [CrossRef]
71. Figueiredo, V.P.; Junior, E.S.L.; Lopes, L.R.; Simões, N.F.; Penitente, A.R.; Bearzoti, E.; Vieira, P.M.d.A.; Schulz, R.; Talvani, A. High Fat Diet Modulates Inflammatory Parameters in the Heart and Liver during Acute *T. cruzi* Infection. *Int. Immunopharmacol.* **2018**, *64*, 192–200. [CrossRef]
72. Lizardo, K.; Ayyappan, J.P.; Ganapathi, U.; Dutra, W.O.; Qiu, Y.; Weiss, L.M.; Nagajyothi, J.F. Diet Alters Serum Metabolomic Profiling in the Mouse Model of Chronic Chagas Cardiomyopathy. *Dis. Markers* **2019**, *2019*, 4956016. [CrossRef] [PubMed]
73. de Souza, D.M.S.; de Paula Costa, G.; Leite, A.L.J.; de Oliveira, D.S.; de Castro Pinto, K.M.; Farias, S.E.B.; Simões, N.F.; de Paiva, N.C.N.; de Abreu Vieira, P.M.; da Silva, C.A.M.; et al. A High-Fat Diet Exacerbates the Course of Experimental *T. cruzi* Infection That Can Be Mitigated by Treatment with Simvastatin. *Biomed Res. Int.* **2020**, *2020*, 1230461. [CrossRef] [PubMed]
74. Souza, D.M.S.d.; Silva, M.C.; Farias, S.E.B.; Menezes, A.P.d.J.; Milanezi, C.M.; Lúcio, K.d.P.; Paiva, N.C.N.; Abreu, P.M.d.; Costa, D.C.; Pinto, K.M.d.C.; et al. Diet Rich in Lard Promotes a Metabolic Environment Favorable to *T. cruzi* Growth. *Front. Cardiovasc. Med.* **2021**, *8*, 667580. [CrossRef]
75. Zaki, P.; Domingues, E.L.; Amjad, F.M.; Narde, M.B.; Gonçalves, K.R.; Viana, M.L.; de Paula, H.; de Lima, W.G.; Huang, H.; Bahia, M.T.; et al. The Role of Fat on Cardiomyopathy Outcome in Mice Models of Chronic *T. cruzi* Infection. *Parasitol Res.* **2020**, *119*, 1829–1843. [CrossRef]
76. Lewis, M.D.; Francisco, A.F.; Taylor, M.C.; Jayawardhana, S.; Kelly, J.M. Host and Parasite Genetics Shape a Link between *T. cruzi* Infection Dynamics and Chronic Cardiomyopathy. *Cell. Microbiol.* **2016**, *18*, 1429–1443. [CrossRef]
77. Santamaría, M.H.; Ríos, L.D.; Corral, R.S. *T. cruzi* Down-Regulates Adiponectin Expression in Mouse Adipocytes via the NFAT Signaling Pathway. *Microbes Infect.* **2021**, *23*, 104757. [CrossRef]
78. González, F.B.; Villar, S.R.; Toneatto, J.; Pacini, M.F.; Márquez, J.; D'Attilio, L.; Bottasso, O.A.; Piwien-Pilipuk, G.; Pérez, A.R. Immune Response Triggered by *T. cruzi* Infection Strikes Adipose Tissue Homeostasis Altering Lipid Storage, Enzyme Profile and Adipokine Expression. *Med. Microbiol. Immunol.* **2019**, *208*, 651–666. [CrossRef] [PubMed]
79. Liang, W.; Qi, Y.; Yi, H.; Mao, C.; Meng, Q.; Wang, H.; Zheng, C. The Roles of Adipose Tissue Macrophages in Human Disease. *Front. Immunol.* **2022**, *13*, 908749. [CrossRef]
80. Nagajyothi, F.; Zhao, D.; Machado, F.S.; Weiss, L.M.; Gary, J.; Desruisseaux, M.S.; Zhao, Y.; Factor, S.M.; Albanese, C.; Teixeira, M.M.; et al. Crucial Role of the Central Leptin Receptor in Murine *T. cruzi* Infection. *J. Infect. Dis.* **2010**, *202*, 1104–1113. [CrossRef] [PubMed]
81. González, F.; Villar, S.; D'Attilio, L.; Leiva, R.; Marquez, J.; Lioi, S.; Beloscar, J.; Bottasso, O.; Perez, A.R. Dysregulated Network of Immune, Endocrine and Metabolic Markers Is Associated to More Severe Human Chronic Chagas Cardiomyopathy. *Neuroimmunomodulation* **2018**, *25*, 119–128. [CrossRef] [PubMed]
82. Ouchi, N.; Shibata, R.; Walsh, K. Targeting Adiponectin for Cardioprotection. *Expert Opin. Ther. Targets* **2006**, *10*, 573–581. [CrossRef]
83. Wang, X.; Zhang, S.; Li, Z. Adipokines in Glucose and Lipid Metabolism. *Adipocyte* **2023**, *12*, 2202976. [CrossRef] [PubMed]
84. Fernandes, F.; Dantas, S.; Ianni, B.M.; Ramires, F.J.A.; Buck, P.; Salemi, V.M.C.; Lopes, H.F.; Mady, C. Leptin Levels in Different Forms of Chagas' Disease. *Braz. J. Med. Biol. Res.* **2007**, *40*, 1631–1636. [CrossRef]
85. Barbosa-Ferreira, J.M.; Mady, C.; Ianni, B.M.; Lopes, H.F.; Ramires, F.J.A.; Salemi, V.M.C.; Grupi, C.J.; Hachul, D.T.; Fernandes, F. Dysregulation of Autonomic Nervous System in Chagas' Heart Disease Is Associated with Altered Adipocytokines Levels. *PLoS ONE* **2015**, *10*, e0131447. [CrossRef] [PubMed]
86. Dabarian, A.L.; Mady, C.; Barbosa-Ferreira, J.M.; Ianni, B.M.; Hotta, V.T.; Ramires, F.J.A.; Lopes, H.F.; Buck, P.d.C.; Pessoa, F.G.; Fonseca, K.C.B.; et al. Dysregulation of Insulin Levels in Chagas Heart Disease Is Associated with Altered Adipocytokine Levels. *Can. J. Physiol. Pharmacol.* **2019**, *97*, 140–145. [CrossRef]
87. Andrade, Z.A.; Silva, H.R. Parasitism of Adipocytes by *T. cruzi*. *Mem. Inst. Oswaldo Cruz* **1995**, *90*, 521–522. [CrossRef] [PubMed]
88. Shoemaker, J.P.; Hoffman, R.V., Jr.; Huffman, D.G. *T. cruzi:* Preference for Brown Adipose Tissue in Mice by the Tulahuen Strain. *Exp. Parasitol.* **1970**, *27*, 403–407. [CrossRef] [PubMed]
89. Ayyappan, J.P.; Nagajyothi, J.F. Diet Modulates Adipose Tissue Oxidative Stress in a Murine Acute Chagas Model. *JSM Atheroscler.* **2017**, *2*, 1030.

90. Zhao, D.; Lizardo, K.; Cui, M.H.; Ambadipudi, K.; Lora, J.; Jelicks, L.A.; Nagajyothi, J.F. Antagonistic Effect of Atorvastatin on High Fat Diet Induced Survival during Acute Chagas Disease. *Microbes Infect.* **2016**, *18*, 675–686. [CrossRef] [PubMed]
91. Popkin, B.M.; Reardon, T. Obesity and the Food System Transformation in Latin America. *Obes. Rev.* **2018**, *19*, 1028–1064. [CrossRef] [PubMed]
92. Geraix, J.; Ardisson, L.P.; Marcondes-Machado, J.; Pereira, P.C.M. Clinical and Nutritional Profile of Individuals with Chagas Disease. *Braz. J. Infect. Dis.* **2007**, *11*, 411–414. [CrossRef] [PubMed]
93. Janoschek, R.; Handwerk, M.; Hucklenbruch-Rother, E.; Schmitz, L.; Bae-Gartz, I.; Kasper, P.; Lackmann, J.W.; Kretschmer, T.; Vohlen, C.; Mesaros, A.; et al. Heterogeneous Effects of Individual High-Fat Diet Compositions on Phenotype, Metabolic Outcome, and Hepatic Proteome Signature in BL/6 Male Mice. *Nutr. Metab.* **2023**, *20*, 8. [CrossRef] [PubMed]
94. Estadella, D.; Oyama, L.M.; Dâmaso, A.R.; Ribeiro, E.B.; Oller Do Nascimento, C.M. Effect of Palatable Hyperlipidic Diet on Lipid Metabolism of Sedentary and Exercised Rats. *Nutrition* **2004**, *20*, 218–224. [CrossRef] [PubMed]
95. Kiran, S.; Kumar, V.; Murphy, E.A.; Enos, R.T.; Singh, U.P. High Fat Diet-Induced $CD8^+$ T Cells in Adipose Tissue Mediate Macrophages to Sustain Low-Grade Chronic Inflammation. *Front. Immunol.* **2021**, *12*, 680944. [CrossRef] [PubMed]
96. Nishimura, S.; Manabe, I.; Nagasaki, M.; Eto, K.; Yamashita, H.; Ohsugi, M.; Otsu, M.; Hara, K.; Ueki, K.; Sugiura, S.; et al. CD8+ Effector T Cells Contribute to Macrophage Recruitment and Adipose Tissue Inflammation in Obesity. *Nat. Med.* **2009**, *15*, 914–920. [CrossRef] [PubMed]
97. Johndrow, C.; Nelson, R.; Tanowitz, H.; Weiss, L.; Nagajyothi, F. *T. cruzi* Infection Results in an Increase in Intracellular Cholesterol. *Microbes Infect.* **2014**, *16*, 337–344. [CrossRef] [PubMed]

pathogens

MDPI

Review

Understanding Host–Pathogen Interactions in Congenital Chagas Disease Through Transcriptomic Approaches

Tatiana M. Cáceres [1], Luz Helena Patiño [1] and Juan David Ramírez [1,2,*]

[1] Centro de Investigaciones en Microbiología y Biotecnología-UR (CIMBIUR), Facultad de Ciencias Naturales, Universidad del Rosario, Bogotá 111221, Colombia; tatiana.caceres@urosario.edu.co (T.M.C.); luzh.patino@urosario.edu.co (L.H.P.)

[2] Molecular Microbiology Laboratory, Department of Pathology, Molecular and Cell-Based Medicine, Icahn School of Medicine at Mount Sinai, New York, NY 10029, USA

* Correspondence: juand.ramirez@urosario.edu.co or juan.ramirezgonzalez@mssm.edu

Abstract: Chagas disease, caused by *Trypanosoma cruzi*, is a parasitic zoonosis with significant health impacts, particularly in Latin America. While traditionally associated with vector-borne transmission, increased migration has expanded its reach into urban and non-endemic regions. Congenital transmission has become a critical route of infection, involving intricate maternal–fetal immune interactions that challenge diagnosis and treatment. This review synthesizes findings from three RNA-seq studies that explore the molecular underpinnings of congenital Chagas disease, emphasizing differentially expressed genes (DEGs) implicated in host–pathogen interactions. The DAVID tool analysis highlighted the overexpression of genes associated with the innate immune response, including pro-inflammatory cytokines that drive chemotaxis and neutrophil activation. Additionally, calcium-dependent pathways critical for parasite invasion were modulated. *T. cruzi* exploits the maternal–fetal immune axis to establish a tolerogenic environment conducive to congenital transmission. Alterations in placental angiogenesis, cellular regeneration, and metabolic processes further demonstrate the parasite's ability to manipulate host responses for its survival and persistence. These findings underscore the complex interplay between the host and pathogen that facilitates disease progression. Future research integrating transcriptomic, proteomic, and metabolomic approaches is essential to unravel the molecular mechanisms underlying congenital Chagas disease, with a particular focus on the contributions of genetic diversity and non-coding RNAs in immune evasion and disease pathogenesis.

Keywords: congenital Chagas disease; transcriptomics; host–pathogen; *Trypanosoma cruzi*

1. Introduction

Chagas disease (CD), caused by the hemoflagellate protozoan *Trypanosoma cruzi* (*T. cruzi*), is a parasitic zoonosis with significant implications for global public health. The World Health Organization (WHO) identifies it as one of the 20 neglected tropical diseases, with a prevalence estimated at 6 to 7 million cases across 21 of the 33 countries in Latin America [1]. Historically, the geographical distribution of CD was predominantly rural, where vector-borne transmission via hematophagous triatomine insects was the principal route of infection. However, anthropogenic factors such as climate change, ecosystem fragmentation, and urbanization have facilitated its spread into urban centers and non-endemic regions, altering its epidemiological pattern.

In addition to vectorial transmission, *T. cruzi* exhibits a capacity for dissemination through non-vectorial mechanisms, including blood transfusions, organ transplantation, ingestion of contaminated food, and accidental exposure in laboratory settings. Among these routes, congenital transmission—from infected mothers to their offspring—has garnered increased attention as a critical pathway contributing to the disease burden [2–4]. This mechanism, exacerbated by the migration of infected populations to non-endemic regions, poses significant diagnostic and therapeutic challenges in countries with limited resources for prenatal and neonatal screening [5,6].

To reduce transmission, vector control programs and blood and organ screening have been implemented in endemic areas, significantly decreasing vector-borne and transfusional transmission. However, congenital transmission remains a major source of new cases. This mechanism presents considerable challenges, mainly due to the difficulty in accurately estimating its global burden, compounded by the lack of representative epidemiological data in many regions and the absence of systematic record-keeping. Current estimates suggest that morbidity is 28.3%, while mortality among affected neonates is 2.2%. These figures likely underestimate the true scope of the problem [7].

Congenital transmission of *T. cruzi* is a complex process involving significant maternal and fetal immune system modulation. Several factors influence the risk of congenital transmission, including maternal factors such as immune status and obstetric history, as well as parasite-related factors such as strain and parasitic load. Some mothers may transmit the parasite in one or more pregnancies, while others do not, and the reasons for this variability are not yet fully understood [8]. During pregnancy, the physiological immunosuppression that prevents fetal rejection may promote the reactivation of parasitemia, as indicated by elevated levels of *T. cruzi*-specific IgM [9]. Mothers who do not transmit the parasite typically exhibit a more pronounced immune response, characterized by elevated levels of proinflammatory cytokines such as TNF-α, IL-1β, and IL-6, which are also detected in their uninfected neonates. In contrast, mothers of neonates with congenital infection demonstrate reduced levels of TNF and IFN-γ, along with diminished expression of key antigen-presenting molecules such as HLA-DR and CD54 [8,10]. These findings underscore the critical role of maternal immune regulation in vertical transmission, highlighting the need for further research to better understand and address this complex process. The parasite employs multiple strategies to invade host cells and evade immune responses, complicating accurate and timely diagnosis and treatment. Furthermore, the use of teratogenic drugs like benznidazole and nifurtimox during pregnancy is contraindicated, allowing neonatal infection to progress to chronic stages of the disease. Consequently, these chronic forms can lead to severe complications, including cardiopathies and gastrointestinal disorders in adulthood [11].

Traditional approaches to studying congenital Chagas disease (CD) are constrained in their ability to elucidate the complex mechanisms underpinning vertical transmission. In contrast, transcriptomics has emerged as a robust and high-resolution tool for analyzing the genetic and molecular dynamics of host–pathogen interactions. This approach allows for the identification of biomarkers and pathways crucial to understanding disease mechanisms and improving diagnostic and therapeutic strategies [12,13]. The utility of transcriptomics is well demonstrated in infectious diseases like tuberculosis, where it has been instrumental in identifying diagnostic markers and differentiating infection states. For instance, the METTL7B gene distinguishes active tuberculosis from latent infection, while the HBA1/2 and HBD genes differentiate *Mycobacterium tuberculosis* from *M. avium* [14]. In respiratory conditions such as pneumonia and chronic obstructive pulmonary disease (COPD) exacerbations, transcriptomic studies have identified over 1600 differentially expressed genes related to inflammation and immune response, advancing diagnostic accuracy and personalized treatment [15]. In the context of congenital CD, transcriptomic approaches offer a framework for investigating critical factors influencing transmission, such as maternal parasitemia, *T. cruzi* genetic diversity, placental molecular interactions, and immune system modulation. By integrating transcriptomic data into multidimensional analyses, researchers can achieve a more comprehensive understanding of the host–pathogen interface [16]. This integration holds promise for refining early diagnostic protocols and developing targeted interventions to mitigate the burden of congenital transmission.

Recent advances in transcriptomic studies have profoundly enriched our understanding of the biological mechanisms driving congenital transmission of *Trypanosoma cruzi*. However, significant knowledge gaps persist regarding the primary mechanisms and immunoregulatory factors that influence fetal infection. Investigations into congenital Chagas disease have employed diverse models, focusing on key pathways such as NF-κB and Toll-like receptor (TLR) signaling, alongside evaluations of placental barrier integrity and its role in parasite invasion [17,18]. Despite these efforts, faithfully replicating the intricate architecture and functional dynamics of the human placenta in experimental systems continues to be a formidable challenge, limiting our ability to fully elucidate the molecular and cellular processes involved. This review provides a structured and comprehensive overview of congenital Chagas disease, beginning with the life cycle and genetic diversity of *Trypanosoma cruzi* and their implications for transmission and pathogenesis. It also examines the mechanisms of interaction between *T. cruzi* and the placenta, setting the stage for a deeper analysis of trophoblast infections. The primary objective is to leverage transcriptomic analyses to identify and characterize the critical biological processes, molecular pathways, and cellular mechanisms affected during infection, with particular emphasis on elucidating the host–pathogen interactions that facilitate congenital transmission.

2. Life Cycle and Genetic Diversity of *T. cruzi*: Implications for Transmission and Pathogenesis

The life cycle of *T. cruzi* is intricate and involves two primary hosts: an insect vector and a mammalian host. It begins when an infected insect vector feeds on the blood of a mammalian host, simultaneously defecating and releasing metacyclic trypomastigotes (MTs), the infective form of the parasite, into its feces. The MTs enter the host through the wound caused by the insect bite, typically through mechanical transmission during scratching. Once inside the host, the MTs invade cells near the entry site and transform into amastigotes. Within these host cells, amastigotes actively proliferate, and after several rounds of replication, they differentiate into Cell-derived trypomastigotes (CDTs), which are then released into the bloodstream [19–21] (Figure 1).

This complex life cycle involves multiple stages, each of which plays a crucial role in the parasite's ability to persist, replicate, and transmit between hosts. The genetic diversity of *T. cruzi* adds a layer of complexity, influencing the parasite's ability to adapt to different environments and evade the immune response, and contributes to variations in disease outcomes across different regions and individuals. Understanding these processes is essential for unraveling the mechanisms underlying transmission and pathogenesis, particularly in congenital CD [22,23].

In an infected pregnant woman, there are two possible scenarios for the transmission of *T. cruzi* to the fetus: reactivation of the infection and direct transmission.

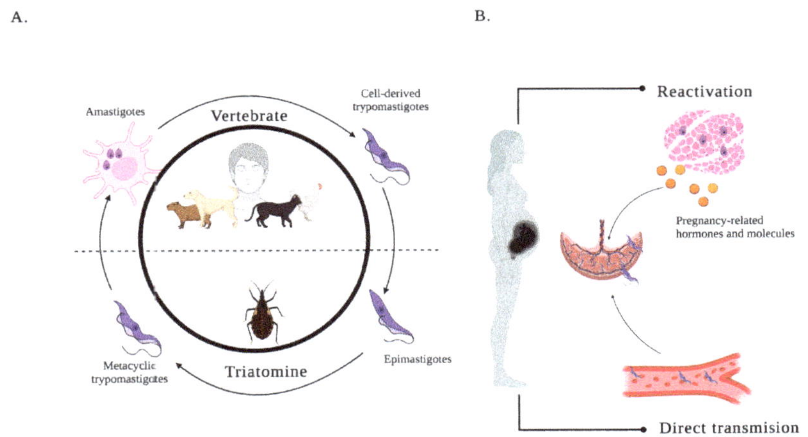

Figure 1. Life cycle. (**A**) The life cycle of *T. cruzi* alternates between vertebrate hosts and the triatomine insect vector. In vertebrates, metacyclic trypomastigotes invade host cells, where they differentiate into intracellular amastigotes, the replicative stage. Amastigotes multiply and transform into cell-derived trypomastigotes, which are released into the bloodstream to infect new cells or be ingested by a triatomine during a blood meal. In the triatomine, trypomastigotes differentiate into epimastigotes in the midgut, where they replicate and eventually transform into infective metacyclic trypomastigotes in the rectal ampoule, completing the cycle. (**B**) Mechanisms of congenital transmission of *T. cruzi*. During pregnancy, *T. cruzi* can be transmitted to the fetus through two mechanisms: (1) reactivation of infection: pregnancy-related hormones stimulate amastigotes in maternal tissues to transform into cell-derived trypomastigotes (CDTs), which are released into the maternal bloodstream. These CDTs can cross the placenta, enter fetal circulation, and spread the infection to fetal organs. (2) Direct transmission: CDTs circulating in the maternal blood invade trophoblastic cells in the placenta, crossing the placental barrier to enter the fetal bloodstream and infecting fetal tissues.

2.1. Reactivation of the Infection

Amastigotes present in maternal tissues can be stimulated by pregnancy-related hormones and molecules, leading to their transformation into cell-derived trypomastigotes (CDTs). This reactivation causes the release of CDTs into the maternal bloodstream. These CDTs can then reach the chorionic villi of the placenta, where they can enter the fetal circulation [24,25].

2.2. Direct Transmission

Circulating CDTs in the maternal blood may meet the chorionic villi, which are in the intervillous space of the placenta. Upon contact with these structures, the CDTs invade the trophoblastic cells that line the villi and cross the placental barrier to enter the fetal blood vessels. In fetal circulation, the CDTs infect nearby cells, where they transform back into amastigotes. During this replicative phase, the amastigotes multiply and are subsequently released as CDTs into the fetal bloodstream. From there, they migrate and spread the infection to various fetal organs and tissues [26,27].

This dual transmission mechanism underscores the complexity of *T. cruzi* interaction with the maternal–fetal immune environment, with significant implications for the pathogenesis of congenital CD.

2.3. Genetic Diversity

A prominent feature of *T. cruzi* is its remarkable plasticity, reflected in its extensive genetic variability and intense intraspecific phenotypic diversity. While it was long assumed that this diversity was solely the result of clonal reproduction, recent research has uncovered the occurrence of genetic recombination events *in vitro* and the presence of hybridization in natural populations [28]. These findings have fundamentally challenged the longstanding hypothesis of strict clonality, revealing a greater degree of genetic complexity in *T. cruzi*. This paradigm shift has necessitated a revision of the parasite's nomenclature, leading to the adoption of a classification system that more accurately represents the phenotypic and genotypic diversity observed within its populations [29,30].

This framework introduced the concept of Discrete Typing Units (DTUs), defined as groups of populations that are more genetically like each other than any other group and are characterized by shared molecular markers. Six DTUs, labeled TcI through TcVI, have been identified and are widely distributed throughout the Americas [31,32]. These DTUs exhibit distinct variations in virulence factors, tissue tropism, distribution among vectors and reservoirs, host immune responses, and clinical manifestations. These differences highlight potential associations between specific traits and individual DTUs [33].

However, the correlation between the genetic diversity of *T. cruzi* and the clinical manifestation of the disease remains challenging to establish. While differential tissue tropism has been observed in some strains, other studies have failed to show a consistent relationship between the parasite's genetics and the clinical forms of the disease [33,34]. This underscores the need to investigate the behavior of different isolates, particularly in co-infection contexts, as interactions between strains could influence the course of the infection [35]. Additionally, it is essential to develop tools that can assess genetic variability and gene expression within parasite populations to understand their role in pathogenesis better. Without a clear understanding of the mechanisms behind these differences, this area of research remains a critical focus in eco-epidemiology and clinical pathophysiology. The complexity of these interactions complicates the ability to generalize host interactions with different DTUs and establish a unified invasion model [36,37].

Genotyping studies on congenital *T. cruzi* transmission have identified several DTUs, including TcI, TcII, TcIII, TcV, and TcVI. Most of these studies have been conducted in endemic regions of South America, such as Argentina, Bolivia, and Chile, where the TcII, TcV, and TcVI genotypes are most common in both pregnant women and adults with chronic CD. Among these, TcV is particularly notable for its high prevalence, being found in 80–100% of congenital transmission cases in countries like Argentina, Bolivia, southern Brazil, Chile, and Paraguay. In contrast, TcI and TcII are less frequently observed in this context [36,38–42]. Notably, congenital *T. cruzi* infections, including severe and sometimes fatal cases, have also been reported in non-endemic countries such as Spain, Italy, the USA, Japan, and Switzerland [43–51]. These cases, however, likely represent only a fraction of the true burden, as the lack of systematic programs for detecting and treating congenital Chagas disease in non-endemic regions limits the recognition and management of these infections.

In a study by Antinori, congenital transmission was reported within a family cluster in Italy, originating from a Bolivian immigrant family. Genotyping identified the presence of TcII and TcV, highlighting the relevance of these DTUs in non-endemic contexts, particularly among migrant populations [52]. The identification of potential links between genetic diversity, transmission risk, and the progression of congenital CD is an emerging area of research. Although the distribution of DTUs varies by country, these associations may significantly influence transmission rates and the severity of congenital CD [53,54]. Moreover, the increasing number of congenital cases reported outside endemic areas underscores the importance of understanding how genetic diversity affects transmission dynamics in diverse epidemiological settings.

Few studies have examined the behavior of DTUs in experimental congenital infections, typically focusing on just one or two *T. cruzi* genotypes. To understand how the parasite's intrinsic characteristics affect congenital transmission, it is crucial to study a wider range of strains that represent all DTUs, considering their diverse biological properties. Some studies have revealed significant genetic variability within the same DTU, which is associated with congenital transmission. For example, Burgos and colleagues identified distinct minicircle signatures in *T. cruzi* strains from the same lineage in the Chaco province of Argentina [38]. Similarly, Herrera and colleagues identified mixed infections of TcI, TcII, and TcV in Argentina, Honduras, and Mexico. They also found a significant association between the parasite haplotypes and congenital transmission [55].

Furthermore, the interaction between *T. cruzi* subpopulations and the host significantly influences the course of congenital infection by modulating the immune response in tissues. Multiple studies have reported alterations in gene expression in the placenta during infection, underscoring the critical role of the local immune response in the placenta in determining the risk of congenital transmission [56,57].

It is essential to deepen our understanding of congenital *T. cruzi* transmission, a multifactorial process influenced by both parasite and host characteristics, which can lead to severe outcomes and high neonatal mortality rates. Strengthening global surveillance of this transmission route is a priority, as is expanding research to include a broader range of strains and geographical regions [35]. This will provide a more comprehensive understanding of the risk factors and help develop effective prevention and intervention strategies to reduce the burden of the disease.

3. Mechanisms of Interaction Between *T. cruzi* and the Placenta

The placenta is a temporary organ that connects the mother's and fetus's circulatory systems, exchanging nutrients and gases for fetal development. In addition to its physical connection role, the placenta produces hormones, peptides, and steroids that regulate various processes critical for the success of the pregnancy [18,58,59]. It also performs metabolic functions, modulates the maternal immune response to tolerate the fetus, and facilitates maternal–fetal communication. Importantly, the placenta serves as a protective barrier, preventing the transmission of pathogens and harmful molecules that could threaten fetal health [58,60]. This tissue consists of chorionic villi, which are in direct contact with maternal blood in the intervillous space. These villi are covered by two layers of trophoblastic cells: an outer layer of syncytiotrophoblasts (STs), a multinucleated, non-replicating layer, and an inner layer of cytotrophoblasts (CTs), a replicating cell layer [61]. Beneath the trophoblast lies the basal lamina and villous stroma (EV), which contains vascular endothelium, fibroblastic cells, and macrophages [62,63].

However, microorganisms such as *Toxoplasma gondii*, cytomegalovirus, rubella, herpesvirus, *Listeria monocytogenes*, *Treponema pallidum*, *T. cruzi*, and others have developed complex immune evasion mechanisms [64,65]. These pathogens possess a range of proteins and enzymes that allow them to bypass the placental barrier. They have evolved to disrupt placental functions, promoting their invasion and facilitating vertical transmission to the fetus. By modulating the immune response, destroying placental cells, and altering the physical and molecular barriers of the placenta, these pathogens can access the fetal environment, leading to severe complications, including malformations, developmental delays, or even fetal death [64,66]. The ability of these microorganisms to bypass placental protection emphasizes the complexity of congenital infections and highlights the need to understand the specific mechanisms of each pathogen. This knowledge is essential for developing more effective prevention and treatment strategies to reduce risks and improve outcomes for both mothers and affected fetuses.

The invasion of *T. cruzi* into host cells is a complex, multifaceted process involving dynamic interactions between the parasite and host cells. The parasite attaches to host cells through specific receptors on its surface. *T. cruzi* presents various molecules on its membrane that interact differently with host cell components and the extracellular matrix (ECM), aiding its invasion. During infection, the parasite is recognized by Toll-like receptors (TLRs) on host cells, which detect pathogen-associated molecular patterns (PAMPs). In the case of *T. cruzi*, several TLRs, including TLR-2, TLR-4, TLR-7, and TLR-9, are involved in this recognition. As these receptors are found in the chorionic villi of the placenta, they are believed to play a key role in the immune response to the infection, particularly in the congenital transmission of the parasite [67–69]. Upon pathogen recognition, trophoblastic cells trigger an innate immune response characterized by the release of cytokines, chemokines, reactive oxygen species (ROS), nitric oxide (NO), and antimicrobial peptides. In response to *T. cruzi* infection, a turnover of syncytiotrophoblasts is induced. This process involves the TLR-2 receptor, which subsequently activates caspase 8 and the MAPK ERK1/2 signaling pathway, leading to the elimination of the pathogen adhering to the surface cell layer [70].

Significant changes in the ECM have been observed in the placentas of women infected with *T. cruzi*. Histopathological findings include the destruction and detachment of syncytiotrophoblasts in chorionic villus explants, selective disorganization of the basal lamina, and alterations in type I collagen within the connective tissue [71,72]. Proteomic studies have shown that *T. cruzi* induces changes in the ECM by either increasing or decreasing the expression of ECM components through several mechanisms: (i) destruction of ECM components via protease activity, (ii) induction of matrix metalloproteinases (MMP-2 and MMP-9), (iii) modulation of signaling pathways, and (iv) reorganization of the cytoskeleton [63,73,74]. These events suggest that the reorganization of the ECM plays a crucial role in regulating inflammatory and immune responses in pregnant women. Together, these alterations contribute to the destabilization of the placental barrier, allowing the parasite to pass through to the fetus. Additionally, the presence of *T. cruzi* triggers an increase in human chorionic gonadotropin levels, which activates the MAPK ERK1/2 signaling pathway. This activation promotes cell proliferation and differentiation, potentially serving as a compensatory mechanism to help maintain the stability of the placental barrier [75].

One of the key strategies used by *T. cruzi* to evade the host immune response is the expression of its pleiotropic protein, calreticulin (TcCalr), which is mainly localized in the flagellar region of the parasite. This protein recruits C1q, a component of the complement system that contains the dimeric serine proteases C1s and C1r. TcCalr's primary function in this context is to prevent the activation of the complement cascade, thus aiding in immune evasion. Additionally, TcCalr acts as a molecular mimicry mechanism, as the TcCalr/C1q complex is recognized by host cells as a signal for phagocytosis, thereby increasing the parasite's infectivity (Figure 2) [76]. This mechanism holds particular significance in the context of congenital transmission of CD, as elevated levels of calreticulin (Calr) have been observed during pregnancy. Furthermore, women who develop pre-eclampsia exhibit even higher levels of Calr compared to those without this complication. These findings underscore the critical role of calreticulin in the host–parasite interaction during congenital transmission of *T. cruzi* and suggest its potential as a biomarker for pre-eclampsia or as a therapeutic target [77,78].

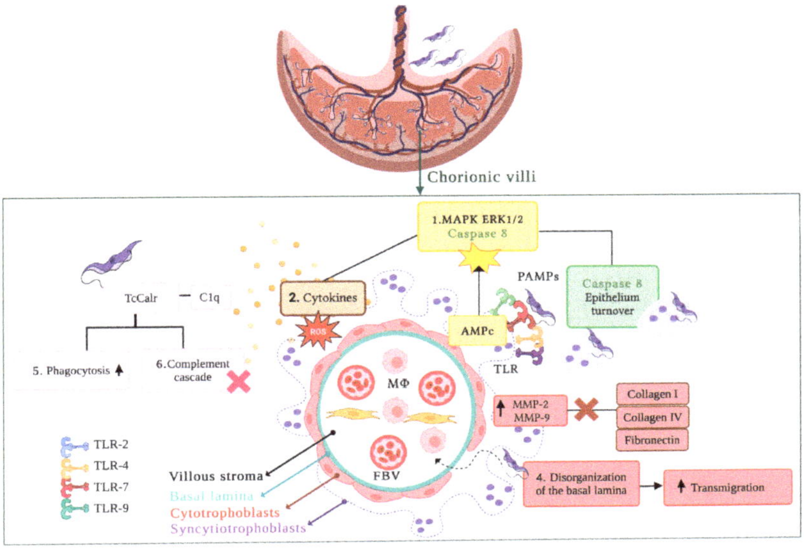

Figure 2. Host–pathogen interactions during congenital *T. cruzi* infection. This figure illustrates the interactions between the *T. cruzi* parasite and chorionic villi, emphasizing key molecular and cellular mechanisms: the interaction between surface molecules of *T. cruzi* and Toll-like receptors (TLRs) on trophoblastic cells initiates signaling cascades that increase cyclic AMP (cAMP) levels and activate the MAPK/ERK1/2 pathway (1). This pathway orchestrates multiple cellular responses, including the production of pro-inflammatory cytokines such as IL-6 and TNF-α, along with the generation of reactive oxygen species (ROS) (2). These inflammatory mediators exacerbate placental tissue damage and create a microenvironment conducive to parasite persistence. Additionally, *T. cruzi* infection triggers the activation of caspase-8, which facilitates the detachment of infected trophoblastic cells. This detachment contributes to cell turnover (3) and forms structural discontinuities in the placental barrier, allowing deeper parasite infiltration. The parasite also induces the overexpression of matrix metalloproteinases (MMP-2 and MMP-9), enzymes that degrade key components of the basal lamina, such as collagen types I and IV and fibronectin. This degradation disrupts the basal lamina's structural integrity (4), enhancing parasite transmigration toward fetal tissues. Finally, *T. cruzi* calcireticulin (TcCRT) binds to C1q, a component of the complement system. This interaction promotes parasite opsonization, increasing its uptake by host cells (5). Concurrently, TcCRT disrupts the classical complement pathway, impairing the host's immune response and facilitating parasite survival (6).

The construction of specific transcriptional profiles is crucial for investigating the biological mechanisms underlying infectious disease development. This approach enables the large-scale analysis of dynamic changes in gene expression, allowing for the identification of thousands of genes while prioritizing those involved in critical processes such as cellular invasion, immune evasion, and disease progression [79]. The identification of differentially expressed genes (DEGs) is a critical first step in understanding how pathogens disrupt host systems. When combined with transcriptional profiling, pathway analysis offers a deeper understanding of the molecular and biological process alterations associated with disease. Tools such as Gene Set Enrichment Analysis (GSEA) and databases like KEGG and Reactome are essential for mapping these interactions. These integrated approaches not only pinpoint altered genes but also reveal the functional networks in which they are involved, providing insights into the mechanisms underlying inflammation, cellular stress responses, and tissue microenvironment remodeling [79,80].

These approaches have been essential for identifying biomarkers and proposing new therapeutic targets [81]. In the context of infectious diseases, genes involved in cytokine signaling, energy metabolism, and apoptotic pathways represent promising targets for developing personalized treatments. Integrating transcriptional profiles with pathway analyses not only enhances our understanding of the underlying pathological mechanisms but also supports the design of more targeted interventions. Additionally, combining transcriptomic data with other omics approaches, such as proteomics and metabolomics, provides a more comprehensive view of disease biology. This multidimensional approach opens new avenues in translational medicine, driving research toward innovative strategies for diagnosing and treating infectious diseases [82].

The construction of transcriptional profiles using advanced techniques such as microarrays and RNA-seq, followed by comprehensive bioinformatics analysis, enables the study of gene expression changes in the human placenta induced by interaction with the *T. cruzi* parasite. This approach not only facilitates the identification of molecular pathways involved in disease pathogenesis but also contributes to uncovering the specific mechanisms of congenital transmission and the host's immune response to infection.

In this context, Castillo (2018) performed a transcriptional profiling analysis using microarrays on human placental explants infected ex vivo with *T. cruzi* (TcII). The analysis of differentially expressed genes (DEGs) revealed alterations in three main categories: (i) immune functions and inflammatory response, highlighting complement system components like CD46 and C1q, as well as the overexpression of Toll-like receptors (TLRs), including TLR-2, TLR-4, TLR-7, TLR-8, and TLR-9; (ii) extracellular matrix (ECM) remodeling, with up-regulation of genes such as ADAM12, ADAMTSL3, MMP10, MMP-2, and MMP-9; and (iii) placental development, nutrient transport, and hormonal function. The latter group represented the most overexpressed genes in infected tissues, including beta-1-glycoproteins, GH2, CSH1, and CSH2. These alterations may disrupt placental function, impairing the proper supply of nutrients and oxygen to the developing fetus, which could lead to complications such as growth restriction or pregnancy loss [56].

A subsequent study utilized RNA sequencing (RNA-seq) to analyze term placentas from both healthy women and women infected with *T. cruzi*, aiming to uncover the molecular mechanisms underlying placental dysfunction observed in pregnant women with chronic CD. Principal component analysis (PCA) revealed distinct clustering patterns based on serological infection status. Among the differentially expressed genes (DEGs), KIF12, a member of the kinesin superfamily involved in microtubule cytoskeleton dynamics, was the most prominently expressed transcript in the seropositive (SP) group. This was followed by HLA-G, which plays a key role in mediating immune tolerance during pregnancy, and PRG2, a cytotoxin involved in antiparasitic defense mechanisms and the inhibition of pregnancy-associated plasma protein A (PAPPA) [59]. Additionally, the seropositive group exhibited down-regulation of immune-modulatory elements, such as IL1F10 and kisspeptin (KISS1), both of which are involved in inhibiting chemotaxis and promoting the secretion of human chorionic gonadotropin (hCG). A negative regulation of hCG subunits, including CGB5, was also observed [83].

Several of the differentially expressed genes mentioned above have been previously linked to conditions such as pre-eclampsia, intrauterine growth restriction, spontaneous abortion, and recurrent pregnancy loss. These include the reduced expression of KISS1 and the subunits of human chorionic gonadotropin (CGB5), as well as the increased expression of neurokinin B (TAC3) and PAPPA [84,85]. Additionally, the dysregulation of glutamine and glutathione synthesis pathways observed in Juiz's study aligns with the findings from the proteomic analysis of the human placenta conducted by Jin in 2017. This study highlights that not only the inflammatory response but also the dysfunction of glutathione metabolism contributes to the development of pre-eclampsia [83,86].

Considering the genetic diversity of *T. cruzi*, Juiz conducted a preliminary study to explore potential differences in host–pathogen interaction mechanisms in congenital CD across different parasite genotypes. The study aimed to examine how the TcI (K98, a myotropic clone) and TcVI (VD, a strain isolated from a congenital case) genotypes modulate the genetic response in the placental environment of mice during experimental chronic infection.

The results showed a higher number of differentially expressed genes (DEGs) in the TcVI genotype, with a greater tropism for placental tissue compared to the TcI genotype. However, functional analysis of the DEGs revealed no significant differences in the activation of innate immune pathways. On the other hand, cellular signaling analysis identified differential regulation in protein kinase and G protein-coupled receptor pathways, suggesting that the negative regulation of these pathways may be linked to the lower infection levels observed in the K98 group [57].

Faral and colleagues evaluated transcriptional and histological changes in placentas infected with congenital isolates (VT), medium-virulence strains (MV), and high-virulence strains (HV) of *T. cruzi*. The VT strains exhibited nearly undetectable parasitemia in BALB/cJ mice and 100% survival, indicating a silent phenotype and low virulence. In contrast, HV strains caused the highest parasitemia, reaching the endpoint of infection in all mice around day 19 post-infection. Despite differences in parasitemia and lethality, VT strains reached the same organs as HV strains and generated a similar immune response, primarily Th1, with the production of IFN-γ, IL-12, TNF-α, and IL-6.

Regarding congenital transmission, the MV strain failed to cross the placenta, while HV strains, although transmitted, did so at lower levels than VT strains. Additionally, HV strains altered the regulation of pregnancy-specific glycoproteins, reducing their levels in the placenta, a change not observed with VT strains. This finding was confirmed histologically, as placentas infected with HV strains showed greater tissue damage compared to those infected with VT strains. In terms of transcriptional profiles, there was a significant disparity in the number of differentially expressed genes (DEGs), with the HV strain showing greater regulation, with 2507 DEGs compared to 408 DEGs in the VT strains. Functionally, the HV strain exhibited overexpression of elements related to inflammation, cellular immune response, and ribosomal proteins, while the VT strains showed down-regulation of processes such as mitosis, the cell cycle, DNA replication, and the DNA damage response.

Based on these findings, the researchers propose two possible scenarios for congenital transmission: (i) exacerbated changes in placental gene expression related to inflammation, as seen in the HV strain, and (ii) induction of an immune response that does not harm the host or interfere with placental function, as seen in the VT strains. These scenarios reflect distinct invasion strategies, with congenital isolates being more efficient due to their low replication and high placental tropism. Although HV strains manage to cross the placental barrier, this process may be associated with tissue damage caused by exacerbated inflammation and imbalance in pregnancy-related molecules. This approach highlights the importance of performing comparative genomic analysis between congenital and high-virulence strains to identify the molecular mechanisms involved in adaptation during congenital transmission [87].

3.1. Enrichment Analysis of Transcriptomic Profiles Using DAVID

We analyzed three RNA-seq studies published between 2010 and 2024 to investigate transcriptomic insights into congenital Chagas disease (CD). Relevant articles were identified using keywords such as "RNA-seq", "CONGENITAL CHAGAS DISEASE", "TRYPANOSOMA CRUZI", "TRANSCRIPTOMICS", and "PLACENTAL". The search algorithm used to identify these studies is outlined in Figure S1. Studies unrelated to congenital CD or those lacking data on differentially expressed genes (DEGs) were excluded from the analysis.

The DEGs, along with their corresponding fold change values and cutoff points of >2 for up-regulated genes and <−2 for down-regulated genes, were extracted from the Supplementary Materials of the selected studies. These datasets were then processed using the DAVID tool for functional annotation and enrichment analysis. This approach enabled the identification of key biological pathways, cellular processes, and molecular functions associated with the congenital transmission of *T. cruzi*. The findings from these analyses provide valuable insights into the transcriptomic profiles of *T. cruzi* infection in both in vivo and ex vivo models of congenital transmission. Overall, the studies focused on identifying biological functions and metabolic pathways linked to the genes of interest (Figure 3).

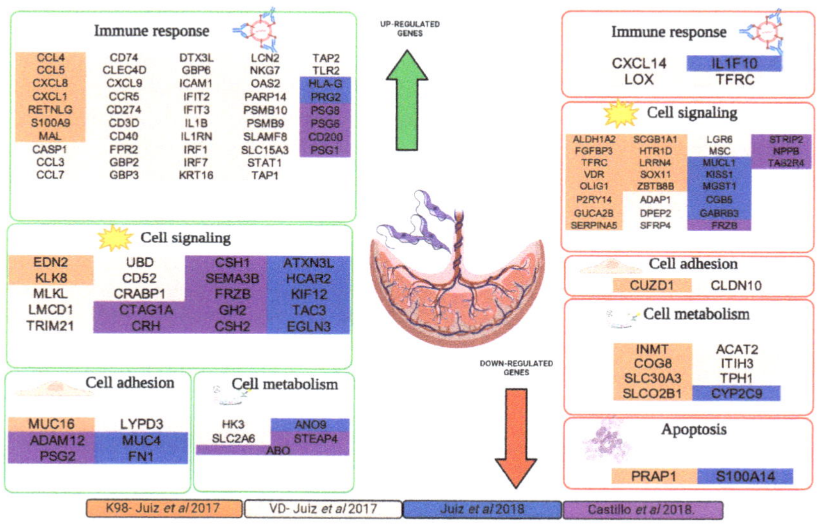

Figure 3. Functional categories of differentially expressed genes in the placental environment during *Trypanosoma cruzi* infection. This figure highlights the functional categories of differentially expressed genes (DEGs) involved in the interaction between *Trypanosoma cruzi* and the placental environment, which plays a critical role in the congenital transmission of Chagas disease. Up-regulated genes are shown in green boxes, while down-regulated genes are in red boxes, illustrating the biological processes activated or suppressed during infection. Background colors within the boxes correspond to the sources of the transcriptomic data, providing a clear link to the original studies and integrating findings across multiple investigations to offer a comprehensive overview of the key transcriptomic changes induced by *T. cruzi* [56,57,83].

In Juiz's study, the up-regulation of genes involved in the immune response was evident, emphasizing the activation of innate immunity and the production of pro-inflammatory cytokines. Conversely, the K98 model displayed a predominance of secreted immune molecules and an intensified pro-inflammatory response characterized by the activation of pathways linked to host defense. Functional ontology analyses highlighted key processes, such as the immune response mediated by antimicrobial peptides, which are essential for the initial containment of pathogens.

Among these processes, interleukin-8 (IL-8) emerged as a pivotal pro-inflammatory cytokine that plays a critical role in neutrophil chemotaxis and activation. IL-8 also regulates essential functions, including neutrophil trafficking, cell adhesion, and placental development. In trophoblastic cells, IL-8 promotes migration and invasion while stimulating progesterone secretion. These cells rely on a finely tuned balance of cytokine signaling to maintain their functionality. However, dysregulation of these signals can result in pathological conditions such as pre-eclampsia [88]. These findings underscore the critical role of innate immune responses as the first line of defense against infections, with IL-8 playing a central role in regulating and coordinating these processes.

In VD, the over-regulation of immune processes is prominent, with inflammatory pathways playing a central role in its pathogenesis. Specifically, the cellular response to interferon type II (IFIT2) is crucial for macrophage activation and the promotion of microbicidal mechanisms. However, the overproduction of this mediator during the chronic phase can exacerbate inflammatory damage, contributing to the tissue deterioration associated with the disease [89,90]. Responses mediated by tumor necrosis factor (TNF-α) and interleukin-1 (IL-1) play a crucial role in balancing the containment of *T. cruzi* and tissue protection. While the activation of these cytokines is beneficial for controlling the infection, their dysregulation can lead to significant inflammatory damage. In this context, IL-1RN, an interleukin-1 receptor antagonist, regulates inflammation by preventing overexposure to inflammatory stimuli that could result in tissue damage [91–93]. This balance mediated by IL-1RN is essential for moderating the immune response, promoting both the control of the infection and the prevention of chronic inflammatory complications.

In line with this mechanism, an over-regulation of CASP1 was observed, a key protease in the activation of caspase-1, responsible for the maturation of interleukins IL-1β and IL-18, which are crucial for the immune response. As part of the inflammasome, CASP1 triggers the inflammatory response during infection, playing a critical role in defense against intracellular pathogens such as *T. cruzi*. In chronic Chagas myocardiopathy, T-cell-mediated inflammation and autoimmune mechanisms contribute to myocardial damage and dysfunction of the cardiac conduction system, highlighting the importance of the CASP1 pathway in the development of this disease [91,94].

Additionally, over-regulated processes such as intracellular calcium homeostasis and the regulation of GTPase activity were observed, both of which play fundamental roles in membrane trafficking and exocytosis. GTPases act as key regulators of various cellular processes, controlling vesicle transport from intracellular compartments. In collaboration with SNARE proteins, GTPases promote efficient membrane fusion during exocytosis. They also participate in calcium-dependent pathways, such as lysosomal exocytosis, to respond to damage in the plasma membrane [95]. This mechanism is linked to the invasion processes described by the parasite, in which it induces damage to the plasma membrane, triggering calcium-dependent lysosomal exocytosis [96–98]. These lysosomes fuse with nascent parasitophorous vacuoles, allowing the parasite to escape into the cytosol and transform into replicative amastigotes. This coordinated process emphasizes the crucial role of GTPases in maintaining cellular homeostasis and responding to the dynamic demands of the cellular environment. The parasite enhances its invasion and survival within host cells by modulating these processes.

On the other hand, another key component in the immune response was identified: signaling mediated by TLRs, which play an essential role in innate immunity by detecting various organisms, including bacteria, fungi, parasites, and viruses [99,100]. In humans, 10 functional TLRs are expressed not only in immune cells, such as macrophages, dendritic cells, and B lymphocytes, but also in non-immune cells, including fibroblasts and epithelial cells [101]. These receptors are crucial for the initial detection of *T. cruzi*, the recruitment of phagocytes, and the activation of inflammatory cytokines [102,103]. However, their dysregulated activation can contribute to the development of inflammatory and autoimmune pathologies, a phenomenon commonly observed in other parasitic and inflammatory diseases, such as leishmaniasis and tuberculosis [104–106]. This immune imbalance reflects shared mechanisms of tissue damage and underscores the immune modulation strategies employed by *T. cruzi* to evade the host's response, leading to progressive tissue damage associated with chronic inflammation [107,108].

The findings reported by Juiz (2018) and Castillo (2018) highlight the presence of immunomodulatory processes and mechanisms regulating the placental microenvironment in the context of congenital CD. Among these, the overexpression of the HLA-G gene stands out. This gene encodes a non-classical molecule of the major histocompatibility complex (MHC) and plays a key role in maternal–fetal immune tolerance. In the literature, elevated levels of HLA-G in mothers infected with *Plasmodium* have been associated with increased susceptibility to infection. This phenomenon suggests that modulation of HLA-G could influence the progression of *T. cruzi* infection and the maternal immune system's ability to control the disease and protect the fetus [56,83,109].

In Castillo's study (2018), the over-regulation of genes from the PSG (pregnancy-specific glycoproteins) family, such as PSG1, PSG6, and PSG8, is reported. These genes play a key role in modulating the immune system during pregnancy. These molecules affect dendritic cells, promoting a balanced immune response that, on one hand, protects against infections, and on the other, supports the tolerance necessary for a successful pregnancy [110,111]. This dynamic suggests that *T. cruzi* could promote a tolerogenic environment that facilitates its congenital transmission, exploiting these maternal immune mechanisms to evade the host's immune response and ensure its survival.

In the reviewed studies, the down-regulation of processes associated with cellular signaling is observed, suggesting *T. cruzi*'s intervention in various biological pathways. Among the proposed strategies, the following stands out.

3.2. Immune Evasion and Modulation of the Maternal–Fetal Immune Response by T. cruzi

The down-regulation of genes related to the immune response, such as SERPINA5, reflects *T. cruzi*'s ability to interfere with key processes of the immune system. SerpinA5, which is involved in various physiological roles, such as host defense, particularly against HIV, still requires further research to clarify its mechanisms of action in these contexts [112,113]. On the other hand, the vitamin D receptor (VDR), known for its significant impact on the immune system, plays a crucial role in modulating immune responses in various physiological and pathological processes [114]. The combination of the down-regulation of SERPINA5 and the disruption of vitamin D signaling suggests that *T. cruzi* employs complex strategies to modulate the activation of key immune mechanisms, promoting a tolerant environment that could facilitate its survival and transmission while evading the host's immune response.

3.3. Disruption of Cellular Repair and Tissue Homeostasis

The down-regulation of genes such as OLIG1, which is essential for the differentiation and maturation of oligodendrocyte precursor cells, suggests that *T. cruzi* may interfere with cellular repair and tissue homeostasis mechanisms [115,116]. Additionally, SOX11, a key transcription factor in the regulation of nervous system development and cellular differentiation, particularly in myelination processes and injury response, indicates a disruption in crucial developmental and cellular repair processes [117]. Furthermore, FGFBP3, a protein essential for regulating fibroblast growth factors (FGFs), modulates key processes such as cell proliferation, differentiation, and migration. FGFBPs, like BP1, are critical for the activation of FGFs, regulating processes such as development, angiogenesis, and tissue repair [118]. The down-regulation of these genes, involved in cellular differentiation, tissue growth, and repair, could interfere with cellular regeneration mechanisms, impairing the host's ability to repair damaged tissues.

3.4. Alteration of Hormonal and Metabolic Signaling

The down-regulation of the KISS1 gene, which encodes kisspeptin, a group of peptide fragments, has been implicated in regulating various key physiological processes such as placental angiogenesis and trophoblast invasiveness, modulating the activity of matrix metalloproteinases and promoting placental remodeling. Disruption of this pathway could contribute to pregnancy instability, as a decrease in kisspeptin expression has been observed in women with recurrent miscarriages. Furthermore, studies suggest that measuring serum kisspeptin levels could serve as a useful biomarker for predicting pregnancy viability [84,119,120]. In addition, the LGR6 gene plays a crucial role in regulating cellular differentiation and proliferation in stem-cell-dependent tissues such as bones and skin, highlighting its importance in tissue regeneration. Alterations in LGR6 expression can create an imbalanced physiological environment, promoting parasite adaptation while interfering with tissue homeostasis [121,122]. Both KISS1 and LGR6 play pivotal roles in regulating essential biological processes, such as placental angiogenesis, trophoblast invasiveness, tissue regeneration, and stem cell differentiation. The down-regulation of these genes suggests that *T. cruzi* may exploit these pathways to manipulate the host's physiological environment, thereby promoting its survival and replication. By disrupting key processes such as hormonal signaling, cellular proliferation, and tissue homeostasis, the parasite may create a favorable environment that enhances immune evasion and supports congenital transmission. These findings indicate that *T. cruzi* employs multiple strategies to manipulate the host's immune response and critical cellular processes, exploiting vulnerabilities to ensure its survival and transmission [123].

A decrease in metabolic, biosynthetic, and macromolecular transport processes has been observed. These alterations in infected and immune cells could arise from the dysregulation of various metabolic pathways. One example is COG8, which is essential for vesicular trafficking in the Golgi apparatus. This protein participates in the formation of SNARE complexes, which are crucial for retrograde transport between the Golgi and other cellular compartments, ensuring the proper localization of glycosylation enzymes in the Golgi. Furthermore, COG8 is involved in biogenesis and vesicular trafficking, highlighting its importance in maintaining cellular functionality [124,125]. Similarly, the transporters SLC30A3 and SLCO2B1 are involved in the transport of zinc from the cytosol to the extracellular space or subcellular organelles, as well as in the transport of steroid and thyroid hormones, which are essential for pregnancy and fetal development, respectively. The placenta regulates the transport of nutrients, hormones, and drugs through ABC and SLC transporters in the syncytiotrophoblast. According to the literature, placental inflammation can alter the expression of these transporters, suggesting direct interference in nutrient uptake and intracellular transport functionality, thus affecting their activity and fetal exposure. This phenomenon has been reported in cases of premature births. Moreover, alterations in CYP2C9, which plays a key role in the metabolism of xenobiotics and endogenous molecules, could reflect an impact on the host's adaptive metabolic capacity [126]. These conditions not only compromise the normal metabolic balance of cells but also affect the regulation of essential hormones and molecules during pregnancy, potentially having implications for fetal development and pregnancy-related complications [127,128].

Additionally, the down-regulation of apoptosis-related genes was observed, such as PRAP1 (Proline-rich acidic protein 1), which plays a crucial role in protecting cells from damage induced by oxidative stress and radiation. Its main function appears to be activation in response to DNA damage, intervening in the p53 signaling pathway to protect cells and improve their viability [129]. On the other hand, S100A14, which belongs to the S100 protein family—small calcium-binding proteins—acts as both an intracellular and extra-cellular signaling molecule. These proteins modulate a wide range of intracellular functions by altering their subcellular localization and interacting with specific target proteins involved in cell growth, differentiation, motility, apoptosis, and cell cycle regulation [130]. This information suggests direct interference with programmed cell death mechanisms. This phenomenon may represent a strategy employed by the parasite to prevent the elimination of infected cells, thereby ensuring its survival in the maternal–fetal environment. Inhibition of apoptosis would create a cellular environment favorable for parasite replication while impeding the host's ability to remove damaged or infected cells. Collectively, these processes could help explain clinical observations such as fetal growth retardation in mothers infected with *T. cruzi* and the emergence of gestational disorders like pre-eclampsia [131].

In total, these factors contribute to the disruption of the delicate molecular balance in the chorionic villi, leading to destabilization and loss of structure, thereby creating an environment conducive to the parasite's invasion. These changes, combined with the down-regulation of apoptotic processes, not only facilitate congenital transmission but also enable the persistence of the parasite within the tissue.

4. MicroRNAs and Their Role in Post-Transcriptional Regulation During Parasitic Infections

To understand the underlying molecular mechanisms in the pathophysiology of the disease and the complex interactions between *T. cruzi* and its host, it is crucial to consider the parasite's ability to modulate host gene expression through non-coding RNAs, such as microRNAs (miRNAs). These miRNAs play a fundamental role in the post-transcriptional regulation of mRNA and protein expression, influencing the molecular interaction networks that govern the host's response to infection [132]. Additionally, miRNAs are being investigated as non-invasive biomarkers due to their easy detection in biological samples, offering significant advantages in terms of accessibility and diagnostic accuracy [133,134].

In humans, chromosome 19 contains the majority of miRNAs, many of which are specifically expressed in the placenta, underscoring their importance in tissue-specific expression profiles [135]. These miRNAs are essential in critical cellular processes during infection, including development, cell proliferation and differentiation, apoptosis, metabolism, and modulation of the immune response [136,137].

A study by Medina et al. (2020) [138] examined the effects of *T. cruzi* and Toxoplasma gondii on the miRNA profile in human placental explants, revealing significant overexpression of three miRNAs: miR-21, miR-146a/b, and miR-210. miR-21, which plays a role in cell proliferation, differentiation, migration, and apoptosis regulation, showed increased expression, potentially linked to physiological changes during pregnancy. The down-regulation of miR-21 in the placenta has been associated with suboptimal fetal growth and may have important implications for fetal development [139,140].

Regarding miR-210, differential expression was observed between normotensive placentas and those with severe pre-eclampsia, suggesting its role as a mediator of mitochondrial dysfunction in pre-eclampsia and a potential marker of hypoxia [141]. In the context of *T. cruzi* infection, miR-210 was overexpressed in infected explants, likely due to the activation of NF-kB pathways, which regulate the transcription of this miRNA [138].

On the other hand, miR-146a, which is induced by NF-kB, plays a crucial role in regulating the placental inflammatory response by preventing dysregulated inflammation. Down-regulation of this miRNA promotes the invasion of *T. cruzi* and *T. gondii*, resulting in an increased parasitic load [142,143]. Moreover, a notable overexpression of miR-512-3p was observed, which inhibits the caspase 8 inhibitor (c-FLIP), thereby enhancing caspase 8 activity. This up-regulation of caspase 8 activity could act as a protective mechanism against infection by promoting the removal of infected or damaged cells, thereby minimizing potential adverse effects on pregnancy [70,138].

In a further analysis, Medina et al. (2022) [144] examined the placental response to *T. cruzi* infection, revealing alterations in miRNA profiles that closely resembled those observed during trophoblast differentiation. The study focused on two miRNAs, miR-512-3p and miR-515-5p, which have opposing effects. While the up-regulation of miR-512-3p promotes cellular turnover, the up-regulation of miR-515-5p inhibits differentiation by directly repressing the transcription factor hGCM-1. miR-515-5p has been proposed as a potential biomarker for pre-eclampsia when overexpressed [145]. It was found that the positive regulation of miR-512-3p and the negative regulation of miR-515-5p promote cell turnover and differentiation, influencing the balance of local defense mechanisms in the placenta against the parasite.

In contrast, inhibition of miR-512-3p and increased levels of miR-515-5p resulted in an increase in parasite DNA load. This suggests that by interfering with trophoblast differentiation through these two miRNAs, one of the initial innate immunity mechanisms in the placental layer is compromised [144]. It is essential to note that the interactions between these molecules are not yet fully understood, and they may work synergistically or regulate other established processes. Therefore, it is crucial to conduct targeted studies to identify the specific microRNAs involved in *T. cruzi* infection and clarify their regulatory roles in this context. These investigations will offer a deeper understanding of the placental response mechanisms to the infection.

5. Future Directions

It is essential to conduct studies that deepen our understanding of the pathogenesis and host response during congenital *T. cruzi* infection, focusing on transcriptomic, proteomic, and metabolomic aspects. Transcriptomic studies are particularly valuable as they provide insights into the changes in gene expression during congenital transmission. These studies offer a detailed understanding of the regulatory mechanisms and molecular pathways involved in the host's response, which can guide subsequent proteomic and metabolomic investigations [146,147].

Integrating protein and metabolite profiles will help clarify how the products of these transcriptomic changes regulate immune responses, and the biological processes involved in transmission. Taken together, these multidisciplinary approaches will not only provide a more comprehensive view of congenital CD pathogenesis but also reveal new opportunities to identify therapeutic targets and biomarkers and develop targeted intervention strategies [81].

However, designing these studies requires careful consideration of the complexity and multifactorial nature of this transmission mechanism. Various factors must be thoroughly evaluated, as they may significantly impact the results and interpretation of the findings. One crucial factor is the selection of the *T. cruzi* genotype. With multiple strains exhibiting variations in virulence, pathogenicity, and tissue tropism, interactions with host cells can differ, affecting immune signaling and response [57,148]. Therefore, selecting the appropriate genotype that mirrors the key characteristics of congenital infection is essential for obtaining relevant and applicable results.

Furthermore, choosing the correct animal or cell model is critical, as different models can replicate specific aspects of congenital infection, providing valuable insights into the mechanisms at play [149,150]. Each model has its strengths and limitations, and the choice should align with the study's objectives. Primates are particularly valuable due to their close resemblance to human placental structures, making them ideal for studying the pathogenesis of congenital transmission [24,151]. However, ethical concerns and high maintenance costs often limit their use. Domestic dogs, while having a distinct type of placentation, are more accessible and offer opportunities to investigate transmission dynamics and immune responses [152]. These species also serve as potential reservoirs, especially in endemic regions, emphasizing their relevance in research.

Rodent models, particularly mice and guinea pigs, are commonly used to study congenital transmission of *T. cruzi*. In mice, infection during pregnancy results in fetal malformations and growth retardation, with the parasite detectable in fetal tissues [153–155]. However, mice are not ideal for studying trophoblast invasion and vascular remodeling, which are critical in fetal growth restriction and pre-eclampsia in pregnant women [156]. Guinea pigs, with a longer gestation period and more human-like fetal development, offer a more suitable model. Infected guinea pigs transmit *T. cruzi* to all offspring, leading to tissue damage and growth retardation, with PCR confirming the presence of the parasite in both the placenta and fetuses [157,158].

The structural and physiological differences in placentas among larger mammals provide valuable opportunities for comparative research, supporting studies of human infection and epidemiological investigations in regions where these species serve as reservoirs or are in close contact with humans [24]. Incorporating advanced molecular techniques like transcriptomics and proteomics enhances understanding of *T. cruzi* interactions with the host during gestation, revealing critical processes such as immune evasion and placental remodeling that facilitate transmission. The genetic variability of *T. cruzi* strains, including Discrete Typing Units (DTUs) like TcI, TcII, and TcV, influences infectivity, virulence, and tissue tropism—key factors in congenital transmission. Including these diverse DTUs in experimental studies allows researchers to explore strain-specific interactions with the host and placenta, providing deeper insights into the epidemiological patterns of congenital Chagas disease and the factors driving its transmission.

Another key consideration is the selection of exposure times and parasitemia levels. Variations in these factors can affect infection dynamics and the molecular and cellular responses [159]. Therefore, establishing experimental protocols that accurately reflect the conditions of congenital transmission is essential. Finally, the choice of omics analysis is also critical. This review emphasizes the value of dual RNA-seq analysis, which enables the evaluation of gene expression modulation in both the host and the pathogen, providing a comprehensive picture of signaling and virulence factors involved in congenital transmission [160,161]. This approach also allows for the examination of differences in invasion and signaling mechanisms, which may differ in various tissues, and helps identify key factors contributing to the development of congenital CD.

Once these key elements are identified, validation through proteomic tools can further solidify the findings, providing a foundation for selecting therapeutic targets and developing strategies to interrupt transmission.

The development of single-cell analysis has gained significant relevance due to its ability to reveal hidden cellular heterogeneity, which is overlooked by traditional approaches that analyze combined cell populations. These variations play a crucial role in the diagnosis and treatment of diseases, enabling the identification of specific responses to therapies [162,163]. Single-cell technologies, by allowing highly sensitive investigations, are fundamental for personalized medicine and more precise diagnoses. In the case of congenital Chagas disease, this approach facilitates the study of the interaction between trophoblasts and *T. cruzi*, uncovering key mechanisms in disease progression, such as immune evasion and placental modulation, thus opening new therapeutic avenues.

Moreover, there is a need to further explore additional regulatory elements, such as non-coding RNAs, which play essential roles in post-transcriptional regulation and contribute to fundamental biological pathways during gestation and infection. MicroRNAs may play a pivotal role in the pathophysiology of congenital Chagas disease transmission. Identifying and characterizing the specific miRNAs involved could provide valuable insights into the molecular mechanisms and signaling pathways at play. These microRNAs could also serve as potential biomarkers for early diagnosis, prognosis, and severity assessment of congenital infection.

Supplementary Materials: The following supporting information can be downloaded at: https://www.mdpi.com/article/10.3390/pathogens14020106/s1. Figure S1. Review for identifying and categorizing research articles on congenital Chagas disease with a focus on RNA-seq and transcriptomic data. This algorithm outlines a review approach used to identify and categorize research articles related to congenital Chagas disease, specifically focusing on RNA-seq and transcriptomic data. The process includes retrieving studies from databases like PubMed, excluding non-congenital research, and analyzing articles that provide data on differentially expressed genes (DEGs), including available Fold Change values.

Funding: This research was funded by Dirección de Investigación e Innovación from Universidad del Rosario.

Institutional Review Board Statement: Not applicable.

Data Availability Statement: Not applicable.

Acknowledgments: We thank Vicerrectoría de Investigación e Innovación from Universidad del Rosario.

Conflicts of Interest: The authors declare no conflicts of interest.

References

1. World Health Organization. Chagas disease in Latin America: An epidemiological update based on 2010 estimates. *Wkly. Epidemiol. Rec.* **2015**, *90*, 33–44.
2. Lidani, K.C.F.; Andrade, F.A.; Bavia, L.; Damasceno, F.S.; Beltrame, M.H.; Messias-Reason, I.J.; Sandri, T.L. Chagas Disease: From Discovery to a Worldwide Health Problem. *Front. Public Health* **2019**, *7*, 166. [CrossRef] [PubMed]
3. Rassi, A.; de Rezende, J.M. American trypanosomiasis (Chagas disease). *Infect. Dis. Clin. N. Am.* **2012**, *26*, 275–291. [CrossRef] [PubMed]
4. Steverding, D. The history of Chagas disease. *Parasites Vectors* **2014**, *7*, 317. [CrossRef]
5. Gascon, J.; Bern, C.; Pinazo, M.J. Chagas disease in Spain, the United States and other non-endemic countries. *Acta Trop.* **2010**, *115*, 22–27. [CrossRef]
6. Schmunis, G.A. Epidemiology of Chagas disease in non-endemic countries: The role of international migration. *Mem. Inst. Oswaldo Cruz* **2007**, *102* (Suppl. S1), 75–85. [CrossRef] [PubMed]
7. Matthews, S.; Tannis, A.; Puchner, K.P.; Bottazzi, M.E.; Cafferata, M.L.; Comandé, D.; Buekens, P. Estimation of the morbidity and mortality of congenital Chagas disease: A systematic review and meta-analysis. *PLoS Neglected Trop. Dis.* **2022**, *16*, e0010376. [CrossRef]
8. Cevallos, A.M.; Hernández, R. Chagas' Disease: Pregnancy and Congenital Transmission. *BioMed Res. Int.* **2014**, *2014*, 401864. [CrossRef] [PubMed]
9. Pinazo, M.J.; Espinosa, G.; Cortes-Lletget, C.; Posada ED, J.; Aldasoro, E.; Oliveira, I.; Muñoz, J.; Gállego, M.; Gascon, J. Immunosuppression and Chagas disease: A management challenge. *PLoS Neglected Trop. Dis.* **2013**, *7*, e1965. [CrossRef]
10. Cardoni, R.L.; García, M.M.; De Rissio, A.M. Proinflammatory and anti-inflammatory cytokines in pregnant women chronically infected with *Trypanosoma cruzi*. *Acta Trop.* **2004**, *90*, 65–72. [CrossRef]
11. Rios, L.; Campos, E.E.; Menon, R.; Zago, M.P.; Garg, N.J. Epidemiology and pathogenesis of maternal-fetal transmission of *Trypanosoma cruzi* and a case for vaccine development against congenital Chagas disease. *Biochim. Biophys. Acta Mol. Basis Dis.* **2020**, *1866*, 165591. [CrossRef]

12. Rivara-Espasandín, M.; Palumbo, M.C.; Sosa, E.J.; Radío, S.; Turjanski, A.G.; Sotelo-Silveira, J.; Porto, D.F.D.; Smircich, P. Omics data integration facilitates target selection for new antiparasitic drugs against TriTryp infections. *Front. Pharmacol.* **2023**, *14*, 1136321. [CrossRef]
13. Brouwer, I.; Lenstra, T.L. Visualizing transcription: Key to understanding gene expression dynamics. *Curr. Opin. Chem. Biol.* **2019**, *51*, 122–129. [CrossRef] [PubMed]
14. Jia, Q.; Wu, Y.; Huang, Y.; Bai, X. New genetic biomarkers from transcriptome RNA-sequencing for *Mycobacterium tuberculosis* complex and *Mycobacterium avium* complex infections by bioinformatics analysis. *Sci. Rep.* **2024**, *14*, 17385. [CrossRef] [PubMed]
15. Bertrams, W.; Griss, K.; Han, M.; Seidel, K.; Klemmer, A.; Sittka-Stark, A.; Hippenstiel, S.; Suttorp, N.; Finkernagel, F.; Wilhelm, J.; et al. Transcriptional analysis identifies potential biomarkers and molecular regulators in pneumonia and COPD exacerbation. *Sci. Rep.* **2020**, *10*, 241. [CrossRef]
16. Kemmerling, U.; Osuna, A.; Schijman, A.G.; Truyens, C. Congenital Transmission of *Trypanosoma cruzi*: A Review About the Interactions Between the Parasite, the Placenta, the Maternal and the Fetal/Neonatal Immune Responses. *Front. Microbiol.* **2019**, *10*, 1854. [CrossRef] [PubMed]
17. Silberstein, E.; Chung, C.C.; Debrabant, A. The transcriptome landscape of 3D-cultured placental trophoblasts reveals activation of TLR2 and TLR3/7 in response to low *Trypanosoma cruzi* parasite exposure. *Front. Microbiol.* **2023**, *14*, 1256385. [CrossRef] [PubMed]
18. Díaz-Luján, C.; Triquell, M.F.; Castillo, C.; Hardisson, D.; Kemmerling, U.; Fretes, R.E. Role of placental barrier integrity in infection by *Trypanosoma cruzi*. *Acta Trop.* **2016**, *164*, 360–368. [CrossRef]
19. Zuma, A.A.; dos Santos Barrias, E.; de Souza, W. Basic Biology of *Trypanosoma cruzi*. *Curr. Pharm. Des.* **2021**, *27*, 1671–1732. [CrossRef] [PubMed]
20. Teixeira, D.E.; Benchimol, M.; Crepaldi, P.H.; de Souza, W. Interactive Multimedia to Teach the Life Cycle of *Trypanosoma cruzi*, the Causative Agent of Chagas Disease. *PLoS Neglected Trop. Dis.* **2012**, *6*, e1749. [CrossRef] [PubMed]
21. Martín-Escolano, J.; Marín, C.; Rosales, M.J.; Tsaousis, A.D.; Medina-Carmona, E.; Martín-Escolano, R. An Updated View of the *Trypanosoma cruzi* Life Cycle: Intervention Points for an Effective Treatment. *ACS Infect. Dis.* **2022**, *8*, 1107–1115. [CrossRef] [PubMed]
22. Coura, J.R.; Borges-Pereira, J. Chagas disease: 100 years after its discovery. A systemic review. *Acta Trop.* **2010**, *115*, 5–13. [CrossRef] [PubMed]
23. Ferri, G.; Edreira, M.M. All Roads Lead to Cytosol: *Trypanosoma cruzi* Multi-Strategic Approach to Invasion. *Front. Cell. Infect. Microbiol.* **2021**, *11*, 634793. [CrossRef]
24. Avalos-Borges, E.E.; Rios, L.E.; Jiménez-Coello, M.; Ortega-Pacheco, A.; Garg, N.J. Animal Models of *Trypanosoma cruzi* Congenital Transmission. *Pathogens* **2022**, *11*, 1172. [CrossRef] [PubMed]
25. Kemmerling, U.; Bosco, C.; Galanti, N. Infection and invasion mechanisms of *Trypanosoma cruzi* in the congenital transmission of Chagas' disease: A proposal. *Biol. Res.* **2010**, *43*, 307–316. [CrossRef] [PubMed]
26. Bustos, P.L.; Milduberger, N.; Volta, B.J.; Perrone, A.E.; Laucella, S.A.; Bua, J. *Trypanosoma cruzi* Infection at the Maternal-Fetal Interface: Implications of Parasite Load in the Congenital Transmission and Challenges in the Diagnosis of Infected Newborns. *Front. Microbiol.* **2019**, *10*, 1250. [CrossRef]
27. Megli, C.J.; Coyne, C.B. Infections at the maternal–fetal interface: An overview of pathogenesis and defence. *Nat. Rev. Microbiol.* **2022**, *20*, 67–82. [CrossRef]
28. Cruz-Saavedra, L.; Caceres, T.; Ballesteros, N.; Posada-Forero, B.; Ramírez, J.D. Differential expression of meiosis and homologous recombination-related genes in the life cycle of *Trypanosoma cruzi*. *Parasitol. Res.* **2023**, *122*, 1747–1757. [CrossRef] [PubMed]
29. Gaunt, M.W.; Yeo, M.; Frame, I.A.; Stothard, J.R.; Carrasco, H.J.; Taylor, M.C.; Mena, S.S.; Veazey, P.; Miles, G.A.J.; Acosta, N.; et al. Mechanism of genetic exchange in American trypanosomes. *Nature* **2003**, *421*, 936–939. [CrossRef]
30. Machado, C.A.; Ayala, F.J. Nucleotide sequences provide evidence of genetic exchange among distantly related lineages of *Trypanosoma cruzi*. *Proc. Natl. Acad. Sci. USA* **2001**, *98*, 7396–7401. [CrossRef] [PubMed]
31. Velásquez-Ortiz, N.; Herrera, G.; Hernández, C.; Muñoz, M.; Ramírez, J.D. Discrete typing units of *Trypanosoma cruzi*: Geographical and biological distribution in the Americas. *Sci. Data* **2022**, *9*, 360. [CrossRef]
32. Zingales, B.; Andrade, S.; Briones, M.R.S.; Campbell, D.; Chiari, E.; Fernandes, O.; Guhl, F.; Lages-Silva, E.; Macedo, A.; Machado, C.; et al. A new consensus for *Trypanosoma cruzi* intraspecific nomenclature: Second revision meeting recommends TcI to TcVI. *Mem. Inst. Oswaldo Cruz* **2009**, *104*, 1051–1054. [CrossRef] [PubMed]
33. Zingales, B.; Bartholomeu, D.C. *Trypanosoma cruzi* genetic diversity: Impact on transmission cycles and Chagas disease. *Mem. Inst. Oswaldo Cruz* **2022**, *117*, e210193. [CrossRef]
34. Jiménez, P.; Jaimes, J.; Poveda, C.; Ramírez, J.D. A systematic review of the *Trypanosoma cruzi* genetic heterogeneity, host immune response and genetic factors as plausible drivers of chronic chagasic cardiomyopathy. *Parasitology* **2019**, *146*, 269–283. [CrossRef] [PubMed]

35. Silvestrini, M.M.A.; Alessio, G.D.; Frias, B.E.D.; Júnior, P.A.S.; Araújo, M.S.S.; Silvestrini, C.M.A.; de Melo, G.E.B.A.; Martins-Filho, O.A.; Teixeira-Carvalho, A.; Martins, H.R.; et al. New insights into *Trypanosoma cruzi* genetic diversity, and its influence on parasite biology and clinical outcomes. *Front. Immunol.* **2024**, *15*, 1342431. [CrossRef]
36. Messenger, L.A.; Miles, M.A.; Bern, C. Between a bug and a hard place: *Trypanosoma cruzi* genetic diversity and the clinical outcomes of Chagas disease. *Expert Rev. Anti. Infect. Ther.* **2015**, *13*, 995–1029. [CrossRef] [PubMed]
37. Miles, M.A.; Llewellyn, M.S.; Lewis, M.D.; Yeo, M.; Baleela, R.; Fitzpatrick, S.; Gaunt, M.W.; Mauricio, I.L. The molecular epidemiology and phylogeography of *Trypanosoma cruzi* and parallel research on Leishmania: Looking back and to the future. *Parasitology* **2009**, *136*, 1509–1528. [CrossRef] [PubMed]
38. Burgos, J.M.; Altcheh, J.; Bisio, M.; Duffy, T.; Valadares, H.M.S.; Seidenstein, M.E.; Piccinali, R.; Freitas, J.M.; Levin, M.J.; Macchi, L.; et al. Direct molecular profiling of minicircle signatures and lineages of *Trypanosoma cruzi* bloodstream populations causing congenital Chagas disease. *Int. J. Parasitol.* **2007**, *37*, 1319–1327. [CrossRef]
39. Corrales, R.M.; Mora, M.C.; Negrette, O.S.; Diosque, P.; Lacunza, D.; Virreira, M.; Brenière, S.F.; Basombrio, M.A. Congenital Chagas disease involves *Trypanosoma cruzi* sub-lineage IId in the northwestern province of Salta, Argentina. *Infect. Genet. Evol.* **2009**, *9*, 278–282. [CrossRef] [PubMed]
40. Garcia, A.; Ortiz, S.; Iribarren, C.; Bahamonde, M.I.; Solari, A. Congenital co-infection with different *Trypanosoma cruzi* lineages. *Parasitol. Int.* **2014**, *63*, 138–139. [CrossRef] [PubMed]
41. Virreira, M.; Alonso-Vega, C.; Solano, M.; Jijena, J.; Brutus, L.; Bustamante, Z.; Truyens, C.; Schneider, D.; Torrico, F.; Carlier, Y.; et al. Congenital Chagas disease in Bolivia is not associated with DNA polymorphism of *Trypanosoma cruzi*. *Am. J. Trop. Med. Hyg.* **2006**, *75*, 871–879. [CrossRef]
42. Virreira, M.; Alonso-Vega, C.; Truyens, C.; Jijena, J.; Brutus, L.; Carlier, Y.; Torrico, F.; Svoboda, M. Comparison of *Trypanosoma cruzi* lineages and levels of parasitic DNA in infected mothers and their newborns. *Am. J. Trop. Med. Hyg.* **2007**, *77*, 102–106. [CrossRef] [PubMed]
43. Lazarte, R.A.; Litman-Mazo, F.; Crewalk, J.-A.; Keim, D.E.; Baram, M.; Klassen-Fischer, M.K.; Massif, D.; Zabielski, M.L.; Colon, A.; Sockwell, D.; et al. Congenital Transmission of Chagas Disease—Virginia, 2010. *MMWR Morb. Mortal. Wkly. Rep.* **2012**, *61*, 477–479.
44. Jackson, Y.; Myers, C.; Diana, A.; Marti, H.-P.; Wolff, H.; Chappuis, F.; Loutan, L.; Gervaix, A. Congenital transmission of Chagas disease in Latin American immigrants in Switzerland. *Emerg. Infect. Dis.* **2009**, *15*, 601–603. [CrossRef]
45. Alarcón, A.; Morgan, M.; Montgomery, S.P.; Scavo, L.; Wong, E.C.C.; Hahn, A.; Jantausch, B. Diagnosis and Treatment of Congenital Chagas Disease in a Premature Infant. *J. Pediatr. Infect. Dis. Soc.* **2016**, *5*, e28–e31. [CrossRef] [PubMed]
46. Flores-Chávez, M.; Faez, Y.; Olalla, J.M.; Cruz, I.; Gárate, T.; Rodríguez, M.; Blanc, P.; Cañavate, C. Fatal congenital Chagas' disease in a non-endemic area: A case report. *Cases J.* **2008**, *1*, 302. [CrossRef] [PubMed]
47. Imai, K.; Maeda, T.; Sayama, Y.; Mikita, K.; Fujikura, Y.; Misawa, K.; Nagumo, M.; Iwata, O.; Ono, T.; Kurane, I.; et al. Mother-to-Child Transmission of Congenital Chagas Disease, Japan. *Emerg. Infect. Dis.* **2014**, *20*, 146–148. [CrossRef]
48. Martínez Ortiz, A.; Herranz Aguirre, M.; Guibert Valencia, J.; Ezpeleta Baquedano, C. Enfermedad de Chagas neonatal de transmisión vertical en países no endémicos: El uso de la PCR en el diagnóstico: Ventajas sobre técnicas convencionales. *An. Sist. Sanit. Navar.* **2013**, *36*, 115–118. [CrossRef]
49. Riera, C.; Guarro, A.; EL Kassab, H.; Jorba, J.M.; Castro, M.; Angrill, R.; Gállego, M.; Fisa, R.; Martin, C.; Lobato, A.; et al. Congenital transmission of *Trypanosoma cruzi* in Europe (Spain): A case report. *Am. J. Trop. Med. Hyg.* **2006**, *75*, 1078–1081. [CrossRef]
50. Bargiggia, G.; Ruggeri, M.; Ortalli, G.; Gabrielli, S.; Rodari, P.; D'Antiga, L.; Farina, C. Congenital Chagas disease in a Bolivian newborn in Bergamo (Italy). *Infez. Med.* **2018**, *26*, 93–96.
51. Basile, L.; Ciruela, P.; Requena-Méndez, A.; Vidal, M.J.; Dopico, E.; Martín-Nalda, A.; Sulleiro, E.; Gascon, J.; Jané, M.; The Working Group of Congenital Chagas Disease in Catalonia. Epidemiology of congenital Chagas disease 6 years after implementation of a public health surveillance system, Catalonia, 2010 to 2015. *Eurosurveillance* **2019**, *24*, 1900011. [CrossRef]
52. Antinori, S.; Galimberti, L.; Grande, R.; Ricaboni, D.; Sala, S.; Giacomet, V.; Colombo, V.; Corbellino, M.; Angheben, A.; Giacomelli, A.; et al. Family cluster of Chagas disease among Bolivian immigrants in Italy: High rate of maternal-fetal transmission. *Travel Med. Infect. Dis.* **2022**, *49*, 102370. [CrossRef]
53. Klein, M.D.; Proaño, A.; Noazin, S.; Sciaudone, M.; Gilman, R.H.; Bowman, N.M. Risk factors for vertical transmission of Chagas disease: A systematic review and meta-analysis. *Int. J. Infect. Dis.* **2021**, *105*, 357–373. [CrossRef]
54. Sánchez, L.V.; Ramírez, J.D. Congenital and oral transmission of American trypanosomiasis: An overview of physiopathogenic aspects. *Parasitology* **2013**, *140*, 147–159. [CrossRef]
55. Herrera, C.; Truyens, C.; Dumonteil, E.; Alger, J.; Sosa-Estani, S.; Cafferata, M.L.; Gibbons, L.; Ciganda, A.; Matute, M.L.; Zuniga, C.; et al. Phylogenetic Analysis of *Trypanosoma cruzi* from Pregnant Women and Newborns from Argentina, Honduras, and Mexico Suggests an Association of Parasite Haplotypes with Congenital Transmission of the Parasite. *J. Mol. Diagn.* **2019**, *21*, 1095–1105. [CrossRef]

56. Castillo, C.; Carrillo, I.; Libisch, G.; Juiz, N.; Schijman, A.; Robello, C.; Kemmerling, U. Host-parasite interaction: Changes in human placental gene expression induced by *Trypanosoma cruzi*. *Parasites Vectors* **2018**, *11*, 479. [CrossRef]
57. Juiz, N.A.; Solana, M.E.; Acevedo, G.R.; Benatar, A.F.; Ramirez, J.C.; da Costa, P.A.; Macedo, A.M.; Longhi, S.A.; Schijman, A.G. Different genotypes of *Trypanosoma cruzi* produce distinctive placental environment genetic response in chronic experimental infection. *PLoS Neglected Trop. Dis.* **2017**, *11*, e0005436. [CrossRef] [PubMed]
58. Burton, G.J.; Fowden, A.L. The placenta: A multifaceted, transient organ. *Philos. Trans. R. Soc. Lond. B Biol. Sci.* **2015**, *370*, 20140066. [CrossRef]
59. Weyer, K.; Glerup, S. Placental Regulation of Peptide Hormone and Growth Factor Activity by proMBP1. *Biol. Reprod.* **2011**, *84*, 1077–1086. [CrossRef]
60. Syme, M.R.; Paxton, J.W.; Keelan, J.A. Drug transfer and metabolism by the human placenta. *Clin. Pharmacokinet.* **2004**, *43*, 487–514. [CrossRef] [PubMed]
61. Benirschke, K.; Burton, G.J.; Baergen, R.N. Basic Structure of the Villous Trees. In *Pathology of the Human Placenta*; Benirschke, K., Burton, G.J., Baergen, R.N., Eds.; Springer: Berlin/Heidelberg, Germany, 2012; pp. 55–100. [CrossRef]
62. Arora, N.; Sadovsky, Y.; Dermody, T.S.; Coyne, C.B. Microbial vertical transmission during human pregnancy. *Cell Host Microbe* **2017**, *21*, 561–567. [CrossRef]
63. Castillo, C.; López-Muñoz, R.; Duaso, J.; Galanti, N.; Jaña, F.; Ferreira, J.; Cabrera, G.; Maya, J.; Kemmerling, U. Role of matrix metalloproteinases 2 and 9 in ex vivo *Trypanosoma cruzi* infection of human placental chorionic villi. *Placenta* **2012**, *33*, 991–997. [CrossRef]
64. Ostrander, B.; Bale, J.F. Congenital and perinatal infections. *Handb. Clin. Neurol.* **2019**, *162*, 133–153.
65. Carlier, Y.; Truyens, C.; Deloron, P.; Peyron, F. Congenital parasitic infections: A review. *Acta Trop.* **2011**, *121*, 55–70. [CrossRef]
66. Costa, M.L.; de Moraes Nobrega, G.; Antolini-Tavares, A. Key Infections in the Placenta. *Obstet. Gynecol. Clin. N. Am.* **2020**, *47*, 133–146. [CrossRef]
67. Cerbán, F.M.; Stempin, C.C.; Volpini, X.; Carrera Silva, E.A.; Gea, S.; Motran, C.C. Signaling pathways that regulate *Trypanosoma cruzi* infection and immune response. *Biochim. Biophys. Acta Mol. Basis Dis.* **2020**, *1866*, 165707. [CrossRef]
68. Liempi, A.; Castillo, C.; Duaso, J.; Droguett, D.; Sandoval, A.; Barahona, K.; Hernández, A.; Galanti, N.; Maya, J.D.; Kemmerling, U.; et al. *Trypanosoma cruzi* induces trophoblast differentiation: A potential local antiparasitic mechanism of the human placenta? *Placenta* **2014**, *35*, 1035–1042. [CrossRef]
69. Tarleton, R.L. Immune system recognition of *Trypanosoma cruzi*. *Curr. Opin. Immunol.* **2007**, *19*, 430–434. [CrossRef]
70. Carrillo, I.; Droguett, D.; Castillo, C.; Liempi, A.; Muñoz, L.; Maya, J.D.; Galanti, N.; Kemmerling, U. Caspase-8 activity is part of the BeWo trophoblast cell defense mechanisms against *Trypanosoma cruzi* infection. *Exp. Parasitol.* **2016**, *168*, 9–15. [CrossRef]
71. Duaso, J.; Rojo, G.; Cabrera, G.; Galanti, N.; Bosco, C.; Maya, J.D.; Morello, A.; Kemmerling, U. *Trypanosoma cruzi* induces tissue disorganization and destruction of chorionic villi in an ex vivo infection model of human placenta. *Placenta* **2010**, *31*, 705–711. [CrossRef]
72. Duaso, J.; Yanez, E.; Castillo, C.; Galanti, N.; Cabrera, G.; Corral, G.; Maya, J.D.; Zulantay, I.; Apt, W.; Kemmerling, U. Reorganization of Extracellular Matrix in Placentas from Women with Asymptomatic Chagas Disease: Mechanism of Parasite Invasion or Local Placental Defense? *J. Trop. Med.* **2011**, *2012*, e758357. [CrossRef] [PubMed]
73. Calvet, C.M.; Oliveira, F.O.R.; Araújo-Jorge, T.C.; Pereira, M.C.S. Regulation of extracellular matrix expression and distribution in *Trypanosoma cruzi*-infected cardiomyocytes. *Int. J. Med. Microbiol.* **2009**, *299*, 301–312. [CrossRef]
74. Gutierrez, F.R.S.; Lalu, M.M.; Mariano, F.S.; Milanezi, C.M.; Cena, J.; Gerlach, R.F.; Santos, J.E.; Torres-Dueñas, D.; Cunha, F.Q.; Schulz, R.; et al. Increased activities of cardiac matrix metalloproteinases matrix metalloproteinase (MMP)-2 and MMP-9 are associated with mortality during the acute phase of experimental *Trypanosoma cruzi* infection. *J. Infect. Dis.* **2008**, *197*, 1468–1476. [CrossRef]
75. Droguett, D.; Carrillo, I.; Castillo, C.; Gómez, F.; Negrete, M.; Liempi, A.; Muñoz, L.; Galanti, N.; Maya, J.D.; Kemmerling, U. *Trypanosoma cruzi* induces cellular proliferation in the trophoblastic cell line BeWo. *Exp. Parasitol.* **2017**, *173*, 9–17. [CrossRef]
76. Ramírez-Toloza, G.; Sosoniuk-Roche, E.; Valck, C.; Aguilar-Guzmán, L.; Ferreira, V.P.; Ferreira, A. *Trypanosoma cruzi* Calreticulin: Immune Evasion, Infectivity, and Tumorigenesis. *Trends Parasitol.* **2020**, *36*, 368–381. [CrossRef]
77. Crawford, K.E.; Kalionis, B.; Stevenson, J.L.; Brennecke, S.P.; Gude, N.M. Calreticulin has opposing effects on the migration of human trophoblast and myometrial endothelial cells. *Placenta* **2012**, *33*, 416–423. [CrossRef]
78. Gu, V.Y.; Wong, M.H.; Stevenson, J.L.; Crawford, K.E.; Brennecke, S.P.; Gude, N.M. Calreticulin in human pregnancy and pre-eclampsia. *Mol. Hum. Reprod.* **2008**, *14*, 309–315. [CrossRef]
79. Kim, K.; Zakharkin, S.O.; Allison, D.B. Expectations, Validity, and Reality in Gene Expression Profiling. *J. Clin. Epidemiol.* **2010**, *63*, 950–959. [CrossRef]
80. Yuryev, A. Contextual data integration in drug discovery. *Expert Opin. Drug Discov.* **2012**, *7*, 659–666. [CrossRef]
81. Quezada, H.; Guzmán-Ortiz, A.L.; Díaz-Sánchez, H.; Valle-Rios, R.; Aguirre-Hernández, J. Omics-based biomarkers: Current status and potential use in the clinic. *Bol. Med. Hosp. Infant. Mex.* **2017**, *74*, 219–226.

82. Tounta, V.; Liu, Y.; Cheyne, A.; Larrouy-Maumus, G. Metabolomics in infectious diseases and drug discovery. *Mol. Omics* **2021**, *17*, 376–393. [CrossRef]
83. Juiz, N.A.; Torrejón, I.; Burgos, M.; Torres, A.M.; Duffy, T.; Cayo, N.M.; Tabasco, A.; Salvo, M.; Longhi, S.A.; Schijman, A.G. Alterations in Placental Gene Expression of Pregnant Women with Chronic Chagas Disease. *Am. J. Pathol.* **2018**, *188*, 1345–1353. [CrossRef]
84. Oride, A.; Kanasaki, H.; Mijiddorj, T.; Sukhbaatar, U.; Ishihara, T.; Kyo, S. Regulation of kisspeptin and gonadotropin-releasing hormone expression in rat placenta: Study using primary cultures of rat placental cells. *Reprod. Biol. Endocrinol.* **2015**, *13*, 90. [CrossRef]
85. Paredes, V.; Espinoza-Caicedo, J.A.; Salazar-Pousada, D.; Escobar, G.S.; Pérez-López, F.R.; Chedraui, P. Lower placental growth factor and higher free β-hCG and PAPP-A levels in the fetal circulation of near-term pregnancies complicated with severe preeclampsia. *Gynecol. Endocrinol.* **2017**, *33*, 79–81. [CrossRef]
86. Jin, X.; Xu, Z.; Cao, J.; Shao, P.; Zhou, M.; Qin, Z.; Liu, Y.; Yu, F.; Zhou, X.; Ji, W.; et al. Proteomics analysis of human placenta reveals glutathione metabolism dysfunction as the underlying pathogenesis for preeclampsia. *Biochim. Biophys. Acta Proteins Proteom.* **2017**, *1865*, 1207–1214. [CrossRef]
87. Faral-Tello, P.; Greif, G.; Romero, S.; Cabrera, A.; Oviedo, C.; González, T.; Libisch, G.; Arévalo, A.P.; Varela, B.; Verdes, J.M.; et al. *Trypanosoma cruzi* Isolates Naturally Adapted to Congenital Transmission Display a Unique Strategy of Transplacental Passage. *Microbiol. Spectr.* **2023**, *11*, e02504-22. [CrossRef]
88. Brünnert, D.; Piccenini, S.; Ehrhardt, J.; Zygmunt, M.; Goyal, P. Sphingosine 1-phosphate regulates IL-8 expression and secretion via S1PR1 and S1PR2 receptors-mediated signaling in extravillous trophoblast derived HTR-8/SVneo cells. *Placenta* **2015**, *36*, 1115–1121. [CrossRef]
89. Boehm, U.; Klamp, T.; Groot, M.; Howard, J.C. Cellular responses to interferon-gamma. *Annu. Rev. Immunol.* **1997**, *15*, 749–795. [CrossRef]
90. Ng, C.T.; Fong, L.Y.; Abdullah, M.N.H. Interferon-gamma (IFN-γ): Reviewing its mechanisms and signaling pathways on the regulation of endothelial barrier function. *Cytokine* **2023**, *166*, 156208. [CrossRef]
91. Cruz-Robles, D.; Chávez-González, J.P.; Cavazos-Quero, M.M.; Pérez-Méndez, O.; Reyes, P.A.; Vargas-Alarcón, G. Association between IL-1B and IL-1RN gene polymorphisms and Chagas' disease development susceptibility. *Immunol. Invest.* **2009**, *38*, 231–239. [CrossRef] [PubMed]
92. de Oliveira, D.S.; Leite, A.L.J.; Pedrosa, T.C.F.; Mota, L.W.R.; de Paula Costa, G.; de Souza, D.M.S.; Perucci, L.O.; Talvani, A. Insights into IL-33 on inflammatory response during in vitro infection by *Trypanosoma cruzi*. *Immunobiology* **2022**, *227*, 152243. [CrossRef] [PubMed]
93. Petersen, C.A.; Burleigh, B.A. Role for interleukin-1 beta in *Trypanosoma cruzi*-induced cardiomyocyte hypertrophy. *Infect. Immun.* **2003**, *71*, 4441–4447. [CrossRef]
94. Clipman, S.J.; Henderson-Frost, J.; Fu, K.Y.; Bern, C.; Flores, J.; Gilman, R.H. Genetic association study of NLRP1, CARD, and CASP1 inflammasome genes with chronic Chagas cardiomyopathy among *Trypanosoma cruzi* seropositive patients in Bolivia. *PLoS ONE* **2018**, *13*, e0192378. [CrossRef]
95. Ireton, K.; Van Ngo, H.; Bhalla, M. Interaction of microbial pathogens with host exocytic pathways. *Cell. Microbiol.* **2018**, *20*, e12861. [CrossRef]
96. Fernandes, M.C.; Corrotte, M.; Miguel, D.C.; Tam, C.; Andrews, N.W. The exocyst is required for trypanosome invasion and the repair of mechanical plasma membrane wounds. *J. Cell Sci.* **2015**, *128*, 27–32. [CrossRef]
97. Fernandes, M.C.; Andrews, N.W. Host Cell Invasion by *Trypanosoma cruzi*: A Unique Strategy that Promotes Persistence. *FEMS Microbiol. Rev.* **2012**, *36*, 734–747. [CrossRef]
98. Reddy, A.; Caler, E.V.; Andrews, N.W. Plasma membrane repair is mediated by Ca(2+)-regulated exocytosis of lysosomes. *Cell* **2001**, *106*, 157–169. [CrossRef] [PubMed]
99. Trinchieri, G.; Sher, A. Cooperation of Toll-like receptor signals in innate immune defence. *Nat. Rev. Immunol.* **2007**, *7*, 179–190. [CrossRef] [PubMed]
100. Uematsu, S.; Akira, S. Toll-like receptors and innate immunity. *J. Mol. Med.* **2006**, *84*, 712–725. [CrossRef] [PubMed]
101. Pellegrini, A.; Guiñazu, N.; Giordanengo, L.; Cano, R.C.; Gea, S. The role of Toll-like receptors and adaptive immunity in the development of protective or pathological immune response triggered by the *Trypanosoma cruzi* protozoan. *Future Microbiol.* **2011**, *6*, 1521–1533. [CrossRef]
102. Campos, M.A.; Gazzinelli, R.T. *Trypanosoma cruzi* and its components as exogenous mediators of inflammation recognized through Toll-like receptors. *Mediat. Inflamm.* **2004**, *13*, 139–143. [CrossRef] [PubMed]
103. Piccinini, A.M.; Midwood, K.S. DAMPening inflammation by modulating TLR signalling. *Mediat. Inflamm.* **2010**, *2010*, 672395. [CrossRef] [PubMed]
104. Domingo-Gonzalez, R.; Prince, O.; Cooper, A.; Khader, S.A. Cytokines and Chemokines in *Mycobacterium tuberculosis* Infection. *Microbiol. Spectr.* **2016**, *4*, 4–5. [CrossRef] [PubMed]

105. Maspi, N.; Abdoli, A.; Ghaffarifar, F. Pro- and anti-inflammatory cytokines in cutaneous leishmaniasis: A review. *Pathog. Glob. Health* **2016**, *110*, 247–260. [CrossRef] [PubMed]
106. Saha, A.; Roy, S.; Ukil, A. Cytokines and Signaling Networks Regulating Disease Outcomes in Leishmaniasis. *Infect. Immun.* **2022**, *90*, e0024822. [CrossRef] [PubMed]
107. Carrera-Silva, E.A.; Guiñazu, N.; Pellegrini, A.; Cano, R.C.; Arocena, A.; Aoki, M.P.; Gea, S. Importance of TLR2 on Hepatic Immune and Non-Immune Cells to Attenuate the Strong Inflammatory Liver Response During *Trypanosoma cruzi* Acute Infection. *PLoS Neglected Trop. Dis.* **2010**, *4*, e863. [CrossRef]
108. Kawai, T.; Akira, S. The role of pattern-recognition receptors in innate immunity: Update on Toll-like receptors. *Nat. Immunol.* **2010**, *11*, 373–384. [CrossRef] [PubMed]
109. Sadissou, I.; D'almeida, T.; Cottrell, G.; Luty, A.; Krawice-Radanne, I.; Massougbodji, A.; Moreau, P.; Moutairou, K.; Garcia, A.; Favier, B.; et al. High plasma levels of HLA-G are associated with low birth weight and with an increased risk of malaria in infancy. *Malar. J.* **2014**, *13*, 312. [CrossRef]
110. Martínez, F.F.; Knubel, C.P.; Sánchez, M.C.; Cervi, L.; Motrán, C.C. Pregnancy-specific glycoprotein 1a activates dendritic cells to provide signals for Th17-, Th2-, and Treg-cell polarization. *Eur. J. Immunol.* **2012**, *42*, 1573–1584. [CrossRef] [PubMed]
111. Thomas, J.; Klebanov, A.; John, S.; Miller, L.S.; Vegesna, A.; Amdur, R.L.; Bhowmick, K.; Mishra, L. CEACAMS 1, 5, and 6 in disease and cancer: Interactions with pathogens. *Genes Cancer* **2023**, *14*, 12–29. [CrossRef]
112. Van Raemdonck, G.; Zegels, G.; Coen, E.; Vuylsteke, B.; Jennes, W.; Van Ostade, X. Increased Serpin A5 levels in the cervicovaginal fluid of HIV-1 exposed seronegatives suggest that a subtle balance between serine proteases and their inhibitors may determine susceptibility to HIV-1 infection. *Virology* **2014**, *458–459*, 11–21. [CrossRef] [PubMed]
113. Yang, H.; Geiger, M. Cell penetrating SERPINA5 (ProteinC inhibitor, PCI): More questions than answers. *Semin. Cell Dev. Biol.* **2017**, *62*, 187–193. [CrossRef] [PubMed]
114. Bikle, D.D. Vitamin D Regulation of Immune Function. *Curr. Osteoporos. Rep.* **2022**, *20*, 186–193. [CrossRef] [PubMed]
115. Dai, J.; Bercury, K.K.; Ahrendsen, J.T.; Macklin, W.B. Olig1 function is required for oligodendrocyte differentiation in the mouse brain. *J. Neurosci.* **2015**, *35*, 4386–4402. [CrossRef]
116. Sabo, J.K.; Heine, V.; Silbereis, J.C.; Schirmer, L.; Levison, S.W.; Rowitch, D.H. Olig1 is required for noggin-induced neonatal myelin repair. *Ann. Neurol.* **2017**, *81*, 560–571. [CrossRef] [PubMed]
117. Tsang, S.M.; Oliemuller, E.; Howard, B.A. Regulatory roles for SOX11 in development, stem cells and cancer. *Semin. Cancer Biol.* **2020**, *67 Pt 1*, 3–11. [CrossRef] [PubMed]
118. Tassi, E.; Garman, K.A.; Schmidt, M.O.; Ma, X.; Kabbara, K.W.; Uren, A.; Tomita, Y.; Goetz, R.; Mohammadi, M.; Wilcox, C.S.; et al. Fibroblast Growth Factor Binding Protein 3 (FGFBP3) impacts carbohydrate and lipid metabolism. *Sci. Rep.* **2018**, *8*, 15973. [CrossRef]
119. Colak, E.; Ozcimen, E.E.; Erinanç, O.H.; Tohma, Y.A.; Ceran, M.U. Is placental KISS-1 expression associated with first trimester abortion spontaneous? *Obstet. Gynecol. Sci.* **2020**, *63*, 490–496. [CrossRef] [PubMed]
120. Jayasena, C.N.; Abbara, A.; Izzi-Engbeaya, C.; Comninos, A.N.; Harvey, R.A.; Maffe, J.G.; Sarang, Z.; Ganiyu-Dada, Z.; Padilha, A.I.; Dhanjal, M.; et al. Reduced Levels of Plasma Kisspeptin During the Antenatal Booking Visit Are Associated With Increased Risk of Miscarriage. *J. Clin. Endocrinol. Metab.* **2014**, *99*, E2652–E2660. [CrossRef]
121. Doherty, L.; Wan, M.; Peterson, A.; Youngstrom, D.W.; King, J.S.; Kalajzic, I.; Hankenson, K.D.; Sanjay, A. Wnt-associated adult stem cell marker Lgr6 is required for osteogenesis and fracture healing. *Bone* **2023**, *169*, 116681. [CrossRef] [PubMed]
122. de Lau, W.; Peng, W.C.; Gros, P.; Clevers, H. The R-spondin/Lgr5/Rnf43 module: Regulator of Wnt signal strength. *Genes Dev.* **2014**, *28*, 305–316. [CrossRef]
123. de Souza, W.; de Carvalho, T.M.U.; Barrias, E.S. Review on *Trypanosoma cruzi*: Host Cell Interaction. *Int. J. Cell Biol.* **2010**, *2010*, 295394. [CrossRef]
124. Laufman, O.; Freeze, H.H.; Hong, W.; Lev, S. Deficiency of the Cog8 subunit in normal and CDG-derived cells impairs the assembly of the COG and Golgi SNARE complexes. *Traffic* **2013**, *14*, 1065–1077. [CrossRef]
125. Willett, R.; Kudlyk, T.; Pokrovskaya, I.; Schönherr, R.; Ungar, D.; Duden, R.; Lupashin, V. COG complexes form spatial landmarks for distinct SNARE complexes. *Nat. Commun.* **2013**, *4*, 1553. [CrossRef] [PubMed]
126. Sangkuhl, K.; Claudio-Campos, K.; Cavallari, L.; Agundez, J.; Whirl-Carrillo, M.; Duconge, J.; Del Tredici, A.L.; Wadelius, M.; Botton, M.R.; Woodahl, E.L.; et al. PharmVar GeneFocus: CYP2C9. *Clin. Pharmacol. Ther.* **2021**, *110*, 662–676. [CrossRef]
127. Ana, Y.; Rojas Marquez, J.D.; Fozzatti, L.; Baigorrí, R.E.; Marin, C.; Maletto, B.A.; Cerbán, F.; Radi, R.; Piacenza, L.; Stempin, C. An exacerbated metabolism and mitochondrial reactive oxygen species contribute to mitochondrial alterations and apoptosis in CD4 T cells during the acute phase of *Trypanosoma cruzi* infection. *Free Radic. Biol. Med.* **2021**, *163*, 268–280. [CrossRef] [PubMed]
128. MacIver, N.J.; Michalek, R.D.; Rathmell, J.C. Metabolic regulation of T lymphocytes. *Annu. Rev. Immunol.* **2013**, *31*, 259–283. [CrossRef] [PubMed]

129. Wolfarth, A.A.; Liu, X.; Darby, T.M.; Boyer, D.J.; Spizman, J.B.; Owens, J.A.; Chandrasekharan, B.; Naudin, C.R.; Hanley, K.Z.; Robinson, B.S.; et al. Proline-Rich Acidic Protein 1 (PRAP1) Protects the Gastrointestinal Epithelium From Irradiation-Induced Apoptosis. *Cell. Mol. Gastroenterol. Hepatol.* **2020**, *10*, 713–727. [CrossRef] [PubMed]
130. Jin, Q.; Chen, H.; Luo, A.; Ding, F.; Liu, Z. S100A14 stimulates cell proliferation and induces cell apoptosis at different concentrations via receptor for advanced glycation end products (RAGE). *PLoS ONE* **2011**, *6*, e19375. [CrossRef]
131. Truyens, C.; Mjihdi, K.; Lambot, M.A.; Rivera, M.T.; Noël, J.C.; Carlier, Y. Effects of acute and chronic *Trypanosoma cruzi* infection in pregnant mice. *Rev. Soc. Bras. Med. Trop.* **2005**, *38* (Suppl. S2), 68–72. [PubMed]
132. Ladomery, M.R.; Maddocks, D.G.; Wilson, I.D. MicroRNAs: Their discovery, biogenesis, function and potential use as biomarkers in non-invasive prenatal diagnostics. *Int. J. Mol. Epidemiol. Genet.* **2011**, *2*, 253–260.
133. Acuña, S.M.; Floeter-Winter, L.M.; Muxel, S.M. MicroRNAs: Biological Regulators in Pathogen-Host Interactions. *Cells* **2020**, *9*, 113. [CrossRef] [PubMed]
134. Salehi, M.; Sharifi, M. Exosomal miRNAs as novel cancer biomarkers: Challenges and opportunities. *J. Cell. Physiol.* **2018**, *233*, 6370–6380. [CrossRef]
135. Sadovsky, Y.; Mouillet, J.F.; Ouyang, Y.; Bayer, A.; Coyne, C.B. The Function of TrophomiRs and Other MicroRNAs in the Human Placenta. *Cold Spring Harb. Perspect. Med.* **2015**, *5*, a023036. [CrossRef] [PubMed]
136. Bartel, D.P. MicroRNAs: Genomics, biogenesis, mechanism, and function. *Cell* **2004**, *116*, 281–297. [CrossRef] [PubMed]
137. Zheng, Y.; Cai, X.; Bradley, J.E. microRNAs in parasites and parasite infection. *RNA Biol.* **2013**, *10*, 371–379. [CrossRef]
138. Medina, L.; Castillo, C.; Liempi, A.; Guerrero-Muñoz, J.; Rojas-Pirela, M.; Maya, J.D.; Prieto, H.; Kemmerling, U. *Trypanosoma cruzi* and Toxoplasma gondii Induce a Differential MicroRNA Profile in Human Placental Explants. *Front. Immunol.* **2020**, *11*, 595250. [CrossRef] [PubMed]
139. Chaiwangyen, W.; Ospina-Prieto, S.; Photini, S.M.; Schleussner, E.; Markert, U.R.; Morales-Prieto, D.M. Dissimilar microRNA-21 functions and targets in trophoblastic cell lines of different origin. *Int. J. Biochem. Cell Biol.* **2015**, *68*, 187–196. [CrossRef]
140. Maccani, M.A.; Padbury, J.F.; Marsit, C.J. miR-16 and miR-21 expression in the placenta is associated with fetal growth. *PLoS ONE* **2011**, *6*, e21210. [CrossRef]
141. Muralimanoharan, S.; Maloyan, A.; Mele, J.; Guo, C.; Myatt, L.G.; Myatt, L. MIR-210 modulates mitochondrial respiration in placenta with preeclampsia. *Placenta* **2012**, *33*, 816–823. [CrossRef]
142. Hou, J.; Wang, P.; Lin, L.; Liu, X.; Ma, F.; An, H.; Wang, Z.; Cao, X. MicroRNA-146a Feedback Inhibits RIG-I-Dependent Type I IFN Production in Macrophages by Targeting TRAF6, IRAK1, and IRAK2. *J. Immunol.* **2009**, *183*, 2150–2158. [CrossRef] [PubMed]
143. Taganov, K.D.; Boldin, M.P.; Chang, K.J.; Baltimore, D. NF-κB-dependent induction of microRNA miR-146, an inhibitor targeted to signaling proteins of innate immune responses. *Proc. Natl. Acad. Sci. USA* **2006**, *103*, 12481–12486. [CrossRef] [PubMed]
144. Medina, L.; Guerrero-Muñoz, J.A.; Liempi, A.I.; Castillo, C.; Ortega, Y.; Sepúlveda, A.; Salomó, F.; Maya, J.D.; Kemmerling, U. Ex Vivo Infection of Human Placental Explants by *Trypanosoma cruzi* Reveals a microRNA Profile Similar to That Seen in Trophoblast Differentiation. *Pathogens* **2022**, *11*, 361. [CrossRef] [PubMed]
145. Zhang, M.; Muralimanoharan, S.; Wortman, A.C.; Mendelson, C.R. Primate-specific miR-515 family members inhibit key genes in human trophoblast differentiation and are upregulated in preeclampsia. *Proc. Natl. Acad. Sci. USA* **2016**, *113*, E7069–E7076. [CrossRef]
146. Raghavachari, N.; Garcia-Reyero, N. Overview of Gene Expression Analysis: Transcriptomics. *Methods Mol. Biol.* **2018**, *1783*, 1–6.
147. Wang, Z.; Gerstein, M.; Snyder, M. RNA-Seq: A revolutionary tool for transcriptomics. *Nat. Rev Genet.* **2009**, *10*, 57–63. [CrossRef] [PubMed]
148. Cencig, S.; Coltel, N.; Truyens, C.; Carlier, Y. Fertility, gestation outcome and parasite congenital transmissibility in mice infected with TcI, TcII and TcVI genotypes of *Trypanosoma cruzi*. *PLoS Neglected Trop. Dis.* **2013**, *7*, e2271. [CrossRef] [PubMed]
149. Bode, C.J.; Jin, H.; Rytting, E.; Silverstein, P.S.; Young, A.M.; Audus, K.L. In Vitro Models for Studying Trophoblast Transcellular Transport. *Methods Mol. Med.* **2006**, *122*, 225–239. [PubMed]
150. Drewlo, S.; Baczyk, D.; Dunk, C.; Kingdom, J. Fusion assays and models for the trophoblast. *Methods Mol. Biol.* **2008**, *475*, 363–382. [PubMed]
151. Li, M.; Brokaw, A.; Furuta, A.M.; Coler, B.; Obregon-Perko, V.; Chahroudi, A.; Wang, H.-Y.; Permar, S.R.; Hotchkiss, C.E.; Golos, T.G.; et al. Non-human Primate Models to Investigate Mechanisms of Infection-Associated Fetal and Pediatric Injury, Teratogenesis and Stillbirth. *Front. Genet.* **2021**, *12*, 680342. [CrossRef] [PubMed]
152. Rodríguez-Morales, O.; Ballinas-Verdugo, M.A.; Alejandre-Aguilar, R.; Reyes, P.A.; Arce-Fonseca, M. *Trypanosoma cruzi* connatal transmission in dogs with Chagas disease: Experimental case report. *Vector Borne Zoonotic Dis.* **2011**, *11*, 1365–1370. [CrossRef]
153. Rossant, J.; Cross, J.C. Placental development: Lessons from mouse mutants. *Nat. Rev. Genet.* **2001**, *2*, 538–548. [CrossRef] [PubMed]
154. Alarcón Mendoza, M.E.; Ruiz Rangel, G.; Lugo de Yarbuh, A.; Briceida Guillén, C.; Moreno Balza, E.A.; Pérez Aguilar, M.C.; Cruz, J. Congenital *Trypanosoma cruzi* transmission in pups of Wistar rats with acute Chagas infection. *Rev. Soc. Venez. Microbiol.* **2010**, *30*, 114–120.

155. Moreno, E.A.; Rivera, I.M.; Moreno, S.C.; Alarcón, M.E.; Lugo-Yarbuh, A. Vertical transmission of *Trypanosoma cruzi* in Wistar rats during the acute phase of infection. *Investig. Clin.* **2003**, *44*, 241–254. [PubMed]
156. Malassiné, A.; Frendo, J.L.; Evain-Brion, D. A comparison of placental development and endocrine functions between the human and mouse model. *Hum. Reprod. Update* **2003**, *9*, 531–539. [CrossRef]
157. Carter, A.M. Animal models of human pregnancy and placentation: Alternatives to the mouse. *Reproduction* **2020**, *160*, R129–R143. [CrossRef]
158. Sherlock, Í.A.; Muniz, T.M. Transmissão do *Trypanosoma cruzi* em três gerações de Cavia porcellus sem a participação de triatomíneos. *Rev. Soc. Bras. Med. Trop.* **1976**, *10*, 27–29. [CrossRef]
159. Libisch, M.G.; Rego, N.; Robello, C. Transcriptional Studies on *Trypanosoma cruzi*—Host Cell Interactions: A Complex Puzzle of Variables. *Front. Cell. Infect. Microbiol.* **2021**, *11*, 692134. [CrossRef] [PubMed]
160. Westermann, A.J.; Barquist, L.; Vogel, J. Resolving host-pathogen interactions by dual RNA-seq. *PLoS Pathog.* **2017**, *13*, e1006033. [CrossRef] [PubMed]
161. Westermann, A.J.; Gorski, S.A.; Vogel, J. Dual RNA-seq of pathogen and host. *Nat. Rev. Microbiol.* **2012**, *10*, 618–630. [CrossRef]
162. Chen, W.; Li, S.; Kulkarni, A.S.; Huang, L.; Cao, J.; Qian, K.; Wan, J. Single Cell Omics: From Assay Design to Biomedical Application. *Biotechnol. J.* **2020**, *15*, e1900262. [CrossRef]
163. Papalexi, E.; Satija, R. Single-cell RNA sequencing to explore immune cell heterogeneity. *Nat. Rev. Immunol.* **2018**, *18*, 35–45. [CrossRef] [PubMed]

pathogens

MDPI

Article

Albaconazole Polymeric Nanocapsules for Treating *Trypanosoma cruzi* Infections

Cristina Maria de Barros [1,2], Vanja Maria Veloso [3], Margareth Spangler Andrade [4], José Mário Carneiro Vilela [4], Maria Alice de Oliveira [1], Marta de Lana [2,5], Maria Terezinha Bahia [2,6] and Vanessa Carla Furtado Mosqueira [1,2,3,*]

1 Laboratory of Pharmaceutics and Nanotechnology, School of Pharmacy, Federal University of Ouro Preto, Campus Universitário Morro do Cruzeiro, Ouro Preto 35400-000, MG, Brazil
2 Post-Graduation Program in Biological Sciences (NUPEB), Exact and Biological Sciences Institute (ICEB), Federal University of Ouro Preto, Ouro Preto 35400-000, MG, Brazil
3 Department of Pharmacy, School of Pharmacy, Federal University of Ouro Preto, Ouro Preto 35400-000, MG, Brazil
4 Laboratory of Nanoscopy, FIEMG SENAI Technological Center—Regional Department of Minas Gerais, Belo Horizonte 31035-536, MG, Brazil
5 Department of Clinical Analysis, School of Pharmacy, Federal University of Ouro Preto, Ouro Preto 35400-000, MG, Brazil
6 Parasitic Diseases Laboratory, School of Medicine, Federal University of Ouro Preto, Ouro Preto 35400-000, MG, Brazil
* Correspondence: mosqueira@ufop.edu.br

Abstract: The therapeutic management of Chagas disease requires new medicines because the standard-of-care drugs available induce adverse effects and have limited efficacy. In this study, we developed a formulation of albaconazole (ABZ) loaded in biodegradable polymeric nanocapsules (NCs). Free ABZ and ABZ-loaded NCs were similarly active against the Y strain and inactive against the Colombian strain epimastigotes of *Trypanosoma cruzi*. Infected mice were given ABZ in different doses and treatment schedules by oral, SC, and IM routes during the acute phase of infection. Free ABZ taken orally reduced parasitemia and suppressed mortality; however, all the animals maintained patent parasitemia during and after treatment. ABZ-NCs increased anti-*T. cruzi* effects ($p < 0.05$), inducing negative parasitemia during treatment in most of the tested regimens. The parasitemia level was also significantly reduced after treatment with ABZ-NCs during the acute phase of the disease, and relapses were delayed compared with the free ABZ treatment. Once- and twice-daily doses were similarly effective, demonstrating that the NCs prolonged the ABZ-NC residence time. Free ABZ and ABZ-NCs did not prevent infection, ABZ seemed to have suppressive effects on *T. cruzi* growth, and encapsulation prolonged this suppression. The analysis of the in vivo results indicated that the NCs significantly improved the safety of ABZ in the mouse model, suggesting that the increased ABZ-NC dosage regimen merits further efficacy and pharmacokinetic evaluations.

Keywords: albaconazole; UR-9825; nanoparticles; efficacy; toxicity; formulation; dose–response; dosage regimen

1. Introduction

Chagas disease (American Trypanosomiasis) is caused by the protozoan *Trypanosoma cruzi*. It is estimated that approximately 6 million individuals have chronic infection with this pathogen, mostly in Latin America [1,2]. Chagas disease (CD) has a significant social impact in Latin America and is a neglected tropical disease [3]. In CD endemic areas, the only drug available for treatment is benznidazole (BZN). However, treatment with

 https://doi.org/10.3390/pathogens14040319

BZN has serious side effects in addition to low cure rates during the chronic phase of the infection [2,4]. In clinical trials performed during the acute phase of the infection, the cure rate obtained was 40% to 76% [5]. Long-term studies on chemotherapy with BZN for children and adolescents with short-term chronic infections have also shown significant success [6,7]. In different regions of Bolivia, however, BZN treatment in children and adolescents induced cure rates of only 0% to 5.4% [8]. No effective treatment exists for the established chronic form of the disease, which is currently the most common clinical presentation in both endemic and non-endemic areas [5,9].

New active pharmaceutical ingredients (APIs) with low toxicity and high cure rates in the acute and chronic phases of CD are urgently needed to control this infection, which is spreading in various countries and continents [1,2,4,10]. A new class of synthetic azole molecules with antifungal activity, such as posaconazole, albaconazole (ABZ), voriconazole, TAK-187, and D0870, has been investigated as an alternative approach to chemotherapy in Chagas disease, because they act as selective inhibitors of lanosterol C-14α-demethylase (CYP51 or Erg11p), a key enzyme implicated in ergosterol synthesis [10–13]. This is a common pathway for the fungi and protozoa of the Trypanosomatidae family, including *T. cruzi* [14]. Triazoles have variable affinity for CYP51, and their action as CYP51 inhibitors depletes ergosterol, causing methylated sterols to accumulate in the cell membrane, which disrupts membrane fluidity, inhibits growth, or induces cell death [14].

Among the azole candidates, ABZ (UR-9825, Group Uriach) [12–16] is a triazole antifungal with a broad spectrum of activity and high oral bioavailability in humans [17–19]. ABZ's chemical structure is shown in Figure 1. Apart from its potent effects against life-threatening and invasive fungal infections [20,21], this drug is also active against *T. cruzi* in vitro and in vivo [22,23].

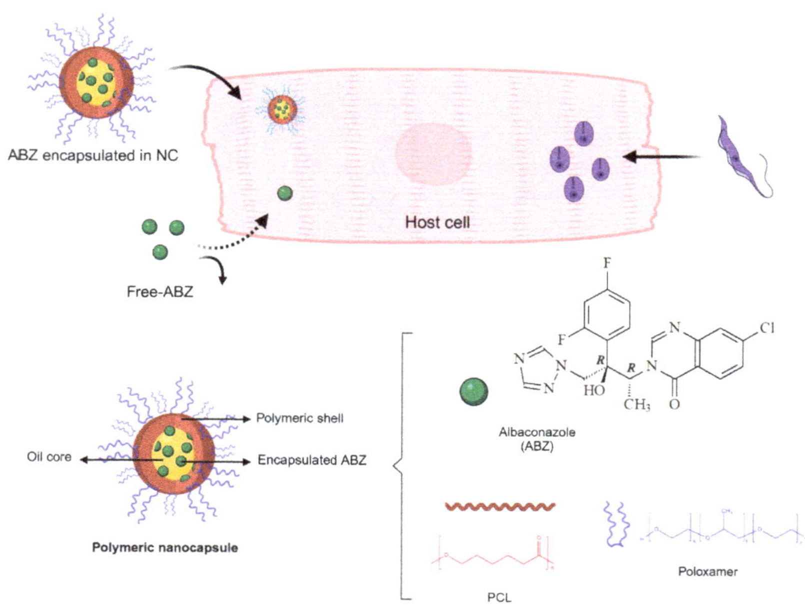

Figure 1. Schematic representation of nanocapsule structure, the chemical structure of ABZ, and the strategy used to deliver ABZ inside *T. cruzi*-infected host cells. Created in BioRender's web-based software 2025. https://BioRender.com/i26s954 (accessed on 21 February 2025).

However, ABZ has solubility limitations. It is a lipophilic molecule (clog $P_{o/w}$ 2.56) and has a short plasma half-life in mice (<1 h) [12,13,15]. Thus, the development of ABZ as an anti-*T. cruzi* candidate is hampered, since it is challenging to evaluate its efficacy in a preclinical murine experimental model of *T. cruzi* infection due to its fast elimination and the inability to detect it in plasma 6 h post-administration [12]. ABZ efficacy has been investigated in dogs infected with the Y and Berenice-78 strains of *T. cruzi*, where it showed promising results in suppressing parasitemia and increasing survival. Cure rates of 25% and 100% were obtained among animals infected with strain Y and treated with 1.5 mg/kg/day of ABZ for 60 and 90 days, respectively [23]. Furthermore, repeated toxicity studies in rats for 28 days with 100 mg/kg twice a day and 250 mg/kg once a day showed a low incidence of toxic effects [24,25].

More recently, ABZ was evaluated in clinical trials (phase II/III), namely, NCT00199264, NCT00509275, and NCT00730405, as an antifungal [17–19,24,26–28]. These studies revealed that ABZ has good pharmacokinetic and oral bioavailability profiles for a once-a-week dosing schedule during 9 months of treatment to control mycoses in humans [19]. ABZ's antifungal effects were dose-dependent [19]. In this phase II clinical trial testing different oral doses (100, 200, 300, and 400 mg/kg), ABZ's toxicological profile showed that it was prone to causing gastrointestinal events such as nausea and diarrhea. However, it was well tolerated, with mild–moderate adverse effects occurring in less than 3% of patients [19].

ABZ has already been formulated as tablets and capsules, and clinical trials comparing these dosage forms demonstrated different bioavailabilities [29]. Its poor water solubility and slow dissolution rate impaired the development of parenteral formulations, and, for this reason, there are no intravenous and subcutaneous injectable formulations of this drug.

Nanocapsules (NCs) represent a dosage form that has demonstrated outstanding therapeutic properties by reducing the toxic effects of anti-*T. cruzi* drugs [30], modifying biodistribution [31], and improving efficacy [32]. This delivery system has been used for encapsulating different lipophilic drugs [33]. We recently reported that NCs effectively interact with infected host cells [34]. After being internalized by them, NCs can deliver the encapsulated drug to the vicinity of intracellular *T. cruzi* amastigotes. NCs have also been shown to interact with blood trypomastigotes [34].

The association of ABZ with polymeric biodegradable NCs is envisaged in the present study, aiming to provide prolonged release, as well as circumventing toxicity issues related to azole derivatives, such as QTc prolongation and arrhythmia [35–37]. NCs have been shown to be effective in reducing both in vitro and in vivo drug-induced cardiotoxicity [38,39]. These expected NC properties may offer relief for patients in the chronic phase of Chagas disease by potentially reducing adverse effects. Furthermore, parenteral (SC, IV and IM) dosage forms for treating *T. cruzi* infections are not currently available. NCs can be administered by SC, IM, IV, and oral routes. Another possible useful application for this delivery system would be to control serious cases, such as those caused by oral infections mediated by contaminated food [40].

The present study reports the development of biodegradable ABZ-loaded NCs, chosen due to the ability of their oily core to entrap lipophilic compounds (Figure 1). The preclinical pharmaceutical development and the physicochemical characterization of ABZ-NCs, in terms of surface charge, hydrodynamic diameter, polydispersity, drug loading, and encapsulation yield, are also reported in this study. A morphological analysis of these new formulations was also conducted by atomic force microscopy (AFM) and scanning electron microscopy (SEM). These techniques were able to image NCs and their surface features, supplying high-resolution information in nanoscale dimensions. An in vitro biological evaluation using acellular cultures of *T. cruzi* epimastigotes was carried out. Furthermore, the dose-response curve of the parasitemia in mice infected with the Y strain of *T. cruzi* (Y strain) and treated with ABZ-NCs or free ABZ was also investigated using a murine model. We aimed to obtain a slow release of ABZ into the tissues to reduce the frequency of administration and the number of doses. The in vivo toxicity induced by scaling doses of ABZ has also been investigated following IV administration.

2. Materials and Methods

2.1. Materials

Lecithin, the soy phosphatidylcholine ~70% purity (Epikuron®170), was acquired from Lucas Meyer (Hamburg, Germany), and poloxamer 188 MW 8400 g/mol (Synperonic PE/F68) was provided by ICI Surfactants (Cleveland, UK). Poly-ε-caprolactone polymer (PCL; Mn 42,500 g/mol, Đ 1.529 at 25 °C) was purchased from Sigma (Sigma-Aldrich Co., St. Louis, MO, USA). Benznidazole (2-nitroimidazol-(N-benzyl-2-nitro-1-imidazolacetamida) was provided by Roche S.A. (Buenos aires, BAL, Argentina). Medium-chain triglycerides (MCTs) (Miglyol 810N) were a donation from IOI OleoChemical (GmbH, Hamburg, Germany). ABZ (UR-9825), the (1*R*,2*R*)-7-chloro-3-[2-(2,4-difluorophenyl)-2-hydroxy-1-methyl-3-(1H-1,2,4-triazol-1-yl)propyl]quinazolin-4(3H)-one, were kindly given by Javier Bartrolí (Uriach & Cia, Barcelona, Spain) to Prof. J. Urbina (Centro de Biofísica y Bioquímica, Miranda state, Venezuela) and Prof. M.T. Bahia. All other solvents (acetone, *N*,*N*-dimethyl-acetamide, dimethyl sulfoxide, glucose, methanol, and polyethylene glycol 300) were analytical-grade substances and used without further purification. The water was purified by reverse osmosis using the Symplicity® System 185 device (Millipore, Burlington, MA, USA).

2.2. Preparation of ABZ Nanocapsules and Solutions

The interfacial deposition method for a preformed polymer followed by the removal of the solvent (nanoprecipitation method) described by Fessi [41] was used to prepare the NCs. The organic phase consisted of 0.6% (*wt*/*v*) polycaprolactone, 0.75% (*wt*/*v*) lecithin, 2.5% (*v*/*v*) MCT oil, and different concentrations of ABZ (0.5, 1.0, and 5.0 mg/mL). The components of the organic phase were dissolved in 10 mL of acetone under magnetic stirring (model PC-200, Corning, NY, USA) at 30 °C at 250 rpm. The organic solution was transferred into the aqueous phase containing 0.75% *wt/v* of poloxamer 188 using a syringe. The mixture was then maintained for 10 min under magnetic stirring (500 rpm). Finally, the colloidal dispersion obtained was concentrated using a rotavapor Laborota 4000/4001 Heidolph (Heidolph Instruments GmbH, Schwabach, Germany), and its final volume was reduced to 10 mL after the complete removal of acetone and part of the water under reduced pressure. Blank NCs, without ABZ, was produced by the same procedure. The preparation process is schematically represented in Figure 2A.

A free ABZ solution was prepared at 2.5 mg/mL. ABZ was dissolved in a mixture composed of (1:2:7) parts of *N*,*N*-dimethyl-acetamide (DMA), PEG 300, and glucose (5% *wt*/*v*). ABZ was first dissolved in DMA, and, afterwards, PEG 300 was added to the solution. The isotonic glucose solution was slowly incorporated into the DMA/PEG 300 solution under low agitation (MS1 vortex apparatus, IKA Works, Campinas, SP, Brazil).

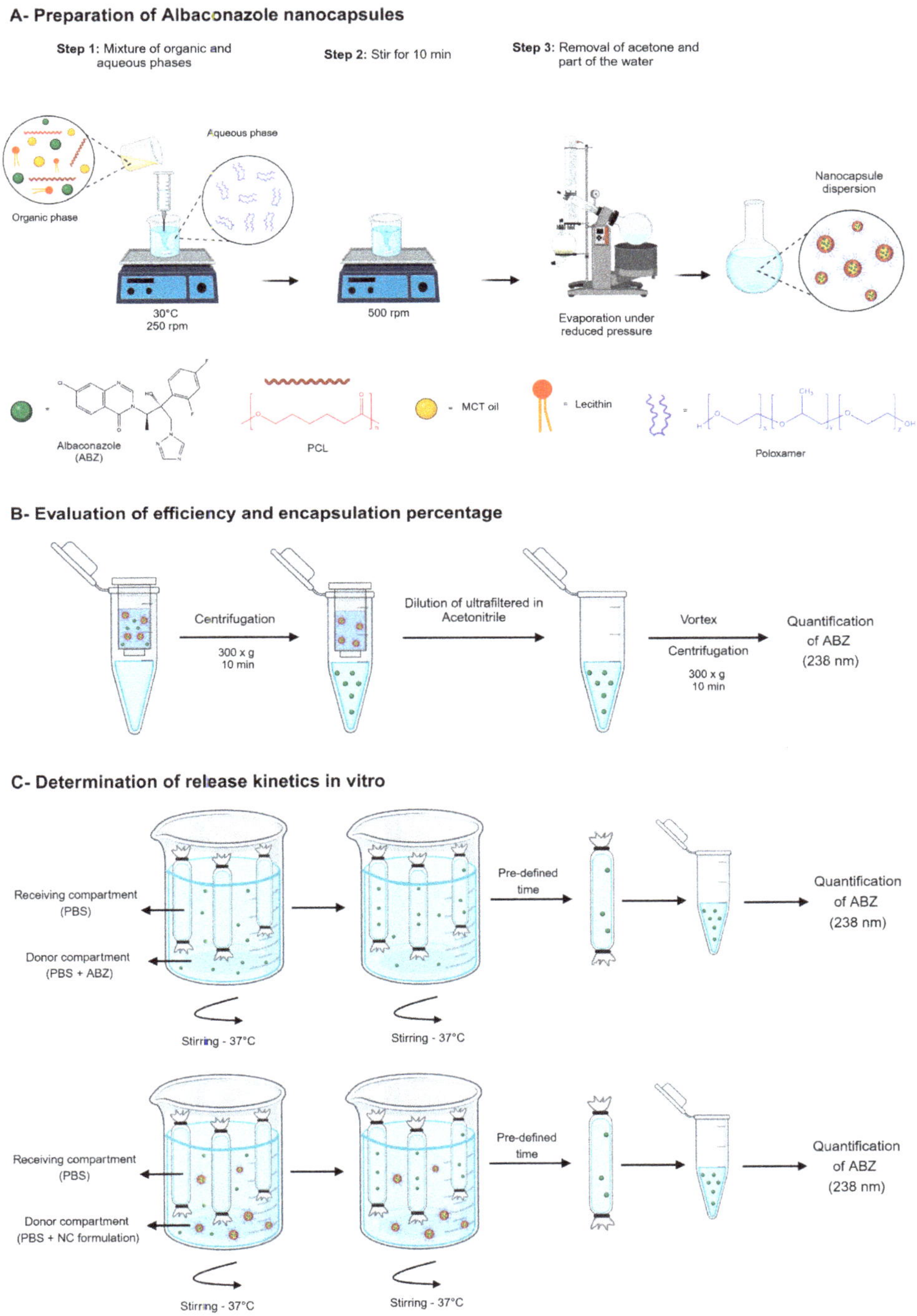

Figure 2. Schematic representation of the preparation method for nanocapsules (**A**); the ultrafiltration/centrifugation method used to separate ABZ loaded in nanocapsules from unloaded ABZ (**B**); and the method of reverse dialysis (**C**) used to study the dissolution/release rate of NCs.

2.3. ABZ Quantification by Ultraviolet Spectrometry

An ultraviolet (UV)-based methodology was developed and validated to quantify free ABZ and ABZ loaded in NCs (Supplementary Materials). The ABZ UV–visible spectrum (Helios α, ThermoSpectronic UV-VIS Spectrophotometer, Pittsburgh, PA, USA) was obtained (Figure S1, Supplementary Materials), and the wavelength of 238 nm was selected to measure ABZ concentrations (details in Supplementary Materials). The ABZ calibration curve (standard curve) was built, containing eight points of known ABZ concentrations in acetonitrile (0.25; 0.5; 1.0; 2.5; 5.0; 7.5; 10.0, and 25 μg/mL). The selectivity, specificity, matrix effect (presence of NCs at 1 and 10%), linearity, repeatability, precision, and accuracy were determined according to the acceptance criteria established by the Analytical and Bioanalytical Methods Validation Guide (ANVISA/MS, BRAZIL, 2003) [42]. The method was linear in the range of 0.25 to 25 mg/mL ($r^2 = 0.9986$ and equation $y = 0.0972x + 0.0226$) and was shown to be selective, precise, and accurate in quantifying free ABZ and ABZ in the presence of NC excipients. The method's intra- and inter-day coefficient of variation were within the maximum established limits. The method was, therefore, considered validated for determining ABZ, and it was applied to quantify ABZ loading and encapsulation efficiency, as well as to quantifying the in vitro release of ABZ from the NCs. Detailed information about this method's development and validation is given in the Supplementary Material Figures and Tables.

2.4. Polymeric Nanocapsule Characterization

The NCs' mean hydrodynamic diameter (Dh) and polydispersity index (PDI) were determined by dynamic light scattering (DLS) at 25 °C with an angle of 90°. NCs were first diluted in ultrapure water. The zeta potential was determined by electrophoretic light scattering in suitable cuvettes, with samples diluted 1:500 times in 1 mM NaCl solution (f(Ka) = 1.5). Both analyses were conducted using a Zetasizer 3000HS Advanced (Malvern Instruments, Worcestershire, UK). The values presented in the tables and graphs correspond to the mean and standard deviations of a minimum of 3 different batches of NCs with 10 readings per sample, determined in triplicate.

2.5. Atomic Force Microscopy and Scanning Electron Microscopy Analysis

The morphological examination of the different nanoparticle formulations was carried out using two different techniques. The SEM images were obtained on a JEOL JSM 5510 microscope at 20 kV, after covering with gold for the shadowing of the samples using sputter equipment. The AFM images were collected using a Dimension 3000 device, monitored by the Nanoscope IIIa controller (Digital Instruments, Santa Barbara, CA, USA), using silicon probes of length 228 μm, with a resonance frequency of 75–98 kHz, constant force of 29–61 N/m, and radius of curvature of 5 nm to 10 nm. To capture images by AFM, 5 μL of each sample was deposited on the atomically flat surface of a freshly cleaved mica, and excess moisture was removed with an argon flow. The images were obtained in intermittent contact mode (tapping mode). The scan was carried out at a speed of 1 Hz and a resolution of 512 × 512 pixels. The quantitative analysis of the images was accomplished using the "section of analyses" applicative of the system (v.5.30r3.sr3), and a minimum of 10 images of each sample were analyzed to ensure the reproducibility of the results. They were presented as the mean and standard deviation of 50 analyzed particles. Additional sizes and the polydispersion measurements of nanoparticles were assessed and calculated by Quantikov 92-99®, a microstructural analyzer for Windows.

2.6. Evaluation of Efficiency and Encapsulation Percentages

The encapsulation efficiency and drug loading were determined following the measurement of the total amount of ABZ present in the NCs, the amount of non-encapsulated ABZ (present in the external aqueous phase of the final dispersion), and the total amount of ABZ used to prepare the formulation. The amount of non-encapsulated ABZ was determined after the ultrafiltration–ultracentrifugation of ABZ-loaded NCs using Microcon filters (Amicon®, 100,000 Da, Millipore, Burlington, MA, USA). ABZ-loaded NCs were centrifuged at 300× *g* for 10 min (Centrifuge 5415 D, Eppendorf), as represented in Figure 2A. A 100 µL ultrafiltrate aliquot was collected and diluted under vortex agitation in 1 mL of acetonitrile, and the ABZ concentration was determined by UV spectroscopy at 238 nm using the methodology described above. The total amount of ABZ present in the NCs dispersion was determined by diluting 100 µL of NCs in 10 mL of acetonitrile and homogenized under vortex agitation to disrupt the NC structure (MS1, IKA Brazil, Campinas, SP, Brazil). After that, the samples were centrifuged at 300× *g*, and the amount of ABZ present in the supernatant was quantified by UV spectroscopy at 238 nm. All analyses were carried out in triplicate, with three readings collected for each sample. NC formulations were prepared containing 0.5, 1.0, and 5.0 mg/mL of ABZ. The percentage of loading was calculated as the difference between the total amount of ABZ in the final NC dispersion and the amount of non-encapsulated ABZ (present in the ultrafiltrate) divided by the total amount of ABZ quantified in the NC dispersion × 100. The ABZ encapsulation efficiency is the amount encapsulated divided by the ABZ weighted to prepare NCs × 100, considering losses in the process. The payload is the mass ratio of encapsulated ABZ to the NC total mass of excipients.

2.7. Determination of ABZ Release In Vitro

The assessment of in vitro free ABZ dissolution and ABZ release profiles from the NCs was performed according to the equilibrium reverse dialysis technique as described by Magalhães et al. (1995) [43]. Briefly, 0.5 mg of an ABZ amorphous powder or 5 mL of NCs was directly diluted in 195 mL of a 0.9% NaCl solution (under sink conditions), where 6 dialysis bags with pores of 12000–14000 Da containing 1 mL of the same 0.9% NaCl solution had previously been immersed. The dialysis bags were kept in equilibrium at 37 °C with the external saline solution for 2 h, before adding the NCs to the external medium. At pre-defined time intervals (0.25, 0.5, 1.0, 2.0, 4.0, and 10.0 h), one of the dialysis bags was removed from the container, and the amount of ABZ present in the bag was determined by UV spectrophotometry at 238 nm, as represented in Figure 2B. The release experiments were carried out in a water bath at 37 °C with constant stirring (Bath Dubnoff mod.144, Fanem, Brazil). The experiment was conducted in triplicate, and each sample was quantified in duplicate.

2.8. In Vitro Activity of Albaconazole Nanocapsules (Epimastigote Inhibition Assay)

The Colombian and Y strains were used to determine the in vitro activity of ABZ NCs against epimastigote forms of *T. cruzi*. Parasites were cultured at 28 °C in liver infusion tryptose (LIT) medium supplemented with (10% *v*/*v*) of fetal bovine serum (FBS) in a biochemical oxygen demand incubator. Cultures were started at a cell density of 2×10^6 epimastigotes/mL, with the drug added 72 h later. Cell density was assessed by counting in a Newbauer chamber. Free ABZ was added as a DMSO solution. The DMSO final concentration did not exceed 0.1% (*v*/*v*). As previously determined in a pilot protocol carried out by our group, this DMSO concentration had no effect on epimastigote proliferation. The experiment was conducted in quadruplicate wells and repeated twice ($n = 8$).

2.9. *Trypanosoma cruzi* Strain and Evolutionary Forms

The Y strain is Tc II DTU (Discrete Typing Unity) of *T. cruzi* [44,45]. It was used to infect mice and simulate the acute phase of the murine disease. The Y strain is partially resistant to BZN [46,47]. The Y strain induces high parasitemia in mice, has a fast infection course, and has high animal mortality rates around the 11th to 13th day post-infection. These characteristics make this an excellent model for the in vivo screening of potential drugs with activity against *T. cruzi*. Trypomastigote forms of *T. cruzi* were collected on the day of parasitemia peak from the blood of experimentally infected mice.

2.10. Experimental Animals and Ethics

Thirty-day-old female albino *Swiss* mice, weighing around 20 g, from the Central Animal Facility of the Federal University of Ouro Preto, were used. All procedures were conducted in accordance with the guidelines of the National Council of Animal Experimentation Control (CONCEA), following approved protocols CEUA/UFOP n°2015/50, 2012/70/(UFOP/n°2004/08, n°2005/07). During the experiments, the animals were kept in the Central Animal Facility and had free access to food and water ad libitum.

2.11. Maximal Tolerated Albaconazole Dose in Healthy Mice

The maximum tolerated dose (MTD) of free ABZ and ABZ-NCs was evaluated using non-infected mice weighing approximately 20 $\pm$ 2 g. Surviving animals were monitored for 72 h. Free ABZ solution and NC formulations were administered via the intravenous route. The maximal volume administered to the animals was 0.2 mL, except when larger volumes were required, where the total volume was divided into applications of 0.2 mL at 20-min intervals. To carry out the DMT experiments, groups of 10 mice were treated with different doses of free ABZ (25, 30, 35, and 40 mg/kg) or encapsulated ABZ (80, 120, and 200 mg/kg). A group of five animals was used for treatment with 500 mg/kg of encapsulated ABZ. All the formulations were freshly prepared, and the free ABZ solution and ABZ-NC concentrations were 2.5 mg/mL and 5.0 mg/mL, respectively.

2.12. Efficacy Evaluation in the Acute Phase in Infected Mice

In this study, the murine *T. cruzi* experimental model of acute infection with the Y strain was used [48]. This strain is partially sensitive to benznidazole [46,47]. Animals were inoculated intraperitoneally with 10,000 blood trypomastigotes per animal. The blood trypomastigote inoculum came from successive passages in mice. The infected animals used in the in vivo experiments were evaluated daily to determine survival. Parasitemia was assessed daily by fresh blood examination (FBE) until complete parasitemia clearance. Blood was collected from the animal's tail vein. The number of parasites was counted according to the technique described by Brener [46] under a light microscope by counting 50 fields on a 22 $\times$ 22 mm microscope slide. The counted number of parasites was used to build the parasitemia curve. Surviving animals were sacrificed 180 days after infection. For the post-treatment cure criteria, a fresh blood test and blood culture were performed according to Chiari's technique (1989) [49].

2.13. Treatment

After confirming parasitemia (approximately 4 days post-inoculation), the formulations were administered for 20 consecutive days [46]. Each experimental group was composed of 10 animals. Control groups were treated with 5% (*wt*/*v*) glucose solution, blank NCs, free ABZ solution, and blank solution. A positive control group was treated by gavage with oral benznidazole (100 mg/kg/day). Different dosages, routes of administration, and treatment durations were utilized to evaluate the effects on parasitemia. Some groups were treated by gavage with controls and 20 mg/kg of ABZ coarse suspension for 20 consecutive days, and other groups were treated by the subcutaneous (SC) route with increasing doses (20, 40, 80, and 120 mg/kg) for the same duration. Another group was treated with 20 mg/kg of ABZ-loaded NCs for 20 consecutive days by the intramuscular (IM) route. The maximal volume administered was 0.2 mL. When a volume larger than 0.2 mL was required, the total volume was divided into applications of 0.2 mL administered at 20-min intervals.

2.14. Statistical Analysis

The hydrodynamic diameter, zeta potential, ABZ efficiency of encapsulation of NCs, and release data points were compared using an unpaired Student's *t*-test using the EpiInfo analysis program version 6.04. Comparisons between the pre-patent and patency periods, peak parasitemia (during and after treatment), and average survival times were performed using Student's *t*-test. To compare the parasitemia curves, the non-parametric Kolmogorov and Smirnov test was used, which compares the area under the parasitemia curve between two samples. Comparisons were made between the curves obtained from untreated and treated mice, as well as between the curves obtained from mice treated with the different formulations. One-way ANOVA with Dunn's multiple-comparisons post-test (95% confidence level) was used for in vitro activity data using Prisma® 8.0 software. The differences were considered significant with $p < 0.05$ and a 95% confidence level.

3. Results

3.1. Nanocapsule Characterization

The NC size measurement was performed by two techniques: DLS in a liquid medium and AFM on dried samples dispersed on mica plates. The pH of NC dispersion and the zeta potential of NC surface containing different concentrations of ABZ were also evaluated, and the results are shown in Table 1. The analysis was performed using freshly prepared formulations.

Table 1. Physicochemical properties of albaconazole nanocapsules.

Formulation	ABZ mg/mL	Hydrodynamic Diameter ± SD [2] (nm)	PDI [3]	AFM [4] Mean Size ± DP [1] (nm)	Zeta Potential ± SD (mV) *	pH ± SD [2]
Blank NCs	0	171.4 ± 0.8	0.127 ± 0.022	336 ± 144	−50.1 ± 1.2	6.69 ± 0.01
0.5 NCs	0.5 *	201.5 ± 0.4	0.096 ± 0.019	195 ± 54	−49.2 ± 3.4	6.44 ± 0.08
1 NCs	1.0 *	225.9 ± 2.1	0.193 ± 0.022	-	−59.5 ± 5.1	7.02 ± 0.04
5 NCs	5.0 [5]	155.9 ± 0.3	0.130 ± 0.004	250 ± 57	−50.5 ± 3.0	7.10 ± 0.20

[1] EE, encapsulation efficiency; [2] SD, standard deviation; [3] PDI, Polydispersion Index, [4] AFM, atomic force microscopy; [5] macroscopic crystallization of ABZ after 7 days post-preparation. Statistical analysis was performed using Student's *t*-test, comparing ABZ-NCs with blank NCs (* $p < 0.05$).

The hydrodynamic diameter of ABZ-NCs increased significantly ($p < 0.05$) with increased ABZ concentrations; however, after loading 5 mg/mL, the maximal payload seemed to be surpassed. After some days, the ABZ crystals grew in the dispersion. All the formulations containing 0 to 1 mg/mL of ABZ were monodisperse, considering a polydispersity index lower than 0.3 during the first week after preparation, and only 5 mg/mL showed instabilities after this storage time. The NC sizes increased until reaching 1 mg/mL, compared to blank NCs. At 5 mg/mL, there is a reduction in mean sizes, probably due to the unloaded ABZ macroscopic crystal that precipitates out of the nanostructure. As a result, 5.0 mg/mL NCs of ABZ show a reduced size, and the PDI is closer to the value of blank-NCs, since the macroscopic crystals of ABZ are not detected by equipment that has a maximum particle detection limit of 3 μm. The pH values of liquid dispersion vary with increasing ABZ concentrations due to its alkaline character. The zeta potential of NCs is negative without significant changes ($p > 0.05$), ranging from −49.2 to −59.5 mV, indicating a low association of ABZ with the surface of NCs. Thus, ABZ is probably associated with the oily nucleus of the NCs, interfering little in the zeta potential. It seems that the optimal loading concentration of ABZ is between 1 and 5 mg/mL.

Scanning electron microscopy (SEM) was used to observe the morphology of NCs containing 1 mg ABZ/mL (Figure 3). Spherical nanostructures are clearly shown. However, this method was not demonstrated to be suitable since the coating of the nanostructures with gold was not uniformly distributed, causing deformation in the structures and even the partial melting of the sample under the electron beam. The NCs can be observed to be slightly deformed, although a three-dimensional spherical appearance is evident. Thus, the analysis of the morphological features of NCs was performed by AFM, since this technique requires no covering of samples (Figure 3A–E). AFM analysis corroborated the physicochemical analysis in the liquid media, indicating that, probably, concentrations of ABZ above 1.0 mg/mL are adsorbed on the NC surface, reaching a saturation level from which the drug precipitates in the external aqueous media in the form of ABZ crystals. In Figure 3, showing height and phase images collected from NC samples containing 0–5 mg of ABZ/mL, nanometric crystals can be observed around NCs (Figure 3D,E). These images show ABZ crystallization from NC nuclei. The ABZ solution allowed to evaporate deposited on mica produced the same standard of microcrystal structures. The crystals and amorphous structures observed in the ABZ-loaded NCs were not observed in blank NCs. The supersaturation of ABZ in the dehydrated samples upon AFM analysis probably induced the rapid crystallization of ABZ, as also observed in NC samples of 5.0 mg/mL in a liquid medium. The schematic representation in Figure 3 is based on the AFM results. In Table 2, the mean geometric diameter measured using AFM images (topographic profiles, Figure 3E) for NC samples was larger than the diameters in the liquid media. This hypothesis is in agreement with observations of NC flattening on the mica surface reported previously [50].

Table 2. Albaconazole association with poly-ε-caprolactone nanocapsules.

ABZ Formulation	ABZ mg/mL (Feed)	ABZ in NCs (mg/mL ± SD [2]) (Real)	EE [1]	% Drug Loading *	Payload # (μg/mg)
0.5 NCs	0.5	0.36 ± 0.18	71.4 ± 4.1	99.79± 0.38	7.83
1 NCs	1.0	1.17± 0.03	117 ± 3.6	94.27± 0.15	25.43
5 NCs	5.0	1.39 ± 0.32	27.9± 1.9 *	94.18± 0.32	30.22 *

[1] EE, encapsulation efficiency; [2] SD, standard deviation; * in freshly prepared NC formulations; # calculations and definitions are provided in the methodology.

The results presented in Table 2 show high encapsulation efficiency for this process, with the highest percentage for 1 mg/mL. The drug loading, the difference in the final NC dispersion between the drug loaded and unloaded, is also very high. These results indicate that ABZ has more affinity for the oily core of NCs than for the external aqueous medium and that it therefore associates better with NCs due to their high lipophilicity. The estimated clog *P* is 2.564 (computed by XLog*P* 3 3.0, PubChem release 14 October 2021). The encapsulation efficiency of ABZ at 5 mg/mL is significantly reduced ($p < 0.05$) in relation to other formulations. Thus, values greater than 1 mg/mL of ABZ probably saturate the system, resulting in the precipitation of ABZ in the external medium, which is responsible for low encapsulation efficiency values at 5 mg/mL concentration (Figure 3D). This result is in agreement with the AFM analysis (Figure 3, SEM image).

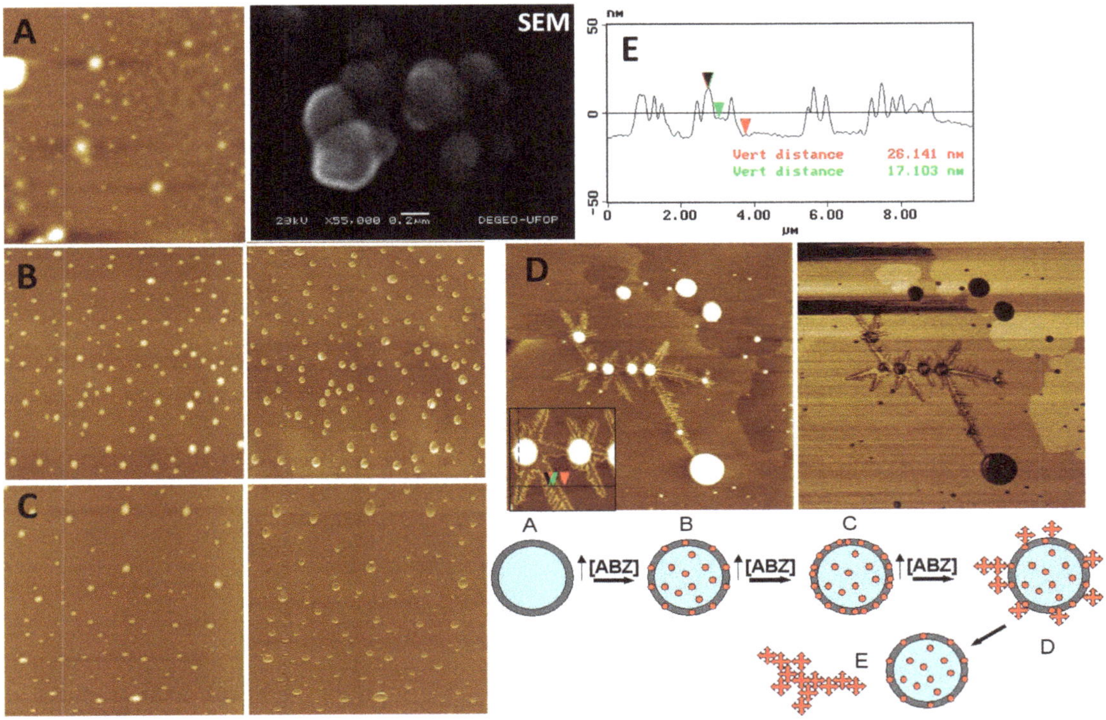

Figure 3. Scanning electron microscopy image (SEM) of ABZ-NCs (1 mg/mL) (in black). Atomic force microscopy (AFM) images showing height (on the left) and phase (on the right), showing (**A**) blank NCs (scan size, 40 μm × 40 μm); (**B**) ABZ-NCs, 0.5 mg/mL (scan size, 40 μm × 40 μm); (**C**) ABZ-NCs, 5.0 mg/mL (scan size, 40 μm × 40 μm); (**D**) ABZ-NCs, 5.0 mg/mL stored for a few hours, showing spherical structures spread on mouse plates surrounded by crystals (scan size, 40 μm × 40 μm); crystal growth, with image insert (scan size, 10 μm × 10 μm) of one crystal size measurement; and (**E**) measurement of ABZ crystals formed on mica plates using AFM equipment software (v.5.30r3.sr3). In the bottom figure, a schematic representation of the ABZ crystallization hypothesis growing from the NC nucleus is shown, indicating the morphological changes following increased ABZ concentration in NC formulations until supersaturation is reached with the presence of nanometric ABZ crystals.

3.2. *Albaconazole Release Profiles*

Figure 4 shows the results for the release profile of ABZ from NCs. The method used to determine the release profile (reverse dialysis) in vitro of encapsulated ABZ was efficiently used in this experiment (Figure 2C).

The release of ABZ from NCs to the external medium was relatively fast, incomplete, and biphasic. From 0 to 1 h, the release profile of the 0.5, 1.0, and 5.0 mg/mL ABZ formulations followed the same pattern, with a *"burst"* release effect for all the NCs (approximately 15%). After 120 min, the formulations containing 0.5 and 1 mg/mL of ABZ showed similar release rates with faster release compared with 5 mg/mL of ABZ. Around 40% of ABZ was released into the external media in less than 4 h. It is observed that after 2 h, a slower profile phase started and continued until it reached a plateau. Up to 10 h, the release of ABZ was incomplete in this media. The reprecipitation of ABZ appeared to occur in the release medium with 5 mg/mL NC formulation after 120 min, which reduced the concentration of ABZ in the release medium. This is also in accordance with the results of encapsulation efficiency (Table 2) and AFM analysis (Figure 3). Free ABZ dissolution was slowed in the first hour in this medium. However, the dissolution continued until it reached 66% in 10 h. NCs were able to retain ABZ even under sink conditions.

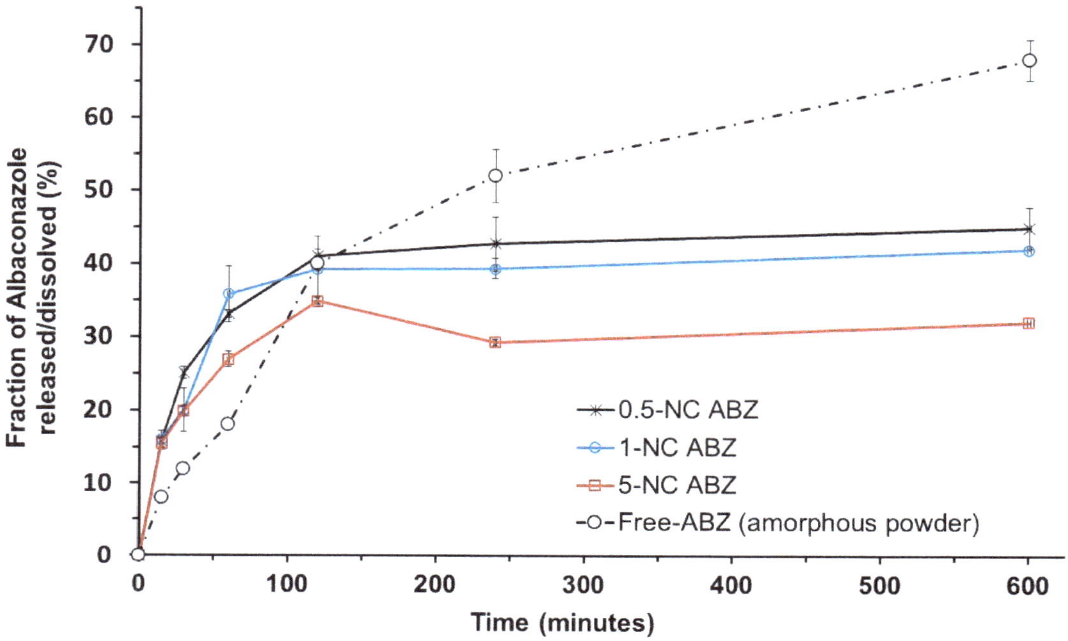

Figure 4. Dissolution profile and release profiles of ABZ from polymeric PCL nanocapsules for three ABZ concentrations: 0.5, 1, and 5 mg/mL. The experiment was performed in saline at 37 °C under sink conditions (20% saturation solubility) by the reverse dialysis method. Each point represents mean values and standard deviations.

3.3. Albaconazole Activity In Vitro Against T. cruzi Epimastigotes

The effects of ABZ against epimastigote forms of *T. cruzi*, Y and Colombian strains, are shown in Figure 5. An inhibitory action of ABZ was observed against the Y strain ($p < 0.05$), compared to untreated control cultures and cultures treated with blank NCs. This inhibitory action began 24 h after the addition of ABZ and was not dose-dependent, as no marked differences were observed between 1 and 3 µM. No differences were observed in the activity of free or encapsulated ABZ regarding culture growth. Free ABZ and ABZ-NCs were ineffective in controlling cell growth when added to cultures of the Colombian strain, even when higher doses were tested (3 µM), confirming the resistance of this strain [46,51].

3.4. General Toxicity Evaluation by Intravenous Route

According to the graph of the percentage of survival versus the concentration of ABZ administered via the IV route represented in Figure S3 (Supplementary Materials), the LD_{50} of free ABZ is 30.4 mg/kg. It was not possible to experimentally obtain the LD_{50} value for ABZ encapsulated in NCs, as even at high concentrations of the drug, no deaths were observed within 72 h (LD_{50} > 500 mg/kg). Therefore, encapsulation proved to be a very useful alternative for reducing the general toxicity of ABZ, since DL_{50} was increased more than 15-fold, as can be seen in Table 3. It shows the adverse effects observed in animals after the intravenous administration of different doses of ABZ. It can be observed that with the encapsulation of ABZ, the intensity and number of side effects are significantly reduced ($p < 0.05$). Free ABZ at 40 mg/kg induced death in all animals, and the double dose (80 mg/kg) encapsulated induced no clinically observed adverse effects in 72 h. Only 500 mg/kg of ABZ-NCs induced a significant increase in serious adverse effects, but no death was observed. Thus, nanoencapsulation dramatically reduces the toxicity of ABZ administered intravenously.

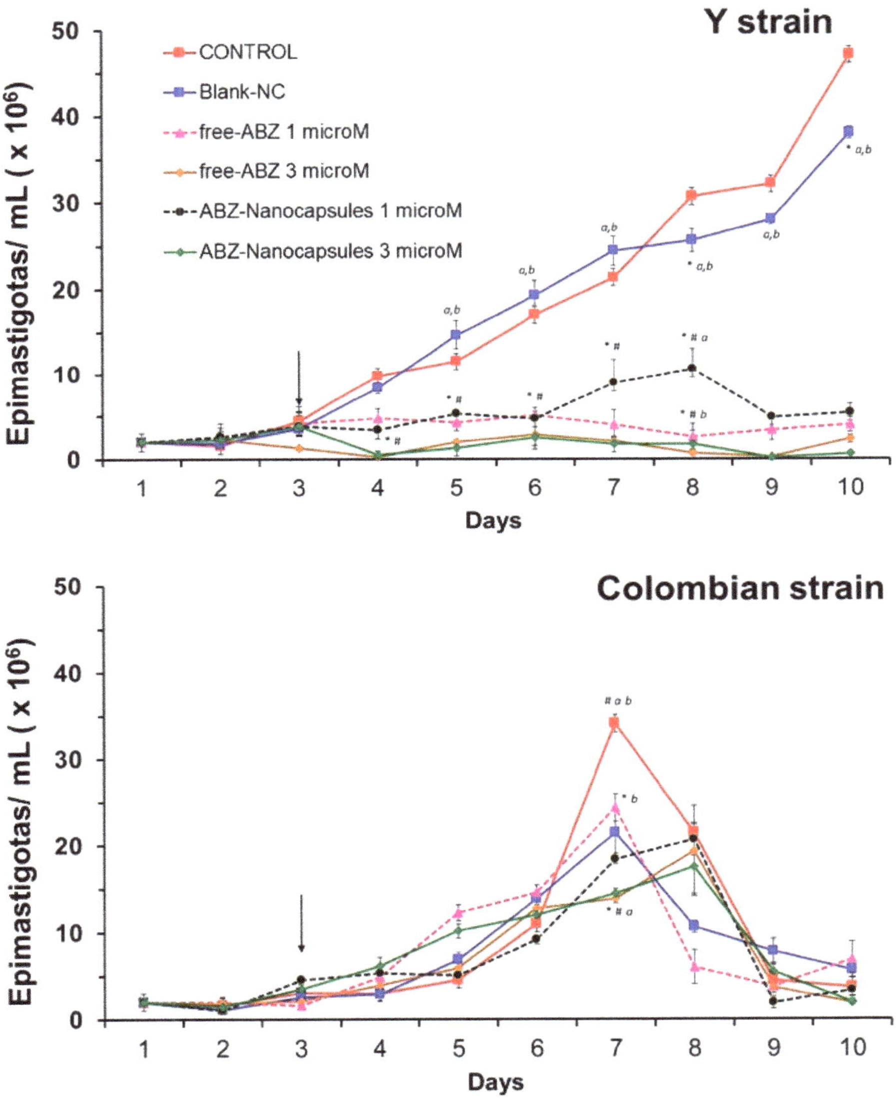

Figure 5. Effects of free ABZ and ABZ-NC in two concentrations (1 and 3 μM) of ABZ on the proliferation of epimastigote forms of the Y and Colombian strains of *T. cruzi*. The arrow indicates the day that formulations were added to the culture media. The experiment was conducted in quadruplicate wells and repeated twice ($n = 8$). The statistical significance ($p < 0.05$) is represented by symbols inside the graphs: * different from control; # different from blank NCs; *a*, different from free -ABZ; *b*, different from ABZ-NCs.

Table 3. Adverse effects caused by intravenous administration of free ABZ or ABZ-loaded nanocapsules.

		Evaluated Parameters			
	Dose	**Ataxia**	**Respiratory Changes**	**Convulsion**	**Survival ***
Free ABZ (intravenous solution) [1]	25 mg/kg	-	-	-	10/10
	30 mg/kg	+++	+++	++	5/8
	35 mg/kg	-	-	++	2/10
	40 mg/kg	-	-	+++	0/9
ABZ NCs [2]	80 mg/kg	-	-	-	10/10
	120 mg/kg	++	+	-	10/10
	200 mg/kg	++	+	-	8/8
	500 mg/kg	+++	+++	-	5/5

[1] ABZ solution (2.5 mg/mL); [2] freshly prepared ABZ nanocapsules (5 mg/mL); * evaluated for 72 h. The symbols represent (+++) strong, (++) moderate, and (+) mild effects.

3.5. *In Vivo Efficacy in Infected Mice*

The efficacy of ABZ in the mouse model infected with the Y strain partially resistant to BNZ was evaluated following different parameters: routes of administration, doses, and dosage regimen. The main results are shown in Table 4.

Table 4. Summary of the effects of treatment with free ABZ and ABZ-loaded nanocapsules on infected mice (Y strain of *T. cruzi*).

Dose (mg/kg/Day)		**Via**	**Survival**	**MST [1] (Days)**	**Patent Period (Days ± SD [2])**	**Maximal Parasitemia Level (×1000) ± SD [2]**		**Negative Parasitological Tests [3]**	
						During ttm.	**After ttm.**	**d 90**	**d 120**
Control (untreated)		SC	0/10	11.4 ± 1.3	8.3 ± 1.5	1418 ± 969	-	-	-
Blank-NCs		SC	0/10	10.2 ± 7.6	7.2 ± 7.6	1427 ± 3082	-	-	-
DMA/PEG 300 solution		SC	0/10	8.4 ± 4.9	5.4 ± 4.9	1976 ± 420	-	-	-
Benznidazole (100 mg/kg/day)		PO	10/10	>60	0	0.5 ± 0 *	nd	10/10	10/10
Free ABZ	20	PO	10/10	>60 *	30 ± 4.8 #	27 ± 29 # *	7.5 ± 4.3	-	-
	20	SC	10/10	>60 *	9.8 ± 12.4 #	0.5 ± 1.6 *	0.5 ± 1.7	-	-
	80	SC	7/10	>60 [4] *	1.4 ± 0.8 #	0 *	0	3/7 #	0/7 #
	120	SC	0/10	4.7 ± 1.6 #	1 ± 0.0	0 *	nd	-	-
ABZ-NCs	20	IM	9/10	>60 [5] *	27.1 ± 3.1 #	318.5 ± 558.5 #	9.0 ± 11.5	-	-
	20	SC	8/10	>60 [6] *	30.1 ± 6.1 #	103.0 ± 104.5 #	12.7 ± 16.9 **	-	-
	40 (2 × 20)	SC	9/10	>60 [7] *	16.5 ± 14.6 #	3.3 ± 5.6 *	3.9 ± 8.6	-	-
	40	SC	10/10	>60 *	22.6 ± 11.9 #	0.5 ± 1.6 *	2.5 ± 3.5	-	-
	80	SC	10/10	>60 *	22.4 ± 12.8 #	4.0 ± 12.6 # *	2.0 ± 2.6	2/10 #	2/10 #
	120	SC	10/10	>60 **	1.1 ± 0.3	0 *	0 **	6/10 # **	6/10 # **

[1] MST, mean survival time; [2] SD, standard deviation; [3] in days after blood culture was performed; [4] animals died on 9th, 10th, or 13th day after inoculation; [5] animal died on 33rd day after inoculation; [6] animals died on 21st and 27th day after inoculation; [7] animal died on 6th day after inoculation; * significantly different from the untreated group ($p < 0.05$); ** significantly different from free ABZ group ($p < 0.05$); # significantly different from BZN group ($p < 0.05$); nd, not determined. PO, per os (oral); SC, subcutaneous; IM, intramuscular.

Blank-NCs and parenteral suspension excipients had no significant effect on infected animals' parasitemia levels ($p > 0.05$) compared to the untreated control that received only isotonic glucose. Benznidazole classical treatment for mice induced 100% survival and negative parasitological tests for 120 days. Free ABZ suspensions were administered via two different routes (PO and SC) with a treatment regimen of 20 uninterrupted days, in addition to being tested via SC at three different doses (20, 80, and 120 mg/kg). Free ABZ administered by the oral route shows no advantageous reduction in parasitemia compared with subcutaneous at the same doses. According to Table 4, the efficacy was significantly higher via SC than oral administration at a dose of 20 mg/kg for the patent period ($p < 0.05$) regarding parasitemia during ($p < 0.05$) and after ($p < 0.05$) the treatment. Survival was 100% for both routes of administration. However, using the SC route, 70% of the animals continued to have negative parasitemia up to 20 days after treatment, while 100% showed patent parasitemia and relapses after oral treatment. At a dose of 80 mg/kg SC, there was a significant reduction in the patent period ($p < 0.02$), without the reactivation of parasitemia after treatment in 100% of the animals. However, survival was 30% lower compared to the 20 mg/kg dose, demonstrating the toxicity of ABZ during long-period SC administration, as the animals died in the second week after starting treatment with negative parasitemia values. There was no significant difference in maximum parasitemia values for surviving animals at the three doses used (20, 80, and 120 mg/kg) via SC ($p > 0.05$) (Table 4). At a dose of 120 mg/kg of free ABZ, there was a significant reduction in the mean survival time ($p < 0.05$), with the animals all dying during the first week of treatment with negative parasitemia. This result indicates that this SC dose of free ABZ was highly toxic to mice, in agreement with previous experiments on the maximal tolerated dose by intravenous administration. The maximum parasitemia, however, was at the same level as the 80 mg/kg dose (Table 4).

Due to the failure of ABZ-NC administration by IM to induce a significant reduction in parasitemia during the patent period, the studies with higher doses of ABZ continued only via the SC route for NCs. The encapsulated ABZ at a dose of 40 mg/kg divided into two daily doses of 20 mg/kg resulted in a longer time for parasitemia reactivation after the end of treatment compared with a single daily dose of 40 mg/kg. However, survival was reduced. At doses of 80 and 120 mg/kg, an increase in efficacy was observed with a significant reduction in the patent period ($p < 0.05$) and the elimination of blood parasites for a period of more than 120 days in 20% and 60% of treated animals, respectively (Table 4). No toxic effects were observed during and after treatment with ABZ-NCs (40–120 mg/kg). This significantly differentiated the formulations, because free ABZ at the same doses induced the death of 30% and 100% of the animals, respectively. The rapid and effective elimination of parasitemia with doses of 80 and 120 mg/kg ABZ-NCs was evident (Figure S4, Supplementary Materials). At the dose of 120 mg/kg ABZ-NCs, there was no reactivation of parasitemia up to 60 days after the end of treatment (Table 4).

4. Discussion

The therapeutic management of Chagas disease requires new drugs, as the standard-of-care medicines available cause serious adverse effects and have limited efficacy. Benznidazole has low efficacy in the chronic phase of the disease, and only this drug is available in many endemic countries of Latin America [2,4]. In the search for new options, in this study, we developed a safe formulation of ABZ loaded into biodegradable polymeric biodegradable nanocapsules (NCs) (Figure 1), which can be administered by oral and parenteral routes (IV, SC, IM). To date, only tablets and conventional gelatin capsule dosage forms have been reported [18], and these formulations have been used only in clinical trials and are not commercially available. Considering the solubility limitations of ABZ (0.001 mg/mL at 30 °C pH 1.2) [52] and its high lipophilicity, we successfully developed a new formulation of ABZ that could be administered parenterally by IM, SC, IV, and even the oral route, composed of biodegradable and biocompatible excipients.

As reported previously, the efficacy of ABZ in the murine model is reduced compared to the efficacy in the dog model [23]. Similarly, herein, parasitological and serological cure in mice treated with ABZ were not obtained in 100% of cases, as previously reported in dogs [23]. Guedes et al. (2004) demonstrated that only a regimen of multiple daily doses was able to improve the therapeutic efficacy in dogs, indicating that ABZ also possesses unfavorable pharmacokinetic properties in this animal model [23].

ABZ appears to have been abandoned in the repurposing strategy against Chagas disease, likely due to its short half-life in animals, in favor of another azole derivative, E-1224, a pro-drug of ravuconazole. Recently, proof-of-concept clinical trials with E-1224 indicated no efficacy [53]. As previously discussed, the rapid suppression of parasitemia and low curative rates of azole derivatives are probably associated with unfavorable pharmacokinetic characteristics in mice that reduce the amount of the drug available in tissues for the eradication of intracellular parasites [11]. For ABZ, a long terminal half-life has been reported in dogs (51 h) and monkeys (24 h) [54]. However, pharmacokinetic studies on ABZ in humans have suggested a prolonged half-life [17,18].

ABZ is a potent lipophilic ergosterol inhibitor that has been studied clinically as an antifungal agent. The inhibition of parasite sterol C14α-demethylase has also been identified as the mechanism of action against *T. cruzi*. Urbina et al. reported potent in vitro activity of ABZ against epimastigotes of the Y strain of *T. cruzi* (MIC 30 nM). In addition, intracellular amastigotes cultured in Vero cells were even more susceptible (IC_{99} of 10 nM and IC_{50} of 1 nM) [22]. Furthermore, no effect was seen on the viability and proliferation of the Vero host cells up to the highest concentration tested (100 nM), indicating a specific antiparasitic activity, as well as a suitable selectivity index (>100). In another study, ABZ induced no reduction in the viability of rat primary hepatocytes at doses up to 40 μM [52]. The PCL NCs developed in this study have been shown to possess low cytotoxicity toward host cells. Their CC_{50} is higher than 248 μg/mL, > 1000 μg/mL, and >1000 μg/mL against the Vero, HepG2, and Caco2 cell lines, respectively, in vitro [55].

More recently, the cardiac toxic effects of supratherapeutic ABZ exposure with multiple and sustained high doses on the ECG parameters of 24 healthy subjects were investigated in clinical trials [17]. Pharmacokinetics studies on humans have indicated that ABZ is rapidly absorbed (T_{max} 5–22.5 h) after an oral dose of 400 mg every 8 h, reaching a mean C_{max} of 11.9 ng/mL after 5 days, and its metabolite, 6-hydroxialbaconazole, reaches a C_{max} of 636 ng/mL. This study concluded that this dosing schedule was well tolerated and the adverse effects were mild, without safety issues [17]. Another clinical study in healthy volunteers demonstrated that capsules had higher bioavailability than tablets, indicating that these two dosage forms were not bioequivalent. This study points out the importance of developing suitable dosage forms for ABZ [18].

In mice, we demonstrated herein that the dosage form can dramatically change the tolerability of ABZ. Nanoencapsulation improves ABZ safety and probably its biopharmaceutical properties, providing prolonged release following the investigated administration routes. The NC formulation is stable and biodegradable, produces reproducible and monodispersed particles, and has suitable sizes and prolonged-release properties. NCs possess an excellent ability to entrap ABZ at a high encapsulation efficiency to be administered by the different parenteral routes tested in this study, as well as by the oral route. In terms of in vitro activity, ABZ exerts suppressive effects against epimastigotes of the Y strain, which is in agreement with previously reported data. However, no significant in vitro effect was observed against the BZN-resistant Colombian strain. This finding is similar to the data obtained by Guedes et al. (2004) against the Berenice-78 strain in vivo, a BZN-sensitive strain [23]. Thus, natural *T. cruzi* resistance to triazole exists, as has been documented previously [56]. No difference was observed between free ABZ and ABZ-NCs against epimastigotes, which is to be expected, as in vitro acellular conditions do not adequately simulate the complexity of in vivo absorption, distribution, metabolization, and elimination. In vitro, the poorly soluble drugs, such as ABZ loaded into nanocapsules, have an incomplete and slow-release rate, as demonstrated here (Figure 4). After SC administration in vivo, the NCs may form a depot of ABZ. In fact, nanocarriers are known to alter drug absorption and distribution steps in vivo. Thus, animal models are more suitable for evaluating the efficacy and toxicity effects of drug encapsulation.

The most recent clinical trials with CYP51 inhibitors indicate that they initially eliminate parasites from blood, but they are not able to maintain this effect after the end of treatment [4,53]. These clinical trials evidence the same effect that we observed in mice with ABZ in this study. As hypothesized by Dumoulin et al. (2020), the anti-*T. cruzi* activity of azoles induces no sterile cure, which opens up the possibility of distinct populations of parasites and/or a heterogeneous environment within a single host underlying treatment failure [57]. The resistance of *T. cruzi* to the azole class of drugs has not yet been reported and merits further investigation.

Effects were markedly different concerning tolerability and reductions in parasitemia relapses in vivo, particularly at doses of 80 and 120 mg/kg, with low levels of parasitemia and a long time to parasitemia reactivation seen with ABZ-NC treatment compared to free ABZ. Protocols for once- and twice-day doses were similarly effective, demonstrating that NCs prolonged ABZ residence time in the body. Therefore, a dose-dependent therapeutic response was observed when ABZ-NCs were administered subcutaneously. These data indicate that the pharmacokinetic profile of the drug was probably altered with the use of NCs. Furthermore, the doses of ABZ required to eliminate parasitemia were significantly higher for NCs than for free ABZ. This may be due to the formation of an NC deposit in the subcutaneous tissue that did not immediately make the concentration of ABZ available to eradicate parasites. It could be suggested, however, that NCs can sustain blood levels of ABZ better than free ABZ, allowing a slower relapse of parasitemia. This hypothesis, which should be confirmed further by biodistribution studies, considers that the NCs can retain ABZ in the injection site, releasing fractions of ABZ, and reducing adverse effects. The in vitro release profile of ABZ from NCs showed that a maximum of 45% of the drug was released in the first 10 h, a result that corroborates our hypothesis. The differences between free ABZ and ABZ-NCs can be attributed to changes in the biodistribution of the drug when administered by the SC route in free or encapsulated form, which merits further investigation.

No parasite elimination was achieved with any formulation in our experimental conditions, indicating a need to carry out future studies with higher doses, as well as experiments in combination with other drugs. ABZ has suppressive effects on *T. cruzi*, and encapsulation prolongs this suppression. From a pharmacological point of view, NCs significantly improved the safety of ABZ in vivo. Hence, appropriate formulations must be developed to control the release of ABZ and optimize the dosage regimen, which may be helpful in the extended treatment regimen for Chagas disease and improve adherence to treatment in patients in the chronic phase.

The most promising effect observed in this work was related to the overall reduced toxicity of ABZ in the encapsulated form. Although a parasitological cure was not obtained in the animals treated with the new ABZ-NC formulations, the results were positive concerning subcutaneous administration, which allowed more sustained action. This effect of NCs promoting prolonged release and reduced toxicity has been observed in different studies [58–60], which attribute the toxicity reduction to the reduced access of NCs to vital organs, such as the heart, reducing, for example, the cardiotoxicity [30,58]. Our results in the present study corroborate this hypothesis, even using another administration route, i.e., SC. Therefore, further in-depth studies on tissue infection in animals treated with different doses and formulations of encapsulated ABZ are necessary.

As Tarlenton (2001) argues, parasite elimination from tissues is one of the treatment goals in the chronic phase of Chagas disease and allows a significant reduction in symptoms and mortality in humans [61]. The present study's data reinforce the importance of developing new dosage forms for this class of drugs to improve biopharmaceutical characteristics.

5. Conclusions

In this study, we developed innovative formulations of ABZ, an investigational triazole, in biodegradable polymeric nanocapsules. Originally, these NCs were able to be delivered by parenteral routes, as well as orally. The use of SC injections emerged as an interesting alternative to administer anti-*T. cruzi* drugs and obtain a depot in this tissue. The ABZ-NCs are stable and monodisperse in size within the nanometrical range. NCs have an adequate ability to load ABZ with high encapsulation efficiency. It was evidenced that NCs dramatically improve in vivo ABZ safety by the intravenous route and SC routes, even at high doses. ABZ-NCs improved mice survival compared to free ABZ, induced the fast elimination of parasitemia, and prolonged the anti-*T. cruzi* effect after the end of treatment in a murine model. This indicates that NCs represent a promising strategy to circumvent the biopharmaceutical limitations of ABZ. The encapsulation of ABZ provides a dosage form that is adaptable to various administration routes. This preliminary preclinical study in mice can pave the way for deeper investigations into its effects on parasite burden in different organs, potential immunological modulation, and further toxicological investigations.

Supplementary Materials: The following supporting information can be downloaded at: https://www.mdpi.com/article/10.3390/pathogens14040319/s1, Figure S1: Ultraviolet spectrum showing the maximum absorption peak of ABZ in acetonitrile (5 mg/mL) at 238 nm; Figure S2: ABZ calibration curve in acetonitrile at a wavelength of 238 nm; R2 = coefficient of determination; Table S1: ABZ average absorbance values as a function of its concentration; Table S2: Results of the calibration curve made with pure ABZ, ABZ + NC at 1% and ABZ + NC at 10%; Table S3: Results of the repeatability test to validate the spectrophotometric assay of ABZ at 238 nm; Table S4: Results of the intermediate precision test, in two days, to validate the spectrophotometric measurement of ABZ at 238 nm; Table S5: Accuracy of the ABZ determination method; Figure S3: Survival after treatment with different doses of Albaconazole coarse suspensions and ABZ-NCs; Figure S4: Parasitemia curves after treatment with different doses of Albaconazole loaded in NCs.

Author Contributions: Conceptualization, C.M.d.B., V.M.V., M.d.L., M.T.B. and V.C.F.M.; Data curation, C.M.d.B., J.M.C.V., M.A.d.O., M.T.B. and V.C.F.M.; Formal analysis, C.M.d.B., V.M.V., M.S.A., J.M.C.V., M.T.B. and V.C.F.M.; Funding acquisition, M.T.B. and V.C.F.M.; Investigation, C.M.d.B., V.M.V., J.M.C.V., M.T.B. and V.C.F.M.; Methodology, C.M.d.B., V.M.V., J.M.C.V., M.S.A., M.T.B. and V.C.F.M.; Project administration, M.T.B., M.d.L. and V.C.F.M.; Software, C.M.d.B., M.T.B. and V.C.F.M.; Supervision, M.T.B. and V.C.F.M.; Validation, C.M.d.B., J.M.C.V., M.A.d.O., M.T.B. and V.C.F.M.; Visualization, C.M.d.B., J.M.C.V., M.A.d.O., M.T.B. and V.C.F.M.; Writing—original draft, C.M.d.B., M.A.d.O., M.T.B. and V.C.F.M.; Writing—review and editing, C.M.d.B., M.A.d.O., M.T.B. and V.C.F.M. All authors have read and agreed to the published version of the manuscript.

Funding: This research received financial support from the UNDP/World Bank/WHO Special Program for Research and Training in Tropical Diseases (ID A00790).

Institutional Review Board Statement: The procedures and experimental conditions were in accordance with COBEA (Brazilian School of Animal Experimentation) guidelines for the use of animals in scientific research and approved by the Ethics Committee in Animal Research at UFOP (number 2012/70; 2004/08 e 2005/07).

Informed Consent Statement: Not applicable.

Data Availability Statement: Data are available from the authors.

Acknowledgments: We are grateful to J. Urbina (Centro de Biofísica y Bioquímica, Venezuela) and Javier Bartrolí (Ulriach & Cia, Spain) for providing albaconazole (UR-9825) to allow the development of the dissertation thesis of C.M. Barros. The authors V.C.F. Mosqueira, M. de Lana, and MT Bahia are productivity research fellows of the National Council for Scientific and Technological Development (CNPq)/Brazil. This investigation also received support from NANOBIOMG Network (grants #00007-14 and #40/11), FAPEMIG universal grant APQ-02576-18. C. M. Barros received an MSc scholarship from CAPES.

Conflicts of Interest: The authors declare no conflicts of interest. The funders had no role in the design of the study, the collection, analysis, or interpretation of data, the writing of the manuscript, or the decision to publish the results.

References

1. PAHO/WHO—Pan American Health Organization. Chagas Disease in the Americas for Public Health Workers-PAHO/WHO. 2022. Available online: https://www.who.int/campaigns/world-chagas-disease-day/2022 (accessed on 10 January 2025).
2. PAHO/WHO. *PAHO/WHO: Guidelines for the Diagnosis and Treatment of Chagas Disease*; PAHO/WHO: Washington, DC, USA, 2019; ISBN 978-92-75-12043-9.
3. Brindley, P.J.; Hotez, P.J.; Kamhawi, S. Revisiting What Constitutes a Neglected Tropical Disease? *PLoS Negl. Trop. Dis.* **2025**, *19*, e0012794. [CrossRef] [PubMed]
4. Torchelsen, F.K.V.D.S.; Mazzeti, A.L.; Mosqueira, V.C.F. Drugs in Preclinical and Early Clinical Development for the Treatment of Chagas's Disease: The Current Status. *Expert Opin. Investig. Drugs* **2024**, *33*, 575–590. [CrossRef] [PubMed]
5. Cançado, J.R. Long term evaluation of etiological treatment of chagas disease with benznidazole. *Rev. Inst. Med. Trop. São Paulo* **2002**, *44*, 29–37. [CrossRef]
6. Andrade, A.L.S.S.; Martelli, C.M.T.; Oliveira, R.M.; Silva, S.A.; Aires, A.I.S.; Soussumi, L.M.T.; Covas, D.T.; Silva, L.S.; Andrade, J.G.; Travassos, L.R.; et al. Short report: Benznidazole efficacy among Trypanosoma cruzi-infected adolescents after a six-year follow-up. *Am. J. Trop. Med. Hyg.* **2004**, *71*, 594–597. [CrossRef] [PubMed]
7. Sosa Estani, S.; Segura, E.L. Treatment of Trypanosoma cruzi infection in the undetermined phase. Experience and current guidelines of treatment in Argentina. *Mem. Inst. Oswaldo Cruz* **1999**, *94*, 363–365. [CrossRef]
8. Yun, O.; Lima, M.A.; Ellman, T.; Chambi, W.; Castillo, S.; Flevaud, L.; Roddy, P.; Parreno, F.; Albajar, V.P.; Palma, P.P. Feasibility, drug safety, and effectiveness of etiological treatment programs for chagas disease in honduras, guatemala, and bolivia: 10-year experience of medecins sans frontieres. *PLoS Negl. Trop. Dis.* **2009**, *3*, e488. [CrossRef] [PubMed]
9. Viotti, R.; Vigliano, C.; Armenti, H.; Segura, E. Treatment of chronic Chagas disease with benznidazole: Clinical and serological evolution of patients with long-term follow-up. *Am. Heart J.* **1994**, *127*, 151–162. [CrossRef]
10. Duschak, V.G. Targets and patented drugs for chemotherapy of Chagas disease in the last 15 years-period. *Recent Pat. Anti-Infect. Drug Discov.* **2016**, *11*, 74–173. [CrossRef]
11. Buckner, F.S.; Urbina, J.A. Recent developments in sterol 14-demethylase inhibitors for Chagas disease. *Int. J. Parasitol. Drugs Drug Resist.* **2012**, *2*, 236–242. [CrossRef]
12. Bartrolí, J.; Turmo, E.; Algueró, M.; Boncompte, E.; Vericat, M.L.; Conte, L.; Ramis, J.; Merlos, M.; García-Rafanell, J.; Forn, J. New azole antifungals. 3. Synthesis and antifungal activity of 3-substituted-4(3H)-quinazolinones. *J. Med. Chem.* **1998**, *41*, 1869–1882. [CrossRef]
13. Sorbera, L.A.; Bartroli, J. Castañer. *Drugs Future* **2003**, *28*, 529–537.
14. Seyedmousavi, S.; Rafati, H.; Ilkit, M.; Tolooe, A.; Hedayati, M.T.; Verweij, P. Systemic Antifungal Agents: Current Status and Projected Future Developments. In *Human Fungal Pathogen Identification*; Methods in Molecular Biology; Lion, T., Ed.; Humana Press: New York, NY, USA, 2017; Volume 1508. [CrossRef]
15. Bartroli, J.; Merlos, M.; Sisniega, H. Overview of albaconazole. *Eur. Infect. Dis.* **2011**, *5*, 88–91.
16. Guillon, R.; Pagniez, F.; Picot, C.; Hédou, D.; Tonnerre, A.; Chosson, E.; Duflos, M.; Besson, T.; Logé, C.; Le Pape, P. Discovery of a Novel Broad-Spectrum Antifungal Agent Derived from Albaconazole. *ACS Med. Chem. Lett.* **2013**, *4*, 288–292. [CrossRef]
17. Dietz, A.J.; Barnard, J.C.; van Rossem, K. A randomized, double-blind, multiple-dose, placebo-controlled, dose escalation study with a 3-cohort parallel group design to investigate the tolerability and pharmacokinetics of albaconazole in healthy subjects. *Clin. Pharmacol. Drug Dev.* **2014**, *3*, 25–33. [CrossRef] [PubMed]
18. van Rossem, K.; Lowe, J.A. A Phase I, randomized, open-label crossover study to evaluate the safety and pharmacokinetics of 400 mg albaconazole administered to healthy participants as a tablet formulation versus a capsule formulation. *Clin. Pharmacol. Adv. Appl.* **2013**, *5*, 23–31. [CrossRef] [PubMed]

19. Sigurgeirsson, B.; van Rossem, K.; Malahias, S.; Raterink, K. A phase II, randomized, double-blind, placebo-controlled, parallel group, dose-ranging study to investigate the efficacy and safety of 4 dose regimens of oral albaconazole in patients with distal subungual onychomycosis. *J. Am. Acad. Dermatol.* **2013**, *69*, 416–425. [CrossRef] [PubMed]
20. Miller, J.L.; Schell, W.A.; Wills, E.A.; Toffaletti, D.L.; Boyce, M.; Benjamin, D.K., Jr.; Bartroli, J.; Perfect, J.R. In vitro and in vivo efficacies of the new triazole albaconazole against *Cryptococcus neoformans*. *Antimicrob. Agents Chemother.* **2004**, *48*, 384–387. [CrossRef]
21. Capilla, J.; Ortoneda, M.; Pastor, F.J.; Guarro, J. In vitro antifungal activities of the new triazole UR-9825 against clinically important filamentous fungi. *Antimicrob. Agents Chemother.* **2001**, *45*, 2635–2637. [CrossRef]
22. Urbina, J.A.; Lira, R.; Visbal, G.; Bartroli, J. In vitro antiproliferative effects and mechanism of action of the new triazole derivative UR-9825 against the protozoan parasite *Trypanosoma* (Schizotrypanum) *cruzi*. *Antimicrob. Agents Chemother.* **2000**, *44*, 2498–2502. [CrossRef]
23. Guedes, P.M.; Urbina, J.A.; Lana, M.; Afonso, L.C.; Veloso, V.M.; Tafuri, W.L.; Machado-Coelho, G.L.; Chiari, E.; Bahia, M.T. Activity of the new triazole derivative albaconazole against *Trypanosoma (Schizotrypanum) cruzi* in dog hosts. *Antimicrob. Agents Chemother.* **2004**, *48*, 4286–4292. [CrossRef]
24. Girmenia, C. New generation azole antifungals in clinical investigation. *Exp. Opin. Investig. Drugs* **2009**, *18*, 1279–1295. [CrossRef] [PubMed]
25. Aperis, G.; Mylonakis, G. Newer triazole antifungal agents: Pharmacology, spectrum, clinical efficacy and limitations. *Expert Opin. Investig. Drugs* **2006**, *15*, 579–602. [CrossRef] [PubMed]
26. Pasqualotto, A.C.; Thiele, K.O.; Goldani, L.Z. Novel triazole antifungal drugs: Focus on isavuconazole, ravuconazole and albaconazole. *Curr. Opin. Investig. Drugs* **2010**, *11*, 165–174. Available online: http://www.ncbi.nlm.nih.gov/pubmed/20112166 (accessed on 10 December 2024).
27. Fávero, M.L.D.; Bonetti, A.F.; Domingos, E.L.; Tonin, F.S.; Pontarolo, R. Oral antifungal therapies for toenail onychomycosis: A systematic review with network meta-analysis toenail mycosis: Network meta-analysis. *J. Dermatol. Treat.* **2022**, *33*, 121–130. [CrossRef]
28. Bartroli, X.; Uriach, J. A clinical multicenter study comparing efficacy and tolerability between five single oral doses of albaconazole and fluconazole 150 mg single dose in acute vulvovaginal candidiasis. In Proceedings of the 45th Interscience Conference on Antimicrobial Agents and Chemotherapy: Abs M-722, Washington, DC, USA, 16–19 December 2005.
29. Girmenia, C.; Finolezzi, E. New-generation triazole antifungal drugs: Review of the Phase II and III trials. *Clin. Investig.* **2011**, *1*, 1577–1594. [CrossRef]
30. Branquinho, R.T.; Roy, J.; Farah, C.; Garcia, G.M.; Aimond, F.; Le Guennec, J.-Y.; Saude-Guimarães, D.A.; Grabe-Guimaraes, A.; Mosqueira, V.C.F.; de Lana, M.; et al. Biodegradable Polymeric Nanocapsules Prevent Cardiotoxicity of Anti-Trypanosomal Lychnopholide. *Sci. Rep.* **2017**, *7*, 44998. [CrossRef]
31. Branquinho, R.T.; Pound-Lana, G.; Marques Milagre, M.; Saúde-Guimarães, D.A.; Vilela, J.M.C.; Spangler Andrade, M.; de Lana, M.; Mosqueira, V.C.F. Increased Body Exposure to New Anti-Trypanosomal Through Nanoencapsulation. *Sci. Rep.* **2017**, *7*, 8429. [CrossRef]
32. Branquinho, R.T.; de Mello, C.G.C.; Oliveira, M.T.; Reis, L.E.S.; de Vieira, P.M.A.; Saúde-Guimarães, D.A.; Mosqueira, V.C.F.; de Lana, M. Lychnopholide in Poly(d,l-Lactide)- block -Polyethylene Glycol Nanocapsules Cures Infection with a Drug-Resistant *Trypanosoma cruzi* Strain at Acute and Chronic Phases. *Antimicrob. Agents Chemother.* **2020**, *64*, 1–13. [CrossRef]
33. Mosqueira, V.C.F.; Mazzeti, A.L.; Bahia, M.T. Chapter 9—Nanomedicines against Chagas disease. In *Applications of Nanobiotechnology for Neglected Tropical Diseases*, 1st ed.; Formiga, F.R., Inamuddin, Severino, P., Eds.; Academic Press: London, UK, 2021; pp. 169–189. [CrossRef]
34. Siqueira, R.P.; Milagre, M.M.; de Oliveira, M.A.; Branquinho, R.T.; Torchelsen, F.K.V.; de Lana, M.; Machado, M.G.C.; Andrade, M.S.; Bahia, M.T.; Mosqueira, V.C.F. In vitro interaction of polyethylene glycol-*block*-poly(*D,L*-lactide) nanocapsule devices with host cardiomyoblasts and *Trypanosoma cruzi*-infective forms. *Parasitol. Res.* **2022**, *121*, 2861–2874. [CrossRef]
35. Okamoto, J.; Fukunami, M.; Kioka, H. Frequent premature ventricular contractions induced by itraconazole. *Circ. J.* **2007**, *71*, 1323–1325. [CrossRef]
36. Philips, J.A.; Marty, F.M.; Stone, R.M.; Koplan, B.A.; Katz, J.T.; Baden, L.R. Torsades de pointes associated with voriconazole use. *Transpl. Infect. Dis.* **2007**, *9*, 33–36. [CrossRef] [PubMed]
37. Tholakanahalli, V.N.; Potti, A.; Hanley, J.F.; Merliss, A.D. Fluconazole-induced torsade de pointes. *Ann. Pharmacother.* **2001**, *35*, 432–434. [CrossRef] [PubMed]
38. Leite, E.A.; Grabe-Guimarães, A.; Guimarães, H.N.; Machado-Coelho, G.L.L.; Barratt, G.; Mosqueira, V.C.F. Cardiotoxicity reduction induced by halofantrine entrapped in nanocapsule devices. *Life Sci.* **2007**, *80*, 1327–1334. [CrossRef] [PubMed]
39. Moreira Souza, A.C.; Grabe-Guimarães, A.; Cruz, J.D.S.; Santos-Miranda, A.; Farah, C.; Teixeira Oliveira, L.; Lucas, A.; Aimond, F.; Sicard, P.; Mosqueira, V.C.F.; et al. Mechanisms of artemether toxicity on single cardiomyocytes and protective effect of nanoencapsulation. *Br. J. Pharmacol.* **2020**, *177*, 4448–4463. [CrossRef]

40. de Sousa, D.R.T.; de Oliveira Guerra, J.A.; Ortiz, J.V.; do Nascimento Couceiro, K.; da Silva e Silva, M.R.H.; Jorge Brandão, A.R.; Guevara, E.; Arcanjo, A.R.L.; de Oliveira Júnior, E.F.; Smith-Doria, S.; et al. Acute Chagas disease associated with ingestion of contaminated food in Brazilian western Amazon. *Trop. Med. Int. Health* **2023**, *28*, 541–550. [CrossRef]
41. Fessi, H.; Puisieux, F.; Devissaguet, J.P.; Ammoury, N.; Benita, S. Nanocapsule formation by interfacial polymer deposition following solvent displacement. *Int. J. Pharm.* **1989**, *55*, R1–R4. [CrossRef]
42. ANVISA; National Health Surveillance Agency Ministry of Health, BRAZIL. Resolution RDC n° 166, of July 24, 2017. Provides Criteria for the Validation of Analytical Methods. Official Gazette of the Federative Republic of Brazil, Brasília, DF, Section 1 25 July. 2017. Available online: https://bvsms.saude.gov.br/bvs/saudelegis/anvisa/2017/rdc0166_24_07_2017.pdf (accessed on 20 July 2018).
43. Santos-Magalhaes, N.S.; Fessi, H.; Puisieux, F.; Benita, S.; Seiller, M. In vitro release kinetic examination and comparative evaluation between submicron emulsion and polylactic acid nanocapsules of clofibride. *J. Microencapsul.* **1995**, *12*, 195–205. [CrossRef]
44. Zingales, B.; Andrade, S.; Briones, M.; Campbell, D.; Chiari, E.; Fernandes, O.; Guhl, F.; Lages-Silva, E.; Macedo, A.; Ma-chado, C.; et al. A New Consensus for *Trypanosoma cruzi* Intraspecific Nomenclature: Second Revision Meeting Recommends TcI to TcVI. *Mem. Inst. Oswaldo Cruz* **2009**, *104*, 1051–1054. [CrossRef]
45. Zingales, B.; Miles, M.A.; Moraes, C.B.; Luquetti, A.; Guhl, F.; Schijman, A.G.; Ribeiro, I. Drug discovery for chagas disease should consider *Trypanosoma cruzi* strain diversity. *Mem. Inst. Oswaldo Cruz* **2014**, *109*, 828–833. [CrossRef]
46. Filardi, L.S.; Brener, Z. Susceptibility and natural resistance of *Trypanosoma cruzi* strains to drugs used clinically in Chagas' disease. *Trans. R. Soc. Trop. Med. Hyg.* **1987**, *81*, 755–759. [CrossRef]
47. Brener, Z. Therapeutic activity and criterion of cure in mice experimentally infected with *Trypanosoma cruzi*. *Rev. Inst. Med. Trop. São Paulo* **1962**, *4*, 89–396.
48. Romanha, A.J.; Castro, S.L.; De Soeiro, M.D.N.C.; Lannes-Vieira, J.; Ribeiro, I.; Talvani, A.; Bourdin, B.; Blum, B.; Olivieri, B.; Zani, C.; et al. In vitro and in vivo experimental models for drug screening and development for Chagas disease. *Mem. Inst. Oswaldo Cruz* **2010**, *105*, 233–238. [CrossRef] [PubMed]
49. Chiari, E.; Dias, J.C.P.; Lana, M.; Chiari, C.A. Hemocultures for the parasitological diagnosis of human chronic Chagas' disease. *Rev. Soc. Bras. Med. Trop.* **1989**, *22*, 19–23. [CrossRef] [PubMed]
50. Leite, E.A.; Vilela, J.M.C.; Mosqueira, V.C.F.; Andrade, M.S. Poly-Caprolactone Nanocapsules Morphological Features by Atomic Force Microscopy. *Microsc. Microanal.* **2005**, *11* (Suppl. S3), 48–51. [CrossRef]
51. Veloso, V.M.; Carneiro, C.M.; Toledo, M.J.O.; Lana, M.; Chiari, E.; Tafuri, W.L.; Bahia, M.T. Variation in Susceptibility to Benznidazole in Isolates Derived from *Trypanosoma cruzi* Parental Strains. *Mem. Inst. Oswaldo Cruz* **2001**, *96*, 1005–1011. [CrossRef]
52. Cao, X.; Sun, Z.; Cao, Y.; Wang, R.; Cai, T.; Chu, W.; Hu, W.; Yang, Y. Design, Synthesis, and Structure-Activity Relationship Studies of Novel Fused Heterocycles-Linked Triazoles with Good Activity and Water Solubility. *J. Med. Chem.* **2014**, *57*, 3687–3706. [CrossRef]
53. Torrico, F.; Gascon, J.; Ortiz, L.; Alonso-Vega, C.; Pinazo, M.-J.; Schijman, A.; Almeida, I.C.; Alves, F.; Strub-Wourgaft, N.; Ribeiro, I. E1224 Study Group. Treatment of adult chronic indeterminate Chagas disease with benznidazole and three E1224 dosing regimens: A proof-of-concept, randomised, placebo-controlled trial. *Lancet Infect. Dis.* **2018**, *18*, 419–430. [CrossRef]
54. Bartroli, J.; Turmo, E.; Algueró, M.; Boncompte, E.; Vericat, M.L.; Conte, L.; Ramis, J.; García-Rafanell, J.; Forn, J. UR-9825: A New Triazole Derivative with Potent, Broad-Spectrum Antifungal Activity. Program and Abstracts of the 37th Interscience Conference on Antimicrobial Agents and Chemotherapy (ICAAC), 1997, abstr. E-67, September, Toronto, Canada. American Soc. for Microbiology, Washington, DC. Available online: https://search.worldcat.org/pt/title/633698851 (accessed on 20 July 2018).
55. Trindade, I.C.; Pound-Lana, G.; Perera, D.G.S.; De Oliveira, L.A.M.; Andrade, M.S.; Vilela, J.M.C.; Postacchini, B.B.; Mosqueira, V.C.F. Mechanisms of interaction of biodegradable polyester nanocapsules with non-phagocytic cells. *Eur. J. Pharm. Sci.* **2018**, *124*, 89–104. [CrossRef]
56. Chen, C.K.; Leung, S.S.; Guilbert, C.; Jacobson, M.P.; McKerrow, J.H.; Podust, L.M. Structural Characterization of CYP51 from *Trypanosoma cruzi* and *Trypanosoma brucei* Bound to the Antifungal Drugs Posaconazole and Fluconazole. *PLoS Negl. Trop. Dis.* **2010**, *4*, e651. [CrossRef]
57. Dumoulin, P.; Vollrath, J.; Tomko, S.S.; Burleigh, B. Glutamine metabolism modulates azole susceptibility in *Trypanosoma cruzi* amastigotes. *eLife* **2020**, *9*, e60226. [CrossRef]
58. Souza, A.C.M.; Mosqueira, V.C.F.; Silveira, A.P.A.; Antunes, L.R.; Richard, S.; Guimarães, H.N.; Grabe-Guimarães, A. Reduced cardiotoxicity and increased oral efficacy of artemether polymeric nanocapsules in *Plasmodium berghei*-infected mice. *Parasitology* **2018**, *145*, 1075–1083. [CrossRef] [PubMed]
59. Mosqueira, V.C.F.; Loiseau, P.M.; Bories, C.; Legrand, P.; Devissaguet, J.-P.; Barratt, G. Efficacy and Pharmacokinetics of Intravenous Nanocapsule Formulations of Halofantrine in *Plasmodium berghei*-Infected Mice. *Antimicrob. Agents Chemother.* **2004**, *48*, 1222–1228. [CrossRef] [PubMed]

60. Mosqueira, V.C.F.; Legrand, P.; Barratt, G. Surface-Modified and Conventional Nanocapsules as Novel Formulations for Parenteral Delivery of Halofantrine. *J. Nanosci. Nanotechnol.* **2006**, *6*, 3193–3202. [CrossRef] [PubMed]
61. Tarlenton, R.L. Parasite persistence in the aetiology of Chagas disease. *Int. J. Parasitol.* **2001**, *31*, 550–554. [CrossRef]

pathogens

MDPI

Review

An Update on Vaccines Against *Trypanosoma cruzi* and Chagas Disease

Nisha J. Garg [1,2]

1 Department of Microbiology & Immunology, University of Texas Medical Branch (UTMB), Galveston, TX 77555-1070, USA; nigarg@utmb.edu
2 Institute for Human Infections and Immunity, University of Texas Medical Branch (UTMB), Galveston, TX 77555-1070, USA

Abstract: Chagas disease (CD) is a global health concern, with no existing therapies to prophylactically treat adults traveling to endemic countries or those who may already be infected with *Trypanosoma cruzi*. The economic burden of Chagas cardiomyopathy and heart failure, due to healthcare costs and lost productivity from premature deaths, provides a strong rationale for investment in the development of immune therapies against CD. Vaccine efficacy is proposed to depend heavily on the induction of a robust Th1 response for the clearance of intracellular pathogens like *T. cruzi*. In this review, updated information on the efforts for vaccine development against CD is provided.

Keywords: Chagas; *Trypanosoma cruzi*; vaccine

1. Introduction

Trypanosoma cruzi (*T. cruzi* or *Tc*) is a kinetoplastid parasite that is the causative agent of Chagas disease (CD). *T. cruzi* is endemic in the Americas, ranging from southern United States to Argentina, due to continuous transmission between triatomine vectors and wildlife and domestic reservoir hosts. Acute blood parasitemia and non-specific somatic symptoms resolve into the subclinical (indeterminate) form of disease within 2–4 months after infection. Decades later, infected individuals may advance to the chronic (determinate) form of CD, involving cardiac, gastrointestinal, and neurological complications [1]. Though under-reported and under-estimated, CD affects 6–8 million people and results in up to 12,000 deaths annually [2]. Approximately 71 million people are exposed to risk of infection and ~28,000 new cases of *T. cruzi* infection occur every year [3]. Recent data suggest that CD results in the loss of 0.7 million disability-adjusted life years [4] and an economic burden of >USD 10 billion [5] per year on the American continent. Even after applying stringent simulation parameters of a 1% infection risk and treatment efficacy of 25%, computational modeling studies suggest that a vaccine against *T. cruzi* costing USD 20/dose would be economically viable and provide net cost savings [5,6]. With an efficacy of >50% and infection risk of 20%, a vaccine costing even >USD 200 per dose is estimated to be economically advantageous [5]. Furthermore, a safe, effective, and reliable vaccine would provide relief from the costs and efforts associated with vector control to reduce parasite dissemination, and the diagnosis and treatment of individuals that are infected in endemic countries.

Upon infection, *T. cruzi* employs a variety of strategies to evade or suppress immune detection and disseminate through the bloodstream to establish active invasion and intracellular replication in a variety of host tissues [7,8]. The host immune system still does a great job in controlling the acute parasitemia, but low-grade tissue infection persists and provides for consistent activation of inflammatory reactions and CD pathogenesis [1].

Several questions as to how parasites persist at low levels remain unanswered: Is it that a few parasites escape from the immune system by changing the dominant antigens on their surface as is noted in *T. brucei*, or does the parasite adopt a different life-cycle stage that is not noted in in vitro cultures and acute in vivo infection? Some studies have suggested that the parasite persists in the gastro-intestinal tract [9] and adipose tissue [10,11] to support a chronic inflammatory state. At least one study documented that a latent form of *T. cruzi* amastigotes, such as that of *Mycobacterium tuberculosis* and *Toxoplasma gondii*, escaped from immune detection and served as so-called persisters [12], though how it differentiated into persister form was not studied. Nevertheless, the finding of low levels of circulating parasites in chronically infected individuals provides strong evidence that *T. cruzi* does come out of dormancy and hiding to maintain an active intracellular/infective life cycle, at least intermittently. These studies provide a basis for recommending drug and immune therapies for parasite control in the indeterminate stage of CD.

Pathogenesis of chronic Chagas disease is attributed to inflammatory immune-mediated myocardial injury, autonomic nervous system derangements, and microvascular disturbances, and has been discussed in several reviews [1,13]. Furthermore, it had been documented that mitochondrial dysfunction of the respiratory chain results in increased generation of reactive oxygen species (ROS) that predispose the heart to oxidative insult during *Tc* infection and chronic disease [14,15]. The sustained occurrence of oxidative adducts was noted in the myocardium of experimental models of CD and in the peripheral blood of rodents and human Chagas subjects [16–18], which was exacerbated by inefficient antioxidant capacity [19,20]. Other studies demonstrating ROS signaling of cytokines and chemokines production in infected cardiomyocytes and murine hearts have provided a potential mechanistic link between ROS generation and chronic inflammation in Chagas cardiomyopathy [15,21]. Readers interested in further understanding the complexity of Chagas disease pathogenesis and efforts toward vaccines, therapies, and drugs development are also referred to excellent recent reviews [22–25].

2. Immunity to *T. cruzi* Infection

The innate immune system is the first to actively elicit a defensive role, followed by the adaptive immune response. It is well documented that *T. cruzi* evades complement pathways to survive in the mammalian host and establish persistent infection [26]. *T. cruzi* surface glycoproteins (mucins), glycophospholipids, and other molecules stimulate the production of cytokines (e.g., IFNγ, TNFα, IL1β, IL6), chemokines (e.g., MCP1, RANTES/CCL5, IP10), and free radicals (e.g., superoxide, nitric oxide, peroxynitrite) in macrophages and cardiomyocytes [27–29]; however, these responses are either delayed or occur at sub-par level and fail to clear the parasite [30,31]. Studies in animal models of *T. cruzi* infection suggest that release of IFNγ and IL12 by natural killer cells and macrophages is essential for the stimulation of adaptive type 1 T cell response [32,33]. $CD4^+$T cells produce Th1 cytokines (IFNγ, IL2) and assist in parasite control through amplification of the phagocytic activity of macrophages, and stimulation of B cells and $CD8^+$ T cells [34,35]. Antibodies produced by B cells must promote opsonization, phagocytosis, and complement-mediated killing of the parasite; and antibodies to the Galα(1,3)Galβ(1,4)GlcNAc epitope of surface expressed mucin glycoproteins exhibit such anti-parasite activity [36]. Finally, *Tc*-specific $CD8^+$T cells, detected in infected mice and humans, contribute to *Tc* control by cytolysis of infected cells and secretion of Th1 cytokines that induce trypanocidal activity [37].

Immunological studies in chronically infected mice and patients have yielded conflicting results. Some reports indicated that immune exhaustion of $CD4^+$ and $CD8^+$ T cells contributes to parasite persistence [38], while others concluded that excessive production of proinflammatory cytokines correlates with tissue damage and clinical disease in chronically infected patients [39]. Recently, it was proposed that antigen-presenting cells producing IL27 contribute to a balanced proinflammatory/anti-inflammatory (IFNγ/IL10) response, leading to reduced inflammatory infiltrate in the myocardium of chagasic mice [40]. Overall, it is safe to conclude from the current literature that an efficient protective response to *T. cruzi* infection requires the combined activities of phagocytes, T helper cells, and cytotoxic T lymphocytes capable of rapidly killing the circulating and intracellular parasites. A sub-par activation of any of these components can result in failure to clear the acute infection and parasite-persistence-associated pathologic events, leading to cardiomyopathy and heart failure in CD.

Based on the above discussed literature, an effective vaccine, whether given prophylactically or therapeutically, is envisioned to: (1) target all parasite forms that circulate in the bloodstream and replicate or hide in tissues, (2) recognize conserved antigens of all clinically relevant parasite lineages and strains that are known to cause infection in mammals, and (3) induce a long-lasting memory immune response that can be recalled to rapidly control/eliminate *T. cruzi* infection.

3. Vaccine Development Against *T. cruzi* and Chagas Disease

Considering the complexity of the antigenic variability in *T. cruzi* populations that leads to diverse immune responses, the goal of achieving sterilizing immunity in which parasite infection and transmission is eliminated by a vaccine is noble and magnanimous but lacks consideration for those that have the least resources and would be maximally benefited by even a partially protective vaccine. Indeed, current efforts have led to the development of candidate vaccines that elicit sufficient immune memory to reduce the parasite load below a threshold level such that a vaccinated individual is protected from chronic Chagas disease. It is highly likely that identification and testing of several candidate vaccines by individual investigators will lead to the selection of a panel of key *T. cruzi* antigens that together will offer a multivalent, highly efficacious vaccine against Chagas disease.

Over the last century, several candidate vaccines have been tested in small animals with varying degree of success in controlling the *T. cruzi* infection and/or tissue damage. Early efforts utilized whole parasites killed by various methods or sub-cellular fractions of the parasite as a vaccine that generated a relevant immune response, evidenced by control of acute challenge infection and survival from lethal infection [41]. Parasite strains attenuated by repeat passage in culture (e.g., TCC) were also used as a vaccine and were shown to provide protection from subsequent challenge with a virulent strain of *Tc* in mice and dogs [42,43]. Researchers also promoted *Trypanosoma rangeli*, which shares significant sequence homology with *T. cruzi* but does not cause disease in humans, as a prophylactic vaccine [44,45]. However, our studies suggest that *T. rangeli* did not offer significant protection from *Tc* infection when used by itself and did not enhance the protective efficacy of a subunit DNA vaccine [46]. With the advancement of genetic engineering technology, investigators have focused on knocking down virulence genes to generate live *Tc* vaccines. Examples include the use of *Tc* deficient in calmodulin-ubiquitin (TulCub8), calreticulin, *Lyt1*, *gp72*, *dhfr-ts*, *Ech1/2*, and other genes as a vaccine, which offered immunity to parasitemia caused by wild-type strains in in-bred and out-bred mice and hamsters [41]. A major limitation of whole organism vaccines is the generation of large quantities of attenuated parasites, without a gain in virulence. Some researchers also remain concerned that an attenuated live vaccine may cause parasitemia and disease in immunocompromised individuals [47]. Overall, concerns about the use of live attenuated vaccines may outweigh the benefits offered for the control of Chagas disease.

3.1. Subunit Vaccine Candidates

Garg et al. utilized recombinant *T. cruzi* expressing a model antigen (chicken ovalbumin) in different cellular compartments to demonstrate that GPI (glycosylphosphatidylinositol)-anchored proteins expressed in the infective and intracellular stages of *T. cruzi* that are released in host cell cytoplasm during parasite differentiation are the most likely source of peptides for immune activation of B and T cells [48]. Equipped with this information, investigators focused on identifying the abundantly expressed surface antigens of the large families, e.g., trans-sialidases, mucin-associated surface proteins (MASP), and glycoprotein 63, as vaccine candidates. Indeed, members of the trans-sialidases superfamily (e.g., TSA1, ASP1/2, ASP9, TS) were recognized by the antibodies and $CD8^+$T lymphocytes in infected mice and humans [49]. Other antigens, including complement regulatory protein (CRP), cruzipain, Tc24, GP82, KMP11, LYT1, paraflagellar rod proteins (PFR), and TC52, were also identified as potential vaccine candidates, because they were recognized by antibodies and IFNγ-producing $CD8^+$T cells in experimental models of *Tc* infection and Chagas patients [50].

We performed a computational screening of the *T. cruzi* sequence database for candidate surface antigens and biological screening to select those that were recognized by IgGs and type 1 $CD8^+$T cells in infected mice, dogs, or humans [51,52]. Of the 11 antigens thus selected, TcG1, TcG2, and TcG4 exhibited additional desirable features, as they were (a) expressed in both infective and intracellular forms of *T. cruzi*, (b) released into host cell cytoplasm during parasite differentiation, and (c) consisted of epitopes presented by the MHC alleles of mice, dogs, and humans [51,52]. Notably, TcG2 and TcG4 were conserved in five of the six *T. cruzi* lineages (80–96% homology) and thus relevant as vaccine candidates for the control of diverse *Tc* isolates circulating in the Americas. Others have employed subtractive proteomics for the selection of immunodominant epitopes and identified MASP1/2 as potential vaccine candidates [53].

3.2. Subunit Vaccines

Investigators have tested the prophylactic and therapeutic efficacy of single- and multi-component candidate vaccines in various experimental models. In general, DNA immunization has been favored due to the ease of production and DNA stability, and because antigen delivery by DNA vaccination was found to be efficient in eliciting antibodies, Th1 cytokines, and $CD8^+$T cell responses to encoded antigens. In parallel studies, adjuvants (e.g., IL12, GMCSF, CD40, HSP70, CpG-ODN, c-di-AMP) were tested to enhance the Th1 responses to defined antigen vaccines.

Initial studies focused on evaluating the vaccine efficacy of antigens of the large families of surface proteins. Members of trans-sialidases family (e.g., ASP1, ASP2, TSA1, TS, SAPA, TSf) have been tested as a DNA or protein vaccine (individually or in combination) and offered a degree of immune resistance to *T. cruzi* infection that correlated with the amount and type of antigen delivered with and without adjuvants (reviewed in [49]). Yet, none of these experimental vaccines elicited sterile immunity, and they have not reached the clinical trial phase. No protection was observed in mice immunized with genes encoding members of the mucin family, though a KLH-conjugated synthetic peptide from the MASP family did offer survival benefits from lethal infection in mice [54]. It can be surmised that the shared epitopes expressed by members of the large families prevent robust activation of the immune system against *Tc*, and a combination of epitopes from other conserved genes would be more useful in stimulating immune effectors against diverse parasite strains.

Subsequently, investigators used a variety of delivery vehicles to test the vaccine efficacy of non-family proteins in experimental models. For example, oral delivery of *Salmonella*-carrying cruzipain DNA (SCz) induced mucosal IgA response, while boosting oral SCz with recombinant cruzipain protein and CpG-ODN produced a strong systemic immune response capable of controlling challenge infection and tissue damage [55]. Vaccines based on PFR2/3 induced IgG2a response; however, only PFR2 conjugated with HSP70 induced IL12 and IFNγ expression and a degree of protection from *Tc* infection [56,57]. Immunization with Tc24, Tcb3, and LYT1 induced cytotoxic T cells; however, only LYT1 provided resistance to lethal infection in mice [58], thus suggesting that antigens recognized by the host immune response after infection may not always be the best candidates for vaccine development. Adenovirus 5 was used as a delivery vehicle by Brazilian scientists; these studies showed that prophylactic Ad-ASP2/Ad-TS elicited robust type 1 cytokines (TNFα, IFNγ) producing CD8$^+$ effector T cell response and 100% survival from lethal challenge infection in Balb/c and C57BL/6 mice [59,60]. The same group showed that therapeutic Ad-ASP2/Ad-TS reprogrammed the immune response such that parasite-persistence-associated oxidative/inflammatory stress, cardiac electrical alterations, and cardiac tissue damage were significantly reduced in chronically infected mice [61]. Ad-ASP2 itself was also found to enhance the inflammatory gene expression profile and decrease amastigote nests in the tissues of mice infected at the time of immunization [62]. Galα(1,3)Galβ(1,4)GlcNAc has been developed as a vaccine candidate in conjugation with human serum albumin with the aim of stimulating anti-α-Gal antibodies, which were shown to provide up to 99.9% protection from tissue parasites in α1,3-galactosyltransferase knockout mice, while non-vaccinated mice exhibited a high degree of tissue parasites, necrotic myocytes, and extensive cardiac inflammation [63].

We have evaluated the protective efficacy of TcG1, TcG2, and TcG4 in mice and dogs, delivering these antigens (individually or in combination) by homologous and heterologous prime/boost approaches (e.g., DNA/DNA, DNA/protein, DNA/Modified Vaccinia Ankara). In all cases, vaccine provided 80–95% protection from acute parasitemia and chronic myocarditis in mice [46,64–66]. The overall extent of protection was primarily associated with elicitation of Th1 cytokine response and CD8$^+$T cell cytolytic activity, while trypanolytic antibodies were not induced by these vaccine candidates. Further studies showed that TcG1 was expendable, and only TcG2/TcG4-based nanovaccine that utilized an antibiotic-free plasmid for delivery was sufficient to offer effective control of infection and disease pathology after primary infection, as well as eliciting a potent recall response potentially capable of controlling the repeat infection [67]. Importantly, the TcG2- and TcG4-encoding nanovaccine-induced immune response to challenge infection subsided after parasite control, and inflammatory infiltrate and oxidative damage were decreased in the tissues of chronically infected, vaccinated mice [68]. As in mice, dogs immunized with the three candidate antigens (TcG1, TcG2, TcG4) also showed significant protection from challenge infection and acute myocarditis [69,70]. TcSP- and TcSSP4-encoding DNA vaccines have also been found to induce anti-parasite IgG2- and IFNγ-producing lymphocytic cell proliferation in dogs, leading to moderate control of tissue parasites and electrocardiographic aberrations [71].

Building on these achievements, in recent studies, investigators have developed chimeric antigens incorporating epitopes of multiple antigens in one vaccine. For example, trivalent cruzipain-TS-ASP2 vaccine (Traspain) showed promising protection from *Tc* infection in mice [72]. We have engineered a bivalent vaccine with or without a string of adjuvants that exhibited prophylactic and therapeutic efficacy, evidenced by >97% control of parasite and tissue damage (unpublished).

Congenital transmission of *T. cruzi* from infected mothers to their newborns has emerged as a major cause of Chagas disease in humans [73]. Experts agree that seropositive women of reproductive age should be treated with benznidazole to potentially prevent congenital transmission [74], and there are no available drugs for the treatment of infected mothers during pregnancy. However, there is only one published report testing the efficacy of a candidate vaccine in pregnant mice. Garg and coworkers found that *Tc* infection resulted in delayed pregnancy, and pregnant female mice gave birth to pups with lower survival rates, decreased birth weight, and slower growth [75]. TcG2- and TcG4-encoding nanovaccine was found to be safe during pregnancy, and it modulated the maternal and placental T cell immune response such that maternal and placental parasite burden was reduced, and vaccine improved the placental integrity and birth/survival rate of newborns of infected mothers [76].

So far, most investigators have devoted efforts to testing the efficacy of their selected candidate vaccines against 1–2 *T. cruzi* isolates or strains of interest in in-bred experimental mouse models that could be established in their laboratories. Considering the diverse geographic and genetic complexity of *T. cruzi* isolates of different lineages, the next phase of studies in vaccine development should address if the candidate vaccine is effective against multiple *T. cruzi* isolates of clinical importance in out-bred animal models or naturally infected animals. It is hoped that candidate antigens that are highly conserved among different parasite lineages and expressed in all life cycle stages of the parasite will be successful in offering broad-range immunity to *T. cruzi* infection.

3.3. Therapeutic Vaccines Against CD

The overall basis for therapeutic vaccines is to enhance multiple immune effector mechanisms to prevent/reduce the parasite's persistence. A therapeutic DNA vaccine encoding Tc52, TSA1, and Tc24 enhanced the number of $CD4^+$ and $CD8^+$ T cells and decreased the parasitemia and mortality in acutely infected mice [77] but failed to arrest cardiomyopathy in chronically infected dogs [78]. Others showed that ASP2- and TS-encoding therapeutic DNA vaccine was not effective in limiting parasitemia or mortality [79] and TSA1 DNA vaccine enhanced myocarditis [80] in infected mice. Subsequently, Tc24 protein was encapsulated in poly(lactic-co-glycolic acid) nanoparticles (with and without E6020 adjuvant), and this vaccine was found to enhance the splenic type 1 $CD8^+$T cell response, such that tissue parasites were reduced when given to mice during acute infection with a highly lethal *T. cruzi* H1 strain, leading to parasite clearance in 60% of chronically infected mice [81,82]. At least one report has recently documented the safety and efficacy of TSA1/Tc24-based therapeutic DNA vaccine in controlling cardiac alterations in a small group of experimentally infected macaques [83].

Considering the limited success of therapeutic vaccines, and the fact that repeat challenge with *Tc* continued to boost the immune responses but did not provide protection from cardiac damage in mice [84], it is proposed that immune therapy against parasite persistence should be adjuvanted with other agents to prevent cardiac damage. Towards this goal, it was found that treatment with benznidazole during the indeterminate phase (when most seropositive patients are identified) controlled the parasite persistence, but did not avert cardiac remodeling and deterioration of ventricular contractility in infected mice and rats. When an antioxidant (phenyl-alpha-tert-butyl nitrone) was added to a benznidazole treatment regimen, mitochondrial function and left ventricular contractile activity were improved in Chagas hearts [85,86]. Likewise, a therapeutic TcG2/TcG4-based DNA vaccine offered better control of the oxidative damage caused by mitochondrial deficiencies of electron transport chain in glutathione-peroxidase-overexpressing mice, while progressive cardiac damage was noted in wildtype mice after TG2/TcG4 therapeutic vaccine [87]. Thus, it is safe to surmise that therapeutic vaccines designed to achieve a rapid stimulation of humoral and cytotoxic immunity to attack persistent parasites, along with adjunct therapies capable of controlling the onset of oxidative insult and mitochondrial deficiencies, would prove to be maximally beneficial in preserving cardiac structure and function in Chagas disease.

4. Future Opportunities and Challenges

Research efforts have led to the identification of several candidate antigens that exhibit promising prophylactic and therapeutic efficacy as a single-component, multi-component, or chimeric vaccine in experimental models of *T. cruzi* infection and Chagas disease. None of the current vaccines offer sterilizing immunity and efforts toward improving vaccine immunogenicity have continued. A recent presentation by Dr. SMR Teixiera and colleagues at the 2024 Brazilian Society of Protozoology meeting deserves mention here: The authors found that TcTS with trans-sialidase enzymatic activity is highly immunogenic but offers low protective efficacy as a vaccine candidate. Further studies using TS deletion mutants led to the findings that the C-terminal repeats known as SAPA (shed acute phase antigen) domain serve as a negative modulator of TS-induced immune response and instead offer virulence to the parasite. Armed with this knowledge, the authors are testing the efficacy of SAPA-deficient TS as mRNA and protein vaccines encapsulated in lipid nanoparticles, and hope that this novel vaccine will be highly protective against Chagas disease. Others reported that incorporation of low-dose benznidazole enhanced the therapeutic efficacy of a bivalent recombinant protein vaccine (TSA1/Tc24) in controlling cardiac fibrosis in chronically infected mice [88]. This study is particularly important because benznidazole itself exhibits lower effectiveness in the chronic phase, likely due to its limited reach to the niches where the parasite hides or its lower effectiveness against the dormant state parasite [89]. Thus, anti-parasite drugs (both old and new) adjuvanted with a vaccine offer a promising approach for controlling Chagas disease.

Development of the next generation of vaccines will likely involve efforts focused on enhancing the in vivo stability of vaccines, such that prolonged antigen presentation and immune memory is achieved. For example, delivery of DNA vaccines encapsulated in cationic liposomes [90] and biopolymers [91], which protect DNA from nucleases, could enhance the stability and expression of antigens carried by DNA vaccines. Liposomes have also been utilized to co-deliver adjuvant such as poly I:C RNA or CpG-ODN to enhance the immunogenicity of DNA and mRNA vaccines [90] and can be designed to carry such adjuvants for DNA vaccines against *T. cruzi*. Other platforms for immunogenic vaccine delivery have also been developed. For example, the assembly of candidate antigens in virus-like particles (VLPs) could enhance antigen presentation and innate and adaptive immunity [92]. T4 bacteriophage can also provide an excellent platform for generating nanoparticle subunit vaccine [93]. The technological knowledge gained from the development of VLP- and T4-phage-based vaccines for high-risk viruses and bacteria could potentially be employed for generating such vaccines for pathogenic parasites. Oral vaccines using tobacco plant for antigen expression exhibited immunologic protection from malaria, toxoplasmosis, and other parasitic diseases [94], and offer a promising perspective for developing cost-effective plant-based vaccines to continue the fight against Chagas disease. Testing the safety and efficacy of candidate vaccines during pregnancy, with the aim of decreasing the impact of congenital Chagas disease, is also urgently needed.

Investigators also need to ensure that the subunit vaccine (along with the delivery vehicle) is safe, elicits recall immune response to protect against repeat infection from the same or different clinically relevant parasite isolates, and is efficacious in reducing the cardiac damage caused by chronic infection in multiple hosts (e.g., mice, dogs, non-human primates). Vaccines that can be delivered orally or via skin patches and stored long-term at room temperature will likely have a better chance of reaching the marketplace. Alternatively, one can envision incorporating *Tc*-specific antigenic epitopes into vaccines for SARS-CoV2 or other high-risk infectious agents to facilitate the vaccination against Chagas disease.

Demonstrating the long-term effectiveness of the current and new anti-parasite vaccines and drugs against a variety of parasite strains of different lineages remains a serious challenge. Such studies are extremely expensive and cannot be carried out with limited availability of resources. Furthermore, *T. cruzi* has a large, diploid genome, in which coding sequence may undergo recombination events to further contribute to antigenic complexity, challenging the development of new vaccines and therapies.

A general challenge for testing the prophylactic efficacy of candidate vaccines in humans is the low incidence rate of disease development over decades. Clinical studies testing the therapeutic efficacy of candidate vaccines in already infected individuals are more feasible; however, the recruitment and long-term follow up required to monitor parasite clearance, immunity, and protection from chronic cardiomyopathy remains a major obstacle. The response to treatment in an infected individual is measured by achievement of negative serology, which can take decades. Some investigators have conducted seminal studies identifying new surrogate markers that can offer a robust readout of vaccine efficacy in the short term and potentially be used in clinical trials testing new drugs and vaccines against CD [95,96].

Finally, it is hoped that the most advanced drugs and vaccines will receive strong support from the academic–private–government partnership and overcome the barriers of costs associated with vaccine testing and production for a neglected tropical disease. Joint collaborative efforts will lead to the development of vaccines that address the antigenic variance of diverse *T. cruzi* lineages circulating in different parts of the Americas, elicit broad and long-lasting immunity to all clinically relevant parasite strains, and offer protection from infection. While testing and developing a vaccine for humans, a vaccine for veterinary use would offer a promising approach to prevent parasite transmission in domestic reservoirs and indirectly protect humans from *T. cruzi* infection.

Funding: NJG has been supported by grants from the National Institute of Allergy and Infectious Diseases (R01AI136031, R44AI172437, R21AI178318) of the National Institutes of Health.

Data Availability Statement: No new data were created or analyzed in this study.

Conflicts of Interest: The author declares no conflict of interest.

References

1. Bonney, K.M.; Luthringer, D.J.; Kim, S.A.; Garg, N.J.; Engman, D.M. Pathology and pathogenesis of chagas heart disease. *Annu. Rev. Pathol. Mech. Dis.* **2020**, *14*, 421–447. [CrossRef]
2. World Health Organization. *Chagas Disease: Control and Elimination*; Report of the Secretariat WHO; UNDP/World Bank/WHO: Geneva, Switerzland, 2010; Available online: http://apps.who.int/gb/ebwha/pdf_files/WHA63/A63_17-en.pdf (accessed on 20 January 2025).
3. Limon-Flores, A.Y.; Cervera-Cetina, R.; Tzec-Arjona, J.L.; Ek-Macias, L.; Sanchez-Burgos, G.; Ramirez-Sierra, M.J.; Cruz-Chan, J.V.; VanWynsberghe, N.R.; Dumonteil, E. Effect of a combination DNA vaccine for the prevention and therapy of *Trypanosoma cruzi* infection in mice: Role of $CD4^+$ and $CD8^+$ T cells. *Vaccine* **2010**, *28*, 7414–7419. [CrossRef] [PubMed]
4. Moncayo, A.; Silveira, A.C. Current epidemiological trends for Chagas disease in Latin America and future challenges in epidemiology, surveillance and health policy. *Mem. Inst. Oswaldo Cruz.* **2009**, *104* (Suppl. S1), 17–30. [CrossRef] [PubMed]
5. Lee, B.Y.; Bacon, K.M.; Bottazzi, M.E.; Hotez, P.J. Global economic burden of Chagas disease: A computational simulation model. *Lancet Infect. Dis.* **2013**, *13*, 342–348. [CrossRef] [PubMed]
6. Lee, B.Y.; Bacon, K.M.; Connor, D.L.; Willig, A.M.; Bailey, R.R. The potential economic value of a *Trypanosoma cruzi* (Chagas disease) vaccine in Latin America. *PLoS Negl. Trop. Dis.* **2010**, *4*, e916. [CrossRef] [PubMed]

7. Gutierrez, F.R.; Guedes, P.M.; Gazzinelli, R.T.; Silva, J.S. The role of parasite persistence in pathogenesis of Chagas heart disease. *Parasite Immunol.* **2009**, *31*, 673–685. [CrossRef]
8. Vellozo, N.S.; Matos-Silva, T.C.; Lopes, M.F. Immunopathogenesis in *Trypanosoma cruzi* infection: A role for suppressed macrophages and apoptotic cells. *Front. Immunol.* **2023**, *14*, 1244071. [CrossRef] [PubMed]
9. Lewis, M.D.; Francisco, A.F.; Jayawardhana, S.; Langston, H.; Taylor, M.C.; Kelly, J.M. Imaging the development of chronic Chagas disease after oral transmission. *Sci. Rep.* **2018**, *8*, 11292. [CrossRef]
10. Combs, T.P.; Nagajyothi; Mukherjee, S.; de Almeida, C.J.; Jelicks, L.A.; Schubert, W.; Lin, Y.; Jayabalan, D.S.; Zhao, D.; Braunstein, V.L.; et al. The adipocyte as an important target cell for *Trypanosoma cruzi* infection. *J. Biol. Chem.* **2005**, *280*, 24085–24094. [CrossRef]
11. Ferreira, A.V.; Segatto, M.; Menezes, Z.; Macedo, A.M.; Gelape, C.; de Oliveira Andrade, L.; Nagajyothi, F.; Scherer, P.E.; Teixeira, M.M.; Tanowitz, H.B. Evidence for *Trypanosoma cruzi* in adipose tissue in human chronic chagas disease. *Microbes Infect.* **2011**, *13*, 1002–1005. [CrossRef] [PubMed]
12. Sanchez-Valdez, F.J.; Padilla, A.; Wang, W.; Orr, D.; Tarleton, R.L. Spontaneous dormancy protects *Trypanosoma cruzi* during extended drug exposure. *Elife* **2018**, *7*, e34039. [CrossRef] [PubMed]
13. Marin-Neto, J.A.; Cunha-Neto, E.; Maciel, B.C.; Simoes, M.V. Pathogenesis of chronic Chagas heart disease. *Circulation* **2007**, *115*, 1109–1123. [CrossRef]
14. Choudhuri, S.; Rios, L.; Vazquez-Chagoyan, J.C.; Garg, N.J. Oxidative stress implications for therapeutic vaccine development against Chagas disease. *Expert. Rev. Vaccines* **2021**, *20*, 1395–1406. [CrossRef]
15. Maldonado, E.; Rojas, D.A.; Urbina, F.; Solari, A. The oxidative stress and chronic inflammatory process in Chagas disease: Role of exosomes and contributing genetic factors. *Oxid. Med. Cell. Longev.* **2021**, *2021*, 4993452. [CrossRef] [PubMed]
16. de Oliveira, T.B.; Pedrosa, R.C.; Filho, D.W. Oxidative stress in chronic cardiopathy associated with Chagas disease. *Int. J. Cardiol.* **2007**, *116*, 357–363. [CrossRef] [PubMed]
17. Lopez, M.; Tanowitz, H.B.; Garg, N.J. Pathogenesis of chronic Chagas disease: Macrophages, mitochondria, and oxidative stress. *Curr. Clin. Microbiol. Rep.* **2018**, *5*, 45–54. [CrossRef]
18. de Carvalho, F.C.T.; de Oliveira, L.R.C.; Gatto, M.; Tasca, K.I.; da Silva, L.D.M.; dos Santos, K.C.; Pierine, D.T.; da Costa, E.A.P.N.; Francisqueti-Ferron, F.V.; dos Santos, R.M.; et al. Oxidative stress evaluation in patients with chronic Chagas disease. *Parasitol. Int.* **2023**, *96*, 102770. [CrossRef]
19. Sanchez-Villamil, J.P.; Bautista-Nino, P.K.; Serrano, N.C.; Rincon, M.Y.; Garg, N.J. Potential role of antioxidants as adjunctive therapy in Chagas disease. *Oxid. Med. Cell. Longev.* **2020**, *2020*, 9081813. [CrossRef]
20. Wan, X.; Garg, N.J. Sirtuin control of mitochondrial dysfunction, oxidative stress, and inflammation in Chagas disease models. *Front. Cell Infect. Microbiol.* **2021**, *11*, 693051. [CrossRef] [PubMed]
21. Choudhuri, S.; Chowdhury, I.H.; Garg, N.J. Mitochondrial regulation of macrophage response against pathogens. *Front. Immunol.* **2020**, *11*, 622602. [CrossRef] [PubMed]
22. Rios, L.; Campos, E.E.; Menon, R.; Zago, M.P.; Garg, N.J. Epidemiology and pathogenesis of maternal-fetal transmission of *Trypanosoma cruzi* and a case for vaccine development against congenital chagas disease. *Biochim. Biophys. Acta Mol. Basis Dis.* **2020**, *1866*, 165591. [CrossRef] [PubMed]
23. Maldonado, E.; Morales-Pison, S.; Urbina, F.; Solari, A. Vaccine design against Chagas disease focused on the use of nucleic acids. *Vaccines* **2022**, *10*, 587. [CrossRef]
24. Pinazo, M.J.; Malchiodi, E.; Ioset, J.R.; Bivona, A.; Gollob, K.J.; Dutra, W.O. Challenges and advancements in the development of vaccines and therapies against Chagas disease. *Lancet Microbe* **2024**, *5*, 100972. [CrossRef]
25. Farani, P.S.G.; Jones, K.M.; Poveda, C. Treatments and the perspectives of developing a vaccine for Chagas disease. *Vaccines* **2024**, *12*, 870. [CrossRef] [PubMed]
26. Ramírez-Toloza, G.; Ferreira, A. *Trypanosoma cruzi* evades the complement system as an efficient strategy to survive in the mammalian host: The specific roles of host/parasite molecules and trypanosoma cruzi calreticulin. *Front. Microbiol.* **2017**, *8*, 1667. [CrossRef] [PubMed]
27. Almeida, I.C.; Camargo, M.M.; Procopio, D.O.; Silva, L.S.; Mehlert, A.; Travassos, L.R.; Gazzinelli, R.T.; Ferguson, M.A. Highly purified glycosylphosphatidylinositols from *Trypanosoma cruzi* are potent proinflammatory agents. *EMBO J.* **2000**, *19*, 1476–1485. [CrossRef]
28. Tanowitz, H.B.; Wen, J.J.; Machado, F.S.; Desruisseaux, M.S.; Robello, C.; Garg, N.J. Trypanosoma cruzi and Chagas disease: Innate immunity, ROS, and cardiovascular system. In *Vascular Responses to Pathogens*; Gavins, F.N.E., Stokes, K.Y., Eds.; Academic Press/Elsevier Inc.: Waltham, MA, USA, 2016; pp. 183–193.

29. Cronemberger-Andrade, A.; Xander, P.; Soares, R.P.; Pessoa, N.L.; Campos, M.A.; Ellis, C.C.; Grajeda, B.; Ofir-Birin, Y.; Almeida, I.C.; Regev-Rudzki, N.; et al. *Trypanosoma cruzi*-infected human macrophages shed proinflammatory extracellular vesicles that enhance host-cell invasion via toll-like receptor 2. *Front. Cell Infect. Microbiol.* **2020**, *10*, 99. [CrossRef]
30. Koo, S.J.; Szczesny, B.; Wan, X.; Putluri, N.; Garg, N.J. Pentose phosphate shunt modulates reactive oxygen species and nitric oxide production controlling *Trypanosoma cruzi* in macrophages. *Front. Immunol.* **2018**, *9*, 202. [CrossRef]
31. Koo, S.J.; Garg, N.J. Metabolic programming of macrophage functions and pathogens control. *Redox Biol.* **2019**, *24*, 101198. [CrossRef] [PubMed]
32. Cerbán, F.M.; Stempin, C.C.; Volpini, X.; Carrera Silva, E.A.; Gea, S.; Motran, C.C. Signaling pathways that regulate *Trypanosoma cruzi* infection and immune response. *Biochim. Biophys. Acta-Mol. Basis Dis.* **2020**, *1866*, 165707. [CrossRef] [PubMed]
33. Macaluso, G.; Grippi, F.; Di Bella, S.; Blanda, V.; Gucciardi, F.; Torina, A.; Guercio, A.; Cannella, V. A review on the immunological response against *Trypanosoma cruzi*. *Pathogens* **2023**, *12*, 282. [CrossRef] [PubMed]
34. Rodrigues, M.M.; Ribeirão, M.; Boscardin, S.B. $CD4^+$ Th1 but not Th2 clones efficiently activate macrophages to eliminate *Trypanosoma cruzi* through a nitric oxide dependent mechanism. *Immunol. Lett.* **2000**, *73*, 43–50. [CrossRef] [PubMed]
35. Hoft, D.F.; Schnapp, A.R.; Eickhoff, C.S.; Roodman, S.T. Involvement of $CD4^+$ Rh1 cells in systemic immunity protective against primary and secondary challenges with *Trypanosoma cruzi*. *Infect. Immun.* **2000**, *68*, 197–204. [CrossRef]
36. Hodžić, A.; Mateos-Hernández, L.; de la Fuente, J.; Cabezas-Cruz, A. A-gal-based vaccines: Advances, opportunities, and perspectives. *Trends Parasitol.* **2020**, *36*, 992–1001. [CrossRef]
37. Acosta Rodríguez, E.V.; Araujo Furlan, C.L.; Fiocca Vernengo, F.; Montes, C.L.; Gruppi, A. Understanding $CD8^+$T cell immunity to *Trypanosoma cruzi* and how to improve it. *Trends Parasitol.* **2019**, *35*, 899–917. [CrossRef]
38. Pérez-Antón, E.; Egui, A.; Thomas, M.C.; Simón, M.; Segovia, M.; López, M.C. Immunological exhaustion and functional profile of $CD8^+$T lymphocytes as cellular biomarkers of therapeutic efficacy in chronic Chagas disease patients. *Acta Trop.* **2020**, *202*, 105242. [CrossRef]
39. Dutra, W.O.; Menezes, C.A.; Magalhaes, L.M.; Gollob, K.J. Immunoregulatory networks in human Chagas disease. *Parasite Immunol.* **2014**, *36*, 377–387. [CrossRef] [PubMed]
40. Medina-Rincón, G.J.; Gallo-Bernal, S.; Jiménez, P.A.; Cruz-Saavedra, L.; Ramírez, J.D.; Rodríguez, M.J.; Medina-Mur, R.; Díaz-Nassif, G.; Valderrama-Achury, M.D.; Medina, H.M. Molecular and clinical aspects of chronic manifestations in Chagas disease: A state-of-the-art review. *Pathogens* **2021**, *10*, 1493. [CrossRef] [PubMed]
41. Solana, J.C.; Moreno, J.; Iborra, S.; Soto, M.; Requena, J.M. Live attenuated vaccines, a favorable strategy to provide long-term immunity against protozoan diseases. *Trends Parasitol.* **2022**, *38*, 316–334. [CrossRef] [PubMed]
42. Revelli, S.; Basombrio, M.A.; Valenti, J.L.; Moreno, H.; Poli, H.; Morini, J.C. Evaluation of an attenuated *Trypanosoma cruzi* strain in rats. Analysis of survival, parasitemia and tissue damage. *Medicina* **1993**, *53*, 39–43. [PubMed]
43. Basombrio, M.A.; Segura, M.A.; Mora, M.C.; Gomez, L. Field trial of vaccination against American Trypanosomiasis (Chagas' disease) in dogs. *Am. J. Trop. Med. Hyg.* **1993**, *49*, 143–151. [CrossRef] [PubMed]
44. Basso, B.; Marini, V. Experimental Chagas disease in balb/c mice previously vaccinated with *T. rangeli*. II. The innate immune response shows immunological memory: Reality or fiction? *Immunobiol.* **2015**, *220*, 428–436. [CrossRef]
45. Basso, B.; Moretti, E.; Fretes, R. Vaccination with *Trypanosoma rangeli* induces resistance of guinea pigs to virulent *Trypanos Cruzi*. *Vet. Immunol. Immunopathol.* **2014**, *157*, 119–123. [CrossRef]
46. Gupta, S.; Salgado-Jimenez, B.; Lokugamage, N.; Vazquez-Chagoyan, J.C.; Garg, N.J. Tcg2/Tcg4 DNA vaccine induces th1 immunity against acute *Trypanosoma cruzi* infection: Adjuvant and antigenic effects of heterologous *T. rangeli* booster immunization. *Front. Immunol.* **2019**, *10*, 1456. [CrossRef] [PubMed]
47. Sanchez-Valdez, F.J.; Perez Brandan, C.; Ferreira, A.; Basombrio, M.A. Gene-deleted live-attenuated *Trypanosoma cruzi* parasites as vaccines to protect against Chagas disease. *Expert. Rev. Vaccines* **2015**, *14*, 681–697. [CrossRef] [PubMed]
48. Garg, N.J.; Nunes, M.P.; Tarleton, R.L. Delivery by *Trypanosoma cruzi* of proteins into the MHC class I antigen processing and presentation pathway. *J. Immunol.* **1997**, *158*, 3293–3302. [CrossRef] [PubMed]
49. da Costa, K.M.; Marques da Fonseca, L.; Dos Reis, J.S.; Santos, M.; Previato, J.O.; Mendonca-Previato, L.; Freire-de-Lima, L. *Trypanosoma cruzi* trans-sialidase as a potential vaccine target against Chagas disease. *Front. Cell Infect. Microbiol.* **2021**, *11*, 768450. [CrossRef] [PubMed]
50. Rios, L.E.; Vazquez-Chagoyan, J.C.; Pacheco, A.O.; Zago, M.P.; Garg, N.J. Immunity and vaccine development efforts against *Trypanosoma cruzi*. *Acta Trop.* **2019**, *200*, 105168. [CrossRef]
51. Bhatia, V.; Sinha, M.; Luxon, B.; Garg, N.J. Utility of *Trypanosoma cruzi* sequence database for the identification of potential vaccine candidates: In silico and in vitro screening. *Infect. Immun.* **2004**, *72*, 6245–6254. [CrossRef]

52. Bhatia, V.; Garg, N.J. Previously unrecognized vaccine candidates control *Trypanosoma cruzi* infection and immunopathology in mice. *Clin. Vaccine Immunol.* **2008**, *15*, 1158–1164. [CrossRef] [PubMed]
53. Albaqami, F.F.; Altharawi, A.; Althurwi, H.N.; Alharthy, K.M. From proteome to candidate vaccines: Target discovery and molecular dynamics-guided multi-epitope vaccine engineering against kissing bug. *Front. Immunol.* **2024**, *15*, 1413893. [CrossRef] [PubMed]
54. Serna, C.; Lara, J.A.; Rodrigues, S.P.; Marques, A.F.; Almeida, I.C.; Maldonado, R.A. A synthetic peptide from *Trypanosoma cruzi* mucin-like associated surface protein as candidate for a vaccine against Chagas disease. *Vaccine* **2014**, *32*, 3525–3532. [CrossRef] [PubMed]
55. Cazorla, S.I.; Becker, P.D.; Frank, F.M.; Ebensen, T.; Sartori, M.J.; Corral, R.S.; Malchiodi, E.L.; Guzman, C.A. Oral vaccination with *Salmonella enterica* as a cruzipain-DNA delivery system confers protective immunity against *Trypanosoma cruzi*. *Infect. Immun.* **2008**, *76*, 324–333. [CrossRef] [PubMed]
56. Morell, M.; Thomas, M.C.; Caballero, T.; Alonso, C.; Lopez, M.C. The genetic immunization with paraflagellar rod protein-2 fused to the hsp70 confers protection against late *Trypanosoma cruzi* infection. *Vaccine* **2006**, *24*, 7046–7055. [CrossRef] [PubMed]
57. Egui, A.; Thomas, M.C.; Morell, M.; Maranon, C.; Carrilero, B.; Segovia, M.; Puerta, C.J.; Pinazo, M.J.; Rosas, F.; Gascon, J.; et al. *Trypanosoma cruzi* paraflagellar rod proteins 2 and 3 contain immunodominant CD8$^+$T-cell epitopes that are recognized by cytotoxic t cells from chagas disease patients. *Mol. Immunol.* **2012**, *52*, 289–298. [CrossRef] [PubMed]
58. Fralish, B.H.; Tarleton, R.L. Genetic immunization with Lyt1 or a pool of trans-sialidase genes protects mice from lethal *Trypanosoma cruzi* infection. *Vaccine* **2003**, *21*, 3070–3080. [CrossRef]
59. Vasconcelos, J.R.; Hiyane, M.I.; Marinho, C.R.; Claser, C.; Machado, A.M.; Gazzinelli, R.T.; Bruna-Romero, O.; Alvarez, J.M.; Boscardin, S.B.; Rodrigues, M.M. Protective immunity against *Trypanosoma cruzi* infection in a highly susceptible mouse strain after vaccination with genes encoding the amastigote surface protein-2 and trans-sialidase. *Hum. Gene Ther.* **2004**, *15*, 878–886. [CrossRef]
60. Vasconcelos, J.R.; Dominguez, M.R.; Neves, R.L.; Ersching, J.; Araujo, A.; Santos, L.I.; Virgilio, F.S.; Machado, A.V.; Bruna-Romero, O.; Gazzinelli, R.T.; et al. Adenovirus vector-induced CD8$^+$T effector memory cell differentiation and recirculation, but not proliferation, are important for protective immunity against experimental *Trypanosoma cruzi* infection. *Hum. Gene Ther.* **2014**, *25*, 350–363. [CrossRef] [PubMed]
61. Pereira, I.R.; Vilar-Pereira, G.; Marques, V.; da Silva, A.A.; Caetano, B.; Moreira, O.C.; Machado, A.V.; Bruna-Romero, O.; Rodrigues, M.M.; Gazzinelli, R.T.; et al. A human type 5 adenovirus-based *Trypanosoma cruzi* therapeutic vaccine re-programs immune response and reverses chronic cardiomyopathy. *PLoS Pathog.* **2015**, *11*, e1004594. [CrossRef] [PubMed]
62. Ribeiro, F.A.P.; Pontes, C.; Machado, A.M.V.; Bruna-Romero, O.; Quintana, H.T.; De Oliveira, F.; De Vasconcelos, J.R.C.; Ribeiro, D. A Therapeutical effects of vaccine from *Trypanosoma cruzi* amastigote surface protein 2 by simultaneous inoculation with live parasites. *J. Cell Biochem.* **2019**, *120*, 3373–3383. [CrossRef]
63. Portillo, S.; Zepeda, B.G.; Iniguez, E.; Olivas, J.J.; Karimi, N.H.; Moreira, O.C.; Marques, A.F.; Michael, K.; Maldonado, R.A.; Almeida, I.C. A prophylactic α-gal-based glycovaccine effectively protects against murine acute chagas disease. *NPJ Vaccines* **2019**, *4*, 13. [CrossRef] [PubMed]
64. Gupta, S.; Garg, N.J. Prophylactic efficacy of TcVac2 against *Trypanosoma cruzi* in mice. *PLoS Negl. Trop. Dis.* **2010**, *4*, e797. [CrossRef]
65. Gupta, S.; Garg, N.J. TcVac3 induced control of *Trypanosoma cruzi* infection and chronic myocarditis in mice. *PLOS ONE* **2013**, *8*, e59434. [CrossRef] [PubMed]
66. Gupta, S.; Garg, N.J. A two-component DNA-prime/protein-boost vaccination strategy for eliciting long-term, protective T cell immunity against *Trypanosoma cruzi*. *PLoS Pathog.* **2015**, *11*, e1004828. [CrossRef] [PubMed]
67. Chowdhury, I.H.; Lokugamage, N.; Garg, N.J. Experimental nanovaccine offers protection against repeat exposures to *Trypanosoma cruzi* through activation of polyfunctional T cell response. *Front. Immunol.* **2020**, *11*, 595039. [CrossRef] [PubMed]
68. Lokugamage, N.; Choudhuri, S.; Davies, C.; Chowdhury, I.H.; Garg, N.J. Antigen-based nano-immunotherapy controls parasite persistence, inflammatory and oxidative stress, and cardiac fibrosis, the hallmarks of chronic Chagas cardiomyopathy, in a mouse model of *Trypanosoma cruzi* infection. *Vaccines* **2020**, *8*, 96. [CrossRef]
69. Aparicio-Burgos, J.E.; Ochoa-Garcia, L.; Zepeda-Escobar, J.A.; Gupta, S.; Dhiman, M.; Martinez, J.S.; de Oca-Jimenez, R.M.; Val Arreola, M.; Barbabosa-Pliego, A.; Vazquez-Chagoyan, J.C.; et al. Testing the efficacy of a multi-component DNA-prime/DNA-boost vaccine against *Trypanosoma cruzi* infection in dogs. *PLoS Negl. Trop. Dis.* **2011**, *5*, e1050. [CrossRef] [PubMed]
70. Aparicio-Burgos, J.E.; Zepeda-Escobar, J.A.; de Oca-Jimenez, R.M.; Estrada-Franco, J.G.; Barbabosa-Pliego, A.; Ochoa-Garcia, L.; Alejandre-Aguilar, R.; Rivas, N.; Penuelas-Rivas, G.; Val-Arreola, M.; et al. Immune protection against *Trypanosoma cruzi* induced by TcVac4 in a canine model. *PLoS Negl. Trop. Dis.* **2015**, *9*, e0003625. [CrossRef]

71. Arce-Fonseca, M.; Ballinas-Verdugo, M.A.; Zenteno, E.R.; Suarez-Flores, D.; Carrillo-Sanchez, S.C.; Alejandre-Aguilar, R.; Rosales-Encina, J.L.; Reyes, P.A.; Rodriguez-Morales, O. Specific humoral and cellular immunity induced by *Trypanosoma cruzi* DNA immunization in a canine model. *Vet. Res.* **2013**, *44*, 15. [CrossRef] [PubMed]
72. Sanchez Alberti, A.; Bivona, A.E.; Cerny, N.; Schulze, K.; Weissmann, S.; Ebensen, T.; Morales, C.; Padilla, A.M.; Cazorla, S.I.; Tarleton, R.L.; et al. Engineered trivalent immunogen adjuvanted with a sting agonist confers protection against *Trypanosoma cruzi* infection. *NPJ Vaccines* **2017**, *2*, 9. [CrossRef]
73. Chancey, R.J.; Edwards, M.S.; Montgomery, S.P. Congenital Chagas disease. *Pediatr. Rev.* **2023**, *44*, 213–221. [CrossRef] [PubMed]
74. Alvarez, M.G.; Vigliano, C.; Lococo, B.; Bertocchi, G.; Viotti, R. Prevention of congenital Chagas disease by benznidazole treatment in reproductive-age women. An observational study. *Acta Trop.* **2017**, *174*, 149–152. [CrossRef] [PubMed]
75. Rios, L.; Lokugamage, N.; Garg, N.J. Effects of acute and chronic *Trypanosoma cruzi* infection on pregnancy outcomes in mice: Parasite transmission, mortality, delayed growth, and organ damage in pups. *Am. J. Pathol.* **2023**, *193*, 313–331. [CrossRef]
76. Rios, L.E.; Lokugamage, N.; Choudhuri, S.; Chowdhury, I.H.; Garg, N.J. Subunit nanovaccine elicited T cell functional activation controls *Trypanosoma cruzi* mediated maternal and placental tissue damage and improves pregnancy outcomes in mice. *NPJ Vaccines* **2023**, *8*, 188. [CrossRef] [PubMed]
77. Dumonteil, E.; Escobedo-Ortegon, J.; Reyes-Rodriguez, N.; Arjona-Torres, A.; Ramirez-Sierra, M.J. Immunotherapy of trypanosoma cruzi infection with DNA vaccines in mice. *Infect. Immun.* **2004**, *72*, 46–53. [CrossRef] [PubMed]
78. Quijano-Hernandez, I.A.; Bolio-Gonzalez, M.E.; Rodriguez-Buenfil, J.C.; Ramirez-Sierra, M.J.; Dumonteil, E. Therapeutic DNA vaccine against trypanosoma cruzi infection in dogs. *Ann. N. Y. Acad. Sci.* **2008**, *1149*, 343–346. [CrossRef]
79. Sanchez-Burgos, G.; Mezquita-Vega, R.G.; Escobedo-Ortegon, J.; Ramirez-Sierra, M.J.; Arjona-Torres, A.; Ouaissi, A.; Rodrigues, M.M.; Dumonteil, E. Comparative evaluation of therapeutic DNA vaccines against trypanosoma cruzi in mice. *FEMS Immunol. Med. Microbiol.* **2007**, *50*, 333–341. [CrossRef] [PubMed]
80. Zapata-Estrella, H.; Hummel-Newell, C.; Sanchez-Burgos, G.; Escobedo-Ortegon, J.; Ramirez-Sierra, M.J.; Arjona-Torres, A.; Dumonteil, E. Control of trypanosoma cruzi infection and changes in t-cell populations induced by a therapeutic DNA vaccine in mice. *Immunol. Lett.* **2006**, *103*, 186–191. [CrossRef] [PubMed]
81. Barry, M.A.; Wang, Q.; Jones, K.M.; Heffernan, M.J.; Buhaya, M.H.; Beaumier, C.M.; Keegan, B.P.; Zhan, B.; Dumonteil, E.; Bottazzi, M.E.; et al. A therapeutic nanoparticle vaccine against trypanosoma cruzi in a balb/c mouse model of chagas disease. *Hum. Vaccin. Immunother.* **2016**, *12*, 976–987. [CrossRef] [PubMed]
82. Barry, M.A.; Versteeg, L.; Wang, Q.; Pollet, J.; Zhan, B.; Gusovsky, F.; Bottazzi, M.E.; Hotez, P.J.; Jones, K.M. A therapeutic vaccine prototype induces protective immunity and reduces cardiac fibrosis in a mouse model of chronic trypanosoma cruzi infection. *PLoS Negl. Trop. Dis.* **2019**, *13*, e0007413. [CrossRef]
83. Dumonteil, E.; Herrera, C.; Marx, P.A. Safety and preservation of cardiac function following therapeutic vaccination against trypanosoma cruzi in rhesus macaques. *J. Microbiol. Immunol. Infect.* **2023**, *56*, 400–407. [CrossRef] [PubMed]
84. Rosas-Jorquera Christian, E.; Sardinha Luiz, R.; Pretel Fernando, D.; Bombeiro André, L.; D'Império Lima Maria, R.; Alvarez José, M. Challenge of chronically infected mice with homologous trypanosoma cruzi parasites enhances the immune response but does not modify cardiopathy: Implications for the design of a therapeutic vaccine. *Clin. Vaccine Immunol.* **2013**, *20*, 248–254. [CrossRef] [PubMed]
85. Wen, J.J.; Bhatia, V.; Popov, V.L.; Garg, N.J. Phenyl-alpha-tert-butyl nitrone reverses mitochondrial decay in acute chagas' disease. *Am. J. Pathol.* **2006**, *169*, 1953–1964. [CrossRef]
86. Wen, J.J.; Bhatia, V.; Popov, V.L.; Garg, N.J. Phenyl-alpha-tert-butyl-nitrone and benzonidazole treatment controlled the mitochondrial oxidative stress and evolution of cardiomyopathy in chronic chagasic rats. *J. Am. Coll. Cardiol.* **2010**, *55*, 2499–2508. [CrossRef]
87. Gupta, S.; Smith, C.; Auclair, S.; Delgadillo Ade, J.; Garg, N.J. Therapeutic efficacy of a subunit vaccine in controlling chronic *Trypanosoma cruzi* infection and chagas disease is enhanced by glutathione peroxidase over-expression. *PLoS ONE* **2015**, *10*, e0130562. [CrossRef] [PubMed]
88. Dzul-Huchim, V.M.; Ramirez-Sierra, M.J.; Martinez-Vega, P.P.; Rosado-Vallado, M.E.; Arana-Argaez, V.E.; Ortega-Lopez, J.; Gusovsky, F.; Dumonteil, E.; Cruz-Chan, J.V.; Hotez, P.; et al. Vaccine-linked chemotherapy with a low dose of benznidazole plus a bivalent recombinant protein vaccine prevents the development of cardiac fibrosis caused by *Trypanosoma cruzi* in chronically-infected balb/c mice. *PLoS Negl. Trop. Dis.* **2022**, *16*, e0010258. [CrossRef] [PubMed]
89. Barrett, M.P.; Kyle, D.E.; Sibley, L.D.; Radke, J.B.; Tarleton, R.L. Protozoan persister-like cells and drug treatment failure. *Nat. Rev. Microbiol.* **2019**, *17*, 607–620. [CrossRef] [PubMed]
90. Schwendener, R.A. Liposomes as vaccine delivery systems: A review of the recent advances. *Ther. Adv. Vaccines* **2014**, *2*, 159–182. [CrossRef]
91. Franck, C.O.; Fanslau, L.; Bistrovic Popov, A.; Tyagi, P.; Fruk, L. Biopolymer-based carriers for DNA vaccine design. *Angew. Chem. Int. Ed. Engl.* **2021**, *60*, 13225–13243. [CrossRef] [PubMed]

92. Kheirvari, M.; Liu, H.; Tumban, E. Virus-like particle vaccines and platforms for vaccine development. *Viruses* **2023**, *15*, 1109. [CrossRef] [PubMed]
93. Tao, P.; Li, Q.; Shivachandra, S.B.; Rao, V.B. Bacteriophage t4 as a nanoparticle platform to display and deliver pathogen antigens: Construction of an effective anthrax vaccine. *Methods Mol. Biol.* **2017**, *1581*, 255–267. [PubMed]
94. Ramos-Vega, A.; Monreal-Escalante, E.; Dumonteil, E.; Banuelos-Hernandez, B.; Angulo, C. Plant-made vaccines against parasites: Bioinspired perspectives to fight against chagas disease. *Expert. Rev. Vaccines* **2021**, *20*, 1373–1388. [CrossRef] [PubMed]
95. Cortes-Serra, N.; Losada-Galvan, I.; Pinazo, M.J.; Fernandez-Becerra, C.; Gascon, J.; Alonso-Padilla, J. State-of-the-art in host-derived biomarkers of Chagas disease prognosis and early evaluation of anti-*Trypanosoma cruzi* treatment response. *Biochim. Biophys. Acta-Mol. Basis Dis.* **2020**, *1866*, 165758. [CrossRef] [PubMed]
96. Choudhuri, S.; Bhavnani, S.K.; Zhang, W.; Botelli, V.; Barrientos, N.; Iñiguez, F.; Zago, M.P.; Garg, N.J. Prognostic performance of peripheral blood biomarkers in identifying seropositive individuals at risk of developing clinically symptomatic Chagas cardiomyopathy. *Microbiol. Spectr.* **2021**, *9*, e0036421. [CrossRef] [PubMed]

pathogens MDPI

Article

New Approaches to the Ecology of *Triatoma sordida* in Peridomestic Environments of an Endemic Area of Minas Gerais, Brazil

Carolina Valença-Barbosa [1,2], Isabel Mayer de Andrade [1], Fellipe Dias Tavares de Simas [1], Ozorino Caldeira Cruz Neto [3], Nilvanei Aparecido da Silva [1], Camila Fortunato Costa [1], Bruno Oliveira Bolivar Moreira [1], Paula Finamore-Araujo [2], Marcus Vinicius Niz Alvarez [4], André Borges-Veloso [1,2], Otacílio da Cruz Moreira [2], Liléia Diotaiuti [1] and Rita de Cássia Moreira de Souza [1,*]

[1] Grupo Triatomíneos, Instituto René Rachou-Fiocruz Minas Gerais, Belo Horizonte 30190-002, MG, Brazil; carolvalencabarbosa@gmail.com (C.V.-B.); belmayer@yahoo.com.br (I.M.d.A.); fellipe.simas@fiocruz.br (F.D.T.d.S.); nilvaneineves1@gmail.com (N.A.d.S.); camilafortunatocosta28@gmail.com (C.F.C.); brunobolivar100@gmail.com (B.O.B.M.); aborgesveloso@gmail.com (A.B.-V.); lileia.diotaiuti@fiocruz.br (L.D.)

[2] Plataforma de Análises Moleculares, Laboratório de Virologia e Parasitologia Molecular, Instituto Oswaldo Cruz/Fiocruz, Rio de Janeiro 21040-360, RJ, Brazil; paulafinamore@gmail.com (P.F.-A.); otacilio@ioc.fiocruz.br (O.d.C.M.)

[3] Centro de Endemias e Zoonoses, Secretaria de Saúde, Espinosa 39510-000, MG, Brazil; ozorinoneto@gmail.com

[4] Instituto de Biotecnologia de Botucatu, Universidade Estadual Paulista, São Paulo 18607-440, SP, Brazil; marcus.alvarez@unesp.br

* Correspondence: rita.souza@fiocruz.br; Tel.: +55-(31)-3349-7760

Abstract: *Triatoma sordida* is a native South American species and the most frequently captured triatomine in artificial environments in Brazil. Although considered a secondary vector of *Trypanosoma cruzi*, it is typically associated with low infection rates. To investigate its role in an endemic area for Chagas disease in northern Minas Gerais, Brazil, we employed a multidimensional approach that combined triatomine capture data with quantitative and qualitative analyses of *T. cruzi*. A total of 1861 *T. sordida* specimens were captured, of which 1455 were examined and 210 (14.4%) were found to be infected with *T. cruzi*. The most prevalent discrete typing unit (DTU) was TcI (80%), followed by TcII (8%), TcV (5%), and TcIII (3%). Molecular techniques provided new insights into the ecology of *T. sordida*, revealing a higher infection rate than previously reported and a parasitic load lower than that observed in other quantified species. Chickens were confirmed as the primary food source, playing an epidemiological role in maintaining infected insects with four *T. cruzi* DTUs. The observed diversity of *T. cruzi* DTUs suggests a lack of environmental segregation, likely due to the extensive movement of various host species between wild and domestic habitats, resulting in overlapping transmission cycles.

Keywords: *Triatoma sordida*; Triatominae; *Trypanosoma cruzi*; Chagas disease; blood meal; discrete typing unit; parasite load; peridomicile; kissing bug

1. Introduction

Triatomines (Hemiptera: Reduviidae) are blood-sucking insects and vectors of the protozoan *Trypanosoma cruzi*, the causative agent of Chagas disease (ChD). It is a neglected tropical disease (NTD) and one of the most significant infectious diseases in Latin America, affecting approximately 6–7 million people [1]. The disease has significant social and economic impacts, as it remains incurable in its chronic phase. Additionally, its acute phase is predominantly asymptomatic, which makes diagnosis even more challenging. The primary measure to reduce the incidence of new infections is by vector control [1].

Although vector control measures have significantly reduced the risk of *T. cruzi* transmission in Brazil through extensive campaigns aimed at eliminating house infestation by the non-native species *Triatoma infestans*, the risk of parasite transmission by autochthonous vectors persists, as they cannot be eradicated [2,3]. More than 150 triatomine species have been formally described, with 67 distributed across the 27 states of Brazil [4–6]. Four species—*Panstrongylus megistus, Triatoma brasiliensis, Triatoma pseudomaculata*, and *Triatoma sordida*—warrant considerable attention due to their epidemiological importance, as they are frequently found in and around human dwellings [7,8].

Triatoma sordida [9] is a native triatomine species in South America, considered synanthropic in Brazil. This taxonomy is under review. Currently, it belongs to the *T. sordida* subcomplex, a monophyletic group, comprising *T. sordida, Triatoma rosai, Triatoma garciabesi, Triatoma jurbergi, Triatoma matogrossensis*, and *Triatoma vandae* [10]. Considering the species *T. sordida, T. patagonica, T. guasayana*, and *T. garciabesi*, cryptic speciation has been suggested [11]. However, some authors argue that the term "cryptic speciation" does not apply, as these species exhibit morphological and/or morphometric differences [12]. It is currently the most commonly captured triatomine in peridomestic Brazilian environments [13–15], where it has been reported to be naturally infected with *T. cruzi* [15]. This species occupies several natural ecotopes, primarily found under the bark of live or dead trees [16] and, to a lesser extent, on the trunks of several palm trees species [17]. Around and near houses, *T. sordida* can form numerous colonies, frequently found colonizing chicken coops and feeding on these animals [16,18,19]. Additionally, *T. sordida* has been found living in association with various mammals, such as rodents and opossums [16,20], which are known as potential wild reservoirs of *T. cruzi* [21]. In this regard, understanding the ecology, behaviour, and infection dynamics of this vector is essential for developing effective control strategies.

Artificial ecotopes, such as the annexes in peridomestic areas, play a crucial role in facilitating interactions between the wild and domestic populations of triatomines and *T. cruzi* genotypes [22,23]. Given that this triatomine species inhabits both natural and artificial environments, the interaction between wild and domestic cycles may be possible [24], posing challenges to control efforts. Furthermore, anthropogenic alterations to the landscape, such as uncontrolled grazing and deforestation, contribute to the concentration of wild mammals and triatomines in peridomestic areas, thereby establishing local transmission cycles of *T. cruzi* [25].

The ecological history of the *T. cruzi* strain distribution has been shaped significantly by mammalian migrations in the Americas [26]. The six parasite genotypes known as discrete typing units (DTUs)—TcI to TcVI—exhibit considerable genetic variability, complicating our understanding of transmission cycles [27,28]. Currently, no associations have been established between parasite genotypes and biological response variables, such as the biome, environment, or host species [29]. Also, it is important to highlight that the parasitemia of *T. cruzi* in mammalian hosts is an important factor in the parasitic infection of vectors [29].

Some mammals associated with synanthropic triatomine species, such as *T. sordida*, act as a link between environments and *T. cruzi* cycles [30,31]. The integration of the extra-domestic population of *T. sordida* into the domestic transmission cycle of the parasite has been reported in Minas Gerais, Brazil [16]. This epidemiological scenario, combined with the high number of specimens captured in households, requires significant efforts from the health system to maintain control. To date, there are no reports on the characterization of *T. cruzi* genotypes in *T. sordida*, and few studies have focused on the blood meal source using precipitin tests [16,20]. Although this method has proven useful, it is either inadequately sensitive or specific [32].

In this study, we employed a multi-dimensional approach combining field survey data of triatomines and both quantitative and qualitative analyses, including blood meal source, parasitic load, and *T. cruzi* DTUs infecting *T. sordida* collected in Espinosa, Minas Gerais, Brazil. Our aim was to advance the understanding of the importance of this vector, highlighting the risk factors associated with potential new *T. cruzi* transmission in domestic environments within this region.

2. Materials and Methods

2.1. Collection of *Triatoma sordida*

Active searches for triatomine insects were conducted in the municipality of Espinosa (14°55′33″ S; 42°49′08″ O), Minas Gerais, Brazil, during routine entomological surveillance surveys. Five localities were investigated, namely Curral Velho, Sítio do Limoeiro, Barro Vermelho, São Pedro, and Coelheiro. The surveys were carried out between 2018 and 2019 by endemic disease control agents in domiciliary units (DUs), including all annexes in the peridomicile. Peridomestic ecotopes were defined as areas surrounding residences, such as shelters for domestic animals, piles of wood and tiles, and other outbuildings [33]. Triatomine insects were captured manually using tweezers. Collected specimens were examined as described below and individually stored at 20 °C in microtubes containing 70% ethanol until DNA extraction.

After the captures, insecticide spraying was carried out in triatomine-infested domiciliary units (intra- and peridomicile) by endemic disease control agents.

2.2. Natural Infection by *Trypanosoma cruzi*

To detect natural infection by *T. cruzi*, rectal content feces were collected by abdominal compression, placed on a slide with saline solution and covered with a glass coverslip (size 22 mm × 22 mm). Samples were examined at the Instituto René Rachou-Fiocruz Minas using a binocular optical microscope at 400× magnification, screening all fields for *T. cruzi*. Rectal contents testing positive for *T. cruzi* by optical microscope were directly cultured in liver infusion tryptose (LIT) medium supplemented with 10% fetal bovine serum and incubated at 38 °C [34]. Cultures were monitored daily for the first seven days using standard light microscopy. Positive cultures were aliquoted and stored at −20 °C until DNA extraction. Additionally, the natural infection index by *T. cruzi* was calculated (number of infected triatomines × 100/number of insects examined).

For *T. cruzi* molecular diagnosis, DNA samples of *T. sordida* were submitted to kDNA-PCR. The set of primers used in the conventional PCR assays were forward 121 (5′-AAATAATGTACGGG(T/G)GAGATGCATGA-3′) and reverse 122 (5′-GGTTCGATT GGGGTTGGTGAATATA-3′), as designed by Sturm et al. [35] and Wincker et al. [36]. The PCR was performed following conditions previously described [31].

2.3. DNA Extraction

The intestine (midgut, hindgut, and rectum) of each *T. sordida* specimen was macerated using a sterile crusher, and genomic DNA was isolated using the DNeasy Qiagen® kit in accordance with the protocol provided by the manufacturer. In addition, DNA was extracted from the positive in vitro cultures. Quantification of the extracted DNA samples was determined using a NanoDrop™ One Microvolume UV-Vis Spectrophotometer (Thermo Scientific, Waltham, MA, USA). All DNA samples were stored at −80 °C until amplification by polymerase chain reaction (PCR).

2.4. *T. cruzi* Genotyping

The molecular characterization of *T. cruzi* DTUs from the DNA of abdominal contents and in vitro culture was performed based on multilocus PCR analysis [37]. For genotype identification, PCRs using three biomarkers (spliced leader, 24 Sα rDNA, and nuclear fragment A10) were conducted. Each biomarker presents a distinct PCR profile, allowing for DTU differentiation. PCR products were electrophoresed on 6% polyacrylamide gels and stained with silver nitrate. The identification of *T. cruzi* DTUs was performed according to the procedures described by the authors [37].

2.5. *T. cruzi* Quantification by Real-Time PCR (qPCR)

Quantification of the parasitic load was performed [38]. The qPCR assays utilized a multiplex Taqman system targeting the *T. cruzi* nuclear satellite DNA (sat-DNA) [39] and the mitochondrial 12S rRNA gene of triatomines [38]. Standard calibration curves for *T. cruzi* and triatomine targets were constructed by serially diluting total DNA obtained from non-infected triatomine intestine samples (adults *T. sordida* from insectary) supplemented with 10^5 *T. cruzi* epimastigotes (Dm28c clone, TcI). The resulting DNA was serially diluted to a range of 10^5 to 0.5 *T. cruzi* equivalents and 5 to 0.002 triatomine intestine unit equivalents.

2.6. Identification of Feeding Sources

DNA samples from intestines of non-starving triatomines were used to identify feeding sources through direct sequencing of the vertebrate 12S rRNA locus (L1085: 5′-CCCAAACTGGGATTAGATACCC-3′; and H11259: 5′-GTTTGCTGAAGATGGCGGTA-3′) [40]. Amplicons were purified using the Wizard® SV gel and PCR Clean-Up System kit (Promega, Madison, WI, USA), then sequenced in both directions using PCR primers. DNA sequencing reactions employed the BigDye® Terminator v.3.1 Cycle Sequencing Kit (Applied Biosystems, Foster City, CA, USA) and were processed on an ABI 3730 Sequencer. Bi-directional sequences were assembled and edited using SeqMan (DNASTAR software package, version 12.3, DNASTAR Inc., Madison, WI, USA). The consensus sequences generated were compared with previously published sequences using BLASTN hosted on the NCBI server (http://www.ncbi.nlm.nih.gov/BLAST) (accessed on 10 February 2025).

2.7. Statistical Analysis

Statistical analysis was performed using R [41] and Prisma software (version 9). Parasite loads between TcI and TcII were compared using the Wilcoxon–Mann–Whitney test. Additionally, the food sources of domestic and wild animals identified in triatomine vectors were analyzed. The Kruskal–Wallis test, followed by Dunn's post hoc test, was applied to evaluate log-transformed parasite loads across different vertebrate blood meal sources, as well as single and mixed *T. cruzi lineages* identified in triatomine vectors and their evolutionary stages. McNemar's test was used to compare the two detection methods (morphological and molecular) for *T. cruzi*.

3. Results

3.1. *T. sordida* Captures and *T. cruzi* Infection Rates

Triatomines were found in 140 (48%) out of 294 domiciles investigated. A total of 1861 specimens were captured across the five localities, with 84 (5%) found indoors and 1777 (95%) in peridomestic environments. Specifically, 1485 (84%) were captured in chicken coops, 114 (6%) in pigsties, 88 (5%) in wood piles, 61 (3%) in tile piles, 14 (0.8%) in bricks and tiles, nine (0.5%) in brick piles, and six (0.3%) in corrals. Colonies of *T. sordida* of varying sizes were recorded, some exceeding 400 specimens. Although stage V nymphs and males predominated, triatomines of all developmental stages were captured, including indoors (Table 1).

Table 1. Number of nymphs (N1–N5) and adults (F, M) of *Triatoma sordida* collected at different locations in the rural region of Espinosa, MG, Brazil (2018–2019), and their infection by *Trypanosoma cruzi* and the respective infection rates (%).

	N1			N2			N3			N4			N5			F			M		
Ecotope	C	E	I	C	E	I	C	E	I	C	E	I	C	E	I	C	E	I	C	E	I
Indoor	18	1	0	12	4	0	20	17	1 (5.9)	17	15	0	12	12	1 (8.3)	3	2	0	2	2	0
Chicken coop	58	1	0	146	31	2 (6.4)	123	111	15 (13.5)	160	153	13 (8.5)	468	442	87 (19.7)	238	190	25 (13.1)	292	225	26 (11.5)
Pigsty	1	1	1 (100)	10	4	1 (25)	8	5	3 (60)	12	10	1 (10)	60	60	14 (23.3)	11	11	1 (9)	12	12	3 (25)
Wood pile	5	2	0	10	2	1 (50)	15	10	1 (10)	24	24	3 (12.5)	27	26	5 (19.2)	5	4	0	2	2	0
Tile pile	1	0	0	3	0	0	5	3	0	4	4	1 (25)	22	21	2 (9.5)	8	7	0	18	18	0
Brick pile	0	0	0	0	0	0	1	1	0	1	1	1 (100)	2	2	0	2	2	1 (50)	3	3	1 (33.3)
Piles of tile/brick	0	0	0	5	0	0	0	0	0	4	3	0	4	4	0	1	1	0	0	0	0
Corral	0	0	0	0	0	0	1	1	0	0	0	0	2	2	0	1	1	0	2	2	0
Total	83	5	1 (20)	186	41	4 (9.7)	173	148	20 (13.5)	222	210	19 (9)	597	569	109 (19.1)	269	218	27 (12.4)	331	264	30 (13.4)

C = number of captured insects; E = number of examined insects by optical microscope and/or PCR; I, (%) = number and percentages of triatomines infected by *T. cruzi*; N1–N5 = first to fifth nymphal instar; F = females; M = males.

Overall, 1455 *T. sordida* were examined by an optical microscope and/or kDNA-PCR, with 210 (14.4%) found to be infected by *T. cruzi* (Table 2). Using only an optical microscope, 1357 fecal samples were examined, with 56 (4%) testing positive for *T. cruzi*. Using the kDNA-PCR technique alone, 985 samples were analyzed, with 201 (20%) showing the expected bands for *T. cruzi*. A comparison of 887 samples analyzed by both an optical microscope and kDNA-PCR revealed that 47 (5%) and 168 (19%) were positive for *T. cruzi*, respectively. McNemar's test indicated a significant difference between the results of these two methods (χ^2 McNemar = 119, p-value < 0.0001).

Table 2. Number of *Triatoma sordida* specimens collected at different ecotopes in the rural region of Espinosa, MG, Brazil (2018–2019), and their infection by *Trypanosoma cruzi* and the respective infection rates (%).

Ecotopes	E	I
Indoor	53	2 (3.8)
Chicken coop	1153	168 (14.6)
Pigsty	103	24 (23.3)
Wood pile	70	10 (14.3)
Tile pile	53	3 (5.7)
Brick pile	9	3 (33.3)
Piles of tile/brick	8	0
Corral	6	0
Total	1.455	210 (14.4)

E = number of examined insects by MO and/or PCR; I (%) = number and percentages of triatomines infected by *T. cruzi*.

Considering the capture environment and ecotopes, the largest infection index was observed in triatomines collected in piles of bricks, followed by those found in pigsties, chicken coops, wood piles, tile piles, and indoors (Table 2).

3.2. *T. cruzi* Genotyping and Detection of Mixed Infections

Out of 201 kDNA-PCR-positive samples of *T. sordida*, we identified the parasite genotype in 143 (71%) samples. The most prevalent DTU was TcI (80%), followed by TcII (8%), TcV (5%), and TcIII (3%). Additionally, we detected mixed infections of TcI and TcII and TcIII and TcV in five (4%) bugs (Table 3). Regarding the DTUs and the capture ecotopes, all DTUs were found in pigsties and chicken coops, which can be attributed to the greater number of samples analyzed from these ecotopes (Table 3).

Table 3. Distribution of *T. cruzi* DTUs and mixed infections in *T. sordida* specimens according to capture ecotopes in the studied areas of the rural region of Espinosa, MG, Brazil.

	TcI	TcII	TcIII	TcV	TcI + TcII	TcI + TcIII	TcI + TcV
Indoor	1	0	0	0	0	0	0
Chicken coop	98	3	3	6	2	1	1
Pigsty	7	9	1	1	0	0	0
Wood pile	6	0	0	0	0	0	0
Tile pile	0	0	1	0	0	0	0
Brick pile	2	0	0	0	0	1	0
Total	114	12	5	7	2	2	1

3.3. *T. cruzi* Parasite Load in *T. sordida* by Real-Time PCR (qPCR)

The quantification of the *T. cruzi* parasite load was performed on 143 *T. sordida* specimens infected with *T. cruzi*, achieving successful quantification in 68 specimens. Of these, one (1%) was in the N2 stage, seven (10%) in N3, three (4%) in N4, 39 (57%) in N5, and 11 (16%) were males and seven (10%) were females. The dynamic range for the *T. cruzi* load quantification in *T. sordida* intestine samples extends from 10^4 to one parasite equivalent, and from 0.0001 to one intestine equivalent. This dynamic range provided linear quantification over a four-log range for *T. cruzi* and six-log range for triatomines, allowing for an accurate standardization of parasite loads. The linearity coefficients (R^2) were 0.99 for both targets. PCR efficiencies were 86.5% for the *T. cruzi* nuclear satellite DNA target and 83.2% for the triatomine 12S ribosomal subunit target, confirming the enhanced performance of the assay. The observed median value for the parasitic load was 10 *T. cruzi*/intestine equivalent (Figure 1).

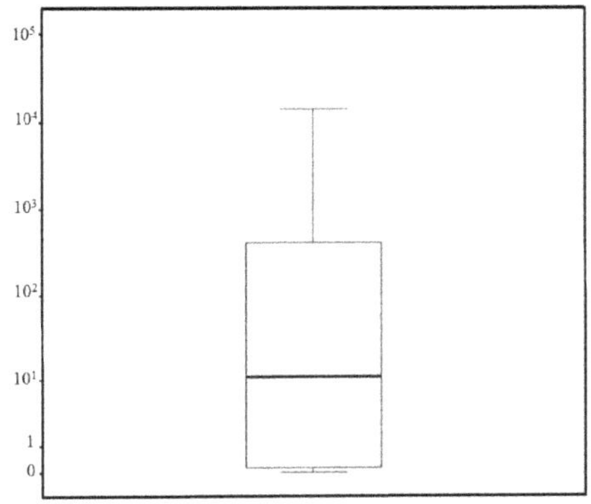

Figure 1. Boxplots of *T. cruzi* parasite load in *T. sordida* captured in the rural region of Espinosa, MG, Brazil.

3.4. Blood Meal Identification in *T. sordida* Specimens

Out of 187 DNA samples tested for blood meal identification, 175 (94%) sequences were obtained. The majority of these were from N5 nymphal instars (62%), followed by adults (18%), N4 (8%), N3 (9%), N2 (0.6%), and N1 (0.6). Additionally, 12 sequences exhibited double peaks in the electropherograms, indicating mixed blood meals. The sequences generated from 174 *T. sordida* showed high identity (98–100%) to sequences of *Gallus gallus* (N = 141, 80%), *Rattus rattus* (N = 14, 8%), *Thrichomys apereoides* (N = 7, 4%), *Sus scrofa* (N = 6, 3%), *Wiedomys cerradensis* (N = 2, 1%), *Canis lupus familiaris* (N = 2, 1%), *Capra hircus* (N = 1, 0.6%), and *Felis catus* (N = 1, 0.6%), available in the NCBI database. However, one sequence returned to a BLAST identity of 97% with a sequence in Genbank (MN206976.1). It is likely that this insect fed on the mustelid *Galictis cuja*, since there are no 12S rRNA sequences for this mammal in the database.

We also assessed the food source for *T. sordida* in relation to the ecotopes where the specimens were captured. As expected, most triatomines that fed on *G. gallus* were collected in chicken coops, although other food sources were also detected in this ecotope. Notably, all insects fed on *R. rattus* were captured in chicken coops, as with other rodents and mammals such as *F. catus*, *W. cerradensis*, *G. cuja*, *T. apereoides*, and *S. scrofa*. Specimens captured indoors had fed primarily on *G. gallus* (Figure 2).

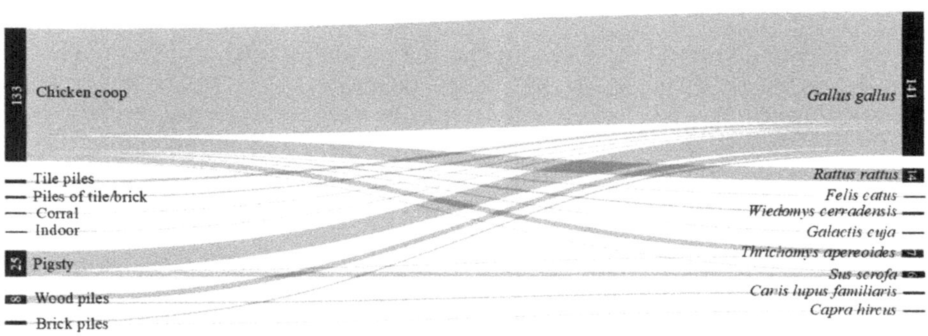

Figure 2. Sankey diagram illustrating the relationships between ecotopes and identified blood meal sources in *T. sordida* collected from rural areas of Espinosa, MG, Brazil.

3.5. Occurrence of DTUs in *T. sordida* After Feeding on Different Hosts

The most common blood meal source in the subsample with determined DTUs was *G. gallus*, accounting for 80%, followed by *R. rattus* (8.9%), *T. apereoides* and *S. scrofa* (3.3% each), *C. lupus familiaris* (2.2%), and *F. catus* and *C. hircus* (1.1% each) (Table 4). For the three *T. sordida* that fed on *W. cerradensis* (N = 2) and *G. cuja* (N = 1), the infecting DTU could not be identified.

Table 4. Number of infections of *T. sordida* with different DTUs after feeding on different hosts.

	TcI	TcII	TcIII	TcV	TcI + TcII	TcI + TcIII
Gallus gallus	48	10	5	6	1	2
Thrichomys apereoides	1	1	0	1	0	0
Capra hircus	1	0	0	0	0	0
Felis catus	1	0	0	0	0	0
Rattus rattus	8	0	0	0	0	0
Sus scrofa	3	0	0	0	0	0
Canis lupus familiaris	2	0	0	0	0	0
Total	64	11	5	7	1	2

3.6. Correlation of DTUs and the Parasite Load in T. sordida

Quantification assays using qPCR revealed parasite loads in 47 *T. sordida* infected with TcI, nine infected with TcII, two infected with TcIII, one infected with TcV, and three infected with mixed DTUs. Due to the small sample sizes, comparisons were only made between TcI and TcII. The highest observed parasite loads were 6.1×10^4, 1.2×10^2, and 48 for TcI, TcII, and mixed infections, respectively. The lowest parasite loads were 0.0001, 0.0004, and 0.008, respectively. Median parasite loads were 22 for TcI, 4.8 for TcII, and 5.8 *T. cruzi*/intestine equivalents for mixed infections. No significant differences were observed in *T. cruzi* load between TcI and TcII (p-value = 0.506) or for mixed infections (p-value = 0.315) (Figure 3).

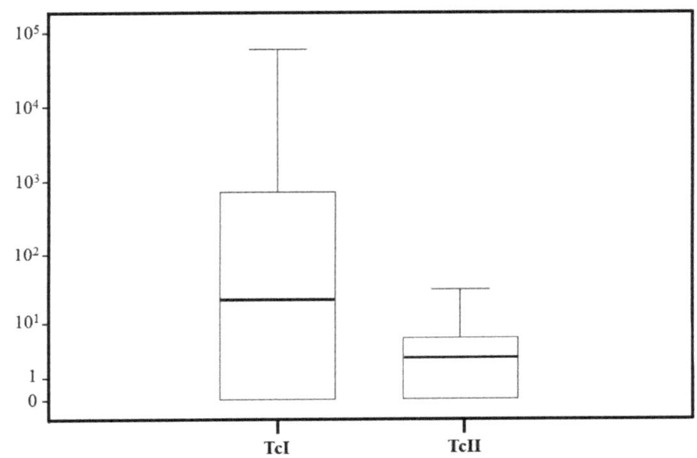

Figure 3. Boxplot comparing estimates of the *T. cruzi* parasite load according to the infecting *T. cruzi* DTU lineage in *T. sordida* from the rural region of Espinosa, MG, Brazil.

3.7. T. cruzi Parasite Load According to the Food Source

A total of 43 triatomines that fed on *G. gallus*, seven on *R. rattus*, four on *Sus scrofa*, two on *C. lupus familiaris*, and one on *T. apereoides* were successfully quantified for parasite load. For statistical analyses, we compared *T. cruzi* parasite loads between *T. sordida* fed on *G. gallus*, *R. rattus*, and *S. scrofa* due to the sample size limitations. Insects fed on *R. rattus* exhibited significantly higher *T. cruzi* loads compared to those fed on *G. gallus* ($p = 0.0007$) (Figure 4). No significant differences were observed in comparisons between *G. gallus* and *S. scrofa* or *S. scrofa* and *R. rattus*. The median *T. cruzi* loads were 0.6, 1.2×10^3, and 4×10^2 parasites/intestine equivalents for *T. sordida* fed on *G. gallus*, *R. rattus*, and *S. scrofa*, respectively (Figure 4).

3.8. T. cruzi Parasite Load According to Triatomine Stage

T. sordida females had significantly higher *T. cruzi* loads compared to N3 (p-value = 0.012), N4 (p-value = 0.018), and N5 (p-value = 0.024) nymphal instars and males (p-value = 0.014). The median values for parasitic loads were 1.8, 0.008, 5.8, 2.8, and 8.9×10^2 *T. cruzi* per intestine equivalents for N3, N4, N5, males, and females, respectively (Figure 5).

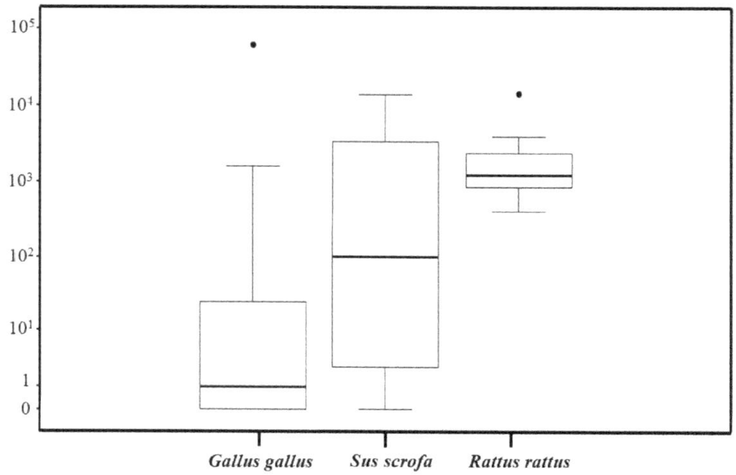

Figure 4. Boxplot comparing estimates of the *T. cruzi* parasite load according to the blood meal sources identified in *T. sordida* from the rural region of Espinosa, MG, Brazil.

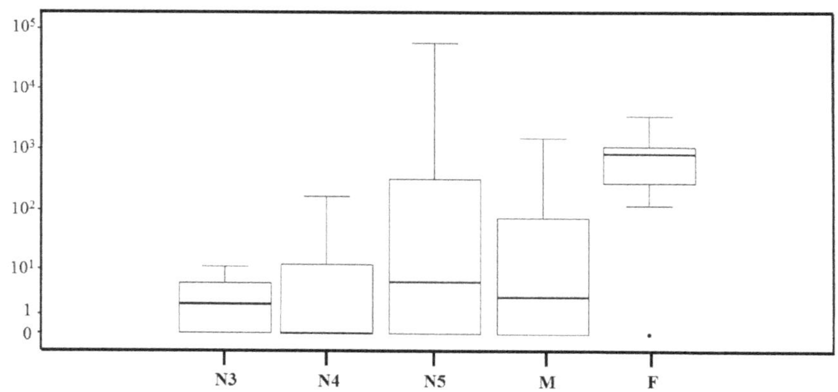

Figure 5. Boxplot comparing the estimate of the *T. cruzi* parasite load according to the development stage of *T. sordida* from the rural region of Espinosa, MG, Brazil.

4. Discussion

Our study provides novel insights into the infections of *T. sordida* by *T. cruzi* in an endemic area of Minas Gerais, Brazil, underscoring the significance of this vector in the peridomestic transmission cycle of the parasite. The principal findings of this study are as follows: (i) a high natural infection rate of *T. cruzi* in *T. sordida*; (ii) a relatively low quantified parasitic load of *T. cruzi*; (iii) specimens that fed on *R. rattus* exhibited the highest *T. cruzi* load per intestine, whereas those that fed on *G. gallus* showed the lowest; (iv) a notable diversity of *T. cruzi* DTU and identified blood meals; (v) no significant difference in *T. cruzi* load between insects infected with TcI and TcII; (vi) females exhibited the highest parasitic load compared to other development stages.

Although *T. sordida* is the most frequently captured triatomine species in Brazil, including in the northern Minas Gerais state, few studies have focused on its natural *T. cruzi* infection rate [13,16,42]. Moreover, due to the low *T. cruzi* infection rate, this vector has been considered to have a low potential for sustaining high rates of *T. cruzi* transmission [42–44]. However, our findings present a different scenario, particularly when considering results obtained using molecular methods. The significant difference between *T. cruzi* detection techniques (optical microscope and kDNA-PCR) was anticipated, as extensively reported in the literature, with parasitic load being a critical factor for accurate diagnosis [39,45,46]. According to Valença-Barbosa et al. [31], triatomine intestinal samples with 10^3 *T. cruzi* per intestine have a 30% chance of being detected as positive by an optical microscope. Considering that the *T. sordida* population examined in our study exhibited a low parasitic load, with a median value of 10 parasites/intestine, false negative results by optical microscope are expected. It is well known that this method tends to underestimate detection rates [39,45,46]. Additionally, it is important to note previous studies evaluating *T. cruzi* infection in this species employed the optical microscope technique, which may explain the lower infection rates reported compared to our findings.

To date, no study has evaluated or defined "high" or "low" parasitic load of *T. cruzi* in different triatomine species and their implications for vector transmission. A few studies have reported a wide range of parasitic load values from 3.02×10^5 to 1.29×10^9, 0.13 to 2.2×10^{10}, and 3.94 to 7.66×10^6 *T. cruzi* per intestine for *Mepraia spinolai* (Chile) [47], *Triatoma melanica* (Brazil) [31], and *Triatoma brasiliensis* (Brazil) [30], respectively. *T. melanica* and *T. sordida* are distributed throughout the northwest of the Minas Gerais state, along the border between Bahia and Minas Gerais, in a transition area between the Caatinga and Cerrado biomes [48]. However, these species occupy different ecotopes. *T. melanica*, which is exclusively sylvatic, is associated with rocky crevices in the Espinhaço mountain range, whereas *T. sordida* in Brazil is arboreal, primarily found under tree bark, and is frequently captured inside and around human dwellings [16].

Our *T. sordida* population exhibited a *T. cruzi* load 10 times lower than that reported for *M. spinolai*. This difference in parasitic load may be attributed to the preferred food sources of these triatomine species. Approximately 84% of *T. sordida* were captured in chicken coops and, consequently, 80% fed on chicken, highlighting their feeding preference for birds, as previously reported [16,49,50]. Chickens are refractory to *T. cruzi* infection, as their serum is resistant to trypomastigotes [51,52]. However, chickens play an important role in the survival and maintenance of triatomine populations already infected by *T. cruzi* [28].

T. sordida is considered a ubiquitous species with a broad ecological range, occupying various ecotopes and feeding on diverse food sources, including rodents and other mammals. In this study, the majority of insects (96%) fed on rodents (*R. rattus*, *W. cerradensis*, and *T. apereoides*) even when captured in chicken coops. These mammals often invade peridomestic areas [53] attracted by chicken food, which may help explain the *T. cruzi* infection rate in this triatomine population. Infestations of *T. sordida* in peridomestic structures in Minas Gerais have frequently been associated with chicken coops, as well as corrals and wood piles [16,43,44], as observed in the present study.

We identified seven different mammals as food sources, including wild and synanthropic rodents, domestic and farm animals (cats, dogs, pigs and goats), and another wild mammal, the lessor grison. Previously, our group reported the presence of food content in *T. melanica*, resembling the mustelid *G. cuja* in the same study area [31]. However, no 12S rRNA sequences for this mammal are available in the database. Most triatomine species are capable of feeding on the blood of a wide diversity of vertebrates, both wild and domestic, which contributes to their adaptation to various habitats [54]. Regarding domestic animals, particularly dogs and cats, they are key reservoir hosts of *T. cruzi*, bridging the gap between the domestic and wild transmission cycles of the parasite [55–57]. Both cats and dogs are epidemiologically important sources of infection for bugs and householders. The movement of domestic cats into wild environments due to human activities underscores the connection between domestic and wild environments [58]. *T. sordida* uses similar native mammals as food sources, much like *T. melanica*, including *T. apereoides*, *W. cerradensis*, and *G. cuja* [31], suggesting intense movement of these mammals between areas. Also noted was this movement of biological elements, including *T. cruzi* reservoirs, driven by anthropogenic actions, which facilitates the reciprocal (sylvatic-peridomestic) circulation of the parasite [58]. For wild mammals, such as rodents, their epidemiological importance as reservoirs of *T. cruzi* is well known [59]. Additionally, all these food sources were identified in specimens from various peridomestic ecotopes, indicating the diversity and intense circulation of animals, which act as a link nature and human-made environments.

Moreover, some *T. sordida* specimens that fed on *R. rattus* exhibited the highest parasitic loads, emphasizing the significant role of these mammals in the transmission cycles of *T. cruzi*. This rodent has been previously identified as an important domiciliary and peridomiciliary reservoir for the parasite [60,61]. As a synanthropic species, *R. rattus* occupies both natural and artificial habitats, facilitating the introduction of *T. cruzi* into transmission cycles near human dwellings. The parasitemia of mammalian hosts is a crucial factor in vector infection [29]. It is important to note that directly associating parasitic load with parasitemia in animals remains challenging due to various influencing factors [28]. However, the dominance of a particular blood source can influence the dynamics of *T. cruzi* transmission cycles [62]. In our study, bugs that fed on chicken exhibited the lowest parasitic loads, highlighting their ability to feed on multiple hosts. While birds do not participate in *T. cruzi* transmission, once triatomines feed on an infected mammal, the infection can persist for a long time [63]. Furthermore, in this study, humans were not identified as a food source for this triatomine species. A plausible explanation for this absence is that most of the insects analyzed were collected in peridomestic ecotopes, where an abundant animal food source is available, while humans are only transient in this environment. Moreover, *T. sordida* is considered a predominantly peridomestic species [18].

This report provides a comprehensive characterization of *T. cruzi* genotypes in *T. sordida* from Brazil. We observed a great diversity of DTU in this triatomine population. However, the parasitic load was not significantly affected by genotypes. TcI was the most prevalent DTU, followed by TcII, TcV, and TcIII.

TcI is the most prevalent genotype in the Americas, predominantly associated with hosts from the orders Didelphimorphia and Rodentia [64]. Moreover, TcI has already been reported infecting *T. sordida* individuals that invade houses from sylvatic environments [24]. Interestingly, we did not detect marsupials as food sources but the eight *R. rattus*, an exotic species whose occurrence is restricted to the artificial environment, identified as a food source for *T. sordida* specimens were infected by TcI. Furthermore, one insect that fed on *T. apereoides* was infected with TcI, while the other two bugs that fed on this rodent were infected with TcII and TcV. Both DTUs were identified much more rarely, with TcII being present in a higher proportion compared to TcV [64]. These DTUs are more often associated with domestic cycles in South America [64,65], though their natural hosts remain poorly defined. In Bolivia, TcII and TcV are the predominant genotypes detected in Rodentia [64]. Regarding the TcIII, this DTU was reported in *T. sordida* from peridomestic environments in Bahia, Brazil, being consistent with our findings [66]. It is important to highlight that TcIII may be underreported in both domestic and wild cycles, because some typing methodologies fail to distinguish between TcIII and TcIV [67].

We observed a significant difference in parasitic load across developmental stages, with females exhibiting a higher *T. cruzi* load compared to N3, N4, N5 instars, and males. Females of *T. melanica* from Espinosa, MG, also exhibited the highest parasitic load when compared to nymphs [31]. The ovarian development and egg production in triatomines are highly dependent on blood meals [68]. *T. sordida* inhabits less stable ecotopes and utilizes a wide range of food sources and allocates more energy to reproduction than maintenance [16,69]. This increased energy investment in reproduction likely elevates the probability of using several food sources, which may contribute to the higher parasitic load in females. However, to prove this correlation, it is necessary to design and develop experimental infection studies under laboratory conditions.

It is important to highlight that *T. sordida* defecates during blood meals, demonstrating its strong vector competence for Chagas disease [70]. Although the parasitic load detected in *T. sordida* was not the highest compared to other triatomine species investigated so far, our findings indicate that this species deserves greater attention due to its high *T. cruzi* infection rate. Furthermore, the diversity of blood meal sources and *T. cruzi* DTUs underscores the intense circulation of different hosts in peridomestic ecotopes, facilitating the overlap between natural and artificial environments. In this regard, the findings of this study have significant implications for vector control, indicating the need to maintain control and surveillance measures, considering the local dynamics of vectors and hosts.

5. Conclusions

The high rates of natural infection by *T. cruzi* observed in *T. sordida*, combined with frequent reports of its prevalence in peridomestic environments, underscore the epidemiological importance of this vector. Despite the parasitic load of *T. sordida* being lower than that of other species previously quantified, this finding emphasizes the underdetection of positive specimens by optical microscope, the commonly employed technique in routine entomological surveillance. This study also confirms the predominance of chickens as food sources and their role in the epidemiological scenario of *T. cruzi* transmission. Although chickens are considered refractory to the parasite, they serve as food sources, sustaining previously infected insects. Wild and synanthropic rodents, along with other mammals identified as food sources, frequently invade peridomestic structures, especially chicken coops, attracted by chicken feed and droppings. Consequently, these animals act as sources of *T. cruzi* infection for the *T. sordida* population. Furthermore, the great diversity of *T. cruzi* DTUs detected in triatomine specimens captured in artificial ecotopes is likely due to the intense circulation of different hosts in wild and domestic habitats, as well as triatomine migration, leading to overlapping DTU transmission cycles.

Author Contributions: Conceptualization, R.d.C.M.d.S., L.D. and C.V.-B.; collection of triatomines, O.C.C.N., R.d.C.M.d.S. and L.D.; *T. cruzi* parasitological diagnosis, I.M.d.A. and R.d.C.M.d.S.; isolation of *T. cruzi*, I.M.d.A.; DNA extraction, C.V.-B., F.D.T.d.S. and C.F.C.; molecular biology, C.V.-B., F.D.T.d.S., C.F.C., N.A.d.S. and B.O.B.M.; formal analysis, C.V.-B., P.F.-A., M.V.N.A., O.d.C.M., L.D. and R.d.C.M.d.S.; validation, all authors; resources, R.d.C.M.d.S. and L.D.; data curation, C.V.-B. and R.d.C.M.d.S.; writing—original draft preparation, C.V.-B.; writing—review and editing, all authors; project administration, R.d.C.M.d.S. and L.D.; supervision, R.d.C.M.d.S. and L.D.; funding acquisition, R.d.C.M.d.S. and L.D. All authors have read and agreed to the published version of the manuscript.

Funding: This research was funded by Fundação de Amparo à Pesquisa do Estado de Minas Gerais (FAPEMIG–APQ-02284-18) and the project "Access to the detection and treatment of CD in the context of primary healthcare in Brazil—IntegraChagas" (VPGDI-017-FIO-20). C.V.B. received postdoctoral fellowships from Conselho Nacional de Desenvolvimento Científico e Tecnológico (152386/2020-3).

Institutional Review Board Statement: Not applicable.

Informed Consent Statement: Not applicable.

Data Availability Statement: Data are included within the article. Sequence data have been deposited in GenBank.

Acknowledgments: The authors would like to thank the Platform for DNA Sequencing Sanger (RPT01E) at the Instituto René Rachou/FIOCRUZ-MG and the Platform for Real-Time PCR (RPT09A) of FIOCRUZ-RJ for their support and the use of their equipment, as well as the PCDCh health workers for their help in capturing the triatomines, especially Bartolomeu Teixeira Lopes.

Conflicts of Interest: The authors declare no conflicts of interest.

References

1. World Health Organization. Available online: https://www.who.int/news-room/fact-sheets/detail/chagas-disease-(american-trypanosomiasis) (accessed on 22 January 2025).
2. Dias, J.C.P. Southern Cone Initiative for the elimination of domestic populations of *Triatoma infestans* and the interruption of transfusion Chagas disease: Historical aspects, present situation, and perspectives. *Memórias Inst. Oswaldo Cruz* **2007**, *102*, 11–18. [CrossRef] [PubMed]
3. Hashimoto, K.; Schofield, C.J. Elimination of *Rhodnius prolixus* in central America. *Parasites Vectors* **2012**, *5*, 45. [CrossRef]
4. Monteiro, F.A.; Weirauch, C.; Felix, M.; Lazoski, C.; Abad-Franch, F. Evolution, systematics, and biogeography of the Triatominae, vectors of Chagas disease. *Adv. Parasitol.* **2018**, *99*, 265–344. [PubMed]
5. Téllez-Rendón, J.; Esteban, L.; Rengifo-Correa, L.; Diaz-Albiter, H.; Huerta, H.; Dale, C. *Triatoma yelapensis* sp. nov. (Hemiptera: Reduviidae) from Mexico, with a key of *Triatoma* species recorded in Mexico. *Insects* **2023**, *14*, 331. [CrossRef] [PubMed]
6. Zhao, Y.; Fan, M.; Li, H.; Cai, W. Review of kissing bugs from China with descriptions of two new species. *Insects* **2023**, *14*, 450. [CrossRef]
7. Coura, J.C. The main sceneries of Chagas disease transmission. The vectors, blood and oral transmissions—A comprehensive review. *Memórias Inst. Oswaldo Cruz* **2014**, *110*, 277–282. [CrossRef]
8. Gurgel-Gonçalves, R.; Komp, E.; Campbell, L.P.; Khalighifar, A.; Mellenbruch, J.; Mendonça, V.J.; Owens, H.L.; Felix, K.D.L.C.; Peterson, A.T.; Ramsey, J.M. Automated identification of insect vectors of Chagas disease in Brazil and Mexico: The Virtual Vector Lab. *PeerJ* **2017**, *5*, e3040. [CrossRef] [PubMed]
9. Stal, C. Monographie der *Gattung conorhinus* und *Verwandten. Berl. Entomol. Z.* **1859**, *3*, 99–117. [CrossRef]
10. Schofield, C.J.; Galvão, C. Classification, evolution, and species groups within the Triatominae. *Acta Trop.* **2009**, *110*, 88–100. [CrossRef]
11. Panzera, F.; Pita, S.; Nattero, J.; Panzera, Y.; Galvão, C.; Chavez, T.; Arias, A.R.; Téllez, L.C.; Noireau, F. Cryptic speciation in the *Triatoma sordida* subcomplex (Hemiptera, Reduviidae) revealed by chromosomal markers. *Parasites Vectors* **2015**, *8*, 495–504. [CrossRef]
12. Garcia, A.C.C.; Oliveira, J.; Cristal, D.C.; Delgado, L.M.G.; Bittinelli, I.F.; Galvão, C.; Britez, N.E.G.; Carrasco, H.J.; Rosa, J.A.; Alevi, K.C.C. Intraspecific and interspecific phenotypic differences confirm the absence of cryptic speciation in *Triatoma sordida* (Hemiptera, Triatominae). *Am. J. Trop. Med. Hyg.* **2020**, *105*, 1759–1766. [CrossRef] [PubMed]
13. Diotaiuti, L.; Azeredo, B.V.M.; Busek, S.C.U.; Fernandes, A.J. Controle do *Triatoma sordida* no peridomicílio rural do município de Porteirinha, Minas Gerais, Brasil. *Rev. Panam. Salud Pública* **1998**, *3*, 21–25. [CrossRef] [PubMed]
14. Gurgel-Gonçalves, R.; Galvão, C.; Costa, J.; Peterson, A.T. Geographic distribution of Chagas disease vectors in Brazil based on ecological niche modeling. *J. Trop. Med.* **2012**, *1*, 705326. [CrossRef] [PubMed]
15. Ribeiro, A.R.; Mendonça, V.J.; Alves, R.T.; Martinez, I.; Araújo, R.F.; Mello, F.; Rosa, J.A. *Trypanosoma cruzi* strains from triatomine collected in Bahia and Rio Grande do Sul, Brazil. *Rev. Saúde Pública* **2014**, *48*, 295–302. [CrossRef]
16. Diotaiuti, L.; Loiola, C.F.; Falcão, P.L.; Dias, J.C.P. The ecology of *Triatoma sordida* in natural environments in two different regions of the state of Minas Gerais, Brazil. *Rev. Inst. Med. Trop. São Paulo* **1993**, *35*, 237–245. [CrossRef] [PubMed]
17. Barretto, M.P.; Carvalheiro, J.R. Estudos sobre reservatórios e vetores silvestres do *Trypanosoma cruzi.* XXVIII: Sobre o encontro de *Triatoma sordida* Stal, 1859 e de *Rhodnius neglectus* Lent, 1954 em ninhos de pássaros da família Furnariidae (Hemiptera, Reduviidae). *Rev. Bras. Biol.* **1968**, *28*, 289–293.
18. Diotaiuti, L.; Paula, O.R.D.; Falcão, L.P.; Dias, P.J. Avaliação do programa de controle vetorial da doença de Chagas em Minas Gerais, Brasil, com referência especial ao *Triatoma sordida. Boletín Oficina Sanit. Panam.* **1995**, *3*, 118.
19. Abad-Franch, F.; Diotaiuti, L.; Gurgel-Gonçalves, R.; Gürtler, R.E. Certifying the interruption of Chagas disease transmission by native vectors: Cui bono? *Memórias Inst. Oswaldo Cruz* **2013**, *108*, 251–254. [CrossRef] [PubMed]
20. Barreto, M.P. Studies on wild reservoirs and vectors of "*Trypanosoma cruzi*". XXI: Observations on the association between reservoirs and vectors, with special reference to the northeast region of the state of São Paulo. *Rev. Bras. Biol.* **1968**, *28*, 481–494.
21. Praça, Y.R.P.; Santiago, P.B.; Charneau, S.; Mandacaru, S.C.; Bastos, I.M.D.; Bentes, K.L.S.; Silva, S.M.M.; Silva, W.M.C.; Silva, I.G.; Sousa, M.V.; et al. An integrative sialomic analysis reveals molecules from *Triatoma sordida* (Hemiptera: Reduviidae). *Front. Cell. Infect. Microbiol.* **2022**, *11*, 798924. [CrossRef]
22. Coura, J.R.; Viñas, P.A.; Junqueira, A.C.V. Ecoepidemiology, short history and control of Chagas disease in the endemic countries and the new challenge for non-endemic countries. *Memórias Inst. Oswaldo Cruz* **2014**, *109*, 856–862. [CrossRef] [PubMed]

23. Barbosa-Silva, A.N.; Câmara, A.C.J.; Martins, K.; Nunes, D.F.; Oliveira, P.I.C.; Azevedo, P.R.M.; Chiari, E.; Galvão, L.M.C. Characteristics of triatomine infestation and natural *Trypanosoma cruzi* infection in the State of Rio Grande do Norte, Brazil. *Rev. Soc. Bras. Med. Trop.* **2016**, *49*, 57–67. [CrossRef] [PubMed]
24. Diotaiuti, L.; Pereira, A.S.; Loiola, C.F.; Fernandes, A.J.; Schofield, J.C.; Dujardin, J.P.; Dias, J.C.P.; Chiari, E. Inter-relation of sylvatic and domestic transmission of *Trypanosoma cruzi* in areas with and without domestic vectorial transmission in Minas Gerais, Brazil. *Memórias Inst. Oswaldo Cruz* **1995**, *90*, 443–448. [CrossRef]
25. Cardozo, M.; Fiad, F.G.; Crocco, L.B.; Gorla, D.E. Effect of habitat fragmentation on rural house invasion by sylvatic triatomines: A multiple landscape-scale approach. *PLoS Neglected Trop. Dis.* **2021**, *15*, e0009579. [CrossRef]
26. Zingales, B.; Miles, M.A.; Campbell, D.A.; Tibayrenc, M.; Macedo, A.M.; Teixeira, M.M.G.; Schijman, A.G.; Llewellyn, M.S.; Lages-Silva, E.; Machado, C.R.; et al. The revised *Trypanosoma cruzi* subspecific nomenclature: Rationale, epidemiological relevance and research applications. *Infect. Genet. Evol.* **2012**, *12*, 240–253. [CrossRef]
27. Fernandes, O.; Mangia, R.H.; Lisboa, C.V.; Pinho, A.P.; Morel, C.M.; Zingales, B.; Campbell, D.A.; Jansen, A.M. The complexity of the sylvatic cycle of *Trypanosoma cruzi* in Rio de Janeiro state (Brazil) revealed by the non-transcribed spacer of the mini-exon gene. *Parasitology* **1999**, *118*, 161–166. [CrossRef]
28. Jansen, A.M.; Xavier, S.C.C.; Roque, A.L.R. *Trypanosoma cruzi* transmission in the wild and its most important reservoir hosts in Brazil. *Parasites Vectors* **2018**, *11*, 502. [CrossRef] [PubMed]
29. Jansen, A.M.; Xavier, S.C.C.; Roque, A.L.R. Landmarks of the knowledge and *Trypanosoma cruzi* biology in the wild environment. *Front. Cell. Infect. Microbiol.* **2020**, *10*, 10. [CrossRef]
30. Valença-Barbosa, C.; Finamore-Araujo, P.; Moreira, O.C.; Vergara-Meza, J.G.; Alvarez, M.V.N.; Nascimento, J.R.; Borges-Veloso, A.; Viana, M.C.; Lilioso, M.; Miguel, D.C.; et al. Genotypic *Trypanosoma cruzi* distribution and parasite load differ ecotypically and according to parasite genotypes in *Triatoma brasiliensis* from endemic and outbreak areas in Northeastern Brazil. *Acta Trop.* **2021**, *222*, 106054. [CrossRef] [PubMed]
31. Valença-Barbosa, C.; Finamore-Araujo, P.; Moreira, O.C.; Alvarez, M.V.N.; Borges-Veloso, A.; Barbosa, S.E.; Diotaiuti, L.; Souza, R.C.M. High parasitic loads quantified in sylvatic *Triatoma melanica*, a Chagas disease vector. *Pathogens* **2022**, *11*, 1498. [CrossRef]
32. Osae, M.; Vezenegho, S.; Spillings, B.; Koekemoer, L. Optimization and validation of a multiplex PCR for identification of mammalian blood meals in malaria vector mosquitoes and time-cost comparison between the PCR and ELISA methods. *Commun. Dis. Surveill. Bull.* **2013**, *11*, 31–34.
33. Arcoverde, W.M.; Lima, J.T.F.; Tauil, P.L.; Silveira, A.C. *Manual de Normas Técnicas da Campanha de Controle da Doença de Chagas*; Ministério da Saúde Brasília: Brasíli, Brazil, 1980.
34. Chiari, E.; Camargo, E.P. Culture and cloning of *Trypanosoma cruzi*. In *Genes and Antigens of Parasites, a Laboratory Manual*; Morel, C.M., Ed.; Fundação Oswaldo Cruz: Rio de Janeiro, Brazil; World Health Organization: Geneva, Switzerland, 1984; pp. 23–26.
35. Sturm, N.R.; Degrave, W.; Morel, C.; Simpson, L. Sensitive detection and schizodeme classification of *Trypanosoma cruzi* cells by amplification of kinetoplast minicircle DNA sequences: Use in diagnosis of Chagas' disease. *Mol. Biochem. Parasitol.* **1989**, *33*, 205–214. [CrossRef]
36. Wincker, P.; Britto, C.; Pereira, J.B.; Cardoso, M.A.; Oelemann, W.; Morel, C.M. Use of a simplified polymerase chain reaction procedure to detect *Trypanosoma cruzi* in blood samples from chronic chagasic patients in a rural endemic area. *Am. J. Trop. Med. Hyg.* **1994**, *51*, 771–777. [CrossRef] [PubMed]
37. Moreira, O.C.; Ramirez, J.C. Genotyping of *Trypanosoma cruzi* from Clinical Samples by Multilocus Conventional PCR. *Methods Mol. Biol.* **2019**, *1955*, 227–238.
38. Moreira, O.C.; Verly, T.; Finamore-Araujo, P.; Gomes, S.A.O.; Lopes, C.M.; Sousa, D.M.; Azevedo, L.R.; Mota, F.F.; D'Avila-Levy, C.M.; Santos-Mallet, J.R.; et al. Development of conventional and real-time multiplex PCR-based assays for estimation of natural infection rates and *Trypanosoma cruzi* load in triatomine vectors. *Parasites Vectors* **2017**, *10*, 404. [CrossRef] [PubMed]
39. Piron, M.; Fisa, R.; Casamitjana, N.; López-Chejade, P.; Puig, L.; Vergés, M.; Gascón, J.; Prat, J.G.; Portús, M.; Sauleda, S. Development of a real-time PCR assay for *Trypanosoma cruzi* detection in blood samples. *Acta Trop.* **2007**, *103*, 195–200. [CrossRef]
40. Kitano, T.; Umetsu, K.; Tian, W.; Osawa, M. Two universal primer sets for species identification among vertebrates. *Int. J. Leg. Med.* **2007**, *21*, 423–427. [CrossRef] [PubMed]
41. The R Project for Statistical Computing. Available online: https://www.r-project.org (accessed on 22 January 2025).
42. Silveira, A.C.; Dias, J.C.P. O controle da transmissão vetorial. *Rev. Soc. Bras. Med. Trop.* **2011**, *44*, 52–63. [CrossRef]
43. Rossi, J.C.N.; Duarte, E.C.; Gurgel-Gonçalves, R. Factors associated with the occurrence of *Triatoma sordida* (Hemiptera: Reduviidae) in rural localities of Central-West Brazil. *Memórias Inst. Oswaldo Cruz* **2015**, *110*, 192–200. [CrossRef]
44. Vianna, E.N.; Guimarães, R.J.P.S.; Souza, C.R.; Gorla, D.; Diotaiuti, L. Chagas disease ecoepidemiology and environmental changes in northern Minas Gerais state, Brazil. *Memórias Inst. Oswaldo Cruz* **2017**, *112*, 760–768. [CrossRef]
45. Vinhaes, M.C.; Oliveira, S.V.; Reis, P.O.; Sousa, A.C.L.; Silva, R.A.; Obara, M.T.; Bezerra, C.M.; Costa, V.M.; Alves, R.V.; Gurgel-Gonçalves, R. Assessing the vulnerability of Brazilian municipalities to the vectorial transmission of *Trypanosoma cruzi* using multi-criteria decision analysis. *Acta Trop.* **2014**, *137*, 105–110. [CrossRef] [PubMed]

46. Minuzzi-Souza, T.T.C.; Nitz, N.; Cuba, C.A.C.; Hagstrom, L.; Hecht, M.M.; Santana, C.; Ribeiro, M.; Vital, T.E.; Santalucia, M.; Know, M.; et al. Surveillance of vector-borne pathogens under imperfect detection: Lessons from Chagas disease risk (mis) measurement. *Sci. Rep.* **2018**, *8*, 151.
47. Saavedra, M.; Bacigalupo, A.; Barrera, M.V.; Vergara, M.J.; Álvarez-Duhart, B.; Martín, C.M.S.; Solís, R.; Cattan, P.E. *Trypanosoma cruzi* infection in the wild Chagas disease vector, *Mepraia spinolai*: Parasitic load, discrete typing units, and blood meal sources. *Acta Trop.* **2022**, *229*, 106365. [CrossRef] [PubMed]
48. Drummond, G.M.; Martins, C.S.; Machado, A.B.M.; Sebaio, F.A.; Antonini, Y. *Biodiversidade em Minas Gerais: Um Atlas para sua conservação*, 2nd ed.; Fundação Biodiversitas: Belo Horizonte, Brazil, 2005.
49. Forattini, O.P.; Ferreira, O.A.; Rabello, E.X.; Barata, J.M.S.; Santos, J.L.F. Aspectos ecológicos da tripanossomíase americana: XVIII-Desenvolvimento e ciclos anuais de colônias de *Triatoma infestans*, *T. sordida* e *Rhodnius neglectus* em ecótopos artificiais, no ambiente peri e extradomiciliar. *Rev. Saúde Pública* **1983**, *17*, 243–262. [CrossRef]
50. Pires, H.H.R.; Borges, E.C.; Andrade, R.E.; Lorosa, E.S.; Diotaiuti, L. Peridomiciliary infestation with *Triatoma sordida* Stal, 1859 in the county of Serra do Ramalho, Bahia, Brazil. *Memórias Inst. Oswaldo Cruz* **1999**, *94*, 147–149. [CrossRef]
51. Kierszenbaum, F.; Ivanyi, J.; Budzko, D.B. Mechanisms of natural resistance to trypanosomal infection. Role of complement in avian resistance to *Trypanosoma cruzi* infection. *Immunology* **1976**, *30*, 1.
52. Minter-Goedbloed, E.; Croon, J.J.A.B. The insusceptibility of chickens to *Trypanosoma* (*Schizotrypanum*) *cruzi*. *Trans. R. Soc. Trop. Med. Hyg.* **1981**, *75*, 350–353. [CrossRef]
53. Sarquis, O.; Borges-Pereira, J.; Cord, J.R.M.; Gomes, T.F.; Cabello, P.H.; Lima, M.M. Epidemiology of Chagas disease in Jaguaruana, Ceará, Brazil. I. Presence of triatomines and index of *Trypanosoma cruzi* infection in four localities of a rural area. *Memórias Inst. Oswaldo Cruz* **2004**, *99*, 263–270. [CrossRef]
54. Gorchakov, R.; Trosclair, L.P.; Wozniak, E.J.; Feria, P.T.; Garcia, M.N.; Gunter, S.M.; Murray, K.O. *Trypanosoma cruzi* infection prevalence and bloodmeal analysis in triatomine vectors of Chagas disease from rural peridomestic locations in Texas, 2013–2014. *J. Med. Entomol.* **2016**, *53*, 911–918. [CrossRef]
55. Gürtler, R.E.; Cecere, M.C.; Fernández, M.D.P.; Vazquez-Prokopec, G.M.; Ceballos, L.A.; Gurevitz, J.M.; Kitron, U.; Cohen, J.E. Key source habitats and potential dispersal of *Triatoma infestans* populations in northwestern Argentina: Implications for vector control. *PLoS Neglected Trop. Dis.* **2014**, *8*, e3238. [CrossRef] [PubMed]
56. Freitas, Y.B.N.; Souza, C.D.S.F.D.; Magalhães, J.M.; Sousa, M.L.R.; d'Escoffer, L.N.; Valle, T.Z.; Gonçalves, T.C.M.; Gil-Santana, H.R.; Kazimoto, T.A.; Amora, S.S.A. Natural infection by *Trypanosoma cruzi* in triatomines and seropositivity for Chagas disease of dogs in rural areas of Rio Grande do Norte, Brazil. *Rev. Soc. Bras. Med. Trop.* **2018**, *51*, 190–197. [CrossRef] [PubMed]
57. Gürtler, R.E.; Cecere, M.C.; Lauricella, M.A.; Cardinal, M.V.; Kitron, U.; Cohen, J.E. Domestic dogs and cats as sources of *Trypanosoma cruzi* infection in rural northwestern Argentina. *Parasitology* **2007**, *134 Pt 1*, 69–82. [CrossRef] [PubMed]
58. Almeida, C.E.; Máximo, M.M.; Pires-Silva, D.; Takiya, D.M.; Valença-Barbosa, C.; Viana, M.C.; Reigada, C.; Iñiguez, A.M.; Harry, M.; Folly-Ramos, E. From molecules to ecosystems: Insights into a network of interactions for a Chagas disease outbreak using *Triatoma brasiliensis* as natural samplers. *Acta Trop.* **2024**, *251*, 107107. [CrossRef] [PubMed]
59. Jansen-Franken, A.M. Reservatórios de Vida Livre. Available online: https://chagas.fiocruz.br/ecologia/reservatorios-de-vida/ (accessed on 22 January 2025).
60. Herrera, L.; Urdaneta-Morales, S. Synanthropic rodent reservoirs of *Trypanosoma* (*Schizotrypanum*) *cruzi* in the valley of Caracas, Venezuela. *Rev. Inst. Med. Trop. São Paulo* **1997**, *39*, 279–282. [CrossRef] [PubMed]
61. Cab-Romero, M.A.; Costa, A.P.; Costa, J.O.J.; Manhães, I.B.O.; Silva, R.E.; Ruiz-Piña, H.A.; Reyes-Novelo, E.A.; Escobedo-Ortegona, J.; Tonhosolo, R. Synanthropic mammals in transmission cycle of *Trypanosoma cruzi* in Yucatán, Mexico. *Braz. J. Glob. Health* **2020**, *1*, 21–28. [CrossRef]
62. Fuentes-Vicente, J.A.; Gutiérrez-Cabrera, A.E.; Flores-Villegas, A.L.; Lowenberger, C.; Benelli, G.; Salazar-Schettino, P.M.; Córdoba-Aguilar, A. What makes an effective Chagas disease vector? Factors underlying *Trypanosoma cruzi*-triatomine interactions. *Acta Trop.* **2018**, *183*, 23–31. [CrossRef] [PubMed]
63. Fuentes-Vicente, J.A.; Santos-Hernández, N.G.; Ruiz-Castillejos, C.; Espinoza-Medinilla, E.E.; Flores-Villegas, A.L.; Alba-Alvarado, M.; Cabrera-Bravo, M.; Moreno-Rodriguez, A.; Vidal-López, D.G. What do you need to know before studying Chagas disease? A beginner's guide. *Trop. Med. Infect. Dis.* **2023**, *8*, 360. [CrossRef]
64. Brenière, S.F.; Waleckx, E.; Barnabé, C. Over Six Thousand *Trypanosoma cruzi* Strains Classified into Discrete Typing Units (DTUs). *PLoS Neglected Trop. Dis.* **2016**, *10*, e0004792. [CrossRef]
65. Zingales, B.; Bartholomeu, D.C. *Trypanosoma cruzi* genetic diversity: Impact on transmission cycles and Chagas disease. *Memórias Inst. Oswaldo Cruz* **2022**, *116*, e210193. [CrossRef] [PubMed]
66. Ribeiro, A.R.; Lima, L.; ALmeida, L.A.; Monteiro, J.; Moreno, C.J.G.; Nascimento, J.D.; Araújo, R.F.; Mello, F.; Martins, L.P.A.; Graminha, M.A.S.; et al. Biological and molecular characterization of *Trypanosoma cruzi* strains from four states of Brazil. *Am. J. Trop. Med. Hyg.* **2018**, *98*, 453. [CrossRef]

67. Fernandes, O.; Souto, R.P.; Castro, J.A.; Pereira, J.B.; Fernandes, N.C.; Junqueira, A.C.V.; Naiff, R.D.; Barrett, T.B.; Degrave, W.M.S.; Zingales, B.S.; et al. Brazilian isolates of *Trypanosoma cruzi* from humans and triatomines classified into two lineages using mini-exon and ribosomal RNA sequences. *Am. J. Trop. Med. Hyg.* **1998**, *58*, 807–811. [CrossRef] [PubMed]
68. Davey, K. The interaction of feeding and mating in the hormonal control of egg production in *Rhodnius prolixus*. *J. Insect Physiol.* **2007**, *53*, 208–215. [CrossRef]
69. Guarneri, A.A.; Pereira, M.H.; Diotaiuti, L. Influence of the blood meal source on the development of *Triatoma infestans*, *Triatoma brasiliensis*, *Triatoma sordida*, and *Triatoma pseudomaculata* (Heteroptera, Reduviidae). *J. Med. Entomol.* **2000**, *37*, 373–379. [CrossRef] [PubMed]
70. Vicente, R.D.; Madeira, F.F.; Borsatto, K.C.; Garcia, A.C.C.; Cristal, D.C.; Delgado, L.M.G.; Bittinelli, I.F.; De Mello, D.V.; Reis, Y.V.; Ravazi, A.; et al. Morphological, cytological and molecular studies and feeding and defecation pattern of hybrids from experimental crosses between *Triatoma sordida* and *T. rosai* (Hemiptera, Triatominae). *Pathogens* **2022**, *11*, 1302. [CrossRef] [PubMed]

pathogens

MDPI

Review

Trypanosoma cruzi/Triatomine Interactions—A Review

Günter A. Schaub

Zoology/Parasitology, Ruhr-University Bochum, Universitätsstr. 150, 44780 Bochum, Germany; guenter.schaub@rub.de

Abstract: This review summarizes the interactions between *Trypanosoma cruzi*, the etiologic agent of Chagas disease, and its vectors, the triatomines, and highlights open questions. Four important facts should be emphasized at the outset: (1) The development of *T. cruzi* strains and their interactions with the mammalian host and the insect vector vary greatly. (2) Only about 10 of over 150 triatomine species have been studied for their interactions with the protozoan parasite. (3) The use of laboratory strains of triatomines makes generalizations difficult, as maintenance conditions influence the interactions. (4) The intestinal microbiota is involved in the interactions, but the mutualistic symbionts, Actinomycetales, have so far only been identified in four species of triatomines. The effects of the vector on *T. cruzi* are reflected in a different colonization ability of *T. cruzi* in different triatomine species. In addition, the conditions in the intestine lead to strong multiplication in the posterior midgut and rectum, with infectious metacyclic trypomastigotes developing almost exclusively in the latter. Starvation and feeding of the vector induce the development of certain stages of *T. cruzi*. The negative effects of *T. cruzi* on the triatomines depend on the *T. cruzi* strain and are particularly evident when the triatomines are stressed. The intestinal immunity of the triatomines responds to ingested blood-stage trypomastigotes of some *T. cruzi* strains and affects many intestinal bacteria, but not all and not the mutualistic symbionts. The specific interaction between *T. cruzi* and the bacteria is evident after the knockdown of antimicrobial peptides: the number of non-symbiotic bacteria increases and the number of *T. cruzi* decreases. In long-term infections, the suppression of intestinal immunity is indicated by the growth of specific microbiota.

Keywords: antibacterial compounds; Chagas disease; interactions; microbiota; mutualistic symbionts; Triatominae; *Trypanosoma cruzi*

1. Introduction

The protozoan parasite *Trypanosoma cruzi* (Kinetoplastida, Trypanosomatidae) is the causative agent of Chagas disease [1,2]. This disease, also known as American trypanosomiasis, is endemic in Latin America, i.e., South and Central America, Mexico, and the southern United States of America. There, its insect vectors, the triatomines, are present. Migration has brought the disease to other countries, particularly to other regions of the United States and Canada, but also to some countries in Europe, the eastern Mediterranean, Africa, and the western Pacific [3–5]. In countries outside America, no triatomines or other insects are infected with *T. cruzi*, but the parasites can be transmitted through blood transfusions and organ transplants. However, the screening of donors varies greatly between countries [6–8]. Also in Germany, blood and organ donors from immigrants from Latin America are not systematically screened [9]. However, *T. cruzi* was found when blood from an immigrant and her daughter was fed to triatomines (Schaub, unpublished). Fifty years ago, the WHO classified Chagas disease as one of the "Big Six" tropical diseases (leprosy, leishmaniasis, trypanosomiasis, malaria, schistosomiasis, and filariasis) and supported investigations within the framework of the "Special Programme for Research and Training in Tropical Diseases" [10]. Today, it is 1 of more than 20 neglected tropical diseases, all characterized by deficiencies in hygiene and public policy, high prevalence in poor populations, difficulties in diagnosis and treatment, and significant economic impact [8].

The fight against Chagas disease is mainly based on insecticide campaigns against the triatomines [11]. Between 1982 and 2010, this led to a sharp decline in the number of infected people from around 20 million to 8 million (summarized by [12]). Since then, however, it has only improved slightly to 6 to 7 million people [2]. Currently, around 30,000 people are infected each year and 10,000 die each year [13]. The new proposed vector control strategies are based on health education programs and control campaigns [14] and not on new methodologies (see Section 6.1). However, re-infestation after the use of insecticides can only be reduced by improvements to houses [15]. The infestation of houses can be detected by screening human sera for antibodies against the vector's saliva [16]. As resistance to insecticides has developed, much greater efforts are needed, including biological control agents such as entomopathogenic fungi [17].

Only mammals are vertebrate hosts of *T. cruzi*. Parasitemia varies greatly between species, and some of them are mainly unaffected by the parasite [18,19]. Curiously, stages of *T. cruzi* that are characteristic of the triatomines develop in the scent glands of opossums [19]. Mammals, and thus also humans, become infected in various ways, mainly vectorially (summarized by [12]). In this way, the parasites in the feces and/or urine of the vectors penetrate mucous membranes or wounds in the skin. Even a small number of metacyclic trypomastigotes are sufficient to penetrate the skin of mice after fecal deposits on the puncture wound of the mouthparts [20]. Oral infections occur through the consumption of infected meat or drinks made from fruit juices or sugar cane that contain feces or crushed triatomines. In the latter case, the risk is increased by the survival of trypanosomes in dead triatomines [21]. The route of infection has a significant impact on the outcome of the infection [22]. Blood transfusions and organ transplants from infected people as well as congenital transmission from mother to child also lead to infections (summarized by [12]).

After infection of humans, the course of the disease varies from person to person, but can be divided into two phases, the initial acute phase and the chronic phase [23–25]. In the acute phase, the flagellate invades all types of host cells that contain a nucleus, as it requires purines from the host for its development. It multiplies intracellularly until the host cell is exhausted and ruptures, and then invades the blood capillaries, where it can usually be easily detected in the blood with a light microscope [26]. Non-specific symptoms such as swelling of the lymph nodes near the infection site and fever occur. The chronic phase begins after about 1–2 months. The parasites are rarely detectable in the blood, but develop in hemocultures or the vectors used for xenodiagnosis [26]. The indeterminate chronic phase is hardly symptomatic and lasts several years to decades [23,24]. In the final chronic phase, organ dysfunction occurs due to the intracellular development and destruction of cells. The heart and intestines enlarge and mega-organs develop. These pathological effects lead to death [24,27].

The development of a vaccine against the parasite is hardly possible, as ethics prohibit experimental infections of humans without the possibility of a consistently successful treatment. For this purpose, only two compounds, benznidazole and nifurtimox, have been available for >50 years (summarized by [12]). They are recommended for the initial acute phase, but often lead to severe side effects. In addition, some *T. cruzi* strains are not affected [28–30]. If congenital infections are detected early, treatment is very successful in almost all children (summarized by [7]). New drugs are currently being tested (summarized by [12]), as well as nanoparticles for the oral administration of benznidazole [31].

In the vector, the triatomine insects, *T. cruzi* develops in the intestinal tract and in the excretory Malpighian tubules (summarized by [32]). In addition to flagellates, fungal and bacterial microbiota, including mutualistic symbionts, colonize the gut. Studies on the interactions between *T. cruzi*, triatomines, and the microbiota offer the opportunity to find new ways to interrupt transmission. The present review is based on the most recent review [12] and provides an updated overview of *T. cruzi*, the vectors, and the microbiota in triatomines and their interactions.

2. *Trypanosoma cruzi*

T. cruzi is a flagellate. All have a kinetoplast as part of a single mitochondrion, subepidermal microtubuli, a surface coat, and a flagellum [23]. The trypanosome multiplies by longitudinal divisions. Rare genetic recombination leads to a predominantly clonal genetic structure of the populations (summarized by [33]). *T. cruzi* strains have been classified by molecular biological methods into six main evolutionary lineages, designated TcI to TcVI and Tcbat (summarized by [33,34]). Often several of them are distributed in a single country (summarized by [35]), e.g., all TcI to TcVI are present in Bolivia, but one is not found in patients and another only in a small group of people [36]. Within each evolutionary lineage, the biological heterogeneity of strains varies greatly [30,37], but the protein profiles of TcI strains correlate with differences in virulence [38]. Many mixed infections with strains from different evolutionary lineages occur in mammals and triatomines [19]. Mixed infections of clones that belong to the same evolutionary lineage can only be detected through labor-intensive cloning of samples from the field and determination of the biological characteristics. Another aspect is the influence of host genetic diversity on disease transmission and severity [39]. Therefore, generalizing the results obtained with a single strain is misleading.

All *T. cruzi* develop four major stages, the non-dividing trypomastigote and the multiplying amastigotes, epimastigotes, and spheromastigotes (summarized by [40]). In mammals, vector-derived metacyclic trypomastigotes invade the host cells and transform into multiplying amastigotes. After the host cell is exhausted, they develop into blood trypomastigotes. These enter the blood capillaries and infect new host cells or circulate in the blood which is ingested by the triatomines. In the vector, they transform into spheromastigotes and epimastigotes and many different intermediate stages. They are heme-auxotrophic and take up heme originating from the digestion of hemoglobin by the insect presumably in the flagellar pocket [41]. Finally, metacyclic trypomastigotes develop [42,43].

3. Vectors

The metacyclic trypomastigotes are transmitted by triatomines which are taxonomically classified to the order Hemiptera, the family Reduviidae (assassin bugs), and the subfamily Triatominae [11]. Around 150 species belong to the Triatominae [44]. *Triatoma rubrofasciata*, the only triatomine species that is distributed worldwide, develops in rat nests in ports of the tropics and subtropics. A few species occur in India, but most are found between the Great Lakes of North America and southern Argentina [11,45,46]. Depending on their habitat, they are divided into sylvatic, peridomestic, and domestic species. Peridomestic triatomines suck the blood of hosts that live in the vicinity of the house, e.g., guinea pigs, chickens, and dogs. Only some species have adapted to house conditions, e.g., *Triatoma infestans* (Figure 1A), *Rhodnius prolixus* (Figure 1B), *Panstrongylus megistus*, and *Triatoma dimidiata* [11]. None of these important domestic species are present in all Latin American countries (summarized by [12,45]).

(A)

(B)

Figure 1. Adults of triatomines. (**A**) Male *Triatoma infestans* (body length: 2.6 cm); (**B**) male *Rhodnius prolixus* (body length: 1.6 cm).

3.1. Development, Attraction, and Blood Ingestion

Triatomines are hemimetabolous insects, and post-embryonic development proceeds through five nymphal stages to the adults [11,45]. All post-embryonic stages feed on blood [47]. Four factors mainly determine the time required for pre-adult development of 5 to 12 months: ambient temperature, availability of hosts, and quality of the blood as well as the supply with mutualistic symbionts. Triatomines feed on all terrestrial vertebrates, mammals, birds, and warm reptiles and amphibians (e.g., [48]). Comparing the effects of feeding on different hosts, *Triatoma pallidipennis* requires a significantly shorter period for pre-adult development and *Triatoma barberi* requires fewer feedings after ingesting blood from mammals than from chickens [49,50]. The blood source—bird or mammal—also significantly influences the morphometry of triatomines (summarized by [51]).

The activity of these nocturnal insects is highest at dawn and dusk [52]. In addition to temperature, exhaled carbon dioxide, skin odor, and visual stimuli also guide the triatomines to the host [52]. There, the proboscis, a sheath that surrounds two pairs of thin mouthparts, is swung forward and pressed against the skin with its tip. Then, the mouthparts are pushed out of the proboscis. While the mandibles cut through the skin, the maxillae penetrate the wound and move within the skin to tap a blood capillary (summarized by [20]). Saliva is pumped into the blood capillary [53]. A probing phase is used to assess food quality and the size of the capillary [54,55]. The rate of blood engorgement is influenced by the host [56]. The nymphs ingest 6 to 12 times their own body weight [11]. Depending on their size, triatomines require 3 to 30 min for complete engorgement, which is then sufficient for development to the next nymphal stage. The nymphs often need to ingest additional amounts of blood in the fifth instar [11].

After molting, the nymphs can take up blood three days later at the earliest, but about three weeks after molting, the number of probings is lower and the amount of ingested blood higher (summarized by [12]). After a complete engorgement, the digestion of fifth-instar nymphs takes more than a month in many species. The subsequent starvation period can last for longer periods, up to a year under optimal humidity conditions. This period varies for the different stages [57–59].

3.2. The Excretory System and the Fate of Ingested Blood

During blood ingestion, the Malpighian tubules begin to produce urine, which flows through the ampullae at the end of the four Malpighian tubules at the border between the midgut and hindgut. Shortly after excretion begins, the ampullae extend their swollen processes far into the rectal lumen [60–62]. They probably absorb ions, water, and other compounds and support a region at the beginning of the rectum, the rectal glands, which differ greatly in the ultrastructure and folding of the cuticle.

The ingested blood passes through the ectodermal foregut, which is lined with cuticle, into the endodermal midgut, which is characterized by microvilli on the lumen side. The first region of the midgut is the short cardia, followed by the highly distensible stomach [11,63]. The third region of the midgut is the small intestine, followed by the ectodermal sac-like rectum [40]. The functions of the intestinal regions are different, especially in the midgut. The cardia protects the mutualistic symbionts (see Section 4.4). In the stomach, blood coagulation is prevented by salivary and stomach-derived anticoagulants. The blood is quickly concentrated by the removal of ions and water [64]. The erythrocytes are lysed and the content has a jelly-like consistency or after feeding on guinea pigs, the hemoglobin crystallizes [65,66]. In the whole midgut, the perimicrovillar membranes develop within hours after feeding, are shed from the microvilli and then degenerate [67]. They cover the contents like the peritrophic membranes in other insects. Within eight days of blood uptake, the stored stomach contents are acidified to a pH of 5.2, followed by an increase to a pH of about 6.5 [68]. Carbohydrates and lipids are digested and absorbed [69]. The activity of aminopeptidases in the stomach contents is higher than in the small intestine [70], possibly digesting leukocytes, plasma proteins, and/or erythrocyte membranes. In a comparison of the two intestinal regions using proteomics, more proteolysis proteins are synthesized in the stomach [71]. These are probably precursors of enzymes that are active in the small intestine, as the digestion of hemoglobin appears to be indicated by a color change from red to brown. According to light and electron microscopy, the functions of the stomach are ion and fluid transport and nutrient storage [72], but the digestion of hemoglobin is discussed [68,73]. Peristaltic movements mix the contents and pass them on to the next region, the small intestine.

There, the contents are liquefied and the hemoglobin is digested or digestion continues, which is indicated by the color change. This region is divided into three parts of similar length by a narrow part in the middle. The anterior half contains more digestive cells than the posterior half and the stomach [74]. Digestion is mainly carried out by two proteases, cathepsins B and L [75]. The activity of cathepsins L and D in the small intestine is higher than in the stomach [70,76]. Finally, the proteins are digested by peptidases, and the aminopeptidases are located between the perimicrovillar membranes [77]. The pH changes after ingestion are similar to those in the stomach, but are somewhat delayed: an initial acidification to pH 5.2 is followed by an increase to around pH 6.5 and pH 5.9 in the posterior part of the small intestine [68]. The residues of digestion are passed into the rectum.

These residues are flushed out after feeding due to the enormous amount of urine. Thereby, the conditions change again [78]. After pH 5.9 in the first dark brown drop of excreta, it rises rapidly to pH 8.4 in the urine and remains at this level for 6 to 24 h after blood ingestion. It then drops to pH 6.2, with individual fluctuations between pH 5.7 and 8.3. The osmolality also changes from about 320 mosmol/kg H_2O in the first drop to about 410 mosmol/kg H_2O in the fourth drop [78]. This is also the value in the rectal contents one day after feeding, followed by about 760 mosmol/kg H_2O two days later. Looking at the change in ion concentration, the rectal contents of unfed insects contain mainly potassium and sulphate, which turns into a sodium chloride solution in the first drops of excreta. Later, the concentrations fluctuate greatly [78]. A change in consistency can also be observed. After flushing out the rectal contents, the urine is clear and then becomes cloudy with yellow-white urate spheres until digestive residues reappear [60,78]. Proteins are hardly present in the urine, but the concentrations are high in the digestive residues stored in the rectum. Nutrients are absorbed before defecation [11].

3.3. Immune System of Triatomines

Like all insects, triatomines have an effective cellular and humoral immune system (e.g., [45,79–82]). It is difficult to assess the importance of immune cells in the hemocel for triatomines, as the liquid diet does not injure the intestinal wall and bacterial penetration through the cuticle can only occur after an attack by a predator. Only some parasites such as *Trypanosoma rangeli* penetrate the intestinal wall, invade the hemocel, and interact with the hemocytes [83]. In the hemocel, there are different types of hemocytes: the phagocytes and others that produce different substances. The ingestion of blood induces the production of antibacterial compounds in the hemocytes, but also in the fat body and in the cells of the small intestine in a systemic reaction [84]. However, the activity of other enzymes for antimicrobial compounds, phenoloxidases, is high in the stomach contents and undetectable in the small intestine [85].

The humoral immunity of triatomines includes many antimicrobial peptides induced by the Toll, IMD (immunodeficiency), JAK/STAT (Janus kinase/signal transducer and activator of transcription), JNK (Jun-N-terminal kinase), and MAPK (mitogen-activated protein kinase) signaling pathways [81,82,86–90]. The pathways react to viruses and many types of bacteria. Non-coding RNAs regulate the immune response during these interactions [91,92]. According to molecular biological data and transcriptomes, several genes encoding at least sixteen antimicrobial compounds of different masses are expressed, some of which are encoded by different genes and exist as multiple isozymes (summarized by [12]). The specific memory that reduces the development of bacteria after a second infection (summarized by [93,94]) has not been studied with bacteria and triatomines. This innate immune priming is active in *Plasmodium* infections of mosquitoes (summarized by [95]). In a comparison of immune memory against two strains of *T. cruzi*, it exists against one strain [96,97], but in one group the mortality rate of 70% within 20 days of infective feeding is surprising.

4. The Microbiota of Triatomines

Triatomines normally ingest sterile blood. However, various bacteria, fungi, and viruses are present in the salivary glands, the Malpighian tubules, and especially in the intestinal tract (summarized by [12]). Before and during blood ingestion, the microbiota first come into contact with the antimicrobial peptides in saliva and then with those in the intestinal contents, which are secreted by the cells of the intestinal wall. Lysis by these compounds has not been studied in bacteria derived from triatomines in the field. The microbiota can act as pathogens, commensals, or mutualistic symbionts [98]. Mutualistic and antagonistic interactions of bacteria regulate the composition of the microbiota [99] but also the antimicrobial compounds of the vector.

4.1. Infection Routes

The microbiome of Triatominae originates from the environment, the host's skin, or the host's blood [100]. Microbes enter the bodies of triatomines in different ways. Some viruses are transmitted transovarially from the female to the offspring (summarized by [101]). Intestinal bacteria and fungi originate from oral infections, e.g., ingestion of water and juices, swallowing air before molting [102], contact of mouthparts with skin before and after blood ingestion, and possibly probing eggshells after hatching (summarized by [12]). Presumably, coprophagy is the most important behavior. Contact with feces is necessary to obtain mutualistic symbionts, but bacteria and fungi present in the environment contaminate the deposited feces.

Coprophagy occurs in all nymphal stages [103], as evidenced by increasing percentages of infections with the coprophagically transmitted homoxenous trypanosomatid *Blastocrithidia triatomae* in groups where the opportunity for coprophagy has been offered since the first nymphal stage. Since this also occurs in the presence of uninfected fifth-instar nymphs, triatomines cannot recognize whether the feces contain symbionts [103]. Coprophagy occurs only after feeding and not during starvation, and only when the nymphs can come into contact with liquid and not dry feces [104]. However, in other studies, *T. infestans* nymphs were not attracted to freshly deposited feces, but rather to volatiles in dry feces within 24 h after feeding [105]. In addition, nymphs leave their shelters to defecate [106]. Therefore, coprophagy requires more detailed studies.

4.2. Microbiota of Triatomines

According to previous culture-dependent identifications of the bacteria, many different species colonize the intestinal tract of triatomines (summarized by [107]). Molecular biological methods determined even higher numbers. Approximately 500 bacterial species are present in and on the posterior segments of the abdomen of field-cultured *T. dimidiata* [108,109]. Since the recent reviews on the microbiota of triatomines [12,46,83,110], only a few additional results have been published, which are considered below.

In molecular biological identifications of bacteria from field-derived *Rhodnius pallescens*, *R. prolixus*, *Panstrongylus geniculatus*, *Psammolestes arthuri*, *Triatoma maculata*, and *T. venosa*, over 90% of gut communities belong to the phyla Proteobacteria, Actinobacteria (including Actinomycetales), Bacterioidetes, and Firmicutes [111,112]. In *Triatoma gerstaeckeri* and *T. sanguisuga*, the majority of the microbiota also belong to four taxa, one of which is the Actinomycetales [113]. These bacteria are not found in all triatomines, but are often found with different genera [114]. However, the identification of the species is missing. Actinobacteria dominate in *Triatoma rubida*, *T. lecticularia*, and *T. protracta* and also in field-derived *Rhodnius ecuadoriensis*, and in the latter species, laboratory colonies contain significantly fewer microbes than samples from the field [100,115]. Such differences between *T. sanguisuga* nymphs from the laboratory and adults from the field result in different metabolic functions of the microbiota [116]. Also, *Triatoma rubrofasciata*, which is distributed worldwide, and *T. maculata*, *T. dimidiata*, *R. pallescens*, *R. prolixus*, *Eratyrus* ssp., and *P. geniculatus* contain high diversity, including Actinobacteria. Some bacterial families are more common in certain species (summarized by [12]).

4.3. Identification of Mutualistic Symbionts

The identification of mutualistic symbionts involves the use of sterile nymphs, infection with individual bacterial species, and the sterile maintenance and feeding of axenic nymphs (details in [110]). Mutual symbionts must establish themselves in the nymphs. Compared to sterile nymphs, the nymphs with the mutualistic symbionts show normal development and reproduction. Because triatomines acquire bacteria from other species grown in the same insectarium and because some of these bacteria are partially symbiotic, identifications of mutualistic symbionts should be based on triatomines from the field (reviewed by [117]).

To date, these symbionts have only been identified from four species of triatomines, but only from two symbionts at the species level [110]. The symbiotic effect of *Rhodococcus rhodnii* in *R. prolixus* has been known for 80 years [118]. Because morphological criteria have been used for identification in the past, all mutualistic symbionts of other triatomines have traditionally been considered *R. rhodnii* (e.g., [119]). Using fifth-instar nymphs and adults from the field, we identified the mutualistic symbiont of three other species of triatomines. The mutualistic symbiont of *T. infestans* is *Rhodococcus triatomae*, previously known as *Nocardia* sp. [120]. It was renamed *R. triatomae* with an almost-complete sequence of the 16S rDNA [121], but without identification as a mutualistic symbiont [122]. The need to determine near-complete sequences points to the difficulty in identification. In *Triatoma sordida* and *P. megistus*, an unnamed *Gordonia rubropertinctus*-like isolate and an unnamed *Rhodococcus equi*-like isolate, respectively, induce normal development [110].

4.4. Development of Mutualistic Symbionts/Bacteria in Triatomines

The development of mutualistic symbionts is strongly linked to the cardia, the short first region of the midgut. It has deep sac-like infoldings with narrow channels opening into the lumen [123]. The dense bacterial populations in these infoldings are protected from the complement factors in the ingested blood. After infection of sterile first-instar nymphs of *T. infestans* and *R. prolixus*, with the mutualistic symbionts *R. triatomae* and *R. rhodnii*, respectively, axenic maintenance and sterile feeding, and dissections of fifth-instar nymphs at different times after feeding, the number of colony-forming mutualistic symbionts per nymph is lower one day after feeding than in the cardia and in the stomach of unfed nymphs [120]. In the cardia of both triatomines, they increase rapidly, up to five and seven days after feeding, respectively, and then remain at this level. In the stomach the course is very similar, although the increase is somewhat delayed. Ten days after feeding, a total of 8×10^8 colony-forming units/fifth-instar nymph are present in *R. prolixus* and 1.8×10^8 in *T. infestans*, 95–99% of which are in the cardia and stomach. In both triatomines, around 99.9% of them are digested during passage through the small intestine. The majority of bacteria in the rectum, 90,000 colony-forming units/rectum, are excreted with digestive residues after blood ingestion [120]. The strong difference between the stomach and small intestine is also evident seven days after feeding fifth-instar nymphs of *R. prolixus* possessing approximately 60×10^8 and 2×10^8 colony-forming unit bacteria, respectively [124] (without separate consideration of mutualistic symbionts and other bacteria). In *T. sordida*, too, fewer actinobacteria colonize the small intestine than the stomach, but only genus levels are taken into account in these metagenome determinations [125]. In contrast to these data, in a metagenome shotgun sequencing approach of *R. prolixus*, the relative abundance of Corynebacteria, which includes *Rhodococcus*, is similar in the stomach and small intestine three days after blood ingestion, but lower in the stomach two and seven days after feeding [126]. There are probably many non-symbiotic Corynebacteria present.

After infecting first-instar nymphs of *R. prolixus* with about a million of different *Rhodococcus* species and maintaining them, avoiding infections with other bacteria, the entire intestine of the subsequent nymphal stages contained about 10^6 genome copies of *R. rhodnii*, but only about 10 of *R. triatomae* and also decreasing concentrations of the other *Rhodococcus* species [127]. With *T. infestans*, exactly the opposite happens, i.e., the mutualistic symbiont of *R. prolixus* does not establish [128].

If we summarize the data on the development of mutualistic symbionts, after feeding there is strong development in the anterior regions of the midgut, cardia, and stomach. There are far fewer mutualistic symbionts in the small intestine and rectum. The sequencing approach highlights the importance of identifying the bacterial species to separate the number of mutualistic symbionts from the number of other bacteria.

4.5. *Functions of Mutualistic Symbionts/Microbiota*

Sterile and aposymbiotic triatomines develop a complex disease syndrome after feeding on guinea pigs, rabbits, or humans. Development is delayed, particularly in the late nymphal stage, and adults rarely develop. In addition, digestion, excretion, and tanning are disturbed, which can partly arise from a reduction in the tracheal system (summarized by [129,130]). After feeding sterile nymphs of *R. prolixus* with blood supplemented with either mutualistic symbionts or B vitamins, development is only slightly delayed and digestion, excretion, and tanning are normal [131]. Functioning as a vitamin B supplier is supported by a metagenome shotgun sequencing approach, as various intestinal bacteria of triatomines have the genes for the synthesis of the various B vitamins [132]. *Bacillus megaterium* has enzymes for the synthesis of these vitamins [133] and colonizes the intestines of field-derived *T. dimidiata* [134].

However, there are arguments against the vitamin B hypothesis. Feeding aposymbiotic *T. infestans* with sterile defibrinated pig blood or on mice and chickens throughout nymphal development avoids aposymbiosis effects that occur after feeding on sheep, cow, and human blood (reviewed by [129]). Unfortunately, the concentration of the various B vitamins in the blood of these animals is unknown. A stronger argument against the vitamin B hypothesis is the normal development of nymphs fed various auxotrophic mutants of *R. rhodnii* that are unable to synthesize certain B-complex vitamins [135]. After feeding on a mixture of blood and all B vitamins, *R. prolixus* nymphs developed more slowly than those infected with the mutualistic symbiont [127,136]. Only *R. rhodnii* and not *R. triatomae*, three other *Rhodococcus* species, *Micrococcus luteus*, and *Escherichia coli* supported the normal growth and reproduction of *R. prolixus* nymphs after feeding with defibrinated rabbit blood [127]. Therefore, these authors concluded that "symbiont B vitamin synthesis is probably a necessary but not sufficient function of gut bacteria".

Are there other compounds that could be provided by the mutualistic symbionts? Since only Actinomycetales are mutualistic symbionts, these are probably peculiarities of these bacteria. All Actinomycetales belong to the Mycolata taxon, in which the cell walls of the bacteria contain mycolic acids [137]. Another compound can be acylglycerols, esters of glycerol and fatty acids, since the enzymes for degradation are only present in the genome of *Rhodococcus* and not in other intestinal bacteria [126].

Summarizing the data on the functions of mutualistic symbionts, the vitamin B hypothesis requires further investigation. Other compounds are conceivable, e.g., mycolic acids and acylglycerols.

4.6. *Intestinal Bacteriolysis*

Bacteriolysis has also already been described in detail [12] and the essential aspects are presented below. When the lyophilized mutualistic symbiont *R. triatomae* is used as a substrate in turbidity assays, the bacteria are not lysed after incubation with homogenates of the stomach and small intestine of unfed *T. infestans* nymphs and those for up to 50 days after feeding [138]. However, when using the Gram-negative bacterium *E. coli* and the Gram-positive *Micrococcus luteus* as substrates, the extracts from the stomach lead to stronger lysis than those from the small intestine. Both intestinal regions of *R. prolixus* show a similar difference [139], which is also reflected in two antibacterial compounds, the midgut glycosidases of *R. prolixus* and the prophenoloxidase activities of *T.* (*M.*) *pallidipennis* [140,141]. Furthermore, the expression levels of genes encoding lysozymes and defensins are higher in the stomach and cardia than in the small intestine of *Triatoma brasiliensis* and *T. infestans* [142–144]. After feeding on a mixture of blood and bacteria, transcripts of lysozyme A, B, defensin C, and prolixicin predominate in the stomach of *R. prolixus*, while mRNAs encoding lysozyme B and prolixicin predominate in the small intestine [139]. The expression rates of prolixicin and three annexins genes in the small intestine of this triatomine are higher than in the stomach [139,145]. In long-term starved *T. infestans*, i.e., 50 days after feeding fifth-instar nymphs, resources are targeted and bacteriolysis of *E. coli* is present in the stomach of adults but is not detectable in the small intestine [138].

In zymographs of stomach and small intestine extracts from unfed *T. infestans* nymphs and those up to 50 days after feeding using nonreducing sodium dodecyl sulfate-polyacrylamide gel electrophoresis, more lysis bands of lyophilized *M. luteus* develop when extracts from the latter region are used. Lytic proteins are mainly present at 15 to 16 kDa, the range of lysozymes, and also at 36 and 40 kDa, but never in the molecular range < 14 kDa, the range of most triatomine antimicrobial proteins [138]. Similar data are available for the saliva of *T. infestans*. In a 24 h incubation of these gels in deionized water to detect the bacteriolytic activity occurring under hypotonic conditions, additional lysis bands appear [146]. This indicates the presence of an undissociated complex of an antimicrobial peptide with another protein, e.g., proteases. Protein complexes are found in the saliva of triatomines and other insects (summarized by [12]).

Summarizing the data of bacteriolysis, the bacteriolytic compounds appear to regulate the non-symbiotic bacteria. Mutualistic symbionts are not lysed by extracts from both midgut regions [138]. The strong development of mutualistic symbionts in the stomach and the decrease in the small intestine are contradictory to the high bacteriolytic activities in the stomach and low activities in the small intestine. Presumably, antimicrobial factors and digestive enzymes, which are linked in complexes, must be separated to digest mutualistic symbionts.

5. Interactions of Triatomines with *Trypanosoma cruzi*

In the following, I will only consider infections via an infected mammalian host. The difficulties in using isolated blood trypomastigotes and metacyclic trypomastigotes as well as epimastigotes have recently been summarized [12].

5.1. Effects of the Vector on Trypanosoma cruzi*—Development of the Parasite in the Vector*

In general, the long coevolution of parasite and vector includes the development of the flagellate in the triatomines, leading to regional coadaptations of both. However, human migration results in both species being introduced to new locations. There, the introduced triatomines are often less susceptible to the local strains of *T. cruzi* than the local native vectors [147–150]. The introduced flagellates may not develop in these vectors and are excreted under unfavorable conditions. Therefore, the species/strain of triatomine and the strain of *T. cruzi* are relevant for establishment in the vector (summarized by [12,43,46,83,148,151]).

Field populations of triatomines regularly contain mixed infections with different *T. cruzi* strains, recognizable by their membership in different evolutionary lineages (e.g., [152,153]). These strains also interact, leading to an improvement or reduction in the population density of a strain [154,155]. Therefore, the main decision before studying the interactions of *T. cruzi* and triatomines should be the choice of system. If they are to reflect the natural scenario, the parasite and vector should come from the same locality. Handling *T. cruzi* is relatively simple, with standardized cyclic infections of vectors and mice and long-term storage of aliquots at −80 °C (summarized by [12]). It is more difficult to maintain the vector for many years without changing the genetic structure of the colonies. A large number of triatomines were to be captured in the field and divided into different groups. After several years of maintenance, the colonies should be mixed and the offspring divided into different groups again. This reduces inbreeding effects. As control experiments, the infections with *T. cruzi* can be compared with the old data. Using this method, we have obtained very similar infections for 20 years.

T. cruzi develops mainly in the intestinal tract [11,156–158]. The Malpighian tubules and their end, the ampullae, appear to be less optimal because they are colonized by a very small number of trypanosomes and not in all nymphs and not in all strains [157,159]. Therefore, the excretory system is rarely considered in infected triatomines. This is even more evident for the cardia. Therefore, the present review focuses on development in the stomach, small intestine, and rectum.

5.1.1. Development of *Trypanosoma cruzi* in the Stomach

After ingesting infectious blood, *T. cruzi* faces conditions that are very different from those of the mammalian host. After temperatures of around 38 °C, pH 7.4 and sufficient glucose as a nutrient in the mammal, the temperature in the stomach changes after a short time to that of the environment in the range of 23 to 30 °C (summarized by [25,160]). The consistency changes quickly and becomes jelly-like or a "serum" with hemoglobin crystals. In *T. infestans*, the pH value is between pH 5.2 and pH 6.5 (summarized by [12]). Glucose is no longer available and *T. cruzi* must use amino acids and lipids derived from the vector's digestion of the blood. However, the flagellate uses carbohydrates and amino acids without a drastic change in its catabolic enzyme levels [161]. In addition, the ingested blood is mixed with the saliva of the triatomine. This appears to affect *T. cruzi* from two evolutionary lineages differently, but requires more detailed studies of activity in the gut (summarized by [12]).

Initial development appears to depend on the combination of *T. cruzi* strain and vector. In one system, all trypomastigotes in the blood are killed in the stomach within four days of ingestion. In another system, individual trypomastigotes are still present three weeks later (summarized by [12]). At one day p.i. of *Dipetalogaster maxima* and *T. (M.) pallidipennis* with a *T. cruzi* strain from Guatemala, many trypanosomes are aggregated. Blood trypomastigotes shorten, become more oval, and one day later develop into spheromastigotes and then into oval epimastigote-like forms, but never into slender epimastigotes [162]. The formation of epimastigotes in the stomach is controversial (summarized by [12]). However, they are present there in long-term infections after molting, presumably originating from the small intestine [104]. After the subsequent ingestion of blood from chickens or uninfected mice, these epimastigotes are killed or survive, respectively.

Summarizing the initial development in the stomach, there are strong system-dependent differences in the proportions of trypanosomes killed. Aggregations occur regularly, often of amastigotes and spheromastigotes. In some systems, precursors of epimastigotes develop.

5.1.2. Development of *Trypanosoma cruzi* in the Small Intestine

In the small intestine of the vector, some factors, temperature and pH changes, are similar to those in the stomach. The jelly-like content is liquefied. There, trypanosomes transform into epimastigotes, which multiply quickly [163]. Presumably, the triatomine's blood digestion provides many nutrients. The vector's digestive enzymes, proteases, lipases, and carbohydrate digestion enzymes, have no effect on the flagellates. The attachment of epimastigotes to the perimicrovillar membranes is controversial (summarized by [12,83]). Since these membranes are only present temporarily, they cannot play a crucial role in development.

The rapid development of the population is evident in our system, where the vector and *T. cruzi* (TcI) originate from the same village in Chile. One week after infection of second-instar nymphs with 8000 to 10,000 blood trypomastigotes per *T. infestans* nymph, approximately 30,000 parasites/nymphs colonize the small intestine [163]. After three to four week starvation periods, the population is reduced slightly before feeding. After feeding subsequent nymphal instars, flagellate populations increase, resulting in approximately 600,000 parasites per fifth-instar nymph [163].

In established infections of *T. infestans*, there are predominantly epimastigotes and various intermediate stages to spheromastigotes, to metacyclic trypomastigotes via spheromastigotes, and very rarely directly to metacyclic trypomastigotes. Then, after periods of starvation of three to four weeks, more round forms develop [163]. Prolonged starvation of fifth-instar nymphs that were originally infected in the first instar leads to a sharp decline in the population, and 60 days after feeding, no more trypanosomes or only small populations are found [57,164]. Since residues of digested blood are present in the small intestine of long-term-starved dead nymphs, the loss of certain compounds probably leads to the death of *T. cruzi*.

Summarizing the development of *T. cruzi* in the small intestine, blood ingestion supports the development of the population. Epimastigotes in particular, but also spheromastigotes, multiply quickly. Morphologically final metacyclic trypomastigotes are only rarely present. Starvation greatly reduces the number of flagellates.

5.1.3. Development of *Trypanosoma cruzi* in the Rectum

The development of *T. cruzi* in the rectum has also recently been reviewed [12,46,83]. Mainly in the rectum, metacyclic trypomastigotes develop [40]. Their number is higher at a maintenance temperature of 30 °C than at 26 and 28 °C [165]. To highlight an extremely important aspect, in the *T. cruzi*/*T. infestans* system from Chile, the number of trypanosomes increases sharply from one nymphal stage to the next and this part is three-times more populated than the small intestine, finally containing 50% metacyclic trypomastigotes [163]. The reason appears to be the possibility of attachment, and approximately two-thirds of the population attach to the cuticle [166,167]. The flagellum attaches with a small hydrophobic area to the wax layer that covers the entire rectal cuticle [61]. This flagellar region contains specific proteins [168]. In *Leishmania*, deletion of the adhesion proteins prevents colonization of the stomodeal valve [169]. In the flagellum of *T. cruzi*, nanotubules are partially dragged when fluids flow and improve adhesion to the cuticle under shear stress [170]. These filaments, along with hemidesmosome-like material, are also present in the enlargements at the attachment site. A hemidesmosome-like attachment plaque between the tip of the flagellum and the substrate develops in many attached trypanosomatids, e.g., in *Crithidia fasciculata* [171]. A specific region of the rectum of triatomines, the four rectal pads, is extremely densely populated with up to four layers of epimastigotes. Epimastigotes in the upper layers attach to the cuticle with the tip of the elongated flagellum. The reason for the preference for the rectal pads, which have lower folds of the cuticle, is probably the connection to the ampullae of the Malpighian tubules, which just cover these pads outside the rectum. In other insects, water and amino acids are heavily absorbed at the rectal pads (summarized by [172]).

Blood ingestion and starvation also affect the population of *T. cruzi* in the rectum. Urine is already produced during blood ingestion, which sometimes leads to excretion before or shortly after cessation of food intake [173–176] (see Section 3.2). The time to start defecation is important because the risk of infection is higher when defecation occurs on the host. This varies greatly between species and between developmental stages [177]; (summarized by [83]). Defecation after leaving the host indicates limited vector capacity (e.g., [174]).

The excreta contain the *T. cruzi* population from the rectal lumen and the unattached population from the rectal wall. These are mainly trypomastigotes that cannot attach due to their short, free flagellum and surface coat (summarized by [40]). Furthermore, metacyclogenesis is induced within four hours after blood ingestion, but only in one precursor stage, the epimastigote [159]. The triggering factors are hemolymph proteins of about 17 kDa, which pass into the urine [40,178]. However, other factors could also support this, e.g., specific digestive compounds, since metacyclogenesis in *R. prolixus* is reduced after the knockdown of an α-glucosidase, which also influences heme detoxification and digestion [179]. The drastic changes in pH and osmolalities (see Section 3.2) can also be relevant. Osmotic changes are controlled by *T. cruzi* through various mechanisms [180–182].

Blood ingestion after prolonged starvation triggers the development of "giant cells", a stage of multiple cell division [183]. When present one day after feeding, their proportion increases to 30 to 50% but they disappear completely between five and ten days after feeding. Low numbers have previously been reported for the small intestine [184].

Starvation without subsequent feeding also influences the population density of *T. cruzi* in *T. infestans* and *R. prolixus* [164,185]. In established infections, approximately 300,000 flagellates colonize the rectum 20 and 30 days after feeding of fourth-instar *T. infestans* nymphs [164]. Starvation for 60 and 90 days reduces the population to 100,000 and 1000 trypanosomes/rectum, respectively, the same level present 120 days after feeding. The reduction 90 days after feeding is not due to defecation, as almost all nymphs stop defecating and store rectal contents after a 60-day starvation period. Considering the different rectal regions in detail using scanning electron microscopy, after a starvation period of 16 weeks most of the rectal regions are colonized by *T. cruzi*. After another four weeks, the flagellates are only attached to the rectal pads, but in all nymphs [186]. In addition to the number of trypanosomes, the percentages of the different development stages also change. Before starvation, 2% of the rectal population are spheromastigotes and 1% are intermediate drop-like forms, but forty days later this number is 30% [164,183].

To summarize the development of *T. cruzi* in the rectum, the population is larger than in the small intestine, presumably due to the attachment of the epimastigotes. The unattached trypanosomes are flushed out through urine production, which is triggered by blood ingestion. Very rapid metacyclogenesis of epimastigotes is induced by hemolymph proteins in urine. After nymphs that have been starved for a long period of time ingest blood, "giant cells" form for a few days. During starvation, trypanosomes progressively decrease in number and the populations contain more spheromastigotes.

5.1.4. Parasite Load in the Whole Intestine

Recent parasite burden studies used PCR-based methods and the whole intestine. Field-caught *Triatoma pseudomaculata, T. brasiliensis, Panstrongylus lutzi, T. sordida,* and *Triatoma vitticeps* contained up to 6×10^{10} *T. cruzi*/intestine units (median 2.3×10^3) [187]. In the qPCR of *T. brasiliensis*, 3.9 to 7.7×10^6 *T. cruzi* per insect were found and in *Triatoma melanica*, TcII-infections resulted in a lower parasitic load compared to TcI and TcIII [152,153]. The difference between TcI and TcII was also evident after the infection of second-instar nymphs and dissection of fifth-instar nymphs using classical counts in Neubauer hemocytometers, resulting in about 1.7×10^6 and 1.4×10^6, respectively, in the whole intestine [164].

5.2. *Effects of Trypanosoma cruzi on Triatomines*

5.2.1. Effects of *Trypanosoma cruzi* on Nymphs and Adults of Triatomines

Several reviews focus on the effects of *T. cruzi* infection on triatomines (e.g., [12,25,32,46,64,83,150,171,188,189]). Summarizing these reviews and considering recent publications, the pathogenicity of *T. cruzi* for the vector is either low or only apparent when triatomines are exposed to adverse conditions. In the laboratory, this may particularly be due to a lack of supply of mutualistic symbionts which is rarely taken into account in publications. In field populations, suboptimal blood sources and starvation, which affect the development of even uninfected triatomines, could support pathogenic effects [58,59,190–192].

Comparing uninfected and infected nymphs in laboratory experiments, some *T. cruzi* strains significantly retard the development of *R. prolixus* nymphs and reduce survival, while infections with other strains have no effect [193]. Considering starvation as a stress factor, fourth- and fifth-instar nymphs of *T. infestans* survive statistically significantly shorter times, up to 17% less, after first-instar infection [57]. This is also evident in *T. (M.) pallidipennis*, but not in *Mepraia spinolai* from the field [141,194]. However, differences in the nutritional status of the field samples must be taken into account. This is lower in infected *M. spinolai* than in uninfected *M. spinolai* [195], perhaps reflecting the differences in the composition of the metabolites in the intestinal tract of uninfected and infected *R. prolixus* [196]. When determining the concentrations of free amino acids and those in peptides, the composition of peptides in the rectal contents of *T. infestans* differs from that of uninfected nymphs (Kollien and Schaub, unpublished), which may be due to proteases in the surface coat of the epimastigotes [197]. In *M. spinolai* collected in the field during the warming period, significantly more *T. cruzi*-infected second- and fourth-instar nymphs than uninfected nymphs molted after two feedings in the laboratory [198]. This is not evident after collection during the cooling season and in the first and third nymphal instar. Therefore, it is difficult to find an explanation.

Studies of the effects of *T. cruzi* on adult longevity and fertility also vary. Under optimal conditions, not all strains of *T. cruzi* affect the fitness and reproduction of the vector ([199]; summarized by [83]). In detailed investigations of *T. infestans*, infected adults copulate earlier and more frequently than uninfected specimens, increasing reproductive efficiency and decreasing life expectancy [200].

To summarize the effects of *T. cruzi* on nymphs and adults, these depend on the parasite–vector system and are relevant to starvation endurance.

5.2.2. Effects of *Trypanosoma cruzi* on the Behavior of Triatomines

In the field, the infection has no influence on the feeding profiles of the different nymphal instars of *M. spinolai*, which select different hosts [201]. In domestic and sylvatic populations of *T. dimidiata*, the abundance of some sensilla types on the antennae is affected by infection [202], but relevance to orientation remains to be investigated. Since the effects on triatomine orientation to the host have recently been reviewed [46,83], this will be briefly presented. Often, infected nymphs respond more quickly to human odors or approach the host more quickly than uninfected nymphs [203]. In *T. infestans* nymphs, parasite load correlates significantly with the number of movements and distances during the photophase [204]. These changes should not affect the risk of transmission, as each nymphal instar requires only one complete engorgement. However, the risk of transmission to an insectivorous host is increased because infected nymphs leave the shelter in an arena for longer and are preyed on by mice more often than uninfected nymphs [205]. As recently highlighted, these behavioral effects can be explained by competition between the trypanosomes and the vector, leading to an earlier starvation response. Therefore, infected nymphs of *Triatoma rubrovaria* consume more blood than uninfected ones [206]. A reduction in weight results in an increase in the dispersal ability of infected female *T. dimidiata* (summarized by [46,83]). The significance of higher levels of negative geotaxis and higher levels of aggregation of infected *T. infestans* in males and females, as well as the impaired aggregation of infected *T. pallidipennis*, is due to a lower concentration of attractive compounds in dry feces [207,208], and the importance is difficult to assess.

To summarize the effects of *T. cruzi* on the behavior, these effects appear to be due to competition between the trypanosomes and the insect for compounds in the blood.

5.2.3. Effects of *Trypanosoma cruzi* on Immunity

The ingestion of *T. cruzi* also triggers a systemic reaction outside the intestine by hemocytes, fat bodies, and other organs (summarized by [32,83]). Detoxifying proteins are immediately induced [209]. The activity of a phospholipase involved in the production of nitric oxide and lipid metabolism, as well as the expression of the corresponding genes in the salivary glands of *R. prolixus*, is significantly reduced in the infected group 3 days after feeding, but not later [210]. *T. cruzi* induces the synthesis of nitric oxide [84]. This is an immediate response at one day p.i. to blood trypomastigotes of *T. cruzi* in the stomach of *T. infestans* [211,212].

The triggering factor of antibacterial activity lies in the surface coat, since the ingestion of blood mixed with blood trypomastigotes or their shed surface coat increases antibacterial activity in the small intestine of *T. infestans* nymphs, but not ingestion of trypomastigotes without a surface layer or epimastigotes [213] (details in [83]). Comparing the expression levels of antimicrobial peptide genes, only some of them are upregulated (summarized by [83]). Expression of a lysozyme gene is increased in *R. prolixus* at 7 and 14 days p.i. with blood trypomastigotes [214] and 20 days p.i. with epimastigotes, that of the gene that encodes a defensin in *T. brasiliensis* [144]. When using different *T. cruzi* strains, the immune responses differ (summarized by [215]). While the expression of genes encoding nitric oxide synthase and phenoloxidase is increased after infection (reviewed by [83,215]), the expression of other genes is reduced, e.g., of defensin C in *Rhodnius neglectus* [216]. Since recombinant defensin kills *T. cruzi* in vitro [217], beneficial downregulation is possible.

The effects of *T. cruzi* on immunity can be summarized as follows: surface coat components of ingested blood trypomastigotes induce it. However, this does not seem to affect the flagellate, but appears to be relevant for the interactions with the intestinal microbiota of the triatomine (see Section 6 below).

6. Interaction of *Trypanosoma cruzi* and the Microbiota of Triatomines

6.1. *Effects of the Microbiota on Trypanosoma cruzi*

Some publications report direct effects of the microbiota on *T. cruzi*. *Serratia marcescens* lyses epimastigotes (summarized by [83]). However, this depends on the strain of the bacterium [218]. Since co-infections of *S. marcescens* and *T. cruzi* occur (e.g., [219]), the significance in natural infections appears to be low. Also, with another bacterium, *Wolbachia*, co-infections with *T. cruzi* occur in the field [111], and an effect of the bacterium on the trypanosome is unlikely (summarized by [12]).

Genetically transformed mutualistic symbionts have been proposed for use in control campaigns [220–225]. These bacteria produce a lepidopteran antibacterial peptide, an antibody fragment, an enzyme disrupting surface glycoconjugates, or dsRNA, all of which kill *T. cruzi* in the intestine. As the release of transformed bacteria into the field is still controversial, no large-scale field studies have been conducted to date.

6.2. Indirect Effects of Trypanosoma cruzi via Inducing and Suppressing Vector Immunity

These interactions are described in several reviews (e.g., [12,46,83,86,226–228]). Blood trypomastigotes induce an upregulation of the synthesis of antimicrobial compounds in the intestine. This impairs the growth of bacteria and fungi and promotes the growth of trypanosomes. The same effect occurs after feeding antibiotics or an inhibitor of nitric oxide synthase [229,230]. In *R. prolixus*, *T. cruzi* infections increase the transcript levels of defensin C and prolixicin, as well as the antibacterial activity in general and specifically the activity of phenoloxidase, but the synthesis of nitric oxide precursors decreases [124,229]. The activation of immunity by *T. cruzi* significantly reduces the abundance of bacteria after feeding in the midgut, but also changes the composition of bacterial species [126] (see Sections 4.2 and 4.4). *T. cruzi* infections suppress specific bacteria and promote other bacteria. This is indicated by the *T. cruzi*-induced synthesis of TiAP; recombinant TiAP has a bacteriostatic effect against Gram-negative *E. coli* and not against the Gram-positive *Micrococcus luteus* [231].

Infected *T. dimidiata* from Colombia contain more Kineosporiaceae, but less Brevibacteriaceae, Dermabacteriaceae, and Enterobacteriaceae [48]. In infected *T. infestans* from the field, species of five bacterial genera are overrepresented and four underrepresented [232]. In field populations of *T. sanguisuga*, certain bacteria are significantly associated with main evolutionary lineages: Bacillales with infections with TcI, Aeromonadales with TcIV, and Burkhodeliales and Enterobacteriales with TcII/V [113]. In specimens of *T. gerstaeckeri* and *T. sanguisuga* from the field, there is no association with the main evolutionary lineages TcI and TcIV [114].

The advantage of this induction of immunity lies in the effect against the competition of bacteria and the flagellate. The knockdown of *T. cruzi*-induced antimicrobial peptides or suppression of immune responses promotes the growth of more bacteria while reducing the number of flagellates compared to the control group [231,233,234]. The specific effect is evident after the silencing of an antibacterial rhamnose-binding lectin from *R. prolixus*, as transcript levels of bacteria are increased, but *T. cruzi* (TcI) is not affected [235].

In field-caught triatomines, infections correlate with a similar or more diverse species composition of the bacteria [109,111,112,219,236,237]. In samples of *T. gerstaeckeri* and *T. sanguisuga* from the field, more Enterobacterales and *Petrimonas* develop in infected specimens [114]. This and the more diverse species composition may indicate immune suppression due to long-term infections with *T. cruzi*: after feeding a mixture of different bacteria or fungi and blood, high numbers of the microorganisms develop only in long-term-infected nymphs and not in uninfected controls [238].

Summarizing these interactions, *T. cruzi* induces as a short-term response the synthesis of antibacterial compounds that kill certain bacteria that might compete with the trypanosomes. Supporting the growth of other bacteria sometimes depends on the main evolutionary lineage of *T. cruzi*. In long-term infections, *T. cruzi* suppresses intestinal immunity, which could be the reason for the over-representation of some bacteria in triatomines from the field.

6.3. Interactions of Trypanosoma cruzi with Mutualistic Symbionts

Only five investigations consider the interactions between the mutualistic symbiont and *T. cruzi*. In *R. prolixus*, which has an established infection with the mutualistic symbiont, more trypanosomes are present at 7 days p.i. than in aposymbiotic nymphs, but fewer *T. cruzi* are present at 21 and 35 days p.i. [239]. The initially stronger development is probably due to support from the mutualistic symbionts: Vitamin B supports the development of the homoxenous flagellate *Blastocrithidia triatomae* in *T. infestans* [240]. After feeding a mixture of blood with cell-culture-derived trypomastigotes and with or without *R. rhodnii* to germ-free first-instar nymphs of *R. prolixus*, the population density of *T. cruzi* is similar in both groups [241]. After a knockdown of the immune component rpRelish, which appears to control the expression of defensin A, the population of *R. rhodnii* in the intestinal regions of *R. prolixus* increases, but the concentrations of *T. cruzi* remain unchanged at 7 and 14 days p.i. [242]. However, the infection of these triatomines with epimastigotes significantly affects the population density of *R. rhodnii* [124]. To investigate the possible effects of trypanosomes on the mutualistic symbionts, the mutualistic symbionts *R. rhodnii* and *R. triatomae* were fed to sterile first-instar nymphs of the respective triatomines, *R. prolixus* and *T. infestans*, followed by axenic maintenance and sterile feedings. After the infection with blood trypomastigotes of *T. cruzi* (TcI) in the fifth-instar and dissections up to 10 days p.i., the population densities of the respective mutualistic symbionts in the cardia, stomach, small intestine, and rectum are similar to those of uninfected nymphs [120].

Summarizing the co-infections of triatomines with *T. cruzi* and the mutualistic symbionts, no long-term support of the trypanosome appears to occur.

7. Suggestions for Future Research

The main suggestion is that investigations should reflect the natural scenario using only blood trypomastigotes to infect triatomines. Should old, well-classified strains of *T. cruzi* be used? These strains were cultured in vitro for many years and lost selection by mammalian or vector hosts. There are probably several "Y" strains with different properties. Therefore, better local strains are used. However, strains of *T. cruzi* not only show an enormous variability in development in experimentally infected mammals, but also in the vector [193,243]. Since it is easier to use a strain that develops rapidly in mammals, such strains are preferred. However, in an identification of *T. cruzi* in *T. infestans* from Chile, only one out of five isolates developed into visible parasitemia in mice without immunosuppression by cyclophosphamide [243]. One of these isolates also multiplied somewhat less intensively in the vector [163] (see Section 5.1.4). Before generalizing, more lower-virulence strains should also be used in investigations of the interactions with the vector.

Taking vectors into account, fewer than about 10 of approximately 150 triatomine species are included in studies of the interactions with *T. cruzi* [215]. Sylvatic species in particular are underrepresented. When establishing triatomines in the laboratory, large quantities should be caught in the field, divided into three to four groups of at least 30 individuals each, and reared separately for a few years. After mixing, the genetic variation should be very similar to that of the original population (see Section 5.1). Important open questions for the vectors are the role of cardia, the concentrations of different carbohydrates in the different regions of the intestine, the digestive processes in the stomach, and more details about the rectal conditions. According to preliminary tests with microelectrodes, which cannot be continued, the oxygen concentrations present are very low (Zinkler and Schaub, unpublished).

Investigations of interactions of flagellate and vector should begin with freshly established triatomine stocks from the field, optimally the population used to isolate *T. cruzi*. The following scheme is optimal: After isolation from field-caught triatomines, mammals should be infected to enable simple sterile in vitro cultivation and the flagellates should be cloned. Then, the main evolutionary lineages should be classified to select different lineages. (Without cloning, different subpopulations develop at different times after infection.) After infection of the progeny of triatomines from which *T. cruzi* originates with epimastigotes and the development of metacyclic trypomastigotes in the rectum, aliquots of feces and urine should be mixed with an antifreeze and stored at −80 °C. Thereby, the infection of experimental mammals in subsequent experiments always begins with identical doses of parasites. Using this standardization, many questions can be solved and potentially related to evolutionary lineages, e.g., the development of slender and stout forms of blood trypomastigotes in the vector [244], the killing of blood trypomastigotes in the stomach of the vector, the development of *T. cruzi* in the presence of guinea pig hemoglobin crystals, the relevance of vector-derived compounds in metacyclogenesis, and the attachment compounds at the tip of the flagellum of epimastigotes. Other important open questions are the role of trypanolytic factors in the stomach (see Section 5.1.1) and the time course of activity of salivary compounds in the intestine. Epimastigotes dominate there and secrete cyclophilin, which inactivates the vector-derived salivary trialysin [245].

Much more information is still missing about the mutualistic symbionts and the interaction with and within the intestinal microbiome. Mutualistic symbionts are only known for the following systems: *R. rhodnii* in *R. prolixus*, R. *triatomae* in *T. infestans*, a *Rhodococcus equi*-like isolate in *P. megistus*, and a *Gordonia* sp. in *T. sordida* [110]. Details on the identification process of mutualistic symbionts and modifications were recently published [12,110] and are not repeated here. In addition to identifying more mutualistic symbionts, the compounds that they release to the triatomines should also be identified, e.g., by biochemical determinations of the various B vitamins in mutualistic symbiont-infected nymphs and feeding membrane fractions and the cytosol without contaminating symbionts to sterile nymphs. Vitamin transporters known for mosquitoes should also be looked for [246].

Since the intestinal tract of triatomines is colonized by many bacterial species, interactions between them and with the mutualistic symbionts should be investigated. Do bacteria from the *Bacillus subtilis* group present in the midgut of field-derived *T.* (*M.*) *pallidipennis* produce antimicrobial compounds [180,247] and alter the composition of the microbiota in triatomines? The presence of a bacterium does not necessarily include colonization, since spores may survive in the insect. After feeding sterile nymphs with blood supplemented with different triatomine-derived bacteria, the majority did not colonize the intestine [122]. Since not all triatomines in the field have mutualistic symbionts, other bacteria seem to act symbiotically. Another possibility are optimal blood donor hosts (see Sections 4.2 and 4.4). The paradoxical situation of high antibacterial activity in the stomach without the lysis of mutualistic symbionts and low antibacterial activity in the small intestine with the lysis of mutualistic symbionts is very interesting. Does lysis occur through complexes of antibacterial factors and digestive enzymes that are activated in the small intestine? (See Section 4.6).

Considering the fascinating interactions of *T. cruzi* and microbiota, the open questions are the identification of the surface compounds of *T. cruzi* that initiate the immune response and the method of the specific modifications by the flagellate. Does *T. cruzi* produce antibacterial compounds or are the effects "only" due to the synthesis of specific antimicrobial compounds of the triatomine? Different evolutionary lineages of *T. cruzi* should be included.

8. Conclusions

Will investigations of the interactions between *T. cruzi*, the triatomines, and the microbiota lead to new approaches to prevent the transmission of Chagas disease by triatomines? New methods could discover new aspects of interactions. They should be based on the natural scenario and use optimal systems. Only with such systems can the pathogenic effect of *T. cruzi* on the vector be quantified and the relevance for transmission assessed.

The small number of parasite/triatomine systems examined so far makes it difficult to generalize because the genetics of *T. cruzi*, the triatomines, and the microbiota are too diverse. Many questions about the interactions of *T. cruzi*, triatomines, and the microbiota require detailed investigation. The accumulation of metagenomic, proteomic, and metabolomic data will provide new insights into these trypanosome–triatomine–microbiota systems, but the enormous amount of data requires a critical review of the relevance.

Funding: The research was funded by the UNDP/World Bank/WHO Special Programme for Research and Training in Tropical Diseases, the Volkswagenstiftung, the German Academic Exchange Service (DAAD), European Community programmes, the Humboldt Foundation and especially the Deutsche Forschungsgemeinschaft.

Institutional Review Board Statement: Not applicable.

Informed Consent Statement: Not applicable.

Data Availability Statement: Detailed data are included in the respective publication.

Acknowledgments: The author would like to thank to thank Jacinta Essling for the correction of the English style of the manuscript.

Conflicts of Interest: The author declares no conflict of interest.

References

1. Chagas, C. Nova trypanozomiaze humana. Über eine neue Trypanosomiasis des Menschen. *Mem. Inst. Oswaldo Cruz* **1909**, *1*, 159–218. [CrossRef]
2. WHO. Chagas Disease (American Trypanosomiasis). Available online: http://www.who.int/health-topics/chagas-disease (accessed on 8 March 2025).
3. Lidani, K.C.F.; Andrade, F.A.; Bavia, L.; Damasceno, F.S.; Beltrame, M.H.; Messias-Reason, I.J.; Sandri, T.L. Chagas Disease: From Discovery to a Worldwide Health Problem. *Front. Public Health* **2019**, *7*, 166. [CrossRef] [PubMed]
4. WHO. Global Distribution of Cases of Chagas Disease, Based on Official Estimates, 2018. 2023. Available online: https://www.who.int/docs/default-source/ntds/chagas-disease/chagas-2018-cases.pdf (accessed on 8 March 2025).
5. Sereno, D.; Oury, B.; Grijalva, M.J. Chagas Disease across the Ages: A Historical View and Commentary on Navigating Future Challenges. *Microorganisms* **2024**, *12*, 1153. [CrossRef] [PubMed]
6. Suárez, C.; Nolder, D.; García-Mingo, A.; Moore, D.A.J.; Chiodini, P.L. Diagnosis and Clinical Management of Chagas Disease: An Increasing Challenge in Non-Endemic Areas. *Res. Rep. Trop. Med.* **2022**, *13*, 25–40. [CrossRef]
7. Cucunubá, Z.M.; Gutiérrez-Romero, S.A.; Ramírez, J.D.; Velásquez-Ortiz, N.; Ceccarelli, S.; Parra-Henao, G.; Henao-Martínez, A.F.; Rabinovich, J.; Basáñez, M.G.; Nouvellet, P.; et al. The Epidemiology of Chagas Disease in the Americas. *Lancet Reg. Health Am.* **2024**, *37*, 100881. [CrossRef]
8. da Gama, A.N.S.; Correia Soeiro, M. *Trypanosoma cruzi* Transmission through Blood Samples and Derivatives: Main Routes, Control Strategies and Recent Advancements in Blood Banks. *Pathogens* **2025**, *14*, 133. [CrossRef] [PubMed]
9. Wirth, M.; Gálvez, R.I.; Jochum, J.; Strauss, R.; Kristensen, K.; Stich, A.; Stegemann, M.; Stahl, P.; Puchner, K.P.; Strasen, J.; et al. Knowledge, Attitudes, Behaviors, and Serological Status Related to Chagas Disease among Latin American Migrants in Germany: A Cross-Sectional Study in Six German Cities. *Front. Cell Infect. Microbiol.* **2023**, *12*, 1047281. [CrossRef]
10. Schaub, G.A.; Wülker, W. Tropische Parasitosen im Programm der Weltgesundheitsorganisation. *Universitas* **1984**, *39*, 71–80. (In German)
11. Schaub, G.A. Kissing Bugs. In *Encyclopedia of Parasitology*, 4th ed.; Mehlhorn, H., Ed.; Springer: Berlin, Germany, 2016; pp. 1400–1403.
12. Schaub, G.A. Interaction of *Trypanosoma cruzi*, Triatomines and the Microbiota of the Vectors—A Review. *Microorganisms* **2024**, *12*, 855. [CrossRef]
13. PAHO. World Chagas Disease Day 2021. Available online: http://www.paho.org/en/news/13-4-2021-70-people-chagas-dont-know-theyre-infected (accessed on 10 March 2025).
14. Vargas-Abasolo, R.; Gutiérrez-Cabrera, A.E.; Cruz-López, L.; Alavez-Rosas, D.; Benelli, G.; Córdoba-Aguilar, A. Chagas Disease Vector Control Strategies: Where we are and where should we Go from here. *Entomol. Gen.* **2023**, *43*, 771–788. [CrossRef]
15. Grijalva, M.J.; Villacís, A.G.; Ocaña-Mayorga, S.; Yumiseva, C.A.; Nieto-Sanchez, C.; Baus, E.G.; Moncayo, A.L. Evaluation of the Effectiveness of Chemical Control for Chagas Disease Vectors in Loja Province, Ecuador. *Vector Borne Zoonotic Dis.* **2022**, *22*, 449–458. [CrossRef]

16. Schwarz, A.; Medrano-Mercado, N.; Billingsley, P.F.; Schaub, G.A.; Sternberg, J.M. IgM-Antibody Responses of Chickens to Salivary Antigens of *Triatoma infestans* as Early Biomarkers for Low-Level Infestation of Triatomines. *Int. J. Parasitol.* **2010**, *40*, 1295–1302. [CrossRef] [PubMed]
17. de Miranda, R.P.R.; Soares, T.K.D.A.; Castro, D.P.; Genta, F.A. General Aspects, Host Interaction, and Application of *Metarhizium* sp. in Arthropod Pest and Vector Control. *Front. Fungal Biol.* **2024**, *5*, 1456964. [CrossRef] [PubMed]
18. Jansen, A.M.; Roque, A.L.R.; Xavier. S.C.C. *Trypanosoma cruzi* Enzootic Cycle: General Aspects, Domestic and Synanthropic Hosts and Reservoirs. In *American Trypanosomiasis Chagas Disease*, 2nd ed.; Telleria, J., Tibayrenc, M., Eds.; Elsevier: Amsterdam, The Netherlands, 2017; pp. 265–282. [CrossRef]
19. Jansen, A.M.; Xavier, S.C.d.C.; Roque, A.L.R. Landmarks of the Knowledge and *Trypanosoma cruzi* Biology in the Wild Environment. *Front. Cell. Infect. Microbiol.* **2020**, *10*, 10. [CrossRef]
20. Waldeck, B.; Schaub, G.A. "Natural Infections" with *Trypanosoma cruzi* via the Skin of Mice: Size of Mouthparts of Vectors and Numbers of Invading Parasites. *Parasitol. Res.* **2022**, *121*, 2033–2041. [CrossRef] [PubMed]
21. Busselman, R.E.; Killets, K.C.; Saunders, A.B.; Hamer, S.A. Viable *Trypanosoma cruzi* Cultured from a Dead *Paratriatoma lecticularia* (Hemiptera: Reduviidae) Encountered in a Large Dog Kennel Environment in South Texas, USA. *J. Med. Entomol.* **2025**, *62*, 225–229. [CrossRef]
22. Gonçalves, K.R.; Mazzeti, A.L.; Nascimento, A.F.D.S.; Castro-Lacerda, J.M.; Nogueira-Paiva, N.C.; Mathias, F.A.S.; Reis, A.B.; Caldas, S.; Bahia, M.T. The Entrance Route: Oral, Mucous, Cutaneous, or Systemic has a Marked Influence on the Outcome of *Trypanosoma cruzi* Experimental Infection. *Acta Trop.* **2022**, *234*, 106581. [CrossRef]
23. Steverding, D. The History of Chagas Disease. *Parasit. Vectors* **2014**, *7*, 317. [CrossRef]
24. Bern, C.; Messenger, L.A.; Whitman, J.D.; Maguire, J.H. Chagas Disease in the United States: A Public Health Approach. *Clin. Microbiol. Rev.* **2020**, *33*, e00023-19. [CrossRef]
25. De Fuentes-Vicente, J.A.; Gutiérrez-Cabrera, A.E.; Flores-Villegas, A.L.; Lowenberger, C.; Benelli, G.; Salazar-Schettino, P.M.; Córdoba-Aguilar, A. What Makes an Effective Chagas Disease Vector? Factors Underlying *Trypanosoma cruzi*-Triatomine Interactions. *Acta Trop.* **2018**, *183*, 23–31. [CrossRef]
26. Meiser, C.K.; Schaub, G.A. Xenodiagnosis. In *Parasitology Research Monographs. Nature Helps... How Plants and Other Organisms Contribute to Solve Health Problems*; Mehlhorn, H., Ed.; Springer: Berlin, Germany, 2011; Volume 1, pp. 273–299.
27. Mehlhorn, H.; Schaub, G.A. Chagas' Disease, Man. In *Encyclopedia of Parasitology*, 4th ed.; Mehlhorn, H., Ed.; Springer: Heidelberg, Germany, 2016; pp. 479–483.
28. Ribeiro, V.; Dias, N.; Paiva, T.; Hagström-Bex, L.; Nitz, N.; Pratesi, R.; Hecht, M. Current Trends in the Pharmacological Management of Chagas Disease. *Int. J. Parasitol. Drugs Drug. Resist.* **2020**, *12*, 7–17. [CrossRef]
29. Gabaldón-Figueira, J.C.; Martinez-Peinado, N.; Escabia, E.; Ros-Lucas, A.; Chatelain, E.; Scandale, I.; Gascon, J.; Pinazo, M.J.; Alonso-Padilla, J. State-of-the-Art in the Drug Discovery Pathway for Chagas Disease: A Framework for Drug Development and Target Validation. *Res. Rep. Trop. Med.* **2023**, *14*, 1–19. [CrossRef] [PubMed]
30. De Fuentes-Vicente, J.A.; Santos-Hernández, N.G.; Ruiz-Castillejos, C.; Espinoza-Medinilla, E.E.; Flores-Villegas, A.L.; de Alba-Alvarado, M.; Cabrera-Bravo, M.; Moreno-Rodríguez, A.; Vidal-López, D.G. What do you Need to Know before Studying Chagas Disease? A Beginner's Guide. *Trop. Med. Infect. Dis.* **2023**, *8*, 360. [CrossRef] [PubMed]
31. Pilicita, V.A.; Sonzogni, A.S.; Allasia, M.; Borra, F.; Minari, R.J.; Gonzalez, V.D.G. Proteins-Based Nanoparticles for Benznidazole Enteric Delivery. *Macromol. Biosci.* **2025**, *25*, e2400338. [CrossRef] [PubMed]
32. Schaub, G.A. Interactions of Trypanosomatids and Triatomines. *Adv. Insect Physiol.* **2009**, *37*, 177–242. [CrossRef]
33. Tibayrenc, M.; Ayala, F.J. Microevolution and Subspecific Taxonomy of *Trypanosoma cruzi*. *Infect. Genet. Evol.* **2022**, *103*, 105344. [CrossRef]
34. Zingales, B.; Bartholomeu, D.C. *Trypanosoma cruzi* Genetic Diversity: Impact on Transmission Cycles and Chagas Disease. *Mem. Inst. Oswaldo Cruz* **2022**, *117*, e210193. [CrossRef]
35. Velásquez-Ortiz, N.; Herrera, G.; Hernández, C.; Muñoz, M.; Ramírez, J.D. Discrete Typing Units of *Trypanosoma cruzi*: Geographical and Biological Distribution in the Americas. *Sci. Data* **2022**, *9*, 360. [CrossRef]
36. Pérez-Cascales, E.; Tellería, J. Unidades discretas de tipificación de *Trypanosoma cruzi*: Una revisión sobre lo que se conoce hasta el momento en Bolivia. *Rev. Arg. Parasitol.* **2024**, *13*, 13–24. (In Spanish)
37. Herreros-Cabello, A.; Callejas-Hernández, F.; Gironès, N.; Fresno, M. *Trypanosoma cruzi*: Genomic Diversity and Structure. *Pathogens* **2025**, *14*, 61. [CrossRef]
38. Telleria, J.; Tibayrenc, M.; Del Salto Mendoza, M.; Seveno, M.; Costales, J.A. Comparative Proteomic Analysis of *Trypanosoma cruzi* TcI Lineage Epimastigotes Unveils Metabolic and Phenotypic Differences between Fast- and Slow-Dividing Strains. *Exp. Parasitol.* **2023**, *252*, 108576. [CrossRef]
39. Tibayrenc, M. Modelling the Transmission of *Trypanosoma cruzi*: The Need for an Integrated Genetic Epidemiological and Population Genomics Approach. *Adv. Exp. Med. Biol.* **2010**, *673*, 200–211. [CrossRef] [PubMed]
40. Kollien, A.H.; Schaub, G.A. The Development of *Trypanosoma cruzi* in Triatominae. *Parasitol. Today* **2000**, *16*, 381–387. [CrossRef]

41. Tevere, E.; Di Capua, C.B.; Chasen, N.M.; Etheridge, R.D.; Cricco, J.A. *Trypanosoma cruzi* Heme Responsive Gene (TcHRG) Plays a Central Role in Orchestrating Heme Uptake in Epimastigotes. *FEBS J.* **2024**, *291*, 1186–1198. [CrossRef] [PubMed]
42. Serrano, A.A.; Schenkman, S.; Yoshida, N.; Mehlert, A.; Richardson, J.M.; Ferguson, M.A. The Lipid Structure of the Glycosylphosphatidylinositol-Anchored Mucin-Like Sialic Acid Acceptors of *Trypanosoma cruzi* Changes during Parasite Differentiation from Epimastigotes to Infective Metacyclic Trypomastigote Forms. *J. Biol. Chem.* **1995**, *270*, 27244–27253. [CrossRef]
43. Schaub, G.A.; Vogel, P.; Balczun, C. Parasite-Vector Interactions. In *Molecular Parasitology—Protozoan Parasites and Their Molecules*; Walochnik, J., Duchêne, M., Eds.; Springer: Wien, Austria, 2016; pp. 431–489.
44. Galvão, C. Taxonomy. In *Triatominae—The Biology of Chagas Disease Vectors*; Guarneri, A.A., Lorenzo, M.G., Eds.; Springer Nature: New York, NY, USA, 2021; pp. 15–38.
45. De Fuentes-Vicente, J.A.; Gutiérrez-Cabrera, A.E. Kissing Bugs (*Triatominae*). In *Encyclopedia of Infection and Immunity*; Rezaei, N., Ed.; Elsevier: Amsterdam, The Netherlands, 2020; Volume 2, pp. 953–970. [CrossRef]
46. Schaub, G.A. An Update on the Knowledge of Parasite-Vector Interactions of Chagas Disease. *Res. Rep. Trop. Med.* **2021**, *12*, 63–76. [CrossRef] [PubMed]
47. Schaub, G.A.; Mehlhorn, H. Insects. In *Encyclopedia of Parasitology*, 4th ed.; Mehlhorn, H., Ed.; Springer: Berlin, Germany, 2016; pp. 1348–1357.
48. Murillo-Solano, C.; López-Domínguez, J.; Gongora, R.; Rojas-Gulloso, A.; Usme-Ciro, J.; Perdomo-Balaguera, E.; Herrera, C.; Parra-Henao, G.; Dumonteil, E. Diversity and Interactions among Triatomine Bugs, their Blood Feeding Sources, Gut Microbiota and *Trypanosoma cruzi* in the Sierra Nevada de Santa Marta in Colombia. *Sci. Rep.* **2021**, *11*, 12306. [CrossRef]
49. Martínez-Ibarra, J.A.; Nogueda-Torres, B.; Meraz-Medina, T.; Diaz-Chavez, R.; Virgen-Cobian, C.J.; Quirarte-Brambila, M. Advantageous Feeding on Different Blood Meal Sources by the Chagas Disease Vector *Triatoma barberi* (Hemiptera: Reduviidae). *J. Med. Entomol.* **2019**, *56*, 1565–1570. [CrossRef]
50. Martínez-Ibarra, J.A.; Nogueda-Torres, B.; Montañez-Valde, O.D.; Michel-Parra, J.G. Influence of Blood Meal Source on the Biological Parameters of *Triatoma pallidipennis* (Heteroptera: Reduviidae) from Mexico. *Med. Vet. Entomol.* **2023**, *37*, 124–131. [CrossRef]
51. Dujardin, J.-P. The Body of Chagas Disease Vectors. *Pathogens* **2025**, *14*, 98. [CrossRef]
52. Lazzari, C.R. The Behaviour of Kissing Bugs. In *Triatominae—The Biology of Chagas Disease Vectors*; Guarneri, A.A., Lorenzo, M.G., Eds.; Springer Nature: New York, NY, USA, 2021; pp. 215–238. [CrossRef]
53. Wenk, P.; Lucic, S.; Betz, O. Functional Anatomy of the Hypopharynx and the Salivary Pump in the Feeding Apparatus of the Assassin Bug *Rhodnius prolixus* (Reduviidae, Heteroptera). *Zoomorphology* **2010**, *129*, 225–234. [CrossRef]
54. Pontes, G.B.; Minoli, S.; Insaurralde, I.O.; de Brito Sánchez, M.G.; Barrozo, R. Bitter Stimuli Modulate the Feeding Decision of a Blood-Sucking Insect via Two Sensory Inputs. *J. Exp. Biol.* **2014**, *217*, 3708–3717. [CrossRef]
55. Sant'Anna, M.R.V.; Soares, A.C.; Araújo, R.N.; Gontijo, N.F.; Pereira, M.H. Triatomines (Hemiptera, Reduviidae) Blood Intake: Physical Constraints and Biological Adaptations. *J. Insect Physiol.* **2017**, *97*, 20–26. [CrossRef] [PubMed]
56. Guarneri, A.A.; Diotaiuti, L.; Gontijo, N.F.; Gontijo, A.F.; Pereira, M.H. Comparison of Feeding Behaviour of *Triatoma infestans*, *Triatoma brasiliensis* and *Triatoma pseudomaculata* in Different Hosts by Electronic Monitoring of the Cibarial Pump. *J. Insect Physiol.* **2000**, *46*, 1121–1127. [CrossRef] [PubMed]
57. Schaub, G.A.; Lösch, P. Parasite/Host-Interrelationships of the Trypanosomatids *Trypanosoma cruzi* and *Blastocrithidia triatomae* and the Reduviid Bug *Triatoma infestans*: Influence of Starvation of the Bug. *Ann. Trop. Med. Parasitol.* **1989**, *83*, 215–223. [CrossRef]
58. Noireau, F.; Dujardin, J.P. Flight and Nutritional Status of Sylvatic *Triatoma sordida* and *Triatoma guasayana*. *Mem. Inst. Oswaldo Cruz* **2001**, *96*, 385–389. [CrossRef] [PubMed]
59. Sarquis, O.; Carvalho-Costa, F.A.; Oliveira, L.S.; Duarte, R.; D.'Andrea, P.S.; de Oliveira, T.G.; Lima, M.M. Ecology of *Triatoma brasiliensis* in Northeastern Brazil: Seasonal Distribution, Feeding Resources, and *Trypanosoma cruzi* Infection in a Sylvatic Population. *J. Vector Ecol.* **2010**, *35*, 385–394. [CrossRef]
60. Wigglesworth, V.B. The Physiology of Excretion in a Blood-Sucking Insect, *Rhodnius prolixus* (Hemiptera; Reduviidae) I. Composition of the Urine. *J. Exp. Biol.* **1931**, *8*, 411–427. [CrossRef]
61. Böker, C.A.; Schaub, G.A. Scanning Electron Microscopic Studies of *Trypanosoma cruzi* in the Rectum of its Vector *Triatoma infestans*. *Z. Parasitenkd.* **1984**, *70*, 459–469. [CrossRef]
62. Maddrell, S.H.P. The Fastest Fluid-Secreting Cell Known: The Upper Malpighian Tubule Cell of *Rhodnius*. *BioEssays* **1991**, *13*, 357–362. [CrossRef]
63. Oliveira, P.L.; Genta, F.A. Blood Digestion in Triatomine Insects. In *Triatominae—The Biology of Chagas Disease Vectors*; Guarneri, A.A., Lorenzo, M.G., Eds.; Springer Nature: New York, NY, USA, 2021; pp. 265–284.
64. Balczun, C.; Meiser, C.K.; Schaub, G.A. Triatomines as Vectors of American Trypanosomiasis. In *Parasitology Research Monographs. Arthropods as Vectors of Emerging Diseases*; Mehlhorn, H., Ed.; Springer: Berlin, Germany, 2012; Volume 3, pp. 275–299. [CrossRef]
65. Lehane, M. Managing the Blood Meal. In *The Biology of Blood Sucking Insects*, 2nd ed.; Lehane, M., Ed.; Cambridge University Press: Cambridge, UK, 2005; pp. 84–115.

66. Smit, J.D.; Guggenheim, R.; Bauer, P.G. Crystallized Hemoglobin in *Rhodnius prolixus* after a Blood Meal on Guinea-Pig. *Experientia* **1983**, *39*, 1335–1338. [CrossRef] [PubMed]
67. Billingsley, P.F.; Downe, A.E.R. Ultrastructural Changes in Posterior Midgut Cells Associated with Blood Feeding in Adult Female *Rhodnius prolixus* Stål (Heteroptera: Reduviidae). *Can. J. Zool.* **1983**, *61*, 2574–2586. [CrossRef]
68. Balczun, C.; Siemanowski, J.; Pausch, J.K.; Helling, S.; Marcus, K.; Stephan, C.; Meyer, H.E.; Schneider, T.; Cizmowski, C.; Oldenburg, M.; et al. Intestinal Aspartate Proteases TiCatD and TiCatD2 of the Haematophagous Bug *Triatoma infestans* (Reduviidae): Sequence Characterisation, Expression Pattern and Characterisation of Proteolytic Activity. *Insect Biochem. Mol. Biol.* **2012**, *42*, 240–250. [CrossRef]
69. Gama, M.d.V.F.; Alexandre, Y.d.N.; Pereira da Silva, J.M.; Castro, D.P.; Genta, F.A. Digestive α-L-Fucosidase Activity in *Rhodnius prolixus* after Blood Feeding: Effect of Secretagogue and Nutritional Stimuli. *Front. Physiol.* **2023**, *14*, 1123414. [CrossRef]
70. Henriques, B.S.; Gomes, B.; Oliveira, P.L.; Garcia, E.S.; Azambuja, P.; Genta, F.A. Characterization of the Temporal Pattern of Blood Protein Digestion in *Rhodnius prolixus*: First Description of Early and Late Gut Cathepsins. *Front. Physiol.* **2021**, *11*, 509310. [CrossRef]
71. Ouali, R.; Vieira, L.R.; Salmon, D.; Bousbata, S. Early Post-Prandial Regulation of Protein Expression in the Midgut of Chagas Disease Vector *Rhodnius prolixus* Highlights New Potential Targets for Vector Control Strategy. *Microorganisms* **2021**, *9*, 804. [CrossRef] [PubMed]
72. Rocha, L.L.; Neves, C.A.; Zanuncio, J.C.; Serrão, J.E. Digestive Cells in the Midgut of *Triatoma vitticeps* (Stal, 1859) in Different Starvation Periods. *C. R. Biol.* **2010**, *333*, 405–415. [CrossRef] [PubMed]
73. Ouali, R.; Bousbata, S. Unveiling the Peptidase Network Orchestrating Hemoglobin Catabolism in *Rhodnius prolixus*. *Mol. Cell. Proteomics.* **2024**, *23*, 100775. [CrossRef]
74. Rocha, L.L.; Neves, C.A.; Zanuncio, J.C.; Serrão, J.E. Endocrine and Regenerative Cells in the Midgut of Chagas' Disease Vector *Triatoma vitticeps* during Different Starvation Periods. *Folia Biol.* **2014**, *62*, 259–267. [CrossRef]
75. Kollien, A.H.; Waniek, P.J.; Nisbet, A.J.; Billingsley, P.F.; Schaub, G.A. Activity and Sequence Characterization of Two Cysteine Proteases in the Digestive Tract of the Reduviid Bug *Triatoma infestans*. *Insect Mol. Biol.* **2004**, *13*, 569–579. [CrossRef]
76. Borges, E.C.; Machado, E.M.M.; Garcia, E.S.; Azambuja, P. *Trypanosoma cruzi*: Effects of Infection on Cathepsin D Activity in the Midgut of *Rhodnius prolixus*. *Exp. Parasitol.* **2006**, *112*, 130–133. [CrossRef]
77. Billingsley, P.F.; Downe, A.E.R. Cellular Localisation of Aminopeptidase in the Midgut of *Rhodnius prolixus* Stål (Hemiptera: Reduviidae) during Blood Digestion. *Cell Tissue Res.* **1985**, *241*, 421–428. [CrossRef]
78. Kollien, A.H.; Grospietsch, T.; Kleffmann, T.; Zerbst-Boroffka, I.; Schaub, G.A. Ionic Composition of the Rectal Contents and Excreta of the Reduviid Bug *Triatoma infestans*. *J. Insect Physiol.* **2001**, *47*, 739–747. [CrossRef] [PubMed]
79. Ratcliffe, N.A.; Whitten, M.M.A. Vector Immunity. In *SGM Symposium 63: Microbe-Vector Interactions in Vector Borne Diseases*; Gillespie, S.H., Osborne, A., Eds.; Cambridge University Press: Cambridge, UK, 2004; pp. 199–262.
80. Müller, U.; Vogel, P.; Alber, G.; Schaub, G.A. The Innate Immune System of Mammals and Insects. In *Contributions to Microbiology*; Egesten, A., Schmidt, A., Herwald, H., Eds.; Karger: Basel, Switzerland, 2008; Volume 15, pp. 21–44.
81. Salcedo-Porras, N.; Lowenberger, C. Immune System of Triatomines. In *Triatominae–The Biology of Chagas Disease Vectors*; Guarneri, A.A., Lorenzo, M.G., Eds.; Springer Nature: New York, NY, USA, 2021; pp. 307–344.
82. Ratcliffe, N.A.; Mello, C.B.; Castro, H.C.; Dyson, P.; Figueiredo, M. Immune Reactions of Vector Insects to Parasites and Pathogens. *Microorganisms.* **2024**, *12*, 568. [CrossRef]
83. Guarneri, A.A.; Schaub, G.A. Interaction of Triatomines with their Bacterial Microbiota and Trypanosomes. In *Triatominae—The Biology of Chagas Disease Vectors*; Guarneri, A.A., Lorenzo, M.G., Eds.; Springer Nature: New York, NY, USA, 2021; pp. 345–386. [CrossRef]
84. Whitten, M.; Sun, F.; Tew, I.; Schaub, G.A.; Soukou, C.; Nappi, A.; Ratcliffe, N. Differential Modulation of *Rhodnius prolixus* Nitric Oxide Activities Following Challenge with *Trypanosoma rangeli*, *T. cruzi* and Bacterial Cell Wall Components. *Insect Biochem. Mol. Biol.* **2007**, *37*, 440–452. [CrossRef] [PubMed]
85. Genta, F.A.; Souza, R.S.; Garcia, E.S.; Azambuja, P. Phenoloxidases from *Rhodnius prolixus*: Temporal and Tissue Expression Pattern and Regulation by Ecdysone. *J. Insect Physiol.* **2010**, *56*, 1253–1259. [CrossRef]
86. Garcia, E.S.; Castro, D.P.; Figueiredo, M.B.; Azambuja, P. Immune Homeostasis to Microorganisms in the Guts of Triatomines (Reduviidae)—A Review. *Mem. Inst. Oswaldo Cruz* **2010**, *105*, 605–610. [CrossRef] [PubMed]
87. Salcedo-Porras, N.; Noor, S.; Cai, C.; Oliveira, P.L.; Lowenberger, C. *Rhodnius prolixus* Uses the Peptidoglycan Recognition Receptor rpPGRP-LC/LA to Detect Gram-Negative Bacteria and Activate the IMD Pathway. *Curr. Res. Insect Sci.* **2021**, *1*, 100006. [CrossRef]
88. Salcedo-Porras, N.; Oliveira, P.L.; Guarneri, A.A.; Lowenberger, C. A Fat Body Transcriptome Analysis of the Immune Responses of *Rhodnius prolixus* to Artificial Infections with Bacteria. *Parasit. Vectors* **2022**, *15*, 269. [CrossRef]

89. Alejandro, A.D.; Lilia, J.P.; Jesús, M.B.; Henry, R.M. The IMD and Toll Canonical Immune Pathways of *Triatoma pallidipennis* are Preferentially Activated by Gram-Negative and Gram-Positive Bacteria, respectively, but Cross-Activation also Occurs. *Parasit. Vectors* **2022**, *15*, 256. [CrossRef]
90. Borsatto, K.C.; Coronado, M.A.; Galvão, C.; Arni, R.K.; Alevi, K.C.C. Transcriptomics Applied to the Study of Chagas Disease Vectors. *Am. J. Trop. Med. Hyg.* **2022**, *106*, 1042–1048. [CrossRef] [PubMed]
91. Satyavathi, V.; Ghosh, R.; Subramanian, S. Long Non-Coding RNAs Regulating Immunity in Insects. *Noncoding RNA* **2017**, *3*, 14. [CrossRef]
92. Moure, U.A.E.; Tan, T.; Sha, L.; Lu, X.; Shao, Z.; Yang, G.; Wang, Y.; Cui, H. Advances in the Immune Regulatory Role of Non-Coding RNAs (miRNAs and lncRNAs) in Insect-Pathogen Interactions. *Front. Immunol.* **2022**, *13*, 856457. [CrossRef] [PubMed]
93. Kurtz, J. Specific Memory within Innate Immune Systems. *Trends Immunol.* **2005**, *26*, 186–192. [CrossRef] [PubMed]
94. Sheehan, G.; Farrell, G.; Kavanagh, K. Immune Priming: The Secret Weapon of the Insect World. *Virulence* **2020**, *11*, 238–246. [CrossRef]
95. Barillas-Mury, C.; Ribeiro, J.M.C.; Valenzuela, J.G. Understanding Pathogen Survival and Transmission by Arthropod Vectors to Pevent Human Disease. *Science* **2022**, *377*, eabc2757. [CrossRef]
96. Carmona-Peña, S.P.; Contreras-Garduño, J.; Castro, D.P.; Manjarrez, J.; Vázquez-Chagoyán, J.C. The Innate Immune Response of Triatomines against *Trypanosoma cruzi* and *Trypanosoma rangeli* with an Unresolved Question: Do Triatomines have Immune Memory? *Acta Trop.* **2021**, *224*, 106108. [CrossRef]
97. Carmona-Peña, S.P.; Vázquez-Chagoyán, J.C.; Castro, D.P.; Genta, F.A.; Contreras-Garduño, J. Benefits and Costs of Immune Memory in *Rhodnius prolixus* against *Trypanosoma cruzi*. *Microb. Pathog.* **2022**, *165*, 105505. [CrossRef]
98. Dillon, R.J.; Dillon, V.M. The Gut Bacteria of Insects: Nonpathogenic Interactions. *Annu. Rev. Entomol.* **2004**, *49*, 71–92. [CrossRef]
99. Douglas, A.E. The Molecular Basis of Bacterial-Insect Symbiosis. *J. Mol. Biol.* **2014**, *426*, 3830–3837. [CrossRef]
100. Tarabai, H.; Floriano, A.M.; Zima, J.; Filová, N.; Brown, J.J.; Roachell, W.; Smith, R.L.; Beatty, N.L.; Vogel, K.J.; Nováková, E. Microbiomes of Blood-Feeding Triatomines in the Context of Their Predatory Relatives and the Environment. *Microbiol. Spectr.* **2023**, *11*, e0168123. [CrossRef] [PubMed]
101. Wang, J.; Gao, L.; Aksoy, S. Microbiota in Disease-Transmitting Vectors. *Nat. Rev. Microbiol.* **2023**, *21*, 604–618. [CrossRef]
102. Da Lage, J.L.; Fontenelle, A.; Filée, J.; Merle, M.; Béranger, J.M.; Almeida, C.E.; Folly Ramos, E.; Harry, M. Evidence that Hematophagous Triatomine Bugs May Eat Plants in the Wild. *Insect Biochem. Mol. Biol.* **2024**, *165*, 104059. [CrossRef] [PubMed]
103. Schaub, G.A.; Jensen, C. Developmental Time and Mortality of the Reduviid Bug *Triatoma infestans* with Differential Exposure to Coprophagic Infections with *Blastocrithidia triatomae* (Trypanosomatidae). *J. Invertebr. Pathol.* **1990**, *55*, 17–27. [CrossRef]
104. Schaub, G.A. Direct Transmission of *Trypanosoma cruzi* between Vectors of Chagas' Disease. *Acta Trop.* **1988**, *45*, 11–19.
105. Lorenzo Figueiras, A.N.; Kenigsten, A.; Lazzari, C.R. Aggregation in the Haematophagous Bug *Triatoma infestans*: Chemical Signals and Temporal Pattern. *J. Insect Physiol.* **1994**, *40*, 311–316. [CrossRef]
106. Lorenzo, M.G.; Lazzari, C.R. The Spatial Pattern of Defaecation in *Triatoma infestans* and the Role of Faeces as a Chemical Mark of the Refuge. *J. Insect Physiol.* **1996**, *42*, 903–907. [CrossRef]
107. Vallejo, G.A.; Guhl, F.; Schaub, G.A. Triatominae—*Trypanosoma cruzi/T rangeli*: Vector-Parasite Interactions. *Acta Trop.* **2009**, *110*, 137–147. [CrossRef] [PubMed]
108. Dumonteil, E.; Ramirez-Sierra, M.J.; Pérez-Carrillo, S.; Teh-Poot, C.; Herrera, C.; Gourbière, S.; Waleckx, E. Detailed Ecological Associations of Triatomines Revealed by Metabarcoding and Next-Generation Sequencing: Implications for Triatomine Behavior and *Trypanosoma cruzi* Transmission Cycles. *Sci. Rep.* **2018**, *8*, 4140. [CrossRef]
109. Orantes, L.C.; Monroy, C.; Dorn, P.L.; Stevens, L.; Rizzo, D.M.; Morrissey, L.; Hanley, J.P.; Rodas, S.G.; Richards, B.; Wallin, K.F.; et al. Uncovering Vector, Parasite, Blood Meal and Microbiome Patterns from Mixed-DNA Specimens of the Chagas Disease Vector *Triatoma dimidiata*. *PLoS Negl. Trop. Dis.* **2018**, *12*, e0006730. [CrossRef]
110. Schaub, G.A. Intestinal Bacteria/Mutualistic Symbionts of Triatomines—A Review. *Mitt. Dtsch. Ges. Allg. Angew. Entomol.* **2020**, *22*, 191–194.
111. Kieran, T.J.; Arnold, K.M.H.; Thomas, J.C.; Varian, C.P.; Saldaña, A.; Calzada, J.E.; Glenn, T.C.; Gottdenker, N.L. Regional Biogeography of Microbiota Composition in the Chagas Disease Vector *Rhodnius pallescens*. *Parasit. Vectors* **2019**, *12*, 504. [CrossRef]
112. Arias-Giraldo, L.M.; Muñoz, M.; Hernández, C.; Herrera, G.; Velásquez-Ortiz, N.; Cantillo-Barraza, O.; Urbano, P.; Ramírez, J.D. Species-Dependent Variation of the Gut Bacterial Communities across *Trypanosoma cruzi* Insect Vectors. *PLoS ONE* **2020**, *15*, e0240916. [CrossRef] [PubMed]
113. Dumonteil, E.; Pronovost, H.; Bierman, E.F.; Sanford, A.; Majeau, A.; Moore, R.; Herrera, C. Interactions among *Triatoma sanguisuga* Blood Feeding Sources, Gut Microbiota and *Trypanosoma cruzi* Diversity in Southern Louisiana. *Mol. Ecol.* **2020**, *29*, 3747–3761. [CrossRef]

114. Mann, A.E.; Mitchell, E.A.; Zhang, Y.; Curtis-Robles, R.; Thapa, S.; Hamer, S.A.; Allen, M.S. Comparison of the Bacterial Gut Microbiome of North American *Triatoma* spp. with and without *Trypanosoma cruzi*. *Front. Microbiol.* **2020**, *11*, 364. [CrossRef] [PubMed]
115. Villacís, J.F.; López-Rosero, A.; Bustillos, J.J.; Cadena, M.; Yumiseva, C.A.; Grijalva, M.J.; Villacís, A.G. Bacterial Microbiota from the Gut of *Rhodnius ecuadoriensis*, a Vector of Chagas Disease in Ecuador's Central Coast and Southern Andes. *Front. Microbiol.* **2024**, *15*, 1464720. [CrossRef] [PubMed]
116. Teal, E.; Herrera, C.; Dumonteil, E. Metabolomics of Developmental Changes in *Triatoma sanguisuga* Gut Microbiota. *PLoS ONE* **2023**, *18*, e0280868. [CrossRef]
117. Gumpert, J. Untersuchungen über die Symbiose von Tieren mit Pilzen und Bakterien. X. Die Symbiose der Triatominen. 2. Infektion symbiontenfreier Triatominen mit symbiontischen und saprophytischen Mikroorganismen und gemeinsame Eigenschaften der symbiontischen Stämme. *Z. Allg. Mikrobiol.* **1962**, *2*, 290–302. [CrossRef]
118. Brecher, G.; Wigglesworth, V.B. The Transmission of *Actinomyces rhodnii* Erikson in *Rhodnius prolixus* Stål (Hemiptera) and its Influence on the Growth of the Host. *Parasitology* **1944**, *35*, 220–224. [CrossRef]
119. Goodchild, A.J.P. The Bacteria associated with *Triatoma infestans* and some other Species of Reduviidae. *Parasitology* **1955**, *45*, 441–448. [CrossRef]
120. Eichler, S.; Schaub, G.A. Development of Symbionts in Triatomine Bugs and the Effects of Infections with Trypanosomatids. *Exp. Parasitol.* **2002**, *100*, 17–27. [CrossRef] [PubMed]
121. Yassin, A.F. *Rhodococcus triatomae* sp. nov., Isolated from a Blood-Sucking Bug. *Int. J. Syst. Evol. Microbiol.* **2005**, *55*, 1575–1579. [CrossRef]
122. Eichler, S.; Reintjes, N.; Jung, M.; Balczun, C.; Dotson, E.; Beard, C.B.; Schaub, G.A. Identification of Bacterial Isolates and the Mutualistic Symbiont of *Triatoma infestans*, Vector of Chagas Disease. 2025; *manuscript in preparation*.
123. Drews, M. Die Cardia von *Triatoma infestans* (Reduviidae; Hemiptera). Untersuchungen zu ihrer Ultrastruktur und zur Lokalisation Symbiontischer Bakterien. Diploma Thesis, University Freiburg, Breisgau, Germany, 1988. (In German).
124. Vieira, C.S.; Waniek, P.J.; Castro, D.P.; Mattos, D.P.; Moreira, O.C.; Azambuja, P. Impact of *Trypanosoma cruzi* on Antimicrobial Peptide Gene Expression and Activity in the Fat Body and Midgut of *Rhodnius prolixus*. *Parasit. Vectors* **2016**, *9*, 119. [CrossRef] [PubMed]
125. Oliveira, J.L.; Cury, J.C.; Gurgel-Gonçalves, R.; Bahia, A.C.; Monteiro, F.A. Field-Collected *Triatoma sordida* from Central Brazil Display High Microbiota Diversity that Varies with Regard to Developmental Stage and Intestinal Segmentation. *PLoS Negl. Trop. Dis.* **2018**, *12*, e0006709. [CrossRef]
126. Eberhard, F.E.; Klimpel, S.; Guarneri, A.A.; Tobias, N.J. Exposure to *Trypanosoma* Parasites Induces Changes in the Microbiome of the Chagas Disease Vector *Rhodnius prolixus*. *Microbiome* **2022**, *10*, 45. [CrossRef]
127. Gilliland, C.A.; Patel, V.; McCormick, A.C.; Mackett, B.M.; Vogel, K.J. Using Axenic and Gnotobiotic Insects to Examine the Role of Different Microbes on the Development and Reproduction of the Kissing Bug *Rhodnius prolixus* (Hemiptera: Reduviidae). *Mol. Ecol.* **2023**, *32*, 920–935. [CrossRef] [PubMed]
128. Jung, M. Auswirkung der Infektion mit verschiedenen Bakterien auf die Entwicklung der Raubwanze *Triatoma infestans* (Reduviidae, Hemiptera). Diploma Thesis, Fakultät für Biologie und Biotechnologie, Ruhr-Universität Bochum, Bochum, Germany, 1995. (In German).
129. Eichler, S.; Schaub, G.A. The Effects of Aposymbiosis and of an Infection with *Blastocrithidia triatomae* (Trypanosomatidae) on the Tracheal System of the Reduviid Bugs *Rhodnius prolixus* and *Triatoma infestans*. *J. Insect Physiol.* **1998**, *44*, 131–140. [CrossRef]
130. Salcedo-Porras, N.; Umaña-Diaz, C.; Bitencourt, R.O.B.; Lowenberger, C. The Role of Bacterial Symbionts in Triatomines: An Evolutionary Perspective. *Microorganisms* **2020**, *8*, 1438. [CrossRef]
131. Wigglesworth, V.B. Symbiotic Bacteria in a Blood-Sucking Insect, *Rhodnius prolixus* Stål (Hemiptera, Triatomidae). *Parasitology* **1936**, *28*, 284–289. [CrossRef]
132. Tobias, N.J.; Eberhard, F.E.; Guarneri, A.A. Enzymatic Biosynthesis of B-Complex Vitamins is Supplied by Diverse Microbiota in the *Rhodnius prolixus* Anterior Midgut Following *Trypanosoma cruzi* Infection. *Comput. Struct. Biotechnol. J.* **2020**, *18*, 3395–3401. [CrossRef]
133. Eppinger, M.; Bunk, B.; Johns, M.A.; Edirisinghe, J.N.; Kutumbaka, K.K.; Koenig, S.S.; Creasy, H.H.; Rosovitz, M.J.; Riley, D.R.; Daugherty, S.; et al. Genome Sequences of the Biotechnologically Important *Bacillus megaterium* Strains QM B1551 and DSM319. *J. Bacteriol.* **2011**, *193*, 4199–42213. [CrossRef] [PubMed]
134. Lopez-Ordonez, T.; Flores-López, C.A.; Montejo-Lopez, R.; Cruz-Hernandez, A.; Conners, E.E. Cultivable Bacterial Diversity in the Gut of the Chagas Disease Vector *Triatoma dimidiata*: Identification of Possible Bacterial Candidates for a Paratransgenesis Approach. *Front. Ecol. Evol.* **2018**, *5*. [CrossRef]
135. Hill, P.; Campbell, J.A.; Petrie, I.A. *Rhodnius prolixus* and its Symbiotic Actinomycete: A Microbiological, Physiological and Behavioural Study. *Proc. R. Soc. Lond. B Biol. Sci.* **1976**, *194*, 501–525. [CrossRef] [PubMed]

136. Lake, P.; Friend, W.G. The Use of Artificial Diets to Determine some of the Effects of *Nocardia rhodnii* on the Development of *Rhodnius prolixus*. *J. Insect Physiol.* **1968**, *14*, 543–562. [CrossRef]
137. de Souza, M.V.N.; Ferreira, M.d.L.; Pinheiro, A.C.; Saraiva, M.F.; de Almeida, M.V.; Valle, M.S. Synthesis and Biological Aspects of Mycolic Acids: An Important Target against *Mycobacterium tuberculosis*. *Sci. World J.* **2008**, *8*, 720–751. [CrossRef]
138. Meiser, C.K.; Pausch, J.K.; Schaub, G.A. Feeding-Induced Changes of Bacteriolytic Activity and of the Pattern of Bacteriolytic Compounds in the Stomach and Small Intestine of the Haematophagous Bug *Triatoma infestans* (Klug, 1834) (Reduviidae, Triatominae). *Parasitologia* **2022**, *2*, 13–26. [CrossRef]
139. Vieira, C.S.; Waniek, P.J.; Mattos, D.P.; Castro, D.P.; Mello, C.B.; Ratcliffe, N.A.; Garcia, E.S.; Azambuja, P. Humoral Responses in *Rhodnius prolixus*: Bacterial Feeding Induces Differential Patterns of Antibacterial Activity and Enhances mRNA Levels of Antimicrobial Peptides in the Midgut. *Parasit. Vectors* **2014**, *7*, 232. [CrossRef]
140. Ribeiro, J.M.C.; Pereira, M.E.A. Midgut Glycosidases of *Rhodnius prolixus*. *Insect Biochem.* **1984**, *14*, 103–108. [CrossRef]
141. González-Rete, B.; Salazar-Schettino, P.M.; Bucio-Torres, M.I.; Córdoba-Aguilar, A.; Cabrera-Bravo, M. Activity of the Prophenoloxidase System and Survival of Triatomines Infected with Different *Trypanosoma cruzi* Strains under Different Temperatures: Understanding Chagas Disease in the Face of Climate Change. *Parasit. Vectors* **2019**, *12*, 219. [CrossRef]
142. Kollien, A.H.; Fechner, S.; Waniek, P.J.; Schaub, G.A. Isolation and Characterization of a cDNA Encoding for a Lysozyme from the Gut of the Reduviid Bug *Triatoma infestans*. *Arch. Insect Biochem. Physiol.* **2003**, *53*, 134–145. [CrossRef]
143. Araújo, C.A.C.; Waniek, P.J.; Stock, P.; Mayer, C.; Jansen, A.M.; Schaub, G.A. Sequence Characterization and Expression Patterns of Defensin and Lysozyme Encoding Genes from the Gut of the Reduviid Bug *Triatoma brasiliensis*. *Insect Biochem. Mol. Biol.* **2006**, *36*, 547–560. [CrossRef] [PubMed]
144. Waniek, P.J.; Jansen, A.M.; Araújo, C.A.C. *Trypanosoma cruzi* Infection Modulates the Expression of *Triatoma brasiliensis* def1 in the Midgut. *Vector-Borne Zoonotic Dis.* **2011**, *1*, 845–847. [CrossRef]
145. Gumiel, M.; de Mattos, D.P.; Vieira, C.S.; Moraes, C.S.; Moreira, C.J.C.; Gonzalez, M.S.; Teixeira-Ferreira, A.; Waghabi, M.; Azambuja, P.; Carels, N. Proteome of the Triatomine Digestive Tract: From Catalytic to Immune Pathways; Focusing on Annexin Expression. *Front. Mol. Biosci.* **2020**, *7*, 589435. [CrossRef]
146. Meiser, C.K.; Klenner, L.; Balczun, C.; Schaub, G.A. Bacteriolytic Activity in Saliva of the Haematophagous *Triatoma infestans* (Reduviidae) and Novel Characterization and Expression Site of a Third Lysozyme. *Arch. Insect Biochem. Physiol.* **2023**, *113*, e22013. [CrossRef] [PubMed]
147. Araújo, C.A.C.; Waniek, P.J.; Jansen, A.M. Development of a *Trypanosoma cruzi* (TcI) Isolate in the Digestive Tract of an Unfamiliar Vector, *Triatoma brasiliensis* (Hemiptera, Reduviidae). *Acta Trop.* **2008**, *107*, 195–199. [CrossRef]
148. Noireau, F.; Diosque, P.; Jansen, A.M. *Trypanosoma cruzi*: Adaptation to its Vectors and its Hosts. *Vet. Res.* **2009**, *40*, 26. [CrossRef] [PubMed]
149. Mejía-Jaramillo, A.M.; Peña, V.H.; Triana-Chávez, O. *Trypanosoma cruzi*: Biological Characterization of Lineages I and II Supports the Predominance of Lineage I in Colombia. *Exp. Parasitol.* **2009**, *121*, 83–91. [CrossRef]
150. Sandoval-Rodríguez, A.; Rojo, G.; López, A.; Ortiz, S.; Saavedra, M.; Botto-Mahan, C.; Cattan, P.E.; Solari, A. Comparing Vector Competence of *Mepraia gajardoi* and *Triatoma infestans* by Genotyping *Trypanosoma cruzi* Discrete Typing Units Present in Naturally Infected *Octodon degus*. *Acta Trop.* **2019**, *190*, 119–122. [CrossRef]
151. Garcia, E.S.; Genta, F.A.; de Azambuja, P.; Schaub, G.A. Interactions between intestinal compounds of triatomines and *Trypanosoma cruzi*. *Trends Parasitol.* **2010**, *26*, 499–505. [CrossRef]
152. Valença-Barbosa, C.; Finamore-Araujo, P.; Moreira, O.C.; Vergara-Meza, J.G.; Alvarez, M.V.N.; Nascimento, J.R.; Borges-Veloso, A.; Viana, M.C.; Lilioso, M.; Miguel, D.C.; et al. Genotypic *Trypanosoma cruzi* Distribution and Parasite Load Differ Ecotypically and According to Parasite Genotypes in *Triatoma brasiliensis* from Endemic and Outbreak Areas in Northeastern Brazil. *Acta Trop.* **2021**, *222*, 106054. [CrossRef]
153. Valença-Barbosa, C.; Finamore-Araujo, P.; Moreira, O.C.; Alvarez, M.V.N.; Borges-Veloso, A.; Barbosa, S.E.; Diotaiuti, L.; de Souza, R.C.M. High Parasitic Loads Quantified in Sylvatic *Triatoma melanica*, a Chagas Disease Vector. *Pathogens* **2022**, *11*, 1498. [CrossRef] [PubMed]
154. Pinto, A.d.S.; de Lana, M.; Bastrenta, B.; Barnabé, C.; Quesney, V.; Noël, S.; Tibayrenc, M. Compared Vectorial Transmissibility of Pure and Mixed Clonal Genotypes of *Trypanosoma cruzi* in *Triatoma infestans*. *Parasitol. Res.* **1998**, *84*, 348–353. [CrossRef] [PubMed]
155. Araújo, C.A.C.; Cabello, P.H.; Jansen, A.M. Growth Behaviour of Two *Trypanosoma cruzi* Strains in Single and Mixed Infections: In Vitro and in the Intestinal Tract of the Blood-Sucking Bug, *Triatoma brasiliensis*. *Acta Trop.* **2007**, *101*, 225–231. [CrossRef] [PubMed]
156. Schaub, G.A.; Böker, C.A. Colonization of the Rectum of *Triatoma infestans* by *Trypanosoma cruzi* Studied by Scanning Electron Microscopy: Influence of Blood Uptake by the Bug. *Parasitol. Res.* **1987**, *73*, 417–420. [CrossRef]
157. Schaub, G.A.; Grünfelder, C.; Zimmermann, D.; Peters, W. Binding of Lectin-Gold Conjugates by Two *Trypanosoma cruzi* Strains in Ampullae and Rectum of *Triatoma infestans*. *Acta Trop.* **1989**, *46*, 291–301. [CrossRef]

158. Ferreira, R.C.; Kessler, R.L.; Lorenzo, M.G.; Paim, R.M.M.; Ferreira, L.L.; Probst, C.M.; Alves-Silva, J.; Guarneri, A.A. Colonization of *Rhodnius prolixus* Gut by *Trypanosoma cruzi* Involves an Extensive Parasite Killing. *Parasitology* **2016**, *143*, 434–443. [CrossRef]
159. Schaub, G.A.; Lösch, P. *Trypanosoma cruzi*: Origin of Metacyclic Trypomastigotes in the Urine of the Vector *Triatoma infestans*. *Exp. Parasitol.* **1988**, *65*, 174–186. [CrossRef]
160. Mwangi, V.I.; Martinez, E.G.; Leda, R.L.; Catunda, M.E.S.L.A.; Dias, A.d.S.; Padron Antonio, Y.; Guerra, M.d.G.V.B. Resisting an Invasion: A Review of the Triatomine Vector (Kissing bug) Defense Strategies against a *Trypanosoma* sp. Infection. *Acta Trop.* **2023**, *238*, 106745. [CrossRef]
161. Urbina, J.A. Intermediary Metabolism of *Trypanosoma cruzi*. *Parasitol. Today* **1994**, *10*, 107–110. [CrossRef]
162. Conzelmann, K.K. Die Entwicklung von *Trypanosoma* (*S.*) *cruzi* Chagas, 1909 (Protozoa: Kinetoplastida) in *Dipetalogaster maxima* Uhler 1894 und *Triatoma pallidipennis* Stal, 1872 (Hemiptera: Triatominae). Diploma Thesis, Faculty Biology, University Tübingen, Tübingen, Germany, 1983. (In German).
163. Schaub, G.A. *Trypanosoma cruzi*: Quantitative Studies of Development of Two Strains in Small Intestine and Rectum of the Vector *Triatoma infestans*. *Exp. Parasitol.* **1989**, *68*, 260–273. [CrossRef]
164. Kollien, A.H.; Schaub, G.A. The Development of *Trypanosoma cruzi* (Trypanosomatidae) in the Reduviid Bug *Triatoma infestans* (Insecta): Influence of Starvation. *J. Eukaryot. Microbiol.* **1998**, *45*, 59–63. [CrossRef] [PubMed]
165. Tamayo, L.D.; Guhl, F.; Vallejo, G.A.; Ramírez, J.D. The Effect of Temperature Increase on the Development of *Rhodnius prolixus* and the Course of *Trypanosoma cruzi* Metacyclogenesis. *PLoS Negl. Trop. Dis.* **2018**, *12*, e0006735. [CrossRef] [PubMed]
166. Kleffmann, T.; Schmidt, J.; Schaub, G.A. Attachment of *Trypanosoma cruzi* Epimastigotes to Hydrophobic Substrates and Use of this Property to Separate Stages and Promote Metacyclogenesis. *J. Eukaryot. Microbiol.* **1998**, *45*, 548–555. [CrossRef]
167. Schmidt, J.; Kleffmann, T.; Schaub, G.A. Hydrophobic Attachment of *Trypanosoma cruzi* to a Superficial Layer of the Rectal Cuticle in the Bug *Triatoma infestans*. *Parasitol. Res.* **1998**, *84*, 527–536. [CrossRef]
168. Won, M.M.; Baublis, A.; Burleigh, B.A. Proximity-Dependent Biotinylation and Identification of Flagellar Proteins in *Trypanosoma cruzi*. *mSphere* **2023**, *8*, e0008823. [CrossRef] [PubMed]
169. Yanase, R.; Pruzinova, K.; Owino, B.O.; Rea, E.; Moreira-Leite, F.; Taniguchi, A.; Nonaka, S.; Sádlová, J.; Vojtkova, B.; Volf, P.; et al. Discovery of Essential Kinetoplastid-Insect Adhesion Proteins and their Function in *Leishmania*-Sand Fly Interactions. *Nat. Commun.* **2024**, *15*, 6960. [CrossRef]
170. Perdomo-Gómez, C.D.; Ruiz-Uribe, N.E.; González, J.M.; Forero-Shelton, M. Extensible Membrane Nanotubules Mediate Attachment of *Trypanosoma cruzi* Epimastigotes under Flow. *PLoS ONE* **2023**, *18*, e0283182. [CrossRef]
171. Denecke, S.; Malfara, M.F.; Hodges, K.R.; Holmes, N.A.; Williams, A.R.; Gallagher-Teske, J.H.; Pascarella, J.M.; Daniels, A.M.; Sterk, G.J.; Leurs, R.; et al. Adhesion of *Crithidia fasciculata* Promotes a Rapid Change in Developmental Fate Driven by cAMP Signaling. *mSphere* **2024**, *9*, e0061724. [CrossRef]
172. Schaub, G.A. The Effects of Trypanosomatids on Insects. *Adv. Parasitol.* **1992**, *31*, 255–319. [CrossRef]
173. Martínez-Ibarra, J.A.; Nogueda-Torres, B.; Montañez-Valde, O.D.; Michel-Parra, J.G.; Ambriz-Santos, M.A. Behavioral Parameters of Six Populations of *Meccus phyllosomus longipennis* (Heteroptera: Reduviidae) from Areas with High and Low Prevalences of *Trypanosoma cruzi* Human Infection. *Insect Sci.* **2021**, *28*, 850–860. [CrossRef]
174. Martínez-Ibarra, J.A.; Grant-Guillén, Y.; Nogueda-Torres, B.; Meraz-Medina, T.; Martínez-Grant, D.M. Life History Data of a *Triatoma protracta nahuatlae*, *T. sinaloensis*, and their Laboratory Hybrids (Hemiptera: Reduviidae). *J. Med. Entomol.* **2024**, *61*, 309–317. [CrossRef] [PubMed]
175. Nogueda-Torres, B.; Montañez-Valdez, O.D.; Michel-Parra, J.G.; Martínez-Grant, D.M.; Martínez-Ibarra, J.A. Biological Parameters of Three Populations of *Triatoma dimidiata s. s.* (Hemiptera: Reduviidae) from Western Mexico. *J. Med. Entomol.* **2021**, *58*, 2114–2123. [CrossRef]
176. Meraz-Medina, T.; Grant-Guillén, Y.; Mercado-Trujillo, J.R.; Nogueda-Torres, B.; Martínez-Ibarra, J.A. Behavioral and Biological Parameters of Six Populations of *Triatoma pallidipennis* (Heteroptera: Reduviidae) from Areas with High and Low Prevalence Rates of *Trypanosoma cruzi* Human Infection. *Acta Trop.* **2024**, *256*, 107259. [CrossRef] [PubMed]
177. Padilla, N.A.; Moncayo, A.L.; Keil, C.B.; Grijalva, M.J.; Villacís, A.G. Life Cycle, Feeding, and Defecation Patterns of *Triatoma carrioni* (Hemiptera: Reduviidae), under Laboratory Conditions. *J. Med. Entomol.* **2019**, *6*, 617–624. [CrossRef]
178. Kleffmann, T. Mechanismen der Anheftung und Induktion der Metazyklogenese von *Trypanosoma cruzi* in *Triatoma infestans*. Ph.D. Thesis, Fakultät für Biologie und Biotechnologie, Ruhr-Universität Bochum, Bochum, Germany, 1999. (In German).
179. Maissner, F.F.; Silva, C.A.O.; Farias, A.B.; Costa, E.P.; Nepomuceno-Silva, J.L.; da Silva, J.R.; Mury, F.B. α-Glucosidase Isoform G Contributes to Heme Detoxification in *Rhodnius prolixus* and its Knockdown Affects *Trypanosoma cruzi* Metacyclogenesis. *Curr. Res. Insect Sci.* **2024**, *6*, 100100. [CrossRef] [PubMed]
180. Jimenez, V.; Miranda, K.; Augusto, I. The Old and the New about the Contractile Vacuole of *Trypanosoma cruzi*. *J. Eukaryot. Microbiol.* **2022**, *69*, e12939. [CrossRef]
181. Docampo, R. Advances in the Cellular Biology, Biochemistry, and Molecular Biology of Acidocalcisomes. *Microbiol. Mol. Biol. Rev.* **2024**, *88*, e0004223. [CrossRef]

182. Augusto, I.; Girard-Dias, W.; Schoijet, A.; Alonso, G.D.; Portugal, R.V.; de Souza, W.; Jimenez, V.; Miranda, K. Quantitative Assessment of the Nanoanatomy of the Contractile Vacuole Complex in *Trypanosoma cruzi*. *Life Sci. Alliance* **2024**, *7*, e202402826. [CrossRef]
183. Kollien, A.H.; Schaub, G.A. *Trypanosoma cruzi* in the Rectum of the Bug *Triatoma infestans*: Effects of Blood Ingestion by the Starved Vector. *Am. J. Trop. Med. Hyg.* **1998**, *59*, 166–170. [CrossRef]
184. Brack, C. Elektronenmikroskopische Untersuchungen zum Lebenszyklus von *Trypanosoma cruzi* unter besonderer Berücksichtigung der Entwicklungsformen im Übertrager *Rhodnius prolixus*. *Acta Trop.* **1968**, *25*, 289–356. (In German)
185. Loshouarn, H.; Guarneri, A.A. Effects of Fasting on the Interplay between Temperature and *Trypanosoma cruzi* Infection on the Life Cycle of the Chagas Disease Vector *Rhodnius prolixus*. *PLoS Negl. Trop. Dis.* **2024**, *18*, e0012665. [CrossRef]
186. Schaub, G.A.; Böker, C.A. Colonization of the Rectum of *Triatoma infestans* by *Trypanosoma cruzi*: Influence of Starvation Studied by Scanning Electron Microscopy. *Acta Trop.* **1986**, *43*, 349–354. [PubMed]
187. Moreira, O.C.; Verly, T.; Finamore-Araujo, P.; Gomes, S.A.O.; Lopes, C.M.; Sousa, D.M.; Azevedo, L.R.; Mota, F.F.; D'Avila-Levy, C.M.; Santos-Mallet, J.R.; et al. Development of Conventional and Real-Time Multiplex PCR-Based Assays for Estimation of Natural Infection Rates and *Trypanosoma cruzi* Load in Triatomine Vectors. *Parasit. Vectors* **2017**, *10*, 404. [CrossRef] [PubMed]
188. Schaub, G.A. Does *Trypanosoma cruzi* Stress its Vector? *Parasitol. Today* **1989**, *5*, 185–188. [CrossRef] [PubMed]
189. Schaub, G.A.; Meiser, C.K.; Balczun, C. Interactions of *Trypanosoma cruzi* and Triatomines. In *Parasitology Research Monographs. Progress in Parasitology*; Mehlhorn, H., Ed.; Springer: Berlin, Germany, 2011; Volume 2, pp. 155–178.
190. Guarneri, A.A.; Pereira, M.H.; Diotaiuti, L. Influence of the Blood Meal Source on the Development of *Triatoma infestans*, *Triatoma brasiliensis*, *Triatoma sordida* and *Triatoma pseudomaculata* (Heteroptera, Reduviidae). *J. Med. Entomol.* **2000**, *37*, 373–379. [CrossRef]
191. Heger, T.J.; Guerin, P.M.; Eugster, W. Microclimatic Factors Influencing Refugium Suitability for *Rhodnius prolixus*. *Physiol. Entomol.* **2006**, *31*, 248–256. [CrossRef]
192. Rolandi, C.; Schilman, P.E. The Costs of Living in a Thermal Fluctuating Environment for the Tropical Haematophagous Bug, *Rhodnius prolixus*. *J. Therm. Biol.* **2018**, *74*, 92–99. [CrossRef]
193. Peterson, J.K.; Graham, A.L.; Dobson, A.P.; Chavez, O.T. *Rhodnius prolixus* Life History Outcomes Differ when Infected with Different *Trypanosoma cruzi* I Strains. *Am. J. Trop. Med. Hyg.* **2015**, *93*, 564–572. [CrossRef]
194. Mc Cabe, A.; Yañez, F.; Pinto, R.; López, A.; Ortiz, S.; Muñoz-San Martin, C.; Botto-Mahan, C.; Solari, A. Survivorship of Wild Caught *Mepraia spinolai* Nymphs: The Effect of Seasonality and *Trypanosoma cruzi* Infection after Feeding and Fasting in the Laboratory. Infect. *Genet. Evol.* **2019**, *71*, 197–204. [CrossRef]
195. Estay-Olea, D.; Correa, J.P.; de Bona, S.; Bacigalupo, A.; Quiroga, N.; San Juan, E.; Solari, A.; Botto-Mahan, C. *Trypanosoma cruzi* could Affect Wild Triatomine Approaching Behaviour to Humans by Altering Vector Nutritional Status: A Field Test. *Acta Trop.* **2020**, *210*, 105574. [CrossRef]
196. Eberhard, F.E.; Klimpel, S.; Guarneri, A.A.; Tobias, N.J. Metabolites as Predictive Biomarkers for *Trypanosoma cruzi* Exposure in Triatomine Bugs. *Comput. Struct. Biotech. J.* **2021**, *19*, 3051–3057. [CrossRef] [PubMed]
197. Cuevas, I.C.; Cazzulo, J.J.; Sánchez, D.O. Gp63 Homologues in *Trypanosoma cruzi*: Surface Antigens with Metalloprotease Activity and a Possible Role in Host Cell Infection. *Infect. Immun.* **2003**, *71*, 5739–5749. [CrossRef]
198. Botzotz, J.; Méndez-Valdés, G.; Ortiz, S.; López, A.; Botto-Mahan, C.; Solari, A. Natural *Trypanosoma cruzi* Infection and Climatic Season Influence the Developmental Capacity in Field-Caught *Mepraia spinolai* Nymphs. *Insects* **2023**, *14*, 272. [CrossRef] [PubMed]
199. Cordero-Montoya, G.; Flores-Villegas, A.L.; Salazar-Schettino, P.M.; Vences-Blanco, M.O.; Rocha-Ortega, M.; Gutiérrez-Cabrera, A.E.; Rojas-Ortega, E.; Córdoba-Aguilar, A. The Cost of being a Killer's Accomplice: *Trypanosoma cruzi* Impairs the Fitness of Kissing Bugs. *Parasitol. Res.* **2019**, *118*, 2523–2529. [CrossRef]
200. Lobbia, P.A.; Rodríguez, C.; Remón, C.; Manteca-Acosta, M. Reproductive Consequences of the Interaction *Trypanosoma cruzi—Triatoma infestans* and its Trade-Off with Survival. *J. Invertebr. Pathol.* **2024**, *206*, 108183. [CrossRef]
201. De Bona, S.; Correa, J.P.; San Juan, E.; Estay-Olea, D.; Quiroga, N.; Bacigalupo, A.; Araya-Donoso, R.; Botto-Mahan, C. Opportunistic or Selective? Stage-Dependent Feeding Behavior in a Wild Vector of Chagas Disease. *Int. J. Parasitol.* **2023**, *53*, 55–64. [CrossRef]
202. May-Concha, I.J.; Escalante-Talavera, M.J.; Dujardin, J.P.; Waleckx, E. Does *Trypanosoma cruzi* (Chagas, 1909) (Kinetoplastida: Trypanosomatidae) Modify the Antennal Phenotype of *Triatoma dimidiata* (Latreille, 1811) (Hemiptera: Triatominae)? *Parasit. Vectors* **2022**, *15*, 466. [CrossRef]
203. Ramírez-González, M.G.; Flores-Villegas, A.L.; Salazar-Schettino, P.M.; Gutiérrez-Cabrera, A.E.; Rojas Ortega, E.; Córdoba-Aguilar, A. Zombie Bugs? Manipulation of Kissing Bug Behavior by the Parasite *Trypanosoma cruzi*. *Acta Trop.* **2019**, *200*, 105177. [CrossRef] [PubMed]
204. Chacón, F.; Muñoz-San Martín, C.; Bacigalupo, A.; Álvarez-Duhart, B.; Solís, R.; Cattan, P.E. *Trypanosoma cruzi* Parasite Load Modulates the Circadian Activity Pattern of *Triatoma infestans*. *Insects* **2022**, *13*, 76. [CrossRef]

205. Marliére, N.P.; Lorenzo, M.G.; Guarneri, A.A. *Trypanosoma cruzi*-Infected *Rhodnius prolixus* Endure Increased Predation Facilitating Parasite Transmission to Mammal Hosts. *PLoS Negl. Trop. Dis.* **2021**, *15*, e0009570. [CrossRef]
206. Verly, T.; Costa, S.; Lima, N.; Mallet, J.; Odêncio, F.; Pereira, M.; Moreira, C.J.C.; Britto, C.; Pavan, M.G. Vector Competence and Feeding-Excretion Behavior of *Triatoma rubrovaria* (Blanchard, 1843) (Hemiptera: Reduviidae) Infected with *Trypanosoma cruzi* TcVI. *PLoS Negl. Trop. Dis.* **2020**, *14*, e0008712. [CrossRef]
207. Depickère, S.; Ramírez-Ávila, G.M.; Deneubourg, J.-L. Alteration of the Aggregation and Spatial Organization of the Vector of Chagas Disease, *Triatoma infestans*, by the Parasite *Trypanosoma cruzi*. *Sci. Rep.* **2019**, *9*, 17432. [CrossRef]
208. Alavez-Rosas, D.; Gutiérrez-Cabrera, A.E.; Cruz-López, L.; Córdoba-Aguilar, A. Lessons to be Popular: The Chemical Basis of Aggregation in *Trypanosoma cruzi*-Infected and Non-infected Chagasic Bugs. *R. Soc. Open Sci.* **2024**, *11*, 231271. [CrossRef]
209. Ouali, R.; Vieira, L.R.; Salmon, D.; Bousbata, S. *Trypanosoma cruzi* Reprograms Mitochondrial Metabolism within the Anterior Midgut of its Vector *Rhodnius prolixus* during the Early Stages of Infection. *Parasit. Vectors* **2024**, *17*, 381. [CrossRef]
210. de Araujo, M.F.C.; Cardoso, L.S.; Pereira, M.H.; Pereira, M.G.; Atella, G.C. *Trypanosoma cruzi* Infection Modulates Secreted Phospholipase A_2 Expression in the Salivary Glands of *Rhodnius prolixus*. *Acta Trop.* **2024**, *257*, 107281. [CrossRef] [PubMed]
211. Buarque, D.S.; Spindola, L.M.; Martins, R.M.; Braz, G.R.; Tanaka, A.S. Tigutcystatin, a Cysteine Protease Inhibitor from *Triatoma infestans* Midgut Expressed in Response to *Trypanosoma cruzi*. *Biochem. Biophys. Res. Comm.* **2011**, *413*, 241–247. [CrossRef] [PubMed]
212. Buarque, D.S.; Braz, G.R.C.; Martins, R.M.; Tanaka-Azevedo, A.M.; Gomes, C.M.; Oliveira, F.A.A.; Schenkman, S.; Tanaka, A.S. Differential Expression Profiles in the Midgut of *Triatoma infestans* Infected with *Trypanosoma cruzi*. *PLoS ONE* **2013**, *8*, e61203. [CrossRef] [PubMed]
213. Pausch, J.K. Characterization of Intestinal Antibacterial Factors of *Triatoma infestans* (Reduviidae, Insecta) and their Interaction with *Trypanosoma cruzi* (Trypanosomatidae, Kinetoplastida). Ph.D. Thesis, Fakultät für Biologie und Biotechnologie, Ruhr-Universität Bochum, Bochum, Germany, 2012.
214. Ursic-Bedoya, R.J.; Nazzari, H.; Cooper, D.; Triana, O.; Wolff, M.; Lowenberger, C. Identification and Characterization of Two Novel Lysozymes from *Rhodnius prolixus*, a Vector of Chagas Disease. *J. Insect Physiol.* **2008**, *54*, 593–603. [CrossRef]
215. Reynoso-Ducoing, O.A.; González-Rete, B.; Díaz, E.; Candelas-Otero, F.N.; López-Aviña, J.A.; Cabrera-Bravo, M.; Bucio-Torres, M.I.; Torres-Gutiérrez, E.; Salazar-Schettino, P.M. Expression of Proteins, Glycoproteins, and Transcripts in the Guts of Fasting, Fed, and *Trypanosoma cruzi*-Infected Triatomines: A Systematic Review. *Pathogens* **2023**, *12*, 1124. [CrossRef]
216. Carvalho-Costa, T.M.; Tiveron, R.D.R.; Mendes, M.T.; Barbosa, C.G.; Nevoa, J.C.; Roza, G.A.; Silva, M.V.; Figueiredo, H.C.P.; Rodrigues, V.; Soares, S.C.; et al. Salivary and Intestinal Transcriptomes Reveal Differential Gene Expression in Starving, Fed and *Trypanosoma cruzi*-Infected *Rhodnius neglectus*. *Front. Cell. Infect. Microbiol.* **2021**, *11*, 773357. [CrossRef]
217. Díaz-Garrido, P.; Cárdenas-Guerra, R.E.; Martínez, I.; Poggio, S.; Rodríguez-Hernández, K.; Rivera-Santiago, L.; Ortega-López, J.; Sánchez-Esquivel, S.; Espinoza, B. Differential Activity on Trypanosomatid Parasites of a Novel Recombinant Defensin Type 1 from the Insect *Triatoma* (*Meccus*) *pallidipennis*. *Insect Biochem. Mol. Biol.* **2021**, *139*, 103673. [CrossRef] [PubMed]
218. da Mota, F.F.; Castro, D.P.; Vieira, C.S.; Gumiel, M.; de Albuquerque, J.P.; Carels, N.; Azambuja, P. In Vitro Trypanocidal Activity, Genomic Analysis of Isolates, and In Vivo Transcription of Type VI Secretion System of *Serratia marcescens* Belonging to the Microbiota of *Rhodnius prolixus* Digestive Tract. *Front. Microbiol.* **2019**, *9*, 3205. [CrossRef]
219. Gumiel, M.; da Mota, F.F.; de Sousa Rizzo, V.; Sarquis, O.; De Castro, D.P.; Lima, M.M.; de Souza Garcia, E.; Carels, N.; Azambuja, P. Characterization of the Microbiota in the Guts of *Triatoma brasiliensis* and *Triatoma pseudomaculata* Infected by *Trypanosoma cruzi* in Natural Conditions Using Culture Independent Methods. *Parasit. Vectors* **2015**, *8*, 245. [CrossRef]
220. Beard, C.B.; Cordon-Rosales, C.; Durvasula, R.V. Bacterial Symbionts of the Triatominae and their Potential Use in Control of Chagas Disease Transmission. *Annu. Rev. Entomol.* **2002**, *47*, 123–141. [CrossRef] [PubMed]
221. Durvasula, R.V.; Sundaram, R.K.; Kirsch, P.; Hurwitz, I.; Crawford, C.V.; Dotson, E.; Beard, C.B. Genetic Transformation of a Corynebacterial Symbiont from the Chagas Disease Vector *Triatoma infestans*. *Exp. Parasitol.* **2008**, *119*, 94–98. [CrossRef] [PubMed]
222. Jose, C.; Klein, N.; Wyss, S.; Fieck, A.; Hurwitz, I.; Durvasula, R. Recombinant Arthrobacter β-1, 3-Glucanase as a Potential Effector Molecule for Paratransgenic Control of Chagas Disease. *Parasit. Vectors* **2013**, *6*, 65. [CrossRef]
223. Taracena, M.L.; Oliveira, P.L.; Almendares, O.; Umaña, C.; Lowenberger, C.; Dotson, E.M.; Paiva-Silva, G.O.; Pennington, P.M. Genetically Modifying the Insect Gut Microbiota to Control Chagas Disease Vectors through Systemic RNAi. *PLoS Negl. Trop. Dis.* **2015**, *9*, e0003358. [CrossRef]
224. Whitten, M.M.; Facey, P.D.; Del Sol, R.; Fernández-Martínez, L.T.; Evans, M.C.; Mitchell, J.J.; Bodger, O.G.; Dyson, P.J. Symbiont-Mediated RNA Interference in Insects. *Proc. Biol. Sci.* **2016**, *283*, 20160042. [CrossRef]
225. Rupawate, P.S.; Roylawar, P.; Khandagale, K.; Gawande, S.; Ade, A.B.; Jaiswal, D.K.; Borgave, S. Role of Gut Symbionts of Insect Pests: A Novel Target for Insect-Pest Control. *Front. Microbiol.* **2023**, *14*, 1146390. [CrossRef]
226. Azambuja, P.; Garcia, E.S.; Ratcliffe, N.A. Gut Microbiota and Parasite Transmission by Insect Vectors. *Trends Parasitol.* **2005**, *21*, 568–572. [CrossRef]

227. Teotônio, I.M.S.N.; Dias, N.; Hagström-Bex, L.; Nitz, N.; Francisco, A.F.; Hecht, M. Intestinal Microbiota—A Modulator of the *Trypanosoma cruzi*-Vector-Host Triad. *Microb. Pathog.* **2019**, *137*, 103711. [CrossRef] [PubMed]
228. Duarte-Silva, E.; Morais, L.H.; Clarke, G.; Savino, W.; Peixoto, C. Targeting the Gut Microbiota in Chagas Disease: What do we Know so Far? *Front. Microbiol.* **2020**, *11*, 585857. [CrossRef] [PubMed]
229. Castro, D.P.; Moraes, C.S.; Gonzalez, M.S.; Ratcliffe, N.A.; Azambuja, P.; Garcia, E.S. *Trypanosoma cruzi* Immune Response Modulation Decreases Microbiota in *Rhodnius prolixus* Gut and is Crucial for Parasite Survival and Development. *PLoS ONE* **2012**, *7*, e36591. [CrossRef] [PubMed]
230. Batista, K.K.S.; Vieira, C.S.; Florentino, E.B.; Caruso, K.F.B.; Teixeira, P.T.P.; da Silva Moraes, C.; Genta, F.A.; de Azambuja, P.; de Castro, D.P. Nitric Oxide Effects on *Rhodnius prolixus*'s Immune Responses, Gut Microbiota and *Trypanosoma cruzi* Development. *J. Insect Physiol.* **2020**, *126*, 104100. [CrossRef]
231. Buarque, D.S.; Gomes, C.M.; Araújo, R.N.; Pereira, M.H.; Ferreira, R.C.; Guarneri, A.A.; Tanaka, A.S. A New Antimicrobial Protein from the Anterior Midgut of *Triatoma infestans* Mediates *Trypanosoma cruzi* Establishment by Controlling the Microbiota. *Biochimie* **2016**, *123*, 138–143. [CrossRef]
232. Waltmann, A.; Willcox, A.C.; Balasubramanian, S.; Mayori, K.B.; Guerrero, S.M.; Sanchez, R.S.; Roach, J.; Pino, C.C.; Gilman, R.H.; Bern, C.; et al. Hindgut Microbiota in Laboratory-Reared and Wild *Triatoma infestans*. *PLoS Negl. Trop. Dis.* **2019**, *13*, e0007383. [CrossRef]
233. Soares, T.S.; Buarque, D.S.; Queiroz, B.R.; Gomes, C.M.; Braz, G.R.; Araújo, R.N.; Pereira, M.H.; Guarneri, A.A.; Tanaka, A.S. A Kazal-Type Inhibitor is Modulated by *Trypanosoma cruzi* to Control Microbiota inside the Anterior Midgut of *Rhodnius prolixus*. *Biochimie* **2015**, *112*, 41–48. [CrossRef]
234. Castro, D.P.; Moraes, C.S.; Gonzalez, M.S.; Ribeiro, I.M.; Tomassini, T.C.B.; Azambuja, P.; Garcia, E.S. Physalin B Inhibits *Trypanosoma cruzi* Infection in the Gut of *Rhodnius prolixus* by Affecting the Immune System and Microbiota. *J. Insect Physiol.* **2012**, *58*, 1620–1625. [CrossRef]
235. Araújo, C.A.C.; Pacheco, J.P.F.; Waniek, P.J.; Geraldo, R.B.; Sibajev, A.; Dos Santos, A.L.; Evangelho, V.G.O.; Dyson, P.J.; Azambuja, P.; Ratcliffe, N.A.; et al. A Rhamnose-Binding Lectin from *Rhodnius prolixus* and the Impact of its Silencing on Gut Bacterial Microbiota and *Trypanosoma cruzi*. *Dev. Comp. Immunol.* **2021**, *114*, 103823. [CrossRef]
236. Rodríguez-Ruano, S.M.; Škochová, V.; Rego, R.O.M.; Schmidt, J.O.; Roachell, W.; Hypša, V.; Nováková, E. Microbiomes of North American Triatominae: The Grounds for Chagas Disease Epidemiology. *Front. Microbiol.* **2018**, *9*, 1167. [CrossRef]
237. Jiménez-Cortés, J.G.; García-Contreras, R.; Bucio-Torres, M.I.; Cabrera-Bravo, M.; López-Jácome, L.E.; Franco-Cendejas, R.; Vences-Blanco, M.O.; Salazar-Schettino, P.M. Bacteria Cultured from the Gut of *Meccus pallidipennis* (Hemiptera: Reduviidae), a Triatomine Species Endemic to Mexico. *Med. Vet. Entomol.* **2021**, *35*, 478–483. [CrossRef] [PubMed]
238. Eichler, S. Interaktionen von Triatominen mit ihren Symbionten und Trypanosomatiden. Ph.D. Thesis, Fakultät für Biologie und Biotechnologie, Ruhr-Universität Bochum, Bochum, Germany, 1998. (In German).
239. Mühlpfordt, H. Der Einfluß der Darmsymbionten von *Rhodnius prolixus* auf *Trypanosoma cruzi*. *Z. Tropenmed. Parasitol.* **1959**, *10*, 314–327. (In German)
240. Jensen, C.; Schaub, G.A. Development of *Blastocrithidia triatomae* (Trypanosomatidae) in *Triatoma infestans* after Vitamin B-Supplementation of the Blood-Diet of the Bug. *Europ. J. Protistol.* **1991**, *27*, 17–20. [CrossRef]
241. de Almeida Dias, F.; Guerra, B.; Vieira, L.R.; Perdomo, H.D.; Gandara, A.C.; Amaral, R.J.; Vollu, R.E.; Gomes, S.A.; Lara, F.A.; Sorgine, M.H.; et al. Monitoring of the Parasite Load in the Digestive Tract of *Rhodnius prolixus* by Combined qPCR Analysis and Imaging Techniques Provides New Insights into the Trypanosome Life Cycle. *PLoS Negl. Trop. Dis.* **2015**, *9*, e0004186. [CrossRef]
242. Mesquita, R.D.; Vionette-Amaral, R.J.; Lowenberger, C.; Rivera-Pomar, R.; Monteiro, F.A.; Minx, P.; Spieth, J.; Carvalho, A.B.; Panzera, F.; Lawson, D.; et al. Genome of *Rhodnius prolixus*, an Insect Vector of Chagas Disease, Reveals Unique Adaptations to Hematophagy and Parasite Infection. *Proc. Natl. Acad. Sci. USA* **2015**, *112*, 14936–14941, Erratum in: *Proc. Natl. Acad. Sci. USA* **2016**, *113*, E1415–E1416. [CrossRef]
243. Schaub, G.A.; Schottelius, J. Identification of Trypanosomes Isolated from Reduviidae from North Chile. *Z. Parasitenkd.* **1984**, *70*, 3–9. [CrossRef]
244. Brener, Z. The Behavior of Slender and Stout Forms of *Trypanosoma cruzi* in the Blood-Stream of Normal and Immune Mice. *Ann. Trop. Med. Parasitol.* **1969**, *63*, 215–220. [CrossRef]
245. Kulkarni, M.M.; Karafova, A.; Kamysz, W.; Schenkman, S.; Pelle, R.; McGwire, B.S. Secreted Trypanosome Cyclophilin Inactivates Lytic Insect Defense Peptides and Induces Parasite Calcineurin Activation and Infectivity. *J. Biol. Chem.* **2013**, *288*, 8772–8784. [CrossRef] [PubMed]

246. Isoe, J.; Riske, B.F.; Dobson, M.E.; Kaylor, H.L.; Brady, J.C.; Debebe, Y.; Saavedra, L.M.; Luckhart, S.; Riehle, M.A. Characterization of the Sodium Multi-Vitamin Transporter in the Mosquito *Anopheles stephensi* and its Capacity to Mobilize Pantothenate and Biotin. *Biomolecules* **2025**, *15*, 59. [CrossRef]
247. Caulier, S.; Nannan, C.; Gillis, A.; Licciardi, F.; Bragard, C.; Mahillon, J. Overview of the Antimicrobial Compounds Produced by Members of the *Bacillus subtilis* Group. *Front. Microbiol.* **2019**, *10*, 302. [CrossRef]

pathogens

Review

The Body of Chagas Disease Vectors

Jean-Pierre Dujardin

UMR INTERTRYP University Montpellier, IRD, CIRAD, F-34398 Montpellier, France; dujjepi@gmail.com

Abstract: Morphometry is an effort to describe or measure the morphology of the body, or parts of it. It also provides quantitative data on the interactions of living organisms with their environment, external or internal. As a discipline, morphometrics has undergone significant developments in the last decade, making its implementation more visual and less laborious. Chagas disease vectors, often referred to by the common name of "kissing bugs", belong to the subfamily Triatominae. Due to their apparent morphological plasticity, they have been the subject of numerous morphometric studies. Most of these have been applied taking into account the particularities of this group of vectors, such as domesticity (synanthropy), food preferences, dispersal ability, insecticide resistance, as well as some taxonomic issues. This brief review over nearly three decades is organized here according to the body organs considered by the authors.

Keywords: Chagas; Triatominae; morphometrics

1. Introduction

In the decades following the discovery of American trypanosomiasis by Carlos Chagas [1], control efforts quickly focused on vector control. On the one hand, there was no known problem of insecticide resistance [2], and on the other hand, infection occurred mainly within human habitats where vectors also bred. Two main strategies were considered, habitat improvement and the application of insecticides in human structures. Habitat improvement, although never convincing as a large-scale solution, is generally a highly desirable measure. However, the use of insecticides over very large areas in a few years has made it possible to sustainably reduce vector transmission. This historic success was possible thanks to major international programs developed successively in the Southern Cone of Latin America, in the Andean region, and in Central America. Thus, some of the most important vectors, *Triatoma infestans* in the Southern Cone and *Rhodnius prolixus* in the Andean region and in Central America, have been dislodged from human structures, leaving today only a few wild foci under surveillance. However, epidemiological surveillance must be maintained not only for them, but also for other species of Triatominae [3]. Indeed, some other species adapt to colonize human spaces, perhaps a collateral damage induced by the elimination of the most synanthropic vectors or a previously undetected natural behavior of some Triatominae [4].

In its past and recent development, the fight against Chagas disease vectors has repeatedly encountered the same two particular problems. One is the level of adaptation of the insect to the human habitat, as well as to that of domestic animals. In the literature on Chagas disease vectors, we almost inevitably find terms such as "domiciled" or "domestic", "peridomestic", "sylvatic", all qualifiers that reflect ecological diversity [5]. The other problem is related to both the ecological and taxonomic diversity of vectors [6,7], a situation that can be recognized in the use of terms such as "main vectors", "secondary vectors", "potential vectors", etc. All species of Triatominae are possible vectors simply because they are hematophagous, but, depending on the level of adaptation of a vector to human habitat, not all of them have the same epidemiological importance. The ability of insects to colonize homes varies according to species, but it can also vary within the same species according to geography [8].

Thus, in many situations, the species identity of the vector and its degree of adaptation to humans are crucial elements to know in the development of control strategies. Many techniques can help to resolve these two questions, but all are relatively costly in terms of men and machines, hindering their use as a routine activity in entomology laboratories. Early morphometric studies applied to these problems showed that helpful answers could be provided quickly and inexpensively by morphometric analyses [9], based on techniques that were easily adopted by any laboratory with a computer and optical devices. These studies were themselves stimulated by an important evolution in the discipline of the morphometric itself [10].

The traditional method of collecting body measurement data is to estimate the linear distance between anatomical landmarks. In entomology textbooks, this method has frequently been applied to a few specimens without giving much indication of intraspecific variability [11]. The multivariate morphometric approach [12] was applied to Triatominae in the 1990s [13–16]. This approach requires larger samples of biological material because it takes into account variation (variance) in measurements. With an acceptable estimate of variances, this approach can produce a synthetic representation of many measurements, thus creating a powerful tool for distinguishing closely related species or groups [17].

In addition to improving discrimination between taxa, the multivariate techniques were also used to tentatively separate the two main properties of the morphometric data: size and shape [18]. However, the different methods to separate these two metric properties could produce different results, and they did not allow to visualize shape differences. These two obstacles could be overcome thanks to the use of anatomical landmark coordinates instead of linear distances between them [10].

Two different statistical approaches use the relative position of coordinates of points: (i) the outline (or pseudo-landmark)- [19] and (ii) landmark (and semi-landmark)-based [20,21] analyses. The former approach, slightly anterior to the latter, is restricted to the situations where landmarks are lacking, or too few. Both techniques belong to the geometric morphometrics (or modern morphometrics) as opposed to linear measurements (or traditional) morphometrics [22], and the most frequently used one currently is the landmark-based method. Application of these modern techniques to different species in many groups of insects, and in other organisms as well, has shown that geometric shape is surprisingly powerful in distinguishing species [23–25], even cryptic [26], calling into question the concept of sibling species [25,27,28].

Moreover, new routes of investigation have been opened regarding the interaction between the environment and the body shape [23], bringing a new tool for the study of phenotypic evolution [28]. Finally, two other virtues of the morphometric approach deserve to be mentioned: (i) its non-traumatic access to the data (image) allowing the use of other techniques on the same sample, including molecular techniques, and (ii) the possibility of exchanging images via the Internet, allowing one to compare own material with that of others (but see Section 4.1).

Initially, and until today, the main application of morphometrics to Chagas disease vectors has been oriented towards both taxonomic and epidemiological questions. But over time, many studies have contributed to the more general question of the evolutionary importance of phenotypic plasticity. In this brief review, I wish not only to highlight the breadth of information provided by modern morphometry applied to the Chagas disease vectors, but also to show, in broad terms, how it achieved this.

2. Methods

All geometric morphometric studies applied to Chagas vectors have used 2D data collected on images of 3D structures. Therefore, a small parallax error is possible. It is certainly not important for the wings of the insect, because the wings are almost 2D structures. However, the heads of Triatominae are closer to 3D structures. A true 3D capture of some body parts is technically possible but requires expensive equipment, which limits applicability to privileged laboratories. This is not the philosophy of morphometry as a tool to support the monitoring of vector control programs, usually under the supervision of many field laboratories with fewer resources.

2.1. Morphometric Techniques

2.1.1. Landmarks

Most of the geometric studies applied to Triatominae use the landmark-based approach [20,29]. In common practice, size and shape are derived from a configuration of landmarks collected on a non-articulated part, often a single organ (see, however, [30]). The relative positions of landmarks contain not only information on size, such as distances between landmarks, but also on shape. Of course, a few landmarks on a wing (or any measurable part of the body) do not completely describe the wing, nor the entire body. But if the landmarks are homologous between individuals, a partial capture of shape is sufficient to allow valid comparisons [28]. Anatomical landmarks are homologous in the sense that they are relocatable points, and according to this criterion, various levels of quality have been recognized (see type I, II and III landmarks, having decreasing levels of precision [21]).

2.1.2. Curves

The use of pseudolandmarks, i.e., points describing closed contours (outline-based approach) [19,31], and the use of semilandmarks, i.e., points describing an open curve between two landmarks [32], are less frequent.

The outline-based approach (pseudolandmarks) is best suited to the situations were no or too few landmarks are available, as, for instance, the eggs [33–35], but it has been adapted also to fit adult bodies [36] or internal wing cells [37]. Historically, the outline-based method is anterior to the landmark-based approach. It also uses the relative positions of coordinates, the ones collected on the contour of the organ (pseudolandmarks), but the statistical analyses are quite different. Since shape is a function of the relative positions of morphological landmarks, the outline-based method is also part of geometric morphometrics [22].

In the case of non-closed contours (open curves), the semilandmark-based method applies [38,39] and, despite the loss of degrees of freedom, receives the same statistical treatment as the landmark-based method [40].

When the idea is to capture a maximum of shape information, the use of pseudolandmarks and/or of semi-landmarks may be combined with landmark-based analysis, as, for instance, in a study on New World versus Old World Triatominae [37].

2.1.3. Softwares

With time, many free software became available, like the TPS series of J. Rohlf [41], the Morpheus from D. Slice (https://morphlab.sc.fsu.edu/ accessed on 16 December 2024), some R packages (geomorph [42], morpho [43]), the MorphoJ software of C. P.

Klingenberg [44], or our own online software XYOM [45]. For the outline-based approach of geometric morphometrics. I consider the R momocs package [46], SHAPE [47], or again XYOM [45]. For a comprehensive list of currently available software, please consult https://www.sbmorphometrics.org/ (accessed on 16 December 2024).

2.2. Morphometric Data

Geometric morphometric data allow a more informative separation, analysis, and visualization of size and shape.

2.2.1. Size

In the landmark-based method, there is a single size estimator, the centroid size (CS). For each configuration of landmarks, the computation of the CS is using the centroid landmark (average X, average Y), using the (squared) distances from this point to each user annotated landmarks [29]. This estimation of size is affected by changes occurring in different orientations, and it is restricted to the configuration of landmarks, thus, of course, to the organ under study. The change in size of an organ is positively correlated to the change in size of the insect's body, but the size of an organ does not necessarily reflect the size of the insect.

In the outline-based approach, the size used to normalize the coefficients of Fourier is an abstract "semi-major axis of the first ellipse" (SMA). It should be used for size comparisons between individuals or groups, but one generally prefers to use the perimeter of the contour which is generally highly correlated to the the SMA, and much more intuitively understandable. This correlation might decrease, however, in a case of a rough, very sinuous contour.

Geometric data produce richer information, but this does not exclude the relevance of linear measurements in certain situations. However, if we want to compare shapes between individuals or groups using such measurements, it is not recommended to use ratios. It is necessary to employ specific multivariate analyses [18].

2.2.2. Shape

In geometric morphometrics, the concept of shape derives from the relative positions of anatomical landmarks. Various works argue that shape variables are polygenic traits [48–50], and that, therefore, their variation under natural conditions could well be explained by genetic drift [24,25,28]. Of course, this does not exclude local environmental factors [51].

The two kinds of shape, landmarks or curves (see above Section 2.1), might reflect also different functional aspects of "shape" (see Section 3.3.2), a question not specifically considered up to now, except for their taxonomic power. In this respect, the two techniques seem to contain an approximately equivalent taxonomic signal [36,37,52], but their statistical treatments are very different. Also very different are the numbers of shape variables produced by these different approaches, as well as the estimation of size. To avoid the "curse of multidimensionality" [53], one technique is to reduce the number of variables by using a representative set of their few first principal components.

2.2.3. Size and Shape

Geometric shape variables are exempt from isometric size change but not from allometric size change. Allometric changes have a direct effect on body proportions (shape). They are produced by different rates of development in different organs or their constituents. Three different levels of allometry are distinguished: static allometry (between individuals of the same age), ontogenetic allometry (due to growth processes), and evolutionary allometry (between taxa) [18].

Thus, part of the geometric shape is a passive change due to variation in size: this part may be called the allometric residue. The effect of this allometric residue could be neglected for interspecific comparisons, where size changes are probably part of the evolutionary divergence between species. But it is important to take it into account, or at least measure it, for conspecific populations [23,28]. Between populations of the same species, size variation is usually due to contingent factors, masking microevolutionary changes reliably detected by shape changes, or interfering with the detection of a possible speciation event [54–56].

Generally, it is relevant to evaluate its importance. The simplest procedure for this is to calculate the linear correlation between size variation and shape variation, using as shape variation the first principal component of the shape variables. To include the totality of shape variables in this estimation, one needs to use a multivariate regression on size. Since it is a parametric test, statistical significance is better evaluated by a non-parametric algorithm. The procedure first calculates the observed values of the vector of regression coefficients (one coefficient for each component of shape), multiplying this vector by its transpose to produce a scalar value (which is the sum of squared coefficients). It then permutes the column of sizes a given number of times; at each permutation, it calculates again the regression coefficients (pseudo-coefficients) and the vector of pseudo-coefficients is multiplied by its transpose. After a given number of permutations, it counts the number of times that the pseudo-values were greater than or equal to the observed one, and it divides this number by the number of permutations. This is the risk of error saying that allometry is significant [57]. Other methods use multivariate analysis of covariance (MANCOVA) of shape variables, considering the compared groups as categorical predictors and centroid size as a covariate [58,59]. Such procedure appears to have been used only once in Triatominae [60].

2.3. Classification Methods

Classification of methods into distinct units (species, ecotypes) is mainly based on shape variables (instead of size), using both unsupervised and supervised analyses. Because the unsupervised methods make no assumption about the pre-established groups of individuals and because they do not distort the Procrustes distances (shape divergence), they are highly recommendable. Among them, the reference one, i.e., principal component analysis (PCA), has been used in all published works on geometric morphometrics applied to Triatominae. Other methods were less frequently or seldom used, such as Hierarchical clustering and Kmeans clustering, respectively [55,61].

The supervised methods try to reproduce, at best, the classification proposed by the user. Among these methods, Linear Discriminant Analysis (LDA) is the most frequently used one. For the morphometrician, the use of LDA suffers from one important criticism: shape variables are modified during the classification process (divided by the groups variances) so that the Mahalanobis distances may depart strongly from the Procrustes distances. To avoid such shape distortion during the classification process, between-group-PCA has been designed [62], later discredited because of unexpected issues when the number of variables was too large relative to the number of groups to classify [63]. LDA is, however, highly powerful, providing relevant classification, but it remains very sensitive to strongly unequal sample sizes [64] and prone to overfitting when the number of input variables is large relative to the smallest group. Not all studies on Triatominae clearly mention the number of variables used as input to LDA.

Both PCA and LDA could be criticized because they are parametric methods, thus depending on various assumptions related to multinormality and variance structure of the data. However, they are robust and generally provide relevant results. In one study, the maximum likelihood method, although a parametric method, has been suggested, performed on the principal components of shape variables [65]. It has the advantage of being applied to univariate (size) or multivariate (shape) data, but it relies heavily on the unrealistic statistical assumptions of normality. Therefore, the statistical significance of the tests should rely on non-parametric techniques (permutations, bootstraps).

To the exception of one study based on the contour of the adult body [36], non-parametric machine learning methods like K-Nearest Neighbors (KNNs), Artificial Neural Network (ANN), Support Vector Machine (SVM), or Random Forest (RF), all techniques that do not affect the Procrustes distances, have not been used in morphometric studies applied to Triatominae [28].

Machine learning attempts to implement an automated identification system [66,67] are not included in this review because not all descriptors are morphometric data, and they are processed directly by neural networks without separately extracting size and shape.

2.4. Reclassification Methods

The usual technique for reclassifying groups proceeds on an individual-by-individual basis: each is reclassified after being removed from the classification model (discriminant analysis, or other). Thus, to reclassify a sample using discriminant analysis (LDA), there are as many LDAs as there are individuals in the sample. And ideally, between each run, the shape variables should be recalculated [68]. The result is a percentage of correct assignments for each pre-established category in the sample.

The question is whether it is possible to account for a correct prediction due to mere chance. Such a special statistic has existed for a long time [69], called total adjusted precision (*TAU*). This *TAU* statistic measures the proportion of correctly assigned individuals in the sample remaining after eliminating (the number of) those correctly assigned by chance. It is valid for any group or species assignment and is calculated as follows:

$$TAU = Nc - (\sum_{i=1}^{G}(Ni * Pi))/(N - \sum_{i=1}^{G}(Ni * Pi)) \quad (1)$$

where N is total sample size; Nc is the total number of cases correctly assigned by the classification method; Pi is the prior probability of group membership in the ith of the G groups; and Ni is the number of cases in the ith group.

2.5. Special Morphometric Comparisons

Size and shape data extracted from linear measurements or coordinate data are generally used to compare distinct taxa. They are also used between different intraspecific entities, such as ecotopes, chromatic variants, geographic areas. In addition, these same techniques are applied to comparisons between sexes or even between different parts of the same individual.

2.5.1. Between Sexes

Sexual size and shape differences are generally studied within species. Sexual size dimorphism was first observed between different ecotopes in *T. infestans*, as well as in *R. domesticus* when moving from field to laboratory conditions [70]. These observations suggested that sexual dimorphism could be a sensitive marker of the environment [71]. Its cause was attributed to an unequal rate of size change between habitats: more pronounced in female specimens, it led to altered sexual dimorphism. Later publications confirmed this observation and included shape in the study of sexual dimorphism (see Section 4.5.1). There are no particular statistics to apply here. Another sexual dimorphism was disclosed and discussed related to the sensory system of the antennae (see Section 3.4), also showing typical changes in the transition from sylvatic to (peri)domestic populations (see Section 4.5.1).

2.5.2. Between Body Sides

To compare the two sides of an insect, i.e., to estimate its symmetry, specialized methods are needed because it is imperative to statistically eliminate measurement error. In fact, quantitative estimation of body symmetry is an important area of morphometrics, and a particular asset: it is a full-fledged application, unmatched by methods using genetic markers. It quantitatively explores the interaction of an organism with its environment, as well as, indirectly, with its internal genetic makeup.

The examination of bilateral symmetry ("matching symmetry", [72,73]) requires bilateral structures, such as wings, antennae, and legs. The statistical procedure can be adapted for an organ which is symmetric in itself ("object symmetry", [73]) like the head, the pronotum, or female genitalia of the Triatominae. For such organs, it is also possible to use the same statistical method as for bilateral symmetry after removing landmarks that are common to both sides [61]. Estimating the bilateral symmetry follows the same statistical procedure for size as for shape (the venation architecture of the wing and/or its contour), adapting, however, the number of degrees of freedom [72,74]: for a comprehensive review of the topic, see [73].

Ideally, the estimation of the symmetry of the body or of an organ requires to minimize annotation errors, so that it should be estimated using, preferably, landmarks of type I (wings, female genitalia) (see Section 2.1.1).

The kind of asymmetry which is explored is fluctuating asymmetry (FA), i.e., the variance of bilateral differences, assumed to exist mainly due to environmental stressors (developmental instability) [75,76]. The source of developmental stress can also be genetic (intense selection processes, inbreeding/exogamy, mutations). For instance, more FA could be demonstrated in insecticide-resistant *T. infestans* versus -susceptible ones [61].

The main hypothesis is that FA increases in cases of stress affecting the development of the insect, whatever the cause of the stress. Under this expression, the most frequent stressing factors are environmental ones: higher population density (more competition for food) [71,77], sublethal doses of insecticides [78], nutrition [79], extreme climatic conditions, flight muscle dimorphism [80] or activity [81], parasitism [82], etc.

For the Chagas disease vectors that have been submitted to symmetry analyses, many of the factors mentioned above are the ones distinguishing sylvatic from domestic populations [71], melanic from non-melanic forms of *T. infestans* [55], good flyers from others [80], insecticide-resistant specimens from susceptible ones [61], and survivors from others [78]. Thus, although not *a priori* a heritable character, the symmetry of an insect is a valuable source of information on its adaptation to the local environment. In the case of vectors that are vectors because of their adaptation to human habitation, this type of information, although not always easy to interpret, can be of great help in understanding their level of domestication and improving vector control strategies (see Section 4.5.1).

2.5.3. Between Body Parts

Less frequently, morphometric data are used to compute the correlation between different body parts: linear correlation for the univariate size estimator [83], two-block partial least-squares method (PLS) [84], or the Escoufier coefficient [85] for the multivariate shape descriptor [35,83,86]. The PLS statistics was applied to estimate the covariation of head and wing shapes as a species specific trait [35]. PLS statistics are also applied to estimate the covariation of organ shape with the size of another organ (wing shape and head size, wing shape and pronotum width, see [86]), not to be confused with the estimation of the allometric residue embedded in shape variables (see Section 2.2.3). The same statistical techniques can be used to correlate shape variation with external parameters such as climate variables [51,86,87].

3. Body

Different stages (eggs, nymphs, adults) and different body organs (heads, wings, pronotum, legs, genitalia) were subjected to 2D geometric morphometric approach.

3.1. Eggs

Traditionally, Triatominae eggs are studied based on linear measurements [88–90], using light microscopy and scanning electron microscopy [91]. The geometric approach to egg morphometry uses light microscopy [33–35,65]. The egg outline can be treated as a closed contour, including or excluding the operculum. The contour-based approach can be used provided that the egg position can be correctly repeated from one individual to another [31]. The operculum could also be treated separately using semi-landmarks describing an open curve between two landmarks [32,40]. Ecuadorian populations of closely related species, *Panstrongylus chinai* and *P. howardi* are clearly distinguishable by egg outlines, but not convincingly by head shape or wing venation [35]. Even slightly deformed nonviable eggs retain significant taxonomic signals [34].

3.2. Nymphs

Because of their epidemiological importance [92] and their more elusive diagnostic characters, nymphs have attracted the interest of morphometricians. Furthermore, because of the incomplete metamorphosis typical of Triatominae, ontogenetic growth could be examined and used as an element of interspecific comparisons.

The projection in a common morphospace of the head shape of successive nymphal stages and of adults allowed to compare the ontogenetic trajectories of four species among the three most important genera of Triatominae: *T. lecticularia*, *T. infestans*, *R. prolixus* and *P. megistus* [93]. A similar approach visualizing the ontogenetic profile of postembryonic head shape development showed that, for *Belminus herreri*, the greatest change occured between the first and second instars [94], while for *T. ryckmani*, largest changes occurred in the first three instars [95].

The head shape of adults and nymphs produced generally concordant results. An intraspecific study (*T. infestans*), however, observed a lower discrimination ability of nymphs between distinct habitats compared to adults [83].

Interspecific comparisons did not observe such a difference between adults and nymphs: the five successive nymphal instars suggested a clear distinction between three of the seven species of the *R. prolixus* complex: *R. marabaensis*, *R. prolixus* and *R. robustus* [96].

3.3. Adults

3.3.1. Body Contour

The body outline excluding legs and antennae was examined as a closed contour (outline-based method). This approach produced better reclassification scores than other methods to distinguish the major evolutionary lineages of *T. dimidiata* [36]. However, it depends on image preprocessing (removal of legs and antennae) which could be an obstacle to its diffusion.

3.3.2. Wings

As in most winged insect groups, the wing (forewing for Triatominae) is the organ of choice for morphometric analysis because it can be easily separated from the body, allowing the rest of the body to be used for other analytical techniques. The nearly two-dimensional structure of the wing reduces digitization errors, at least those due to different orientations as is the case for heads (see Section 3.3.3).

Due to the better repeatability of measurements, wing vein geometry is ideal for exploring bilateral symmetry (see Section 2.5.2), where minimizing measurement errors is of crucial importance.

For the same reason, the wings have been used to compare very close, even cryptic or sibling species, an exercise where wing vein geometry has proven powerful [39,52,97–99] (see also Section 4.2), as well as to compare intraspecific entities, separated by either space [100] or time [101] (see also Section 4.4.1).

The geometric approach applied to the wing can cover different realities: either the architecture of the internal veins or the outline of the wing or its parts (rigid and membranous parts).

The venation pattern is typically analyzed by the landmark-based approach, all of them landmarks of type I [29], i.e., landmarks having the highest level of relocability from one individual to another.

The external outline of the forewing (or the outline of some internal cells [37]) can be approximated by the pseudolandmark technique (outline-based approach). For the Triatominae, the outline of an internal part (rigid, membranous) contains some landmarks so that it can also be approximated by the use of semi-landmarks [37,39]. The shape described by these techniques does not use or uses very few type I landmarks.

It is, however, not certain that these different forms respond in the same way to external factors, in particular by comparing the external contour of the wing and its internal venation architecture. The external contour of the wing (contour-based shape), as well as the size of the wing itself, can present changes visible to the naked eye [102]. Intuitively, these changes mechanically affect the insect's flight capacity. On the contrary, it is less obvious to imagine that invisible or subtle changes in the internal venation of the wing have an impact on flight capacity. For any flying insect, the flight function actually depends on various constituent elements of the wing apparatus as a whole (surface and external dimensions of the wing, shape of the wing contour, dimensions of the eye, spacing of the eyes, presence or activity of wing muscles, antennal sensory system, nutritional status, body weight, etc.). Such a complex system, reflected by changes in body parts other than the wings themselves [80,103], probably does not depend much on subtle variations in the architecture of forewing venation.

In other words, the internal venation architecture variation is not likely to induce by itself a different flying capacity, nor is it likely to have any adaptive value: acording to [104], wing venation, as a by product of natural selection, is almost a neutral character.

3.3.3. Head

The above considerations about shape as landmarks and shape as outline may be important for wings, but not for heads. For heads, indeed, landmarks probably tell the same story as those of the outline, because the landmarks of the head are part of its outline (as captured by the image of the head).

Unlike the internal venation of the wings, the shape of the head seems more likely to vary as an adaptive trait. The head of Triatominae is a composite, multifunctional organ, bearing a proboscis (rostrum), two antennae and two eyes, two ocelli, all structures that allow the insect to evaluate its environment and respond to a wide range of stimuli. Head shape variation, in addition to being affected by genetic drift in isolated populations [38,83], could be more directly related to external parameters such as feeding habits [60,83,105–107], environmental factors detected by the antennae [83,108], or light exposure affecting eye size [27]. Other associations have been highlighted, such as the ability to fly and the separation of the eyes, or simply the size of the head [102,103].

Assuming that the head is attributed a greater correspondence to environmental factors, this organ then turns out to be the one of choice to explore intraspecific variation related to different food sources (see Section 4.6.1) or habitats [83,109], thus leading to the possible detection of different ecotypes [38,108].

In any case, the shape of the head, derived from linear measurements or a geometric approach, is generally distinct between species (see Section 4.2) and has been used as an additional argument for the description of new species [110]. Another advantage of the head, and not the least, is that it allows to include the nymphs in a morphometric study (see Section 3.2).

However, the use of the head alone [110–113] is not encouraged without some precautions. For a morphometrician, the head can be a problematic organ for three reasons: (i) the landmarks describing the geometry of the head are of lower quality (there is no type I landmark), (ii) its volume makes it difficult to offer each specimen the same head orientation, generating errors in the annotation of the landmarks, and (iii) the landmarks actually cover the left and right parts of the body, thus including possible asymmetries (object symmetry) in their intergroup variation. Due to its lack of type I landmarks alone, the head is less recommendable than the wings for the study of asymmetry, where geometric capture must be as rigorous as possible (see Section 2.5.2). Authors generally take the precaution of either averaging the bilateral homologous landmarks or testing only one side of the head.

3.3.4. Head and Wing

A combined morphometric analysis of heads (adults and fifth instar nymphs) and wings was used to study the mobility of *T. infestans* between poultry and goat pens: the authors observed a lack of covariation in shape between the two organs [83]. Subsequent morphometric studies including both heads and wings often highlight the difference in information provided by the two organs (but see [39]), either within species [56,107,109] or between them [97,114].

For the *R. prolixus* complex, heads appeared to be more discriminating than wings [99], while for the *T. sordida* subcomplex, especially the *T. sordida* and *T. garciabesi* comparison, wings were more informative [98,114]. In a phylogenetic study focusing on the *T. brasiliensis* species complex (*T. brasiliensis*, *T. juazeirensis*, *T. sherlocki*, *T. melanica*, *T. lenti*, *T. petrocchiae*), adding *T. melanocephala*, *T. vitticeps*, and *T. tibiamaculata* species, the head shape information did not correlate well with the wings, and the shape variation of the wings showed more agreement with genetic variation [97]. Within the highly morphologically variable and ecologically flexible *R. ecuadoriensis* from Ecuador and Peru, the geometric shapes of head and wings suggested different patterns of variation, with the wing shape closer to the genetic classification [109]. Such discrepancies could be related to the taxa examined, or to the properties of the organ itself (head, wing). A recommended precaution in such a comparison between head and wings is to also include the comparison of measurement errors (see Section 4.1).

3.3.5. Pronotum

In Triatominae, linear measurements of the pronotum (length, width) have traditionally been associated with species descriptions [11,115]. Its study with geometric techniques used the using semilandmarks approach due to the scarcity of valid landmarks. Its use is recent, and publications on this organ combined its information with that of the head [38] or wings [39,98]. As for heads, interference of object symmetry in final shape variables can be avoided by averaging the sides or by considering only one side.

Pronotum shape appears to have good discriminatory power in the *T. sordida* subcomplex. However, the topology (NJ trees) of shape-based classifications, similar when derived from heads or wings, was different when based on the pronotum [98]. The pronotum shape was also effective for distinguishing different haplotypes of *T. pallidipennis* [38].

Since a broad thorax may imply highly developed forewing muscles [116], the pronotum has been included in studies related to flight dispersal ability. Between two geographic lineages of *T. garciabesi*, the size of the pronotum, unlike its shape, showed a different trend than that observed for the head and forewings [39].

Visualization of the body contours of the three haplogroups of *T. dimidiata* suggested that the pronotum is the most variable body part between haplogroups [36].

3.3.6. Legs

Along with the head and pronotum, the legs represent another organ that can be examined in nymphs and adults. The idea is that walking is certainly an active dispersal mechanism, especially for nymphs, and thus likely to play an important role in exchanges between domestic and peridomestic sites [92].

There is a pioneering work that paves the way for this morphometric application where the relative dimensions of legs and leg segments explore the locomotion of *T. infestans* [117]. The authors use both linear measurements (femur, tibia) and the geometric approach (meron, coxa, trochanter) of the right side of *T. infestans*, comparing forelimbs, midlimbs, and hindlimbs, and calculating the correlation between them.

In the future, legs could be a candidate for asymmetry studies, in spite of not having landmarks of type I (see Section 2.5.2).

3.3.7. Female Genitalia

Another morphometric study examined, for the first time, the geometry of female genitalia to explore their variation between genera and species [118]. This is an organ limited to one sex of adults, thus unable to include nymphs, unable to explore sexual dimorphism, but it presents an interesting feature: according to the authors, all landmarks were type I [21]. It is then a possible candidate for reliable object symmetry analyses [73].

3.4. *Antennae*

This organ deserves special attention because of its particular reading, based on the density of micro-organs that line its segments. Traditionally, relative linear dimensions of antennal segments are used in species description [11], but, whether by electron or optical microscope, other antennal characteristics are measurable. The four segments of the antennae are covered with specialized sensilla (setae, basiconic, trichoids with thick and thin walls) capturing chemical, physical, or mechanical signals [119]. Their counting (and length) under the optical microscope produces meristic variables (instead of continuous variables) defining the antennal phenotype (AP). The AP was insect developmental stage specific and distinct between genera and species [120,121] (but see [35]), distinct between haplotypes of *T. dimidiata* [122], and distinct between cytotypes (chromosomal variants) [123] or ecotypes [108] of *T. infestans*.

Since the insect sensory system directly reads its external environment, the AP appears as a powerful tool to study habitat adaptation [83,124], host-searching behavior, and dispersal ability [122], or its possible association with insecticide resistance [61] or with parasitism [125]. Other interesting points inviting to study antennae are that (i) they can be explored in nymphs, (ii) they provide valuable information of sexual differentiation, and (iii) as bilateral structures, antennae could be used to explore the developmental stability of the insect (see Section 2.5.2).

4. Discussion

4.1. The Measurement Error

Measurement error is an often neglected chapter in published studies, at the risk of leading to classification errors [126,127]. Measurement error exists at different stages of morphometric analysis: the mounting technique of specimens or organs, the photography conditions, and the user's skill in collecting landmark coordinates [128]. In general, modern digital photography techniques provide adequate resolution for correct landmark recognition under different conditions. However, it may be convenient to photograph a distortion test chart to ensure that there is no image distortion. Among the sources of measurement error other than those resulting from an incorrect photography step, user intervention is often the most important. The error is usually due to small but systematic differences between different users in the exact location of the same landmarks. These subtle discrepancies can be seriously amplified by the power of a classification analysis such as discriminant analysis [129]. Measurement error can become a significant obstacle to mixing data from different users to enable a larger sample study [129,130]. As a result, User A should not enter their own measurements into a coordinate database collected by User B, and vice versa.

Currently, the recommended method for performing morphometric comparisons is to allow a single user to produce the data (see next paragraph). Even when performed by a single user, the digitization step should be repeated at least once, allowing its accuracy to be measured and error to be reduced by averaging the two digitizations [128].

4.2. Distinguishing Species by Shape

Significant differences in geometric shape between populations, even if free of allometric effects (see Section 2.2.3), should not be the sole argument for deciding the specific status of populations. In fact, no new species of Triatominae have been declared on the basis of morphometric variation alone. Studies describing new species [131] or exploring species homogeneity have generally used morphometric variation as a complementary argument [35,132,133]. Because of the difficulty of clearly defining what a species is, the modern trend in taxonomy is to combine various disciplines to help reach consensus on the specific status of new or questioned taxa [6], an approach now called integrative taxonomy [134,135]. However, thanks to a simple morphometric analysis, certain taxa have been legitimately called into question [136], synonymized [137,138], detected [139], or even provisionally explained [140].

In other insect groups, such as Diptera, geometric shape variation has proven to be a reliable character for detecting or distinguishing morphologically close, cryptic, or even sibling species [23,28], and there is no evidence that this would be different in Triatominae. Morphometric studies based on geometric shape have distinguished cryptic species such as *R. prolixus* and *R. robustus* [141–143], or morphologically close species within the *R. prolixus* complex (11 species) [99], within the *T. brasiliensis* complex [97], within the *T. oliverai* complex [144], within the *T. sordida* subcomplex [98,145], or evolutionary lineages within the *T. dimidiata* complex [27,36] (but see [82]) or the *T. pallidipennis* complex [38]. Most of the papers demonstrating the usefulness of the shape variable in species recognition have been applied to adults. However, its application to juvenile stages has been successfully tested (see Sections 3.1 and 3.2) and could be of high epidemiological interest because the presence of pre-adult stages in the home may suggest the ability of the species to colonize human habitations, an important condition for being a vector.

4.3. Phylogeny?

After a classic morphometric study presenting the level of separation of certain species based on morphometric traits, the temptation is never far away to make a phylogenetic presentation. However, there is no clear basis for expecting parallelism between the phylogenetic signal of shape variables and that of genetic markers [22]. According to theoretical and modelization studies, the phylogenetic signal embedded in landmark locations is not strong enough to provide reliable results [146,147].

And yet, even a simple UPGMA tree derived from the wing venation architecture could be in global agreement with phylogenetic trees derived from genetic characters: a striking example is the clear proximity of the New World wings shapes with the North American ones [37]. Another example is the inclusion of a predator, *Opisthacidius pertinax*, in the UPGMA tree derived from the wing shape of Triatominae [148], as confirmed later in molecular phylogenetic trees [149]. The above-mentioned examples confirm that divergence of form between species contains the effects of large evolutionary events, without, however, reflecting these events alone. Indeed, other examples of phylogenetic hypotheses drawn from either geometric shape [33,96,97,118,144,150] or shape as derived from linear measurements [144,151,152] do not always show complete congruence with related molecular trees.

In fact, the morphogenetic processes underlying morphological changes are not supposed to ensure a linear reflection of genetic changes, which makes morphometric changes unpredictable indicators of underlying genetic changes. In conclusion, to examine the relationship between shape variation and evolutionary events, it is less questionable (and less risky) to fit shape data to a known phylogeny than to estimate a phylogeny from shape data [22,146].

4.4. Exploring Species Homogeneity

Morphometric variation within a species has generally been explored between pre-established groups, such as geographical areas (Section 4.4.1), color varieties (Section 4.4.2), or habitats (Section 4.5.1). Another approach is to explore the structure of a population without apparent subdivision (Section 4.5.2) with its important epidemiological correlate: the probability of reinfestations after control measures (Section 4.5.3).

4.4.1. Geographic Variation

Large-scale geographic studies of populations of genera *Triatoma* and *Rhodnius* observed a progressive reduction in body size from the putative geographic origin and the periphery of the species' range. In *T. infestans*, the decrease in body size was accompanied by a reduction in chromosome heterochromatin blocks [27,153]. Based on these observations in *T. infestans*, new hypotheses were suggested for *R. prolixus* from Central America. Indeed, the analyses in this region showed a decrease in the size of the head of *R. prolixus*, as well as a significantly lower number of RAPD bands relative to Venezuelian populations. In agreement with historical data, these clines suggest that the Central American populations represents a recently imported population, therefore, in principle, easier to eliminate [154].

Morphometric properties (size and shape) using either linear measurements [155] or landmark-based methods are frequently applied with the idea to detect or to reinforce the hypothesis of microevolutionary events in *T. infestans* [55,156], *T. garciabesi* [39,56], *T. guasayana* [56,86,87], *T. nitida* [155] or *Rhodnius ecuadoriensis* [109]. Such studies are always a source of information about the within-species microevolutionary trends, and can detect or suspect possible speciation events like in *Panstrongylus rufotuberculatus* of Bolivia [133], in *R. pallescens* of Colombia and Panama [157], in *T. sordida* [158], *T. patagonica* [159], *T. dimidiata* [27,36], or *T. pallidipennis* [38]. In addition to detection of possible speciation processes, the within-species geographic variation of morphometric properties also has important epidemiological applications (see Section 4.5.2).

4.4.2. Chromatic Forms

Intuitively, morphometric variation should parallel morphological variation. However, between *T. infestans* and *T. melanosoma*, there is a clear lack of parallelism between chromatic variation and wing shape [138], or between chromatic variation and antennal phenotype [137], sufficient to question the validity of *T. melanosoma* as a distinct species from *T. infestans*. In Bolivia, no morphometric evidence was found to consider the "dark morph" as a distinct species from *T. infestans* [160]. In Argentina, chromatic and non-chromatic morphs of *T. infestans* showed significant differences in size and shape but did not question their conspecificity [55].

Shape differences between different conspecific forms are interesting to know from a taxonomic point of view, but they do not tell the whole story. In the morphometrist's toolbox, there are other ways to understand local morphological variation, such as symmetry (see Section 2.5.2) or sexual dimorphism (see Section 2.5.1), not to mention antennal phenotype (see Section 3.4). Consider, for example, the chromatic and non-chromatic forms of species *T. infestans* in domestic and peridomestic habitats of Argentina. They show similar sexual dimorphism in size, but they display different levels of asymmetry: higher in melanic individuals, suggesting a selective advantage of non-melanic individuals in domestic and peridomestic habitats [55].

4.5. Targeting Bugs

The target of a vector control programme must be properly delimited at different geographical scales and, in addition to the choice of effective insecticides, must be prepared for possible reinfestation problems. In this respect, the morphometric approach has been useful.

4.5.1. Domestic Versus Peridomestic or Sylvatic Populations

The apparent distinction between domestic and peridomestic populations [155], or sylvatic [161], of the same species is often examined because of its epidemiological importance. Are these isolated populations or are they connected, jeopardizing habitat control measures [162,163]?

It has been suggested that the size of kissing bugs decreases during the transition from sylvatic or peridomestic to domestic foci of *T. infestans* [70,71]. This transition can be seen as a shift from low to high population density, but also from a wild to an artificial environment. Furthermore, insect size also exhibits seasonal variation that may affect this pattern [101]. To explore this question, morphometric differences between a field population and its laboratory descendants were presented as a model for the changes observed between natural samples of sylvatic and domestic populations.

In *Panstrongylus geniculatus*, the analyses over five generations showed differences in size but not in shape, with the decrease in head size occurring only up to the second generation while the decrease in wing size proceeded up to the fifth generation [164]. A laboratory protocol in *R. pallescens* allowed to reproduce the reduction in size in the first laboratory generation [54] by simulating the apparent changes occurring for a Triatominae in the transition from silvatic (low population density, low feeding frequency) to domestic habitats (higher density, higher feeding frequency). As in *P. geniculatus*, size decreased but no shape differences were observed other than a passive change in shape with size change. The authors emphasized the importance of the interaction between population density and feeding frequency in producing specific and significant changes in insect dimensions [54].

In *R. neglectus*, a reduction in wing size was observed between sylvatic and laboratory conditions [165]. A similar pattern of size reduction was observed in Brazilian *T. infestans* samples [166]. All these observations and experiments support the idea that size changes can be used as a marker of the colonization of human habitations by Triatominae, as recently confirmed in species *T. sordida* from Brazil [167].

To help distinguishing domestic from peridomestic or sylvatic specimens, another morphometric strategy was explored: the sexual dimorphism (see Section 2.5.1). A reduction in sexual dimorphism has been suggested as a signal of established domestication [70,71,165]. This suggestion was frequently verified, to the extent of considering it as an argument to infer the recentness of the domestic populations of *T. venosa* [168].

Another very different morphometric strategy was also suggested to compare domestic and field samples of Triatominae: the measurement of symmetry. A significant asymmetry in the wings of domestic samples was apparent in *T. sordida* from Bolivia, while it was not detectable in its field counterparts [71]. Later studies on Triatominae also suggested the level of asymmetry (actually of the fluctuating asymmetry) as a possible marker of different factors, all related to possible sources of stress, either genetic or environmental ones (see Section 2.5.2).

As already mentioned above (see Section 3.4), the sensory pattern of antennal segments also contributed to characterize sylvatic and (peri)domestic populations [108,124,125].

4.5.2. Biogeographical Islands

Exploring morphometric variation within a species at a broader spatial scale could help design control strategies by detecting isolation, i.e., the absence of exchanges of individuals, between certain territories that would therefore behave like isolated "islands". In this situation, vector control measures could be applied sequentially, one area ("island") after another.

It seems commonly admitted that the adaptation to different ecotopes or microhabitats (ecological pressure) is perhaps the main mechanism that drives the speciation in the subfamily Triatominae [27,71,109]. This adaptation could generate morphological changes, generally qualitative, even before the installation of reproductive isolation [109,169]. But before developing these rapid morphological changes (see Section 4.4.1), the separated populations develop a less visible, but significant, morphometric divergence.

Thus, from the point of view of the morphometrician, but also from the point of view of the epidemiologist, it is necessary to distinguish reproductive isolation from physical isolation (geographical, mainly). Shape characters, in particular the internal wing venation architecture, diverge first because of spatial isolation (genetic drift, see Section 2.2.2). In fact, between isolated conspecific populations, the internal wing venation architecture would be affected because of both genetic drift and environmental factors [56,87,109,170]. In the case of gene exchanges between populations, even allopatric ones, or at the frontier between two biogeographical islands [39], no genetic drift would take place and the internal wing venation architecture would hardly diverge. My working hypothesis has a corollary: in a common environment, the venation architecture diverges between spatially isolated subpopulations, even at a very low spatial scale [100,156,171]. In another insect, *Aedes aegypti*, two isolated isofemale laboratory lines, in spite of a common laboratory environment, could diverge in the geometry of wing venation after less than fifteen generations [172]. Intraspecific studies could help vector control strategy by detecting biogeographical islands, or it could help to understand the origin of reinfestations after control measures (see Section 4.5.3) [28].

4.5.3. Reinfestation Studies

As for linear morphometric measurements [9], the geometric method was applied to the recurring problem of the origin of reinfesting specimens after control measures: In the absence of insecticide resistance, do reinfesting individuals come from a residual population or from untreated neighboring foci? In order to try to answer this question using shape variables or using the antennal phenotype, a few conditions are necessary.

First, the target sample (the pre-treatment population) and neighboring samples must be morphometrically distinguishable [161], at a small [83,100,124,156,173] or at a larger geographic scale [83,108,158,174]. Second, it is assumed that in such a comparison, unexpected, cryptic species would be detected, such as *R. robustus* and *R. prolixus* [175], *R. neglectus* and *R. prolixus* [142], and others.

Furthermore, as this type of study [163,176–178] can compare several generations of insects (before and after control measures), it should be clear to what extent the morphometric shape is transmitted to subsequent generations [162] and to what extent the geometric shape changes over time [101,166,179,180].

Finally, as the statistical procedure consists in assigning a sample (reinfested specimen) to several populations of different generations, the interference of a possible correct prediction by mere chance must be taken into account (see *TAU* statistics in Section 2.4).

4.6. Vector Biology

4.6.1. Feeding Habits

Triatominae are obligate blood-sucking insects, with nymphs, males, and females all feeding on vertebrate blood. Exceptional or occasional predatory behavior has been reported, suggesting their evolutionary origin as predators [181]. Experimental work compared the effect of bird and mammalian (pigeon, guinea pig) blood sources on the development of *T. infestans*, showing significant effects on head shape and size [60,105]. Later experiments confirmed that different blood sources can have different effects on head or body shape in other species, as in *T. williami* [106] or in *T. pallidipennis* [79].

4.6.2. Dispersal Capacity

Dispersal of bugs is possible through flight and walking [92]. When it comes to flight, morphometric studies of Triatominae have not necessarily focused on wings: other aspects of bug morphology have been associated with different flight abilities [102,103].

When comparing wings of natural populations, it must be specified which wing shape is being compared (see Section 3.3.2), but this does not constitute sufficient precision. While not omitting the caveat regarding statistical association and causality, other obstacles inherent in studies of natural populations must also be considered.

(i) To the possible exception of laboratory experiments, there is frequently an important additional effect modifying size and shape between categories: geography or simply spatial isolation. Some studies exploring the flight ability of the bugs are at the same time interesting studies about the geographic variation of metric properties [56,86,87]. In the case of both allopatry and distinct environments (urban, rural, etc.), it might become very difficult to disentangle these interfering factors. An example of a possible solution uses the optimum K of the Kmeans approach and the metric disparities of various samples configurations [61].

(ii) In addition to variation in shape between populations adapted to different environments or microenvironments, variation in size (not just wing size) is frequently observed. Since in such comparisons size often shows significant differences, shape comparisons should take care to be free of allometry (see Section 2.2.3). If this is not possible (or not required, as in [80]), the effect of size on shape should be taken into account in the overall interpretation, or its importance should be illustrated graphically (as in [182]).

There is one morphometric trait that may help to circumvent at least some of these obstacles: insect symmetry (see Section 2.5.2). As an indicator of developmental instability, symmetry may be more related to the environment than to genetic drift. The relationship between increased asymmetry and reduced flight ability was already suspected in a simple laboratory experiment where flight ability was assessed by flight activity [81]. More comprehensive studies have been published recently [55,78,80,179].

4.6.3. Insecticide Susceptibility

This important aspect of insect physiology was approached by tentatively answering two questions: (i) Is there an association between the application of sublethal doses of pyrethroid insecticides and the metric properties of insects [77,78,170]? and (ii) Are resistant specimens detectable morphometrically [61]?

Variation in the fluctuating asymmetry of wings of *T. infestans* collected before and after a community-wide pyrethroid spraying campaign in Argentina suggested the possible effects of a lower population density after spraying or a selective advantage of more symmetrical individuals [77]. A laboratory experiment on *T. infestans* concluded that insects exposed to a sublethal dose of deltamethrin had larger, less symmetrical, and less canalized wings [78]. In another laboratory experiment comparing pyrethroid-sensitive and pyrethroid-resistant *T. infestans*, insects previously treated with deltamethrin developed significantly thicker cuticles than the control only in the resistant population and significantly larger wings in both populations [170].

Recently, the morphometric properties of natural populations of resistant *T. infestans* specimens were examined for their wings, heads, and antennae. Various differences were suspected to be specific of insecticide resistance acquisition, including smaller size of the wings and heads, shape differences, and antennal sensory simplification, laying the foundations for the identification of a resistance biomarker in Triatominae [61].

5. Conclusions

Over the last three decades, the morphometric study of Chagas disease vectors has undergone considerable development and a steady increase in the quality of publications. In this group of vectors, morphometrics has been successfully applied not only to clarify species delimitation issues, but also, and perhaps more importantly, to reveal and understand their intraspecific variation. In this way, modern morphometrics applied to Triatominae has maintained a balance between two objectives: to detect and understand the microevolutionary (and beyond) events of an insect and to provide useful information for the development of vector control strategies.

Funding: This research received no external funding.

Institutional Review Board Statement: Not applicable.

Acknowledgments: I warmly thank the colleagues who brought me their enthusiasm and intelligence during my years of stay in Latin America. Their names are found for the most part in the authors of the works cited as references in this review.

Conflicts of Interest: The author declares no conflicts of interest.

References

1. Chagas, C. Nova tripanosomiase humana. *Gaz. Medica Bahia* **1909**, *40*, 453–440.
2. Mougabure-Cueto, G.; Picollo, M.I. Insecticide resistance in vector Chagas disease: Evolution, mechanisms and management. *Acta Trop.* **2015**, *149*, 70–85. [CrossRef]

3. Cavallo, M.J.; Amelotti, I.; Gorla, D.E. Invasion of rural houses by wild Triatominae in the arid Chaco. *J. Vector Ecol.* **2016**, *41*, 97–102. [CrossRef] [PubMed]
4. Dujardin, J.P.; Schofield, C.J. Vector control by surveillance networks: The ECLAT program and Chagas. In *Encyclopedia of Infectious Diseases: Modern Methodologies*; Tibayrenc, M., Ed.; Wiley & Sons: Hoboken, NJ, USA, 2007; Chapter 24, pp. 425–434.
5. Schofield, C.J.; Diotaiuti, L.; Dujardin, J.P. The process of Domestication in Triatominae. *Mem. Inst. Oswaldo Cruz* **1999**, *94* (Suppl. S1), 375–378. [CrossRef] [PubMed]
6. Bargues, M.; Schofield, C.; Dujardin, J.P. 6-Classification and systematics of the Triatominae. In *American Trypanosomiasis Chagas Disease*, 2nd ed.; Telleria, J., Tibayrenc, M., Eds.; Elsevier: London, UK, 2017; pp. 113–143. [CrossRef]
7. de Paiva, V.F.; Belintani, T.; de Oliveira, J.; Galvão, C.; da Rosa, J.A. A review of the taxonomy and biology of Triatominae subspecies (Hemiptera: Reduviidae). *Parasitol. Res.* **2022**, *121*, 499–512. [CrossRef] [PubMed]
8. Noireau, F.; Dujardin, J.P. Biology of Triatominae. In *American Trypanosomiasis. Chagas Disease. One Hundred Years of Research*; Telleria, J., Tibayrenc, M., Eds.: Elsevier: Amsterdam, The Netherlands, 2010; pp. 149–169, ISBN 978-0-12-384876-5.
9. Dujardin, J.P.; Bermúdez, H.; Schofield, C.J. The use of morphometrics in entomological surveillance of silvatic foci of *Triatoma infestans* in Bolivia. *Acta Trop.* **1997**, *66*, 145–153. [CrossRef] [PubMed]
10. Rohlf, F.J.; Marcus, L.F. A revolution in morphometrics. *TREE* **1993**, *8*, 129–132.
11. Lent, H.; Wygodzinsky, P. Revision of the Triatominae (Hemiptera, Reduviidae), and their significance as vectors of Chagas disease. *Bull. Am. Mus. Nat. Hist.* **1979**, *163*, 123–520.
12. Sneath, P.; Sokal, R. Numerical taxonomy. *Nature* **1962**, *193*, 855–860. [CrossRef]
13. Gorla, D.E.; Jurberg, J.; Catalá, S.S.; Schofield, C.J. *Systematics of Triatoma sordida, T. guasayana and T. patagonica (Hemiptera, Reduviidae)*; Memorias do Instituto Oswaldo Cruz: Rio de Janeiro, Brazil, 1993; Volume 88, pp. 379–385.
14. Harry, M. Morphometric variability in the Chagas' disease vector *Rhodnius prolixus*. *Jpn. J. Genet.* **1994**, *69*, 233–250. [CrossRef]
15. Torres, E.; Galíndez, G.; Itamar, M.E.; Araujo, V.; Barazarte, R.; Marquez, J. Multivariate analysis applied to some populations of *Panstrongylus geniculatus* Latreill 1811 (Hemiptera, Reduviidae, Triatominae). *Entomol. Vect.* **1997**, *4*, 191–202.
16. Dujardin, J.P.; Forgues, G.; Torrez, M.; Martínez, E.; Cordoba, C.; Gianella, A. Morphometrics of domestic *Panstrongylus rufotuberculatus* in Bolivia. *Ann. Trop. Med. Parasitol.* **1998**, *92*, 219–228. [CrossRef]
17. Rohlf, F.J. Perspectives on the application of multivariate statistics to taxonomy. *Taxon* **1971**, *20*, 85–90. [CrossRef]
18. Klingenberg, C.P. Multivariate allometry. In *Advances in Morphometrics Proceedings of the 1993 NATO-ASI on Morphometrics*; Marcus, L.F., Corti, M., Loy, A., Naylor, G.J.P., Slice, D., Eds.; Series A, Life Sciences; Plenum Press NATO ASI: New York, NY, USA, 1996; pp. 23–49.
19. Rohlf, F.J.; Archie, J.W. A comparison of Fourier methods for the description of wing shape in mosquitoes (Diptera:Culicidae). *Syst. Zool.* **1984**, *33*, 302–317. [CrossRef]
20. Rohlf, F.J. Rotational fit (Procrustes) methods. In Proceedings of the Michigan Morphometrics Workshop, Ann Arbor, MI, USA, 16–28 May 1988; Rohlf, F., Bookstein, F., Eds.; Special Publication Number 2; The University of Michigan Museum of Zoology: Ann Arbor, MI, USA, 1990; pp. 227–236.
21. Bookstein, F.L. *Morphometric Tools for Landmark Data. Geometry and Biology*; Cambridge University Press: New York, NY, USA, 1991.
22. Rohlf, F.J. Geometric Morphometrics in Systematics. 2000. Available online: https://api.semanticscholar.org/CorpusID:53362120 (accessed on 16 December 2024).
23. Dujardin, J.P.; Slice, D.E. Contributions of Morphometrics to Medical Entomology. In *Encyclopedia of Infectious Diseases*; Tibayrenc, M., Ed.; John Wiley & Sons, Inc.: Hoboken, NJ, USA, 2006; pp. 435–447, ISBN 9780470114209.
24. Dujardin, J.P. Morphometrics applied to Medical Entomology. *Infect. Genet. Evol.* **2008**, *8*, 875–890. [CrossRef]
25. Perrard, A.; Baylac, M.; Carpenter, J.M.; Villemant, C. Evolution of wing shape in hornets: Why is the wing venation efficient for species identification? *J. Evol. Biol.* **2014**, *27*, 2665–2675. [CrossRef]
26. Zúñiga-Reinoso, Á.; Benítez, H.A. The overrated use of the morphological cryptic species concept: An example with *Nyctelia* darkbeetles (Coleoptera: Tenebrionidae) using geometric morphometrics. *Zool. Anz.* **2015**, 255, 47–53. [CrossRef]
27. Dujardin, J.P.; Costa, J.; Bustamante, D.; Jaramillo, N.; Catalá, S. Deciphering morphology in Triatominae: The evolutionary signals. *Acta Trop.* **2009**, *110*, 101–111. [CrossRef] [PubMed]
28. Dujardin, J.P. Chapter 16—Modern Morphometrics of Arthropods: A Phenotypic Approach to Species Recognition and Population Structure. In *Genetics and Evolution of Infectious Diseases*, 3rd ed.; Tibayrenc, M., Ed.; Elsevier: Amsterdam, The Netherlands, 2024; pp. 385–425. [CrossRef]
29. Bookstein, F.L. Introduction to methods for landmark data. In Proceedings of the Michigan Morphometrics Workshop, Ann Arbor, MI, USA, 16–28 May 1988; Rohlf, F.J., Bookstein, F.L., Eds.; Special Publication No. 2; The University of Michigan Museum of Zoology: Ann Arbor, MI, USA, 1990; pp. 216–225.
30. Adams, D.C. Methods for shape analysis of landmark data from articulated structures. *Evol. Ecol. Res.* **1999**, *1*, 959–970.
31. Kuhl, F.P.; Giardina, C.R. Elliptic Fourier features of a closed contour. *Comput. Graph. Image Process.* **1982**, *18*, 236–258. [CrossRef]

32. Bookstein, F.L. Landmark methods for forms without landmarks: Localizing group differences in outline shape. *Med. Image Anal.* **1997**, *1*, 225–243. [CrossRef] [PubMed]
33. Santillán-Guayasamín, S.; Villacís, A.; Grijalva, M.; Dujardin, J. The modern morphometric approach to identify eggs of Triatominae. *Parasit Vectors* **2017**, *10*, 55. [CrossRef] [PubMed] [PubMed Central]
34. Santillan Guayasamin, S.; Villacis, A.G.; Grijalva, M.J.; Dujardin, J.P. Triatominae: Does the shape change of non-viable eggs compromise species recognition? *Parasites Vectors* **2018**, *11*, 543. [CrossRef] [PubMed]
35. Villacís, A.; Dujardin, J.; Panzera, F.; Yumiseva, C.; Pita, S.; Santillán-Guayasamín, S.; Orozco, M.; Mosquera, K.; Grijalva, M. Chagas vectors *Panstrongylus chinai* (Del Ponte, 1929) and *Panstrongylus howardi* (Neiva, 1911): Chromatic forms or true species? *Parasit Vectors* **2020**, *13*, 226. [CrossRef] [PubMed] [PubMed Central]
36. Cruz, D.; Arellano, E.; Denis Ávila, D.; Ibarra-Cerdena, C.N. Identifying Chagas disease vectors using elliptic Fourier descriptors of body contour: A case for the cryptic dimidiata complex. *Parasites Vectors* **2020**, *13*, 332. [CrossRef] [PubMed]
37. Dujardin, J.P.; Pham Thi, K.; Truong Xuan, L.; Panzera, F.; Pita, S.; Schofield, C.J. Epidemiological status of kissing-bugs in South East Asia: A preliminary assessment. *Acta Trop.* **2015**, *151*, 142–149. [CrossRef]
38. Cruz, D.D.; Ospina-Garcés, S.M.; Arellano, E.; Ibarra-Cerdeña, C.N.; Nava-García, E.; Alcalá, R. Geometric morphometrics and ecological niche modelling for delimitation of *Triatoma pallidipennis* (Hemiptera: Reduviidae: Triatominae) haplogroups. *Curr. Res. Parasitol. Vector-Borne Dis.* **2023**, *3*, 100119. [CrossRef]
39. Verly, T.; Pita, S.; Carbajal de la Fuente, A.; Burgueño Rodríguez, G.; Piccinali, R.; Fiad, F.; Ríos, N.; Panzera, F.; Lobbia, P.; Sanchez Casaccia, P.; et al. Relationship between genetic diversity and morpho-functional characteristics of flight-related traits in *Triatoma garciabesi* (Hemiptera: Reduviidae). *Parasites Vectors* **2024**, *17*, 145. [CrossRef]
40. Dujardin, J.P. A template-dependent semilandmarks treatment and its use in medical entomology. *Infect. Genet. Evol.* **2019**, *70*, 197–207. [CrossRef]
41. Rohlf, F. The Tps series of software. *Hystrix* **2015**, *26*, 9–12. [CrossRef]
42. Baken, E.; Collyer, M.; Kaliontzopoulou, A.; Adams, D. geomorph v4.0 and gmShiny: Enhanced analytics and a new graphical interface for a comprehensive morphometric experience. *Methods Ecol. Evol.* **2021**, *12*, 2355–2363. [CrossRef]
43. Schlager, S. Morpho and Rvcg—Shape Analysis in R. In *Statistical Shape and Deformation Analysis*; Zheng, G., Li, S., Szekely, G., Eds.; Academic Press: Cambridge, MA, USA, 2017; pp. 217–256.
44. Klingenberg, C.P. MorphoJ: An integrated software package for geometric morphometrics. *Mol. Ecol. Resour.* **2011**, *11*, 353–357. [CrossRef]
45. Dujardin, S.; Dujardin, J.P. Geometric morphometrics in the cloud. *Infect. Genet. Evol.* **2019**, *70*, 189–196. [CrossRef]
46. Bonhomme, V.; Picq, S.; Gaucherel, C.; Claude, J. Momocs: Outline Analysis Using R. *J. Stat. Softw.* **2014**, *56*, 1–24. [CrossRef]
47. Iwata, H.; Ukai, Y. A computer program package for quantitative evaluation of biological shapes based on elliptic Fourier descriptors. *J. Hered.* **2002**, *93*, 384–385. [CrossRef] [PubMed]
48. Klingenberg, C.P.; Leamy, L.J. Quantitative genetics of geometric shape in the mouse mandible. *Evolution* **2001**, *55*, 2342–2352. [PubMed]
49. Klingenberg, C.P.; Leamy, L.J.; Cheverud, J.M. Integration and modularity of quantitative trait locus effects on geometric shape in the mouse mandible. *Genetics* **2004**, *166*, 1909–1921. [CrossRef] [PubMed]
50. Shrimpton, A.; Robertson, A. The isolation of polygenic factors controlling bristle score in *Drosophila melanogaster*: I. Allocation of third chromosome bristle effects to chromosome sections. *Genetics* **1988**, *118*, 437–443. [CrossRef] [PubMed]
51. Carbajal-de-la Fuente, A.; Piccinali, R.; Porcasi, X.; Marti, G.; Rojas de Arias, A.; Abrahan, L.; Suárez, F.; Lobbia, P.; Medina, G.; Provecho, Y.; et al. Variety is the spice: The role of morphological variation of *Triatoma infestans* (Hemiptera, Reduviidae) at a macro-scale. *Acta Trop.* **2024**, *256*, 107239. [CrossRef] [PubMed]
52. Dujardin, J.P.; Kaba, D.; Solano, P.; Dupraz, M.; McCoy, K.D.; Jaramillo-O, N. Outline-based Morphometrics, an overlooked method in arthropod studies? *Infect. Genet. Evol.* **2014**, *28*, 704–714. [CrossRef]
53. Bellman, R. *Adaptive Control Processes: A Guided Tour*; Princeton Legacy Library, Princeton University Press: Princeton, NJ, USA, 1961.
54. Caro-Riaño, H.; Jaramillo, N.; Dujardin, J.P. Growth changes in *Rhodnius pallescens* under simulated domestic and sylvatic conditions. *Infect. Genet. Evol.* **2009**, *9*, 162–168. [CrossRef] [PubMed]
55. Nattero, J.; Carbajal de la Fuente, A.L.; Piccinali, R.V.; Cardozo, M.; Rodríguez, C.S.; Crocco, L.B. Characterization of melanic and non-melanic forms in domestic and peridomestic populations of *Triatoma infestans* (Hemiptera: Reduviidae). *Parasites Vectors* **2020**, *13*, 47. [CrossRef] [PubMed]
56. Fiad, F.; Cardozo, M.; Nattero, J.; Gigena, G.V. Gorla, D.; Rodríguez, C. Association between environmental gradient of anthropization and phenotypic plasticity in two species of triatomines. *Parasites Vectors* **2024**, *17*, 169. [CrossRef] [PubMed]
57. Good, P. *Permutation Tests. A Practical Guide to Resampling Methods for Testing Hypotheses*; Springer: New York, NY, USA, 2000.
58. Debat, V.; Begin, M.; Legout, H.; David, J.R. Allometric and nonallometric components of *Drosophila* wing shape respond differently to developmental temperature. *Evolution* **2003**, *57*, 2773–2784. [PubMed]

59. Collyer, M.C.; Adams, D.C. Analysis Of Two-State Multivariate Phenotypic Change In Ecological Studies. *Ecology* **2007**, *88*, 683–692. [CrossRef] [PubMed]
60. Nattero, J.; Leonhard, G.; Gürtler, R.E.; Crocco, L.B. Evidence of selection on phenotypic plasticity and cost of plasticity in response to host-feeding sources in the major Chagas disease vector *Triatoma infestans*. *Acta Trop.* **2015**, *152*, 237–244. [CrossRef] [PubMed]
61. Hernández, M.L.; Dujardin, J.P.; Villacís, A.G.; Yumiseva, C.A.; Remón, C.; Mougabure-Cueto, G. Resistance to deltamethrin in *Triatoma infestans* (Hemiptera: Reduviidae): Does it influence the phenotype of antennae, wings, and heads? *Acta Trop.* **2023**, *245*, 106976. [CrossRef] [PubMed]
62. Mitteröcker, P.; Bookstein, F. Linear discrimination, ordination, and the visualization of selection gradients in modern morphometrics. *Evol. Biol.* **2011**, *38*, 100–114. [CrossRef]
63. Bookstein, F.L. Pathologies of Between-Groups Principal Components Analysis in Geometric Morphometrics. *Evolutionary Biology* **2019**, *46*, 271–302. [CrossRef]
64. Pimentel, R.A. An introduction to ordination, principal components analysis and discriminant analysis. In *Ordination in the Study of Morphology, Evolution and Systematics of Insects: Applications and Quantitative Genetic Rationales*; Foottit, R.G., Sorensen, J.T., Eds.; Elsevier: New York, NY, USA, 1992; p. 418.
65. Dujardin, J.; Dujardin, S.; Kaba, D.; Santillán-Guayasamín, S.; Villacís, A.G.; Piyaselakul, S.; Sumruayphol, S.; Samung, Y.; Morales-Vargas, R. The maximum likelihood identification method applied to insect morphometric data. *Zool. Syst.* **2017**, *42*, 46–58.
66. Gurgel-Gonçalves, R., Komp, E.; Campbell, L.P.; Khalighifar, A.; Mellenbruch, J.; Mendonça, V.J.; Owens, H.L.; de la Cruz Felix, K.; Peterson, A.T.; Ramsey, J.M. Automated identification of insect vectors of Chagas disease in Brazil and Mexico: The Virtual Vector Lab. *PeerJ* **2017**, *5*, e3040. [CrossRef] [PubMed]
67. de Miranda, V.L.; de Souza, E.P.; Bambil, D.; Khalighifar, A.; Peterson, A.T.; de Oliveira Nascimento, F.A.; Gurgel-Gonçalves, R.; Abad-Franch, F. Cellphone picture-based, genus-level automated identification of Chagas disease vectors: Effects of picture orientation on the performance of five machine-learning algorithms. *Ecol. Inform.* **2024**, *79*, 102430. [CrossRef]
68. Kitthawee, S.; Dujardin, J. The *Diachasmimorpha longicaudata* complex in Thailand discriminated by its wing venation. *Zoomorphology* **2016**, *135*, 323–332. [CrossRef]
69. Klecka, W.R. Discriminant analysis. In *Sage University Paper Series on Quantitative Applications in the Social Sciences*; No. 07-019; Sage: Beverly Hills, CA, USA, 1980.
70. Dujardin, J.P.; Steindel, M.; Chavez, T.; Martínez, E.; Schofield, C.J. Changes in the sexual dimorphism of Triatominae in the transition from natural to artificial habitats. *Memórias Inst. Oswaldo Cruz* **1999**, *94*, 565–569. [CrossRef]
71. Dujardin, J.P.; Panzera, P.; Schofield, C.J. Triatominae as a model of morphological plasticity under ecological pressure. *Memórias Inst. Oswaldo Cruz* **1999**, *94*, 223–228. [CrossRef] [PubMed]
72. Klingenberg, C.P.; McIntyre, G. Geometric morphometrics of developmental instability: Analyzing patterns of fluctuating asymmetry with Procrustes methods. *Evolution* **1998**, *52*, 1363–1375. [CrossRef] [PubMed]
73. Klingenberg, C.P. Analyzing Fluctuating Asymmetry with Geometric Morphometrics: Concepts, Methods, and Applications. *Symmetry* **2015**, *7*, 843–934. [CrossRef]
74. Claude, J. *Morphometrics with R*; Springer Science+Business Media, LLC: New York, NY, USA, 2008; 330p, ISBN 978-0-387-77789-4.
75. Palmer, A.R.; Strobeck, C. Fluctuating Asymmetry: Measurement, Analysis, Patterns. *Annu. Rev. Ecol. Syst.* **1986**, *17*, 391–421. [CrossRef]
76. Van Dongen, S.V. Fluctuating asymmetry and developmental instability in evolutionary biology: Past, present and future. *J. Evol. Biol.* **2006**, *19*, 1727–1743. [CrossRef] [PubMed]
77. Nattero, J.; Piccinali, R.V.; Gaspe, M.S.; Gürtler, R.E. Fluctuating asymmetry and exposure to pyrethroid insecticides in *Triatoma infestans* populations in northeastern Argentina. *Infect. Genet. Evol.* **2019**, *74*, 103925. [CrossRef] [PubMed]
78. Nattero, J.; Mougabure-Cueto, G.; Debat, V.; Gürtler, R.E. Phenotypic plasticity, canalisation and developmental stability of *Triatoma infestans* wings: Effects of a sublethal application of a pyrethroid insecticide. *Parasites Vectors* **2021**, *14*, 355. [CrossRef]
79. Gutiérrez-Cabrera, A.; Badillo, R.; Gonzalez, L.; Ospina-Garcés, S.; Córdoba-Aguilar, A. Body shape and fluctuating asymmetry following different feeding sources and feeding time in a triatomine, *Triatoma pallidipennis* (Stal, 1892). *Infect. Genet. Evol.* **2022**, *98*, 105199. [CrossRef] [PubMed]
80. Nattero, J.; Piccinali, R.; Fiad, F.; Cano, F.; Carbajal-de-la Fuente, A. Relationship between flight muscle dimorphism and wing morphometry in *Triatoma infestans* (Klug, 1834) (Hemiptera, Reduviidae, Triatominae). *Front. Ecol. Evol.* **2023**, *11*, 1211219. [CrossRef]
81. Soares, R.; Dujardin, J.P.; Schofield, C.; Diotaiuti, L. Wing symmetry and flight activity. *Mem. Inst. Oswaldo Cruz* **1996**, *91*, 143.
82. Nouvellet, P.; Ramirez-Sierra, M.J.; Dumonteil, E.; Gourbière, S. Effects of genetic factors and infection status on wing morphology of *Triatoma dimidiata* species complex in the Yucatán peninsula, Mexico. *Infect. Genet. Evol.* **2011**, *11*, 1243–1249. [CrossRef] [PubMed]

83. Hernández, M.L.; Abrahan, L.B.; Dujardin, J.P.; Gorla, D.E.; Catalá, S.S. Phenotypic Variability and Population Structure of Peridomestic *Triatoma infestans* in Rural Areas of the Arid Chaco (Western Argentina): Spatial Influence of Macro- and Microhabitats. *Vector Borne Zoonotic Dis.* **2011**, *11*, 503–513. [CrossRef] [PubMed]
84. Rohlf, F.J.; Corti, M. Use of two-block partial least-squares to study covariation in shape. *Syst. Biol.* **2000**, *49*, 740–753. [CrossRef]
85. Escoufier, Y. La dépendance de deux aléas vectoriels critères et visualisation. *Rev. Stat. Appliquée* **1973**, *21*, 5–16.
86. Gigena, V.; Rodríguez, C.; Fiad, F.; Hernández, M.; Carbajal-de-la Fuente, A.; Piccinali, R.; Casaccia, P.; Rojas de Arias, A.R.; Lobbia, P.; Abrahan, L.; et al. Phenotypic variability in traits related to flight dispersal in the wing dimorphic species *Triatoma guasayana*. *Parasit Vectors* **2023**, *16*, 8. [CrossRef] [PubMed]
87. Fiad, F.G.; Cardozo, M.; Rodríguez, C.S.; Hernández, M.L.; Crocco, L.B.; Gorla, D.E. Ecomorphological variation of the *Triatoma guasayana* wing shape in semi-arid Chaco region. *Acta Trop.* **2022**, *232*, 106488. [CrossRef] [PubMed]
88. Barata, J.M.S. Aspectos morfológicos de ovos de Triatominae. II. Caracteristicas macroscopicas e exocoriais de dez especies do gênero *Rhodnius* Stal, 1859 (Hemiptera, Reduviidae). *Rev. Saúde Publica* **1981**, *15*, 490–542. [CrossRef] [PubMed]
89. Costa, J.; Barth, O.M.; Marchon Silva, V.; de Almeida, C.E.; FreitasSibajev, M.G.R.; Panzera, F. Morphological studies on the *Triatoma brasiliensis* Neiva, 1911 (Hemiptera, Reduviidae, Triatominae) genital structures and eggs of different chromatic forms. *Memórias Inst. Oswaldo Cruz* **1997**, *92*, 493–498. [CrossRef]
90. Souza, S. *Aspectos Morfológicos e Morfometria de Ovos de Triatoma pseudomaculata Corrêa & Espínola, 1964 (Hemiptera, Reduviidae, Triatominae) No Estado do Ceará*; Instituto Oswaldo Cruz, Fundação Oswaldo Cruz: Rio de Janeiro, Brazil, 2011; p. 73.
91. Obara, M.T.; Barata, J.; da Silva, N.N.; Ceretti Júnior, W.; Urbinatti, P.R.; da Rosa, J.A.; Jurberg, J.; Galvão, C. Morphologic, morphometrical, and histological aspects of the eggs of four species of the genera *Meccus* (Hemiptera, Reduviidae, Triatominae). *Mem. Inst. Oswaldo Cruz* **2007**, *102*, 13–19. [CrossRef] [PubMed]
92. Abrahan, L.B.; Gorla, D.E.; Catalá, S.S. Dispersal of *Triatoma infestans* and other Triatominae species in the arid Chaco of Argentina: Flying, walking or passive carriage? The importance of walking females. *Mem. Inst. Oswaldo Cruz* **2011**, *106*, 232–239. [CrossRef]
93. Patterson, J.S.; Barbosa, S.E.; Feliciangeli, M.D. On the genus *Panstrongylus* Berg 1879: Evolution, ecology and epidemiological significance. *Acta Trop.* **2009**, *110*, 187–199. [CrossRef]
94. Rocha, D.S.; Patterson, J.S.; Sandoval, C.M.; Jurberg, J.; Ângulo, V.M.; Esteban, L.; Galvão, C. Description and ontogenetic morphometrics of nymphs of *Belminus herreri* Lent & Wygodzinsky (Hemiptera: Reduviidae, Triatominae). *Neotrop. Entomol.* **2005**, *34*, 491–497. [CrossRef]
95. Rocha, D.; Dale, C.; da Rosa, J.A.; Galvão, C. Description of nymphs and ontogenetic morphometry of *Triatoma ryckmani* Zeledón & Ponce, 1972 (Hemiptera: Heteroptera: Reduviidae: Triatominae). *EntomoBrasilis* **2020**, *13*, e899. [CrossRef]
96. Lázari Cacini, G.; de Oliveira, J.; Belintani, T.; dos Santos Souza, E.; Olaia, N.; Pinto, M.C.; da Rosa, J. Immature instars of three species of *Rhodnius* Stål, 1859 (Hemiptera, Reduviidae, Triatominae): Morphology, morphometry, and taxonomic implications. *Parasites Vectors* **2022**, *15*, 91. [CrossRef] [PubMed]
97. Oliveira, J.; Marcet, P.L.; Takiya, D.M.; Mendonça, V.J.; Belintani, T.; Bargues, M.D.; Mateo, L.; Chagas, V.; Folly-Ramos, E.; Cordeiro-Estrela, P.; et al. Combined phylogenetic and morphometric information to delimit and unify the *Triatoma brasiliensis* species complex and the Brasiliensis subcomplex. *Acta Trop.* **2017**, *170*, 140–148. [CrossRef] [PubMed]
98. Nattero, J.; Piccinali, R.; Macedo Lopes, C.; Hernández, M.; Abrahan, L.; Lobbia, P.; Rodríguez, C.; Carbajal De La Fuente, A. Morphometric variability among the species of the Sordida subcomplex (Hemiptera: Reduviidae: Triatominae): Evidence for differentiation across the distribution range of *Triatoma sordida*. *Parasites Vectors* **2017**, *10*, 412. [CrossRef] [PubMed]
99. Alvarez, A.; Dale, C.; Galvão, C. Geometric morphometry of the *Rhodnius prolixus* complex (Hemiptera, Triatominae): Patterns of intraspecific and interspecific allometry and their taxonomic implications. *ZooKeys* **2024**, *1202*, 213–228. [CrossRef] [PubMed]
100. Schachter-Broide, J.S.; Dujardin, J.P.; Kitron, U.; Gürtler, R.E. Spatial structuring of *Triatoma infestans* (Hemiptera: Reduviidae) populations from northwestern Argentina using wing geometric morphometry. *J. Med. Entomol.* **2004**, *41*, 643–649. [CrossRef]
101. Schachter-Broide, J.; Gürtler, R.E.; Kitron, U.; Dujardin, J.P. Temporal variations of wing size and shape of *Triatoma infestans* (Hemiptera: Reduviidae) populations from northwestern Argentina using geometric morphometry. *J. Med. Entomol.* **2009**, *46*, 994–1000. [CrossRef]
102. Hernández, M.; Espinoza, J.; Gomez, M.; Gorla, D. Morphological changes associated with brachypterous *Triatoma guasayana* (Hemiptera, Reduvii- dae) and their relationship with flight. *Int. J. Trop. Insect Sci.* **2020**, *40*, 413–421. [CrossRef]
103. Hernández, M.; Dujardin, J.; Gorla, D.; Catalá, S. Can body traits, other than wings, reflect the flight ability of triatominae bugs? *Rev. Soc. Bras. Med. Trop.* **2015**, *48*, 682–691. [CrossRef] [PubMed]
104. Grodnitsky, D.L. *Form and Function of Insect Wings: The Evolution of Biological Structures*; The John Hopkins University Press: Baltimore, MD, USA; London, UK, 2000; p. 99, ISBN 0-8018-6003-2.
105. Nattero, J.; Malerba, R.; Rodríguez, C.; Crocco, L. Phenotypic plasticity in response to food source in *Triatoma infestans* (Klug, 1834) (Hemiptera, Reduviidae: Triatominae). *Infect. Genet. Evol.* **2013**, *19*, 38–44. [CrossRef] [PubMed]

106. Lunardi, R.R.; Benítez, H.A.; Câmara, T.P.; Gomes, L.P.; Arrais-Silva, W.W. Head shape variation in response to diet in *Triatoma williami* (Hemiptera, Reduviidae: Triatominae), a possible Chagas disease vector of legal Amazonia. *Zool. Anz.* **2017**, *267*, 187–193. [CrossRef]
107. Vilaseca, C.; Méndez, M.A.; Pinto, C.F.; Lemic, D.; Benítez, H.A. Unraveling the Morphological Variation of *Triatoma infestans* in the Peridomestic Habitats of Chuquisaca Bolivia: A Geometric Morphometric Approach. *Insects* **2021**, *12*, 185. [CrossRef] [PubMed]
108. Catalá, S.; Dujardin, J.P. Antennal sensilla patterns indicate geographic and ecotopic variability among *Triatoma infestans* (Hemiptera: Reduviidae) populations. *J. Med. Entomol.* **2001**, *38*, 423–428. [CrossRef]
109. Abad-Franch, F.; Monteiro, F.A.; Pavan, M.G.; Patterson, J.S.; Bargues, M.D.; Zuriaga, M.; Aguilar, M.; Beard, C.B.; Mas-Coma, S.; Miles, M.A. Under pressure: Phenotypic divergence and convergence associated with microhabitat adaptations in Triatominae. *Parasites Vectors* **2021**, *14*, 195. [CrossRef] [PubMed]
110. Dorn, P.L.; Justi, S.A.; Dale, C.; Stevens, L.; Galvão, C.; Lima-Cordón, R.; Monroy, C. Description of *Triatoma mopan sp. n.* from a cave in Belize (Hemiptera, Reduviidae, Triatominae). *ZooKeys* **2018**, *775*, 69–95. [CrossRef] [PubMed]
111. Monroy, M.C.; Bustamante, D.M.; Rodas, A.G.; Enriquez, M.E.; Rosales, R.G. Habitats, dispersion and invasion of sylvatic *Triatoma dimidiata* (Hemiptera: Reduviidae: Triatominae) in Peten, Guatemala. *J. Med. Entomol.* **2003**, *40*, 800–806. [CrossRef]
112. Bustamante, D.M.; Monroy, C.; Menes, M.; Rodas, A.; Salazar-Schettino, P.M.; Rojas, G.; Pinto, N.; Guhl, F.; Dujardin, J.P. Metric Variation among Geographic Populations of the Chagas Vector *Triatoma dimidiata* (Hemiptera: Reduviidae: Triatominae) and Related Species. *J. Med. Entomol.* **2004**, *41*, 96–301. [CrossRef]
113. de la Rúa, N.M.; Bustamante, D.M.; Menes, M.; Stevens, L.; Monroy, C.; Kilpatrick, C.W.; Rizzo, D.; Klotz, S.A.; Schmidt, J.; Axen, H.J.; et al. Towards a phylogenetic approach to the composition of species complexes in the North and Central American *Triatoma*, vectors of Chagas disease. *Infect. Genet. Evol.* **2014**, *24*, 157–166. [CrossRef]
114. Gurgel-Gonçalves, R.; Ferreira, J.; Rosa, A.; Bar, M.; Galvão, C. Geometric morphometrics and ecological niche modelling for delimitation of near-sibling triatomine species. *Med. Vet. Entomol.* **2011**, *25*, 84–93. [CrossRef]
115. Lima-Cordón, R.A.; Monroy, M.C.; Stevens, L.; Rodas, A.; Rodas, G.A.; Dorn, P.L.; Justi, S.A. Description of *Triatoma huehuetenanguensis sp. n.*, a potential Chagas disease vector (Hemiptera, Reduviidae, Triatominae). *ZooKeys* **2019**, *820*, 51–70. [CrossRef] [PubMed]
116. Almeida, C.E.; Oliveira, H.L.; Correia, N.; Dornak, L.L.; Gumiel, M.; Neiva, V.L.; Harry, M.; Mendonça, V.J.; Costa, J.; Galvão, C. Dispersion capacity of *Triatoma sherlocki*, *Triatoma juazeirensis* and laboratory-bred hybrids. *Acta Trop.* **2012**, *122*, 71–79. [CrossRef] [PubMed]
117. Mougabure-Cueto, G.; Hernández, M.L.; Gilardoni, J.J.; Nattero, J. Morphometric study of the legs of the main Chagas vector, *Triatoma infestans* (Hemiptera: Reduviidae). *Acta Trop.* **2024**, *255*, 107219. [CrossRef]
118. Belintani, T.; de Paiva, V.F.; de Oliveira, J.; da Rosa, J.A. New in morphometry: Geometric morphometry of the external female genitalia of Triatominae (Hemiptera: Reduviidae). *Acta Trop.* **2022**, *229*, 106383. [CrossRef] [PubMed]
119. Catalá, S.; Schofield, C.J. Antennal sensilla of *Rhodnius*. *J. Morphol.* **1994**, *219*, 193–204. [CrossRef] [PubMed]
120. Catalá, S. Antennal sensilla of Triatominae. A comparative study of five genera. *Int. J. Insect Morphol. Embryol.* **1997**, *26*, 67–73. [CrossRef]
121. Gracco, M.; Catalá, S.S. Inter-specific and developmental differences on the array of antennal chemoreceptors in four species of Triatominae (Hemiptera: Reduviidae). *Mem. Inst. Oswaldo Cruz* **2000**, *95*, 67–74. [CrossRef] [PubMed]
122. May-Concha, I.; Guerenstein, P.G.; Ramsey, J.M.; Rojas, J.C.; Catalá, S. Antennal phenotype of Mexican haplogroups of the *Triatoma dimidiata* complex, vectors of Chagas disease. *Infect. Genet. Evol.* **2016**, *40*, 73–79. [CrossRef]
123. Hernández, M.L.; Abrahan, L.; Moreno, M.; Gorla, D.; Catalá, S. Phenotypic Variability Associated to Genomic Changes in the main vector of Chagas Disease in South America. *Acta Trop.* **2008**, *106*, 60–67. [CrossRef] [PubMed]
124. Abrahan, L.; Hernández, L.; Gorla, D.; Catalá, S.S. Phenotypic diversity of *Triatoma infestans* at the microgeographic level on the Gran Chaco of Argentina and the Andean valleys of Bolivia. *J. Med. Entomol.* **2008**, *45*, 660–666. [CrossRef] [PubMed]
125. May-Concha, I.J.; Escalante-Talavera, M.J.; Dujardin, J.P.; Waleckx, E. Does *Trypanosoma cruzi* (Chagas, 1909) (Kinetoplastida: Trypanosomatidae) modify the antennal phenotype of *Triatoma dimidiata* (Latreille, 1811) (Hemiptera: Triatominae)? *Parasites Vectors* **2022**, *15*, 466. [CrossRef]
126. Rasmussen, P.; Wheeler, W.; Moser, T.; Vine, L.; Sullivan, B.; Rusch, D. Measurements of Canada goose morphology - Sources of error and effects on classification of subspecies. *J. Wildl. Manag.* **2001**, *65*, 716–725. [CrossRef]
127. Jordaens, K.; Van Dongen, S.; Van Riel, P.; Geenen, S.; Verhagen, R.; Backeljau, T. Multivariate morphometrics of soft body parts in terrestrial slugs: Comparison between two datasets, error assessment and taxonomic implications. *Biol. J. Linn. Soc.* **2002**, *75*, 533–542. [CrossRef]
128. Arnqvist, G.; Mårtensson, T. Measurement error in geometric morphometrics: Empirical strategies to assess and reduce its impact on measure of shape. *Acta Zool. Acad. Sci. Hung.* **1998**, *44*, 73–96.
129. Dujardin, J.P.; Kaba, D.; Henry, A.B. The exchangeability of shape. *BMC Res. Notes* **2010**, *3*, 266. [CrossRef]

130. Robinson, C.; Terhune, C.E. Error in geometric morphometric data collection: Combining data from multiple sources. *Am. J. Phys. Anthropol.* **2017**, *164*, 62–75. [CrossRef]
131. Galvao, C.; Patterson, J.S.; Rocha, D.D.; Jurberg, J.; Carcavallo, R.; Rajen, K.; Ambrose, D.P.; Miles, M.A. A new species of Triatominae from Tamil Nadu, India. *Med. Vet. Entomol.* **2002**, *16*, 75–82. [CrossRef] [PubMed]
132. Mendonça, V.; Alevi, K.; Pinotti, H.; Gurgel-Gonçalves, R.; Pita, S.; Guerra, A.; Panzera, F.; Araújo, R.; Azeredo-Oliveir, M.; da Rosa, J. Revalidation of *Triatoma bahiensis* Sherlock & Serafim, 1967 (Hemiptera: Reduviidae) and phylogeny of the *T. brasiliensis* species complex. *Zootaxa* **2016**, *4107*, 239. [CrossRef]
133. Pita, S.; Gómez-Palacio, A.; Lorite, P.; Dujardin, J.; Chavez, T.; Villacís, A.; Galvão, C.; Panzera, Y.; Calleros, L.; Pereyra-Mello, S.; et al. Multidisciplinary approach detects speciation within the kissing bug *Panstrongylus rufotuberculatus* populations (Hemiptera, Heteroptera, Reduviidae). *Mem. Inst. Oswaldo Cruz* **2022**, *116*, e210259. [CrossRef] [PubMed]
134. Alevi, K.C.C.; de Oliveira, J.; Garcia, A.C.C.; Cristal, D.C.; Delgado, L.M.G.; de Freitas Bittinelli, I.; Dos Reis, Y.V.; Ravazi, A.; de Oliveira, A.B.B.; Galvão, C.; et al. *Triatoma rosai* sp. nov. (Hemiptera, Triatominae): A New Species of *Argentinian Chagas* Disease Vector Described Based on Integrative Taxonomy. *Insects* **2020**, *11*, 830. [CrossRef] [PubMed]
135. Alevi, K.C.C.; de Oliveira, J.; da Silva Rocha, D.; Galvão, C. Trends in Taxonomy of Chagas Disease Vectors (Hemiptera, Reduviidae, Triatominae): From Linnaean to Integrative Taxonomy. *Pathogens* **2021**, *10*, 1627. [CrossRef]
136. Carbajal de la Fuente, A.L.; Jaramillo, N.; Barata, J.M.S.; Noireau, F.; Diotaiuti, L. Misidentification of two Brazilian triatomes, *Triatoma arthurneivai* and *Triatoma wygodzinskyi*, revealed by geometric morphometrics. *Med. Vet. Entomol.* **2011**, *25*, 178–183. [CrossRef] [PubMed]
137. Catalá, S.; Torres, M. Similitude of the patterns of sensilla on the antennae of *Triatoma melanosoma* and *Triatoma infestans*. *Ann. Trop. Med. Parasitol.* **2001**, *95*, 287–295. [CrossRef]
138. Gumiel, M.; Catalá, S.; Noireau, F.; de Arias, A.R.; Garcia, A.; Dujardin, J.P. Wing geometry in *Triatoma infestans* (Klug) and *T. melanosoma* Martinez, Olmedo and Carcavallo (Hemiptera: Reduviidae). *Syst. Entomol.* **2003**, *28*, 173–179. [CrossRef]
139. Matias, A.; De la Riva, J.X.; Torrez, M.; Dujardin, J.P. *Rhodnius robustus* in Bolivia identified by its wings. *Mem. Inst. Oswaldo Cruz* **2001**, *96*, 947–950. [CrossRef] [PubMed]
140. Costa, J.; Peterson, A.T.; Dujardin, J.P. Indirect evidences suggest homoploid hybridization as a possible mode of speciation in Triatominae (Hemiptera, Heteroptera, Reduviidae). *Infect. Genet. Evol.* **2008**, *9*, 263–270. [CrossRef]
141. Villegas, J.; Feliciangeli, M.D.; Dujardin, J.P. Wing shape divergence between *Rhodnius prolixus* from Cojedes (Venezuela) and *R. robustus* from Mérida (Venezuela). *Infect. Genet. Evol.* **2002**, *2*, 121–128. [CrossRef] [PubMed]
142. Gurgel-Goncalves, R.; Abad-Franch, F.; Ferreira, J.; Santana, D.; Cuba, C. Is *Rhodnius prolixus* (Triatominae) invading houses in central Brazil ? *Acta Trop.* **2008**, *107*, 90–98. [CrossRef] [PubMed]
143. Márquez, E.J.; Jaramillo-O, N.; Gómez-Palacio, A.; Dujardin, J.P. Morphometric and molecular differentiation of a *Rhodnius robustus*-like form from *R. robustus* Larousse, 1927 and *R. prolixus* Stal, 1859 (Hemiptera, Reduviidae). *Acta Trop.* **2011**, *120*, 103–109. [CrossRef] [PubMed]
144. Noireau, F.; dos Santos, S.M.; Gumiel, M.; Dujardin, J.P.; dos Santos Soares, M.; Carcavallo, R.U.; Galvão, C.; Jurberg, J. Phylogenetic relationships within the oliveirai complex (Hemiptera: Reduviidae: Triatominae). *Infect. Genet. Evol.* **2002**, *2*, 11–17. [CrossRef]
145. Gonzalez-Britez, N.E.; Carrasco, H.J.; Martínez Purroy, C.E.; Feliciangeli, D.; Maldonado, M.; López, E.; Segovia, M.J.; Rojas de Arias, A. Genetic and morphometric variability of *Triatoma sordida* (Hemiptera: Reduviidae) from the eastern and western regions of Paraguay. *Front. Public Health* **2014**, *2*, 149. [CrossRef] [PubMed]
146. Klingenberg, C.P.; Gidaszewski, N.A. Testing and Quantifying Phylogenetic Signals and Homoplasy in Morphometric Data. *Syst. Biol.* **2010**, *59*, 245–261. [CrossRef] [PubMed]
147. Varón-González, C.; Whelan, S.; Klingenberg, C.P. Estimating Phylogenies from Shape and Similar Multidimensional Data: Why It Is Not Reliable. *Syst. Biol.* **2020**, *69*, 863–883. [CrossRef] [PubMed]
148. Schofield, C.J.; Galvão, C. Classificatión, evolutión, and species groups within the Triatominae. *Acta Trop.* **2009**, *110*, 88–100. [CrossRef] [PubMed]
149. Hwang, W.S.; Weirauch, C. Evolutionary History of Assassin Bugs (Insecta: Hemiptera: Reduviidae): Insights from Divergence Dating and Ancestral State Reconstruction. *PLoS ONE* **2012**, *7*, e45523. [CrossRef] [PubMed]
150. Patterson, J.S.; Schofield, C.J.; Dujardin, J.P.; Miles, M.A. Population morphometric analysis of the tropicopolitan bug *Triatoma rubrofasciata* and relationships with Old World species of *Triatoma*: Evidence of New World ancestry. *Med. Vet. Entomol.* **2001**, *15*, 443–451. [CrossRef] [PubMed]
151. Dujardin, J.P.; Chavez, T.; Moreno, J.M.; Machane, M.; Noireau, F.; Schofield, C.J. Comparison of isoenzyme electrophoresis and morphometric analysis for phylogenetic reconstruction of the Rhodniini (Hemiptera: Reduviidae: Triatominae). *J. Med. Entomol.* **1999**, *36*, 653–659. [CrossRef] [PubMed]

152. Almeida, C.E.; Marcet, P.L.; Gumiel, M.; Takiya, D.M.; Cardozo-de Almeida, M.; Pacheco, R.S.; Lopes, C.M.; Dotson, E.M.; Costa, J. Phylogenetic and phenotypic relationships among *Triatoma carcavalloi* (Hemiptera: Reduviidae: Triatominae) and related species collected in domiciles in Rio Grande do Sul State, Brazil. *J. Vector Ecol.* **2009**, *34*, 164–173. [CrossRef] [PubMed]
153. Panzera, F.; Dujardin, J.P.; Nicolini, P.; Caraccio, M.N.; Rose, V.; Tellez, T.; Bermudez, H.; Bargues, M.D.; Mas-Coma, S.; O'Connor, J.E.; et al. Genomic changes of Chagas disease vector, South America. *Emerg. Infect. Dis.* **2004**, *10*, 438–446. [CrossRef]
154. Dujardin, J.P.; Muñoz, M.; Chavez, T.; Ponce, C.; Moreno, J.; Schofield, C.J. The origin of *Rhodnius prolixus* in Central America. *Med. Vet. Entomol.* **1998**, *12*, 113–115. [CrossRef]
155. Monroy, C.; Bustamante, D.M.; Rodas, A.; Rosales, R.; Mejia, M.; Tabaru, Y. Geographic distribution and morphometric differentiation of *Triatoma nitida* Usinger 1939 (Hemiptera: Reduviidae: Triatominae) in Guatemala. *Mem. Inst. Oswaldo Cruz* **2003**, *98*, 37–43. [CrossRef]
156. Piccinali, R.V.; Gaspe, M.S.; Nattero, J.; Gürtler, R.E. Population structure and migration in *Triatoma infestans* (Hemiptera: Reduviidae) from the Argentine Chaco: An integration of genetic and morphometric data. *Acta Trop.* **2023**, *247*, 107010. [CrossRef] [PubMed]
157. Gómez-Palacio, A.; Jaramillo-O, N.; Caro-Riaño, H.; Diaz, S.; Monteiro, F.; Pérez, R.; Panzera, F.; Triana, O. Morphometric and molecular evidence of intraspecific biogeographical differentiation of *Rhodnius pallescens* (Hemiptera: Reduviidae: Rhodniini) from Colombia and Panama. *Infect. Genet. Evol.* **2012**, *12*, 1975–1983. [CrossRef]
158. Vendrami, D.P.; Obara, M.T.; Gurgel-Gonçalves, R.; Ceretti-Junior, W.; Marrelli, M.T. Wing geometry of *Triatoma sordida* (Hemiptera: Reduviidae) populations from Brazil. *Infect. Genet. Evol.* **2017**, *49*, 17–20. [CrossRef]
159. Nattero, J.; Pita, S.; Calleros, L.; Crocco, L.; Panzera, Y.; Rodríguez, C.; Panzera, F. Morphological and Genetic Differentiation within the Southernmost Vector of Chagas Disease: *Triatoma patagonica* (Hemiptera– Reduviidae). *PLoS ONE* **2016**, *11*, e0168853. [CrossRef] [PubMed]
160. Noireau, F.; Flores, R.; Gutierrez, T.; Dujardin, J.P. Detection of silvatic dark morphs of *Triatoma infestans* in the Bolivian Chaco. *Memórias Inst. Oswaldo Cruz* **1997**, *92*, 583–584. [CrossRef]
161. Dujardin, J.P.; Bermudez, H.; Casini, C.; Schofield, C.J.; Tibayrenc, M. Metric differences between silvatic and domestic *Triatoma infestans* (Hemiptera, Reduviidae) in Bolivia. *J. Med. Entomol.* **1997**, *34*, 544–552. [CrossRef] [PubMed]
162. Dujardin, J.P.; Beard, C.B.; Ryckman, R. The relevance of wing geometry in entomological surveillance of Triatominae, vectors of Chagas disease. *Infect. Genet. Evol.* **2007**, *7*, 161–167. [CrossRef] [PubMed]
163. Gaspe, M.S.; Provecho, Y.M.; Piccinali, R.V.; Gürtler, R.E. Where do these bugs come from? Phenotypic structure of *Triatoma infestans* populations after control interventions in the Argentine Chaco. *Mem. Inst. Oswaldo Cruz* **2015**, *110*, 310–318. [CrossRef] [PubMed]
164. Jaramillo, N.; Castillo, D.; Wolff, M. Geometric morphometric differences between *Panstrongylus geniculatus* from field and laboratory. *Mem. Inst. Oswaldo Cruz* **2002**, *97*, 667–673. [CrossRef]
165. Gurgel-Gonçalves, R.; Maeda, M.; Ferreira, J.; Rosa, A.; Cuba, C. Morphometric changes of *Rhodnius neglectus* (Hemiptera: Reduviidae) in the transition from sylvatic to laboratory conditions. *Zoologia (Curitiba)* **2011**, *28*, 5. [CrossRef]
166. Paschoaletto, L.; Santos-Mallet, J.; Dale, C.; Neiva, V.; Macedo Lopes, C.; Costa, J. A 72 years of temporal analysis through geometric morphometrics detects phenotypic variationin populations of *Triatoma infestans* (Klug, 1834). *Acta Biol. Par.* **2020**, *49*, 1-17. [CrossRef]
167. Ribeiro, G., Jr.; Reis, J.; Vaccarezza, F.; Silva, A.C.O.; Lanza, F.C.; Miranda, D.L.P.; Gurgel-Gonçalves, R.; Reis, M.G.D. Sometimes, the size matters: Wing geometric morphometrics as a tool to assess domiciliation by *Triatoma sordida* (Stäl 1859). *Rev. Soc. Bras. Med. Trop.* **2023**, *57*, e007092024. [CrossRef] [PubMed]
168. Vargas, E.; Espitia, C.; Patiño, C.; Pinto, N.; Aguilera, G.; Jaramillo, C.; Bargues, M.D.; Guhl, F. Genetic structure of *Triatoma venosa* (Hemiptera: Reduviidae): Molecular and morphometric evidence. *Mem. Inst. Oswaldo Cruz* **2006**, *101*, 1. [CrossRef]
169. Usinger, R.L.; Wygodzinsky, P.; Ryckman, R.E. The Biosystematics of Triatominae. *Annu. Rev. Entomol.* **1966**, *11*, 309–330. [CrossRef] [PubMed]
170. Nattero, J.; Mougabure-Cueto, G.; Gürtler, R.E. Sublethal effects of a pyrethroid insecticide on cuticle thickness, wing size and shape in the main vector *Triatoma infestans*. *Med. Vet. Entomol.* **2022**, *36*, 397–407. [CrossRef] [PubMed]
171. Gaspe, M.S.; Schachter-Broide, J.; Gurevitz, J.M.; Kitron, U.; Gürtler, R.E.; Dujardin, J.P. Microgeographic Spatial Structuring of *Triatoma infestans* (Hemiptera: Reduviidae) Populations Using Wing Geometric Morphometry in the Argentine Chaco. *J. Med. Entomol.* **2012**, *49*, 504–514. [CrossRef] [PubMed]
172. Jirakanjanakit, N.; Leemingsawat, S.; Dujardin, J.P. The geometry of the wing of *Aedes* (*Stegomyia*) *aegypti* in isofemale lines through successive generations. *Infect. Genet. Evol.* **2008**, *8*, 414–421. [CrossRef] [PubMed]
173. Kamimura, E.; Viana, M.C.; Lilioso, M.; Fontes, F.; Pires da Silva, D.; Valença-Barbosa, C.; Carbajal de la Fuente, A.; Folly-Ramos, E.; Solferin, V.; Thyssen, P.; et al. Drivers of molecular and morphometric variation in *Triatoma brasiliensis (Hemiptera: Triatominae)*: The resolution of geometric morphometrics for populational structuring on a microgeographical scale. *Parasites Vectors* **2020**, *13*, 455. [CrossRef]

174. Dumonteil, E.; Tripet, F.; Ramirez-Sierra, M.J.; Payet, V.; Lanzaro, G.; Menu, F. Assessment of *Triatoma dimidiata* dispersal in the Yucatan peninsula of Mexico by morphometry and microsatellite markers. *Am. J. Trop. Med. Hyg.* **2007**, *76*, 930–937. [CrossRef] [PubMed]
175. Feliciangeli, M.D.; Sanchez-Martin, M.; Marrero, R.; Davies, C.; Dujardin, J.P. Morphometric evidence for a possible role of *Rhodnius prolixus* from palm trees in house re-infestation in the State of Barinas (Venezuela). *Acta Trop.* **2007**, *101*, 169–177. [CrossRef]
176. Ceballos, L.A.; Piccinali, R.V.; Marcet, P.L.; Vazquez-Prokopec, G.M.; Cardinal, M.V.; Schachter-Broide, J.; Dujardin, J.P.; Dotson, E.M.; Kitron, U.; Gürtler, R.E. Hidden Sylvatic Foci of the Main Vector of Chagas Disease *Triatoma infestans*: Threats to the Vector Elimination Campaign? *PLoS Negl. Trop. Dis.* **2011**, *5*, e1365. [CrossRef] [PubMed]
177. Hernández, M.L.; Dujardin, J.P.; Gorla, D.E.; Catalá, S.S. Potential sources of *Triatoma infestans* reinfesting peridomiciles identified by morphological characterization in Los Llanos, La Rioja, Argentina. *Mem. Inst. Oswaldo Cruz* **2013**, *108*, 91–97. [CrossRef] [PubMed]
178. Gaspe, M.S.; Gurevitz, J.M.; Gürtler, R.E.; Dujardin, J.P. Origins of house reinfestation with *Triatoma infestans* after insecticide spraying in the Argentine Chaco using wing geometric morphometry. *Infect. Genet. Evol.* **2013**, *17*, 93–100. [CrossRef]
179. Nattero, J.; Cecere, M.C.; Gürtler, R.E. Temporal variations of fluctuating asymmetry in wing size and shape of *Triatoma infestans* populations from northwest Argentina. *Infect. Genet. Evol.* **2017**, *56*, 133–142. [CrossRef]
180. Paschoaletto, L.; Dale, C.; Neiva, V.; Carbajal de la Fuente, A.; Oliveira, J.; Benítez, H.; Costa, J. Morphological Stasis in Time? A *Triatoma brasiliensis brasiliensis* Study Using Geometric Morphometrics in the Long Run. *Animals* **2022**, *12*, 1362. [CrossRef] [PubMed]
181. Schofield, C.J. Biosystematics of the Triatominae. In *Biosystematic of Haematophagous Insects*; Systematics Assoc. Special Volume 37; Service, M.W., Ed.; Oxford Clarendon Press:Oxford, UK, 1988; pp. 284–312.
182. Vilaseca, C.; Méndez, M.A.; Pinto, C.F.; Benítez, H.A. Assessment of Shape Variation Patterns in *Triatoma infestans* (Klug 1834) (Hemiptera: Reduviidae: Triatominae): A First Report in Populations from Bolivia. *Insects* **2020**, *11*, 274. [CrossRef] [PubMed]

pathogens

MDPI

Review

Mitochondrial DNA Structure in *Trypanosoma cruzi*

Alfonso Herreros-Cabello [1], Francisco Callejas-Hernández [2], Manuel Fresno [1,3,*] and Núria Gironès [1,3,*]

1 Centro de Biología Molecular Severo Ochoa, Consejo Superior de Investigaciones Científicas, Universidad Autónoma de Madrid, Cantoblanco, 28049 Madrid, Spain
2 Bloomberg School of Public Health, Johns Hopkins University, Baltimore, MD 21205, USA
3 Instituto Sanitario de Investigación Princesa, 28006 Madrid, Spain
* Correspondence: mfresno@cbm.csic.es (M.F.); ngirones@cbm.csic.es (N.G.)

Abstract: Kinetoplastids display a single, large mitochondrion per cell, with their mitochondrial DNA referred to as the kinetoplast. This kinetoplast is a network of concatenated circular molecules comprising a maxicircle (20–64 kb) and up to thousands of minicircles varying in size depending on the species (0.5–10 kb). In *Trypanosoma cruzi*, maxicircles contain typical mitochondrial genes found in other eukaryotes. They consist of coding and divergent/variable regions, complicating their assembly due to repetitive elements. However, next-generation sequencing (NGS) methods have resolved these issues, enabling the complete sequencing of maxicircles from different strains. Furthermore, several insertions and deletions in the maxicircle sequences have been identified among strains, affecting specific genes. Unique to kinetoplastids, minicircles play a crucial role in a particular U-insertion/deletion RNA editing system by encoding guide RNAs (gRNAs). These gRNAs are essential for editing and maturing maxicircle mRNAs. In *Trypanosoma cruzi*, although only a few studies have utilized NGS methods to date, the structure of these molecules suggests a classification into four main groups of minicircles. This classification is based on their size and the number of highly conserved regions (mHCRs) and hypervariable regions (mHVRs).

Keywords: *Trypanosoma cruzi*; kinetoplast; maxicircles; minicircles

1. Introduction to the *Trypanosoma cruzi* Kinetoplast

Trypanosoma cruzi (*T. cruzi*) is a unicellular protist responsible for Chagas disease, also known as American Trypanosomiasis. This chronic disease is endemic to Latin America and is classified as a neglected tropical disease [1]. The life cycle of *T. cruzi* is quite intricate, involving a blood-feeding triatomine insect vector and various mammalian hosts. This parasite undergoes several biological stages. In the midgut of the triatomine, non-infective epimastigotes transform into infective metacyclic trypomastigotes. Once inside a host, these trypomastigotes invade cells and differentiate into replicative amastigotes, which later become infective trypomastigotes that enter the bloodstream [2].

T. cruzi exhibits a huge genomic diversity and plasticity, considering both nuclear and mitochondrial genomes [3]. This genetic variability has led to the development of numerous classification methods for the many strains identified. These classifications are based on conserved genetic sequences, including genomic, mitochondrial, and microsatellite DNA. Nowadays, six discrete typing units (DTUs), known as TcI-VI, and a seventh one called TcBat, are the principal method for strain typing. DTUs are defined as "sets of stocks that are genetically more related to each other than to any other stock and that are identifiable by common genetic, molecular or immunological markers" [4,5]. Moreover, researchers

https://doi.org/10.3390/pathogens14010073

have proposed that the genetic variability of the parasite may contribute to the different clinical manifestations of Chagas disease [6–8].

Kinetoplastids display a single large mitochondrion per cell, and their DNA is constituted by a network of concatenated circular molecules known as the kinetoplast or kDNA (or mtDNA), including several maxicircles (20–64 kb) and thousands of minicircles (0.5–10 kb) of different sizes depending on the specie [9–11]. Interestingly, many essential cellular processes take place in the mitochondria, such as oxidative phosphorylation, which is linked to ATP production. This process is carried out by a series of protein complexes in the electron transport chain (ETC), known as complexes I-IV (cI-cIV), and ATP synthase [12]. In trypanosomatids, cII consists entirely of subunits encoded by nuclear DNA. However, complexes I, III, IV, and ATP synthase include at least one subunit encoded by the mitochondrial genome [13].

According to different studies, the metacyclogenesis process in *T. cruzi* involves significant morphological changes affecting cellular structures, such as the DNA configuration within the kinetoplast. In the later stages during the metacyclogenesis, the kinetoplast transitions from a densely packed structure to a more globular and less dense form. Subsequently, the kinetoplast shifts from the anterior end (epimastigotes) to the posterior region of the cell body (metacyclic trypomastigotes) [14]. Furthermore, during the posterior amastigogenesis, after the host-cell invasion, parasites undergo a folding process, bringing their posterior end closer to their anterior end. This folding facilitates the repositioning of the basal body and kinetoplast from this posterior region, typical of trypomastigotes, to the anterior region again, which is characteristic of the amastigote stage [15].

Notably, the topological state of the kinetoplast is regulated by topoisomerases, which are responsible for catenating and decatenating the circular DNA molecules. Topoisomerase II is not expressed in metacyclic trypomastigotes, although its mRNA is present across all stages. This suggests that its mRNA might be stored in a translation-repressed form in trypomastigotes. Also, kinetoplast-associated proteins play a crucial role in the organization of the kinetoplast, and their distribution varies depending on the stage of the parasite [14].

The mitochondrial genome complexity is added to a unique kDNA S-phase in addition to the traditional G1, S, and G2/M stages of the cell cycle. However, many molecular players involved in the later stages, particularly those responsible for kDNA division and segregation, remain unidentified. Recently, a study investigated the role of Aurora kinase genes in *T. cruzi* (TcAUK1, TcAUK2, and TcAUK3) that are involved in the cell cycle. Detailed analysis of TcAUK1 localization revealed that this protein behaves like a canonical chromosome passenger protein, associated with the mitotic spindle during nuclear division. Additionally, during interphase, TcAUK1 is located on both sides of the kinetoplast, and its overexpression in epimastigotes results in a delayed G2-M transition, likely due to its impact on the initiation of kinetoplast duplication [16]. In *Leishmania major*, a single Aurora gene (Lmairk) has been reported [17]; however, its function has not yet been studied. Meanwhile, in *Trypanosoma brucei*, three Aurora genes were identified, but gene-silencing experiments revealed that only TbAUK1 is functional [18].

2. Maxicircle Diversity and Composition

Kinetoplastid maxicircles contain mitochondrial genes similar to those found in other eukaryotes [19]. Their sequences are divided into two primary regions: a coding region (CR), which is highly conserved among different species [20–22], and a divergent or variable region (DR), which is challenging to sequence due to its repetitive nature and variability in length [23].

Recent advancements in third-generation sequencing technologies have uncovered that the length of maxicircle sequences can vary significantly among trypanosomatid species [23]. Short-length sequencing technologies (such as Illumina, and including paired-end) often fall short in assembling the complete maxicircle in *T. cruzi*, leading most studies to focus solely on the coding regions with short flanking sequences [24,25]. Repetitive elements, such as tandem repeats and large duplications, complicate the assembly process and can introduce errors even in coding sequences. However, long-read sequencing technologies, such as PacBio, can mitigate these challenges [26].

The first *T. cruzi* maxicircles were described in 2006 [27] through a comparative analysis of the CL-Brener and Esmeraldo strains. The maxicircle sizes were estimated to be 22 kb and 28 kb, respectively, but a collapsed zone of repetitive sequences hindered their complete assembly. After this analysis, in 2011, the maxicircle coding region (CR) of the SylvioX10 strain was sequenced by PCR amplification of specific fragments, followed by assembly and comparison with the CL-Brener and Esmeraldo maxicircles [28], confirming the synteny of these CRs among different *T. cruzi* clades.

Later, in 2020, Gerasimov et al. utilized PacBio reads to assemble 17 maxicircles of different trypanosomatids, including two strains of *T. cruzi* (Dm28c and TCC), thereby expanding the known collection of divergent region (DR) structures and solving the problems with their repetitive regions [23]. On one hand, the Dm28c maxicircle was the longest, considering all the species (including *Leishmania* and *T. brucei*), with 47.39 kb, and both Dm28c and TCC maxicircles displayed the largest DR, even doubling the size of the other maxicircles. On the other hand, the CRs of all assembled maxicircles were comparable in length and demonstrated a high degree of synteny. Furthermore, in this study, DR organization was also analyzed by identifying two distinct regions, the P5 and P12 elements, based on their proximity to the *nd5* and *12S* genes, respectively. Similar structures have previously been reported in *Trypanosoma lewisi* [29]. The P12 element comprises short, highly repetitive units organized in repeat arrays of varying length and that are very dynamic [30,31], whereas the P5 element is a tandem repeat with a large period, but it has a small number of repetitions. This architecture appears to be a typical feature of the DRs in trypanosomatid maxicircles.

Subsequently, two articles published in 2021 sequenced new maxicircles, significantly advancing our understanding in this field. In the first one, the authors assembled the maxicircles of Y and Bug2148 [32], displaying minimal sequence-length variations in the CRs and supporting the notion that the well-documented differences between strains in the maxicircle are primarily due to the DRs, consistent with the findings of Gerasimov et al. Additionally, for the first time in *T. cruzi*, the researchers observed differences in %GC content between the CR and DR; hence, the DR is highly enriched in AT nucleotides (up to 76%), whereas the CR exhibits a more balanced nucleotide composition (approximately 47.6% GC content).

In the second manuscript from 2021, Berná et al. described high-quality assemblies of maxicircle genomes from the six DTUs of *T. cruzi*, including the resolution of repetitive regions, using long reads from PacBio and Nanopore and post-corrected with Illumina reads [33]. They reassembled the maxicircle of Dm28c and TCC, increasing the DR size in 2–3 kb compared to Gerasimov et al. [23]. Notably, these researchers separated the DR into three different sections: the short repeats (SRs) and long repeats (LRs) and the AT-rich region (AT content 83%). This last one presents a size < 1 kb and is located among the CR and the SR (Figure 1). This fits with the previous observations of Callejas-Hernández et al. and Gerasimov et al. in other *T. cruzi* strains. Regarding the characteristics of these regions, differences were found among DTUs in the length of the SR (from ~2.1 kb to ~6.8 kb), LR (from ~14.3 kb to ~30.3 kb), and the AT-rich region (from ~0.1 kb to ~1 kb).

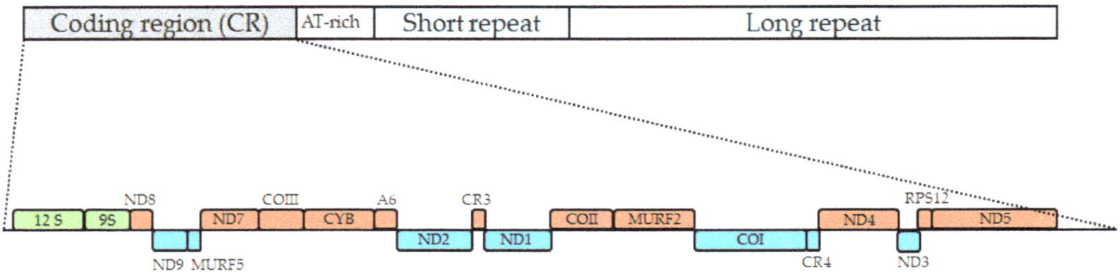

Figure 1. The architecture of the *T. cruzi* maxicircle. The coding region (CR), AT-rich, short-repeat region, and long-repeat region are shown, according to Berná et al., [33] as well as the genes of the CR: green for the ribosomal genes, blue for the genes in the negative strand, and brown for the genes in the positive strand.

Table 1 summarizes all these assembled maxicircles of *T. cruzi* by strains, showing the lengths of both CRs and DRs. Interestingly, the Y strain maxicircle was sequenced twice obtaining different results in the length of both CR and DR, which raises questions about whether they are the same strain or the impact of its maintenance under laboratory conditions between distinct research groups may have on this strain. Also, in the case of the Bug2148 strain, the sequence length is the largest recorded to date at 64.11 kb, indicating a potential unknown level of complexity in some *T. cruzi* strains. The DTU and the origin of the Bug2148 strain are currently under debate. Phylogenetic analysis using the complete coding regions of maxicircles associates it with TcI, although it has been previously classified as TcV [33].

Table 1. Information about the different complete, assembled maxicircles of *T. cruzi*. CR: coding region. DR: divergent/variable region.

Strain	DTU	Length (kb)	CR (kb)	DR (kb)	GenBank Accession	Reference
Dm28c	I	50.48	15.36	35.12	MW421590	[33]
Y	II	38.79	13.85	24.94	MW421591	
MT3663	III	44.19	15.29	28.89	MW567142	
JoseJulio	IV	44.28	15.26	29.02	MW421592	
BolFc10A	V	34.80	17.54	17.27	MW567141	
TCC	VI	42.48	15.34	27.14	MW407947	
Y	II	24.09	15.18	8.91	MW732647	[32]
Bug2148	I/V	64.11	15.18	48.93	MW732648	

Altogether, these genomic studies have been used to establish the complete repertoire of the 20 genes of the *T. cruzi* maxicircle (Figure 1). Among them, 12S rRNA and 9S rRNA encode ribosomal RNA components essential for mitochondrial protein synthesis. COI, COII, and COIII encode subunits of cytochrome c oxidase, which is involved in the electron transport chain and ATP production. The Cyb gene encodes cytochrome b, a component of the bc1 complex in the electron transport chain. Also, there are several genes (ND1, 2, 3, 4, 5, 7, 8, and 9) that encode the NADH dehydrogenase subunits of the electron transport chain, while CR3 and CR4 are cytosine-rich genes that are believed to encode homologous proteins to ND4L and ND6. Finally, RSP12 is the ribosomal protein S12, A6 is the ATPase

6 and the MURFs (Maxicircle Unidentified Reading Frames) genes encode proteins of unknown function, but they could be involved in mitochondrial processes.

However, some modifications have been described for different strains as insertions/deletions of specific genes or sequences. These changes may be common to a given lineage or exclusive of a particular strain. According to Berná et al., the Y strain (TcII) exhibits two deletions of 452 bp and 1071 bp. The first deletion disrupts the *nd7* gene, while the second causes a 5′ deletion of *nd2*, the complete elimination of *cr3*, and a 3′ deletion of *nd1*, leaving only the first 167 nucleotides at the 5′ end [33]. Interestingly, in strains isolated from asymptomatic patients, the *nd7* truncation was also detected [34], demonstrating that it is not exclusive to TcII. While this deletion is not found in either the Esmeraldo or Berenice strains, both strains show a similar deletion impacting the *nd4* gene [27].

In contrast, the BolFc10A strain (TcV) has an interruption of the *nd2* gene, and it contains two insertions of 1408 bp and 1017 bp, separated by 893 bp, belonging to minicircle sequences [33]. The 1408 bp insertion corresponds to an entire minicircle with four conserved regions disrupting the *nd4* gene, while the 1017 bp insertion is a fragment of a minicircle sequence situated in the intergenic region between the *nd4* and *nd3* genes. Despite being different minicircles, both insertions carry the same gRNA for the *nd3* gene. Moreover, a predicted gRNA for the *murf2* gene was found in the SylvioX10 maxicircle between the *nd4* and *cr4* genes, and it was also confirmed in CL Brener, Y, and Bug2148 strains [28,32].

Therefore, all of these modifications affecting, above all, mitochondrial *nd* genes (that encode for NADH dehydrogenase subunits) generate a debate about the existence of a functional complex I in *T. cruzi* [35]. In this parasite, succinate serves as the primary electron donor for the ETC, rather than NADH. Respiration is inhibited by malonate, a competitive inhibitor of mitochondrial succinate dehydrogenase, rather than by complex I inhibitors [36]. Further analysis involving mitochondrial respiration, with distinct strains of *T. cruzi*, is mandatory to clarify the role of the maxicircle genes and the functionality of complex I.

3. Minicircles Diversity and Composition

Minicircles are unique to kinetoplastids and play a key role in a U-insertion/deletion RNA-editing system by encoding guide RNAs (gRNAs) [37–39]. This process is critical for editing the maxicircles mRNAs and their maturation. Regarding its length and composition in *T. cruzi*, since the 1980s, several studies have shown that the set of minicircles had a conserved structure across strains, approximately 1.4 kb in size. These minicircles were organized into four highly conserved regions (mHCRs) of 120 bp, positioned 90 degrees apart, and the same number of 330 bp hypervariable regions (mHVRs) interspersed among the conserved regions [40–45]. The diversity of mHVRs has garnered attention; hence, these regions encode the gRNAs that guide the editing of many mitochondrial mRNAs, transforming these primary transcripts into functional messages [11,46].

Furthermore, minicircles contain three conserved sequence blocks (CSBs). CSB-1 (10 bp) and CSB-2 (8 bp) exhibit lower homology between species, while CSB-3 (12 bp), known as the Universal Minicircle Sequence (UMS), is conserved across most kinetoplastids and serves as the origin of minicircle replication [47]. CSBs number, location, and size vary between species [43,48–50]. Specifically, CSB-3, or UMS, is the binding site for the UMS binding protein (UMSBP), associated with minicircle replication and kDNA segregation [51]. Although this specific protein has been widely studied in other species, such as *Crithidia fasciculata* [52,53], its presence was confirmed in *T. cruzi* by a proteomic study [54]. According to all of the results available, the consensus sequences for each CSB, including

T. cruzi, would be as follows: CSB-1: AGGGGCGTTC; CSB-2: CCCCGTAC; and CSB-3: GGGGTTGGTGTA.

However, two recent studies have changed all of these statements about the minicircles of *T. cruzi*. In the first one, in 2021, the complete repertoire of minicircles of the Y strain was published [32], detecting 286 different minicircles. Contrary to the previous affirmations, Y strain minicircles displayed four distinct groups of size: 42 minicircles between 336 and 376 bp (group 1), 9 minicircles between 698 and 769 bp (group 2), 47 minicircles between 1051 and 1095 bp (group 3) and, finally, the largest group consisted of 188 minicircles between 1357 and 1448 bp (group 4). Thus, the authors proposed a new classification into four groups based on size ranges, with each group separated by approximately 300 bp. Interestingly, these minicircle subtypes had already been documented and suggested in the 1980s [55,56]. Furthermore, analyzing the number and structure of the mHCRs and mHVRs, the authors discovered an increasing number of both regions according to the minicircle size: from one mHCR and one mHVR in the group 1 minicircles, to four mHCRs and four mHVRs in the group 4 minicircles. In addition, in the mHCRs of the Y-strain minicircles, the composition and number of the CSB-1, CSB-2, and CSB-3 were also analyzed. The authors highlighted that the theoretical consensus sequences were larger than the previous ones for *T. cruzi*, featuring more conserved nucleotides at both ends in all the CSBs. Interestingly, small differences, such as changes in the nucleotide proportion in a position or nucleotide insertions, between mHCRs of distinct minicircle groups and between the mHCRs of the same minicircle group were also detected.

In the second one, in 2024, this minicircle organization was confirmed, in which the authors established the minicirculome of 50 clones of TcI populations [57]. They identified a total of 8254 minicircles, with each clone having an average of 165 minicircles. They categorized these minicircles into four distinct groups: group 1 (331–393 bp), group 2 (697–751 bp), group 3 (1059–1098 bp), and group 4 (1409–1556 bp), ranges very similar to the previous research with the Y strain. Table 2 and Figure 2 compile all of these minicircle groups' data. The primary difference is observed in group 4, where TcI clones exhibit a slightly larger size compared to the Y strain (TcII).

Table 2. Group composition of the assembled minicircles of *T. cruzi*. The numbers of mHCRs and mHVRs for each group and DTU are displayed, along with their percentage of the total, based on the data of Callejas-Hernández et al. [32] and Gómez-Palacio et al. [57].

Group	DTU	Size Range (bp)	Number of mHCRs	Number of mHVRs	% of Minicircles
1	TcI	331–393	1	1	12.13
	TcII	336–376	1	1	14.69
2	TcI	697–751	2	2	5.21
	TcII	698–769	2	2	3.15
3	TcI	1059–1098	3	3	10.17
	TcII	1051–1095	3	3	16.43
4	TcI	1409–1556	4	4	72.49
	TcII	1357–1448	4	4	65.73

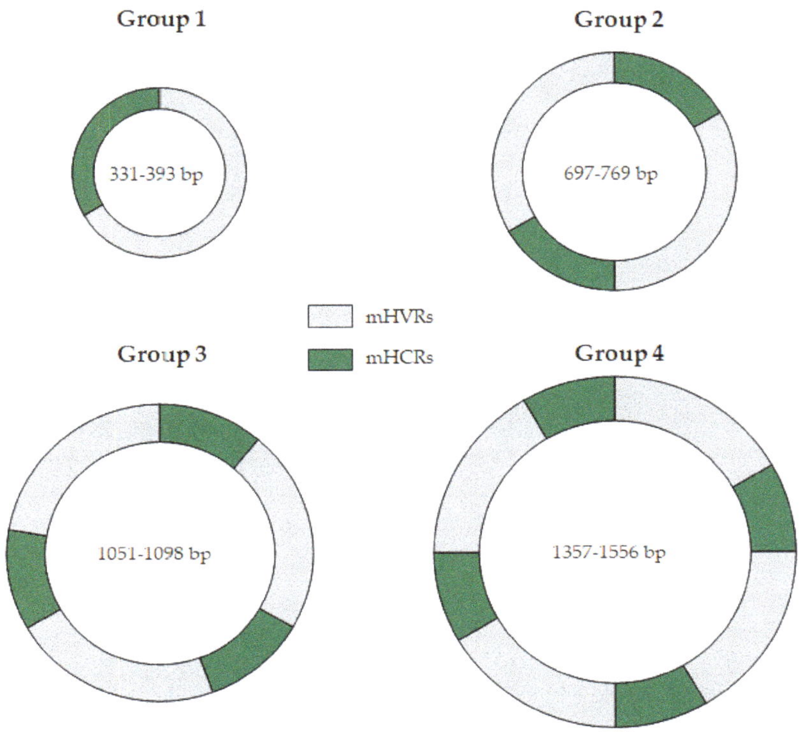

Figure 2. Architecture of the minicircles of TcI and TcII strains of *T. cruzi*. Minicircle groups 1–4 of sizes are shown, along with their ranges, according to the data of Callejas-Hernández et al. [32] and Gómez-Palacio et al. [57]. mHCRs: highly conserved regions (green). mHVRs: hypervariable regions (grey). bp = base pairs.

A first study with the minicircle sequences of the CL-Brener and Esmeraldo strains described 108 potential gRNAs from 248 minicircle mHVRs, and two gRNAs within CL-Brener maxicircle [11]. After that, in 2021, Rusman et al. identified a total of 7,476,003 gRNAs across nine *T. cruzi* strains belonging to the six DTUs [46]. Subsequently, the gRNAs were clustered into gRNA classes based on the editing region and sequence similarity. A total of 1334 gRNA classes were identified, and their number per strain varied, ranging from 104 to 401. Notably, TcV strains had up to three- or four-times fewer gRNA classes than other DTUs. Altogether, it suggests rapid changes in gRNA repertoires and potential gRNA redundancy.

Remarkably, Gómez-Palacio et al. identified a total of 9849 gRNAs in the minicircles of TcI clones specifically [57]. These gRNAs were predicted for 16 out of 18 tested genes, with no gRNAs found for the *nd2* gene in any of the analyzed clones, and only 1 gRNA identified for the *nd7* gene. Moreover, significant differences were observed in the number of predicted gRNAs and the abundance of edited genes among strains. For the most prevalent minicircle group (group 4), a total of 8562 gRNAs were predicted for all edited genes, except *nd2*. In contrast, group 1 had 802 gRNAs, group 2 had 666, and group 3 had 1135. No predicted gRNAs were found for the *nd8* and *nd9* genes in group 1; the *nd9* and *cr4* genes in group 2; and the *nd1*, *nd9*, and *cr4* genes in group 3.

Compared with different *Trypanosoma* species, *T. cruzi* possesses four groups of minicircles, whereas *T. rangeli* displays three distinct groups of minicircle sequences, termed

KP1, KP2, and KP3, based on the number of conserved regions. These classes include one conserved region (KP1), two conserved regions positioned 180 degrees apart (KP2), or four conserved regions located at 90-degree intervals (KP3) [58,59]. Moreover, in *T. cruzi*, minicircle sizes range from 336 to 1448 bp, while *T. rangeli* shows a more consistent size of 1.6–1.8 kb [60]. Also, *Trypanosoma copemani* minicircles are approximately 2048 bp and are categorized into two groups based on the number of conserved and variable regions: G1M1 minicircles (two conserved and two variable regions) and G1M2 minicircles (four conserved and four variable regions) [50]. In *Leishmania*, *Leishmania major* has 97 different minicircles, whereas *Leishmania infantum* has only 49. Despite this, the sizes of their minicircles are quite uniform, ranging from 660 to 876 bp in *L. major* and from 775 to 832 bp in *L. infantum* [22]. Additionally, *Leishmania* minicircles contain only one conserved region, in contrast to the *T. cruzi* model.

4. Conclusions

The kinetoplast, or kDNA, is a network of concatenated circular molecules in kinetoplastids. It consists of a maxicircle (20–64 kb) and thousands of minicircles (0.5–10 kb), with variable sizes among species. In *T. cruzi*, NGS techniques have enabled the complete sequencing of maxicircles from different strains over the past decade. This research has confirmed the CR synteny and the DR diversity, which appears to be divided into three sections. These analyses have also revealed various modifications in maxicircle genes among strains, although much remains to be understood about their origins. Additionally, some aspects, such as the specific roles of certain maxicircle genes, particularly those encoding NADH dehydrogenase subunits, require further investigation.

Recently, there has been a significant shift in understanding the *T. cruzi* minicircles. Initially, it was thought that minicircles were of a single size (1.4 kb), with four mHCRs and four mHVRs. However, recent studies have shown that *T. cruzi* minicircles can be classified into four main size groups, each with an increasing number of mHCRs and mHVRs. This conclusion is based on studies of TcI and TcII strains, and further research is needed to confirm it as a general pattern among strains. However, many questions remain unanswered, such as the origin of these four different size structures of minicircles and the significant diversity of gRNAs found among strains.

Author Contributions: Conceptualization, A.H.-C., N.G. and M.F.; methodology, A.H.-C. and F.C.-H.; formal analysis, A.H.-C. and F.C.-H.; investigation, A.H.-C.; writing—original draft preparation, A.H.-C.; writing—review and editing, A.H.-C., F.C.-H., N.G. and M.F.; supervision, N.G. and M.F.; project administration, N.G. and M.F.; funding acquisition, N.G. and M.F. All authors have read and agreed to the published version of the manuscript.

Funding: This research was funded by "Ministerio de Ciencia e Innovación", grant number PID2021-123389OB-I00.

Institutional Review Board Statement: Not applicable.

Informed Consent Statement: Not applicable.

Conflicts of Interest: The authors declare no conflicts of interest.

References

1. Pérez-Molina, J.A.; Molina, I. Chagas Disease. *Lancet* **2018**, *391*, 82–94. [CrossRef] [PubMed]
2. Echeverria, L.E.; Morillo, C.A. American Trypanosomiasis (Chagas Disease). *Infect. Dis. Clin. N. Am.* **2019**, *33*, 119–134. [CrossRef] [PubMed]
3. Herreros-Cabello, A.; Callejas-Hernández, F.; Gironès, N.; Fresno, M. Trypanosoma Cruzi Genome: Organization, Multi-Gene Families, Transcription, and Biological Implications. *Genes* **2020**, *11*, 1196. [CrossRef]

4. Zingales, B.; Miles, M.A.; Campbell, D.A.; Tibayrenc, M.; Macedo, A.M.; Teixeira, M.M.G.; Schijman, A.G.; Llewellyn, M.S.; Lages-Silva, E.; Machado, C.R.; et al. The Revised Trypanosoma Cruzi Subspecific Nomenclature: Rationale, Epidemiological Relevance and Research Applications. *Infect. Genet. Evol. J. Mol. Epidemiol. Evol. Genet. Infect. Dis.* **2012**, *12*, 240–253. [CrossRef]
5. Barnabé, C.; Mobarec, H.I.; Jurado, M.R.; Cortez, J.A.; Brenière, S.F. Reconsideration of the Seven Discrete Typing Units within the Species Trypanosoma Cruzi, a New Proposal of Three Reliable Mitochondrial Clades. *Infect. Genet. Evol. J. Mol. Epidemiol. Evol. Genet. Infect. Dis.* **2016**, *39*, 176–186. [CrossRef]
6. Manoel-Caetano, F.d.S.; Silva, A.E. Implications of Genetic Variability of Trypanosoma Cruzi for the Pathogenesis of Chagas Disease. *Cad. Saude Publica* **2007**, *23*, 2263–2274. [CrossRef]
7. Belew, A.T.; Junqueira, C.; Rodrigues-Luiz, G.F.; Valente, B.M.; Oliveira, A.E.R.; Polidoro, R.B.; Zuccherato, L.W.; Bartholomeu, D.C.; Schenkman, S.; Gazzinelli, R.T.; et al. Comparative Transcriptome Profiling of Virulent and Non-Virulent Trypanosoma Cruzi Underlines the Role of Surface Proteins during Infection. *PLoS Pathog.* **2017**, *13*, e1006767. [CrossRef]
8. Oliveira, A.E.R.; Pereira, M.C.A.; Belew, A.T.; Ferreira, L.R.P.; Pereira, L.M.N.; Neves, E.G.A.; Nunes, M.d.C.P.; Burleigh, B.A.; Dutra, W.O.; El-Sayed, N.M.; et al. Gene Expression Network Analyses during Infection with Virulent and Avirulent Trypanosoma Cruzi Strains Unveil a Role for Fibroblasts in Neutrophil Recruitment and Activation. *PLoS Pathog.* **2020**, *16*, e1008781. [CrossRef]
9. Maslov, D.A.; Opperdoes, F.R.; Kostygov, A.Y.; Hashimi, H.; Lukeš, J.; Yurchenko, V. Recent Advances in Trypanosomatid Research: Genome Organization, Expression, Metabolism, Taxonomy and Evolution. *Parasitology* **2019**, *146*, 1–27. [CrossRef]
10. Lukes, J.; Guilbride, D.L.; Votýpka, J.; Zíková, A.; Benne, R.; Englund, P.T. Kinetoplast DNA Network: Evolution of an Improbable Structure. *Eukaryot. Cell* **2002**, *1*, 495–502. [CrossRef]
11. Thomas, S.; Martinez, L.I.T.; Westenberger, S.J.; Sturm, N.R. A Population Study of the Minicircles in Trypanosoma Cruzi: Predicting Guide RNAs in the Absence of Empirical RNA Editing. *BMC Genom.* **2007**, *8*, 133. [CrossRef] [PubMed]
12. Vercellino, I.; Sazanov, L.A. The Assembly, Regulation and Function of the Mitochondrial Respiratory Chain. *Nat. Rev. Mol. Cell Biol.* **2022**, *23*, 141–161. [CrossRef] [PubMed]
13. Škodová-Sveráková, I.; Verner, Z.; Skalický, T.; Votýpka, J.; Horváth, A.; Lukeš, J. Lineage-Specific Activities of a Multipotent Mitochondrion of Trypanosomatid Flagellates. *Mol. Microbiol.* **2015**, *96*, 55–67. [CrossRef] [PubMed]
14. Zuma, A.A.; Dos Santos Barrias, E.; de Souza, W. Basic Biology of Trypanosoma Cruzi. *Curr. Pharm. Des.* **2021**, *27*, 1671–1732. [CrossRef] [PubMed]
15. Tomasina, R.; González, F.C.; Cabrera, A.; Basmadjián, Y.; Robello, C. From Trypomastigotes to Trypomastigotes: Analyzing the One-Way Intracellular Journey of Trypanosoma Cruzi by Ultrastructure Expansion Microscopy. *Pathogens* **2024**, *13*, 866. [CrossRef]
16. Fassolari, M.; Alonso, G.D. Aurora Kinase Protein Family in Trypanosoma Cruzi: Novel Role of an AUK-B Homologue in Kinetoplast Replication. *PLoS Negl. Trop. Dis.* **2019**, *13*, e0007256. [CrossRef]
17. Siman-Tov, M.M.; Ivens, A.C.; Jaffe, C.L. Identification and Cloning of Lmairk, a Member of the Aurora/Ipl1p Protein Kinase Family, from the Human Protozoan Parasite Leishmania. *Biochim. Biophys. Acta* **2001**, *1519*, 241–245. [CrossRef]
18. Tu, X.; Kumar, P.; Li, Z.; Wang, C.C. An Aurora Kinase Homologue Is Involved in Regulating Both Mitosis and Cytokinesis in Trypanosoma Brucei. *J. Biol. Chem.* **2006**, *281*, 9677–9687. [CrossRef]
19. Simpson, L. The Mitochondrial Genome of Kinetoplastid Protozoa: Genomic Organization, Transcription, Replication, and Evolution. *Annu. Rev. Microbiol.* **1987**, *41*, 363–382. [CrossRef]
20. Simpson, L.; Neckelmann, N.; de la Cruz, V.F.; Simpson, A.M.; Feagin, J.E.; Jasmer, D.P.; Stuart, K. Comparison of the Maxicircle (Mitochondrial) Genomes of Leishmania Tarentolae and Trypanosoma Brucei at the Level of Nucleotide Sequence. *J. Biol. Chem.* **1987**, *262*, 6182–6196. [CrossRef]
21. Kaufer, A.; Barratt, J.; Stark, D.; Ellis, J. The Complete Coding Region of the Maxicircle as a Superior Phylogenetic Marker for Exploring Evolutionary Relationships between Members of the Leishmaniinae. *Infect. Genet. Evol. J. Mol. Epidemiol. Evol. Genet. Infect. Dis.* **2019**, *70*, 90–100. [CrossRef] [PubMed]
22. Camacho, E.; Rastrojo, A.; Sanchiz, Á.; González-de la Fuente, S.; Aguado, B.; Requena, J.M. Leishmania Mitochondrial Genomes: Maxicircle Structure and Heterogeneity of Minicircles. *Genes* **2019**, *10*, 758. [CrossRef] [PubMed]
23. Gerasimov, E.S.; Zamyatnina, K.A.; Matveeva, N.S.; Rudenskaya, Y.A.; Kraeva, N.; Kolesnikov, A.A.; Yurchenko, V. Common Structural Patterns in the Maxicircle Divergent Region of Trypanosomatidae. *Pathogens* **2020**, *9*, 100. [CrossRef] [PubMed]
24. Messenger, L.A.; Llewellyn, M.S.; Bhattacharyya, T.; Franzén, O.; Lewis, M.D.; Ramírez, J.D.; Carrasco, H.J.; Andersson, B.; Miles, M.A. Multiple Mitochondrial Introgression Events and Heteroplasmy in Trypanosoma Cruzi Revealed by Maxicircle MLST and next Generation Sequencing. *PLoS Negl. Trop. Dis.* **2012**, *6*, e1584. [CrossRef] [PubMed]
25. Carranza, J.C.; Valadares, H.M.S.; D'Avila, D.A.; Baptista, R.P.; Moreno, M.; Galvão, L.M.C.; Chiari, E.; Sturm, N.R.; Gontijo, E.D.; Macedo, A.M.; et al. Trypanosoma Cruzi Maxicircle Heterogeneity in Chagas Disease Patients from Brazil. *Int. J. Parasitol.* **2009**, *39*, 963–973. [CrossRef]

26. Tørresen, O.K.; Star, B.; Mier, P.; Andrade-Navarro, M.A.; Bateman, A.; Jarnot, P.; Gruca, A.; Grynberg, M.; Kajava, A.V.; Promponas, V.J.; et al. Tandem Repeats Lead to Sequence Assembly Errors and Impose Multi-Level Challenges for Genome and Protein Databases. *Nucleic Acids Res.* **2019**, *47*, 10994–11006. [CrossRef]
27. Westenberger, S.J.; Cerqueira, G.C.; El-Sayed, N.M.; Zingales, B.; Campbell, D.A.; Sturm, N.R. Trypanosoma Cruzi Mitochondrial Maxicircles Display Species- and Strain-Specific Variation and a Conserved Element in the Non-Coding Region. *BMC Genom.* **2006**, *7*, 60. [CrossRef]
28. Ruvalcaba-Trejo, L.I.; Sturm, N.R. The Trypanosoma Cruzi Sylvio X10 Strain Maxicircle Sequence: The Third Musketeer. *BMC Genom.* **2011**, *12*, 58. [CrossRef]
29. Lin, R.-H.; Lai, D.-H.; Zheng, L.-L.; Wu, J.; Lukeš, J.; Hide, G.; Lun, Z.-R. Analysis of the Mitochondrial Maxicircle of Trypanosoma Lewisi, a Neglected Human Pathogen. *Parasit. Vectors* **2015**, *8*, 665. [CrossRef]
30. Flegontov, P.N.; Guo, Q.; Ren, L.; Strelkova, M.V.; Kolesnikov, A.A. Conserved Repeats in the Kinetoplast Maxicircle Divergent Region of Leishmania Sp. and Leptomonas Seymouri. *Mol. Genet. Genom. MGG* **2006**, *276*, 322–333. [CrossRef]
31. Flegontov, P.N.; Strelkova, M.V.; Kolesnikov, A.A. The Leishmania Major Maxicircle Divergent Region Is Variable in Different Isolates and Cell Types. *Mol. Biochem. Parasitol.* **2006**, *146*, 173–179. [CrossRef] [PubMed]
32. Callejas-Hernández, F.; Herreros-Cabello, A.; del Moral-Salmoral, J.; Fresno, M.; Gironès, N. The Complete Mitochondrial DNA of Trypanosoma Cruzi: Maxicircles and Minicircles. *Front. Cell. Infect. Microbiol.* **2021**, *11*, 672448. [CrossRef] [PubMed]
33. Berná, L.; Greif, G.; Pita, S.; Faral-Tello, P.; Díaz-Viraqué, F.; Souza, R.D.C.M.D.; Vallejo, G.A.; Alvarez-Valin, F.; Robello, C. Maxicircle Architecture and Evolutionary Insights into Trypanosoma Cruzi Complex. *PLoS Negl. Trop. Dis.* **2021**, *15*, e0009719. [CrossRef] [PubMed]
34. Baptista, C.S.; Vêncio, R.Z.N.; Abdala, S.; Carranza, J.C.; Westenberger, S.J.; Silva, M.N.; Pereira, C.A.d.B.; Galvão, L.M.C.; Gontijo, E.D.; Chiari, E.; et al. Differential Transcription Profiles in Trypanosoma Cruzi Associated with Clinical Forms of Chagas Disease: Maxicircle NADH Dehydrogenase Subunit 7 Gene Truncation in Asymptomatic Patient Isolates. *Mol. Biochem. Parasitol.* **2006**, *150*, 236–248. [CrossRef]
35. Opperdoes, F.R.; Michels, P.A.M. Complex I of Trypanosomatidae: Does It Exist? *Trends Parasitol.* **2008**, *24*, 310–317. [CrossRef]
36. Denicola-Seoane, A.; Rubbo, H.; Prodanov, E.; Turrens, J.F. Succinate-Dependent Metabolism in Trypanosoma Cruzi Epimastigotes. *Mol. Biochem. Parasitol.* **1992**, *54*, 43–50. [CrossRef]
37. Aphasizhev, R.; Aphasizheva, I. Mitochondrial RNA Editing in Trypanosomes: Small RNAs in Control. *Biochimie* **2014**, *100*, 125–131. [CrossRef]
38. Simpson, L.; Sbicego, S.; Aphasizhev, R. Uridine Insertion/Deletion RNA Editing in Trypanosome Mitochondria: A Complex Business. *RNA* **2003**, *9*, 265–276. [CrossRef]
39. Smith, J.T., Jr.; Doleželová, E.; Tylec, B.; Bard, J.E.; Chen, R.; Sun, Y.; Zíková, A.; Read, L.K. Developmental Regulation of Edited CYb and COIII Mitochondrial mRNAs Is Achieved by Distinct Mechanisms in Trypanosoma Brucei. *Nucleic Acids Res.* **2020**, *48*, 8704–8723. [CrossRef]
40. Leon, W.; Frasch, A.C.; Hoeijmakers, J.H.; Fase-Fowler, F.; Borst, P.; Brunel, F.; Davison, J. Maxi-Circles and Mini-Circles in Kinetoplast DNA from Trypanosoma Cruzi. *Biochim. Biophys. Acta* **1980**, *607*, 221–231. [CrossRef]
41. González, A. Nucleotide Sequence of a Trypanosoma Cruzi Minicircle. *Nucleic Acids Res.* **1986**, *14*, 9217. [CrossRef] [PubMed]
42. Macina, R.A.; Sanchez, D.O.; Gluschankof, D.A.; Burrone, O.R.; Frasch, A.C. Sequence Diversity in the Kinetoplast DNA Minicircles of Trypanosoma Cruzi. *Mol. Biochem. Parasitol.* **1986**, *21*, 25–32. [CrossRef] [PubMed]
43. Degrave, W.; Fragoso, S.P.; Britto, C.; van Heuverswyn, H.; Kidane, G.Z.; Cardoso, M.A.; Mueller, R.U.; Simpson, L.; Morel, C.M. Peculiar Sequence Organization of Kinetoplast DNA Minicircles from Trypanosoma Cruzi. *Mol. Biochem. Parasitol.* **1988**, *27*, 63–70. [CrossRef] [PubMed]
44. Shapiro, T.A.; Englund, P.T. The Structure and Replication of Kinetoplast DNA. *Annu. Rev. Microbiol.* **1995**, *49*, 117–143. [CrossRef]
45. Junqueira, A.C.V.; Degrave, W.; Brandão, A. Minicircle Organization and Diversity in Trypanosoma Cruzi Populations. *Trends Parasitol.* **2005**, *21*, 270–272. [CrossRef]
46. Rusman, F.; Floridia-Yapur, N.; Tomasini, N.; Diosque, P. Guide RNA Repertoires in the Main Lineages of Trypanosoma Cruzi: High Diversity and Variable Redundancy Among Strains. *Front. Cell. Infect. Microbiol.* **2021**, *11*, 663416. [CrossRef]
47. Ray, D.S. Conserved Sequence Blocks in Kinetoplast Minicircles from Diverse Species of Trypanosomes. *Mol. Cell. Biol.* **1989**, *9*, 1365–1367.
48. Ponzi, M.; Birago, C.; Battaglia, P.A. Two Identical Symmetrical Regions in the Minicircle Structure of Trypanosoma Lewisi Kinetoplast DNA. *Mol. Biochem. Parasitol.* **1984**, *13*, 111–119. [CrossRef]
49. Sugisaki, H.; Ray, D.S. DNA Sequence of Crithidia Fasciculata Kinetoplast Minicircles. *Mol. Biochem. Parasitol.* **1987**, *23*, 253–263. [CrossRef]
50. Botero, A.; Kapeller, I.; Cooper, C.; Clode, P.L.; Shlomai, J.; Thompson, R.C.A. The Kinetoplast DNA of the Australian Trypanosome, Trypanosoma Copemani, Shares Features with Trypanosoma Cruzi and Trypanosoma Lewisi. *Int. J. Parasitol.* **2018**, *48*, 691–700. [CrossRef]

51. Milman, N.; Motyka, S.A.; Englund, P.T.; Robinson, D.; Shlomai, J. Mitochondrial Origin-Binding Protein UMSBP Mediates DNA Replication and Segregation in Trypanosomes. *Proc. Natl. Acad. Sci. USA* **2007**, *104*, 19250–19255. [CrossRef] [PubMed]
52. Onn, I.; Kapeller, I.; Abu-Elneel, K.; Shlomai, J. Binding of the Universal Minicircle Sequence Binding Protein at the Kinetoplast DNA Replication Origin. *J. Biol. Chem.* **2006**, *281*, 37468–37476. [CrossRef] [PubMed]
53. Abu-Elneel, K.; Robinson, D.R.; Drew, M.E.; Englund, P.T.; Shlomai, J. Intramitochondrial Localization of Universal Minicircle Sequence-Binding Protein, a Trypanosomatid Protein That Binds Kinetoplast Minicircle Replication Origins. *J. Cell Biol.* **2001**, *153*, 725–734. [CrossRef] [PubMed]
54. Herreros-Cabello, A.; Callejas-Hernández, F.; Fresno, M.; Gironès, N. Comparative Proteomic Analysis of Trypomastigotes from Trypanosoma Cruzi Strains with Different Pathogenicity. *Infect. Genet. Evol. J. Mol. Epidemiol. Evol. Genet. Infect. Dis.* **2019**, *76*, 104041. [CrossRef] [PubMed]
55. Frasch, A.C.; Sánchez, D.O.; Stoppani, A.O. Homogeneous and Heterogeneous Mini-Circle Subpopulations in Trypanosoma Cruzi Kinetoplast DNA. *Biochim. Biophys. Acta* **1984**, *782*, 26–33. [CrossRef]
56. Sanchez, D.O.; Frasch, A.C.; Carrasco, A.E.; Gonzalez-Cappa, S.M.; de Isola, E.D.; Stoppani, A.O. Rapid Evolution of Kinetoplast DNA Mini-Circle Subpopulations in Trypanosoma Cruzi. *Mol. Biochem. Parasitol.* **1984**, *11*, 169–178. [CrossRef]
57. Gómez-Palacio, A.; Cruz-Saavedra, L.; Van den Broeck, F.; Geerts, M.; Pita, S.; Vallejo, G.A.; Carranza, J.C.; Ramírez, J.D. High-Throughput Analysis of the Trypanosoma Cruzi Minicirculome (mcDNA) Unveils Structural Variation and Functional Diversity. *Sci. Rep.* **2024**, *14*, 5578. [CrossRef]
58. Recinos, R.F.; Kirchhoff, L.V.; Donelson, J.E. Characterization of Kinetoplast DNA Minicircles in Trypanosoma Rangeli. *Mol. Biochem. Parasitol.* **1994**, *63*, 59–67. [CrossRef]
59. Vallejo, G.A.; Guhl, F.; Carranza, J.C.; Lozano, L.E.; Sánchez, J.L.; Jaramillo, J.C.; Gualtero, D.; Castañeda, N.; Silva, J.C.; Steindel, M. kDNA Markers Define Two Major Trypanosoma Rangeli Lineages in Latin-America. *Acta Trop.* **2002**, *81*, 77–82. [CrossRef]
60. Vallejo, G.A.; Macedo, A.M.; Chiari, E.; Pena, S.D. Kinetoplast DNA from Trypanosoma Rangeli Contains Two Distinct Classes of Minicircles with Different Size and Molecular Organization. *Mol. Biochem. Parasitol.* **1994**, *67*, 245–253. [CrossRef]

pathogens

MDPI

Review

The Functions of Cytokines in the Cardiac Immunopathogenesis of Chagas Disease

Mariana Citlalli de Alba-Alvarado [1,*], Margarita Cabrera-Bravo [1], Edgar Zenteno [2], Paz María Salazar-Schetino [1] and Martha Irene Bucio-Torres [1,*]

[1] Departamento de Microbiología y Parasitología, Facultad de Medicina, Universidad Nacional Autónoma de México, Ciudad México 04510, Mexico; imay@unam.mx (M.C.-B.); pazmar@unam.mx (P.M.S.-S.)
[2] Departamento de Bioquímica, Facultad de Medicina, Universidad Nacional Autónoma de México, Ciudad de México 04510, Mexico; ezenteno@unam.mx
* Correspondence: marianadealba@comunidad.unam.mx (M.C.d.A.-A.); marbu@unam.mx (M.I.B.-T.)

Abstract: Chagas disease is a complex zoonosis. Clinically, it presents in two distinct phases, acute and chronic. The ability of patients to respond to *Trypanosoma cruzi* infection depends on the balance between inflammatory and anti-inflammatory responses, in which cytokines play a key regulatory role. In this review, we discuss the role of cytokines in regulating the host response and as mediators of cardiac injury by inducing profibrotic alterations. The importance of characterizing cytokine profiles as biomarkers of the evolution of cardiac damage in *T.-cruzi*-infected individuals is also emphasized.

Keywords: *Trypanosoma cruzi*; Chagas disease; cardiopathy; cytokines; Th1; Th17

1. Introduction

Chagas disease is a complex zoonosis involving interactions between vector species and wild, domestic, and peridomestic mammals, with several known modes of transmission [1]. In humans, the disease presents with different clinical manifestations as it progresses. The causative agent is the flagellated protozoan *Trypanosoma cruzi*, which is naturally transmitted by the excretions of triatomines, a subfamily of hematophagous hemipterans; this intracellular parasite was described by Dr. Carlos Chagas in Brazil in 1909 [2]. The vector colonizes homes in rural and suburban areas, keeping the infection close to population centers. In these environments, squirrels, opossums, voles, and armadillos participate in transmission as mammalian reservoirs [3].

Chagas disease is characterized by two clinical phases: acute and chronic. In the acute phase, the signs and symptoms are related to the portal of entry of the parasite, including the Romaña–Mazza sign (unilateral bipalpebral edema) or an inoculation chagoma (subcutaneous nodular lesion), often accompanied by nonspecific systemic manifestations (5% of cases), such as fever, adenitis, myalgia, arthralgia, hepatosplenomegaly, and anemia [1]. Atypical or less-common manifestations in this phase include acute cardiomyopathy, meningoencephalitis [4], lymphochagoma, (not reported in Mexico) [5], and skin rash [6] particularly in immunocompromised patients. The chronic phase of Chagas disease can present asymptomatically or symptomatically. The asymptomatic phase, characterized by low or no parasitemia, usually lasts between 10 and 20 years, and is diagnosed by serological methods. In the symptomatic phase, the main effects involve the heart, the digestive system, and the nervous system [7,8].

Approximately 27% of patients develop progressive heart failure evidenced by an incomplete block of the right bundle branch, left anterior hemiblock, and ventricular extrasystoles [8,9]. Echocardiography reveals apical aneurysms in 8.5–55% of cases, intramural thrombi in the left ventricle, and ventricular dysfunction with a reduced ejection fraction [10,11]. In this phase, visceral manifestations are infrequently observed, including a megacolon and megaesophagus with alterations of the nervous system [12,13]. On the other hand, atypical complications such as pericarditis and sudden death [14] can occur in both phases of the disease.

The immune response in the vertebrate host depends on the clinical phase of infection, the presence of cytokine-producing immune cells, and other host genetic factors. With respect to the parasite, factors such as virulence, inoculum size, and the recurrence of reinfection, which is common in areas of high endemicity, are important [15].

Controversy remains regarding the variables involved in the clinical evolution following infection, with 30% of patients progressing to the chronic symptomatic phase, while 70% remain asymptomatic [16]; it can be inferred that the expression of cytokine profiles determined by host genetics, as well as the response to parasite antigens, interact to contribute to this progression.

Cytokine profiles, as key regulators of the immune response, play an important role in controlling parasitism and regulating the inflammatory process in Chagas disease [15–18]. Therefore, this review aims to provide a context for their role in the clinical evolution of the disease as it progresses to chronicity.

2. Role of Cytokines in the Acute Phase of Chagas Disease

Despite the parasite's mechanisms of evading the host immune response, Toll-like receptors (TLRs) are capable of recognizing specific parasite antigens on the *T. cruzi* membrane. These receptors are expressed on phagocytes, macrophages, and cells of the innate response; among these antigens, glycosylphosphatidylinositol (GPI), recognized by TLR2 and TLR6 [19], and glycosylinositol phospholipids (GIPLs), recognized by TLR4, stand out. On the other hand, oligonucleotide sequences with non-methylated CpG motifs of *T. cruzi* DNA are recognized by TLR9, which is highly expressed on dendritic cells [20].

TLR activation initiates a MyD88-dependent signaling cascade that activates transcription factors and promotes the expression of proinflammatory cytokines such as IL-12 and subsequently interferon gamma (IFN-γ). It has been shown experimentally that MyD88-knockout mice are susceptible to acute *T. cruzi* infection [21]. Activation of these receptors also induces various transcription factors that promote cytokine expression, thereby stimulating both innate and adaptive immune responses.

After primary infection, the immune response mainly involves the activation of immune cells such as neutrophils, eosinophils, and phagocytic cells. During the acute phase of infection, neutrophils, one of the most abundant innate cell types, are recruited to the site of infection through the action of IL-17 [22]. In response to parasitic antigens, neutrophils produce extracellular traps (NETS) and proinflammatory cytokines such as tumor necrosis factor-alpha (TNF-α) and IL-1 [23].

During the acute phase, the production of cytokines with proliferative properties, such as the hematopoietic growth factor (IL-3) and IL-2, has also been reported. Such production is extremely important in the acute phase because it stimulates the development of NK cells. NK cells are crucial in the innate response against the parasite because they produce IFN-γ and TNF-α, and these cytokines will lead to the elimination of *T. cruzi* by inducing the activation of macrophages in the early phase of infection [24].

TNF-α promotes inflammation in infected individuals, as well as apoptosis of *T. cruzi*-infected cells. This cytokine, together with IL-6 [25], increases iNOS activation in macrophages, in addition to increasing the expression of chemokines to induce migration of other immune cells to the site of infection, in synergy with IFN-γ [26–28]. Macrophages are phagocytic cells that can polarize into different subsets depending on microenvironmental stimuli, parasite antigens, and signaling through PPARγ [28]. In the acute phase of infection, M1 macrophages play an inflammatory role, stimulating the expression of molecules such as iNOS, NADPH, NO, and ROS, and increasing the expression of MHC class II, CD86, AP-1, and NF-κB. They also produce cytokines such as TNF-α, IL-6, IL-12, and IL-1β [29–31]. This response can lead to parasite destruction, but it can also promote inflammation and tissue damage driven by the response to parasite antigens.

On the other hand, the M2 macrophage subset is associated with tissue remodeling and parasitic phagocytosis. M2 macrophages secrete Th2-type cytokines such as IL-10 and transforming growth factor beta (TGF-β) and express molecules such as CD36, CD163, mannose receptor (MR), AMPK, Arg1, PPARs, and STAT6 [32–34], which are associated with an immunoregulatory response and extracellular matrix remodeling.

The role of monocytes and macrophages is crucial at the onset of infection. Activation of these cells is mediated by proinflammatory Th1-type cytokines that are important for phagocytosis. The major cytokines involved in this acute-phase response are IL-12, IFN-γ, and TNF-α [35,36] (Figure 1). IL-12 appears to be the cytokine initiator and promoter of the acute-phase response against *T. cruzi* [37]; IL-12 is also required for the production of mediators for the transcription and generation of IFN-γ and TNF-α. Some authors have indicated that when the function of this cytokine is blocked by anti-IL-12 antibodies, the survival of blood trypomastigotes during the acute phase is increased because there is no initial proinflammatory response [37]. Although the important role of IL-12 is known, the master control for the development of the Th1-type response depends on the presence of interferon gamma (IFN-γ) [20,38]. During this phase, phagocytic cells or M1 macrophages are activated by IFN-γ, which stimulates the inducible nitric oxide synthase (iNOS) pathway in macrophages; this activation is necessary to lyse parasites in the blood during the acute phase [39], along with macrophage activation and an increase in MHC expression to promote the function of antigen-presenting cells (APC) [17,40] (Figure 1).

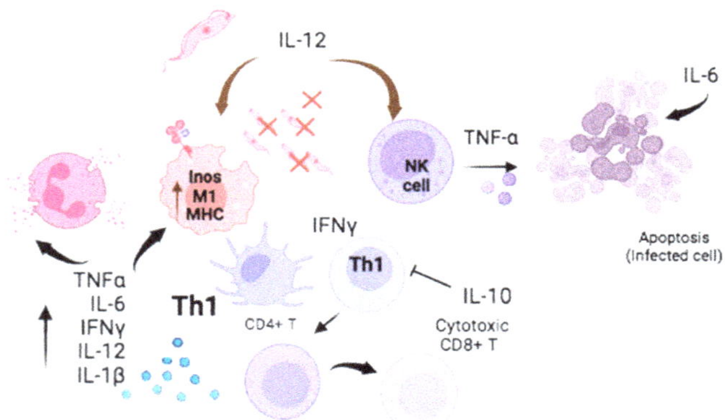

Figure 1. Major cytokines in the acute response. IFN-γ activates phagocytic cells and iNOS, reducing parasitemia and increasing MHC expression, which optimize the antigen presentation by APCs. TNF-α in synergy with IL-6 promotes apoptosis of infected cells and activates neutrophils and eosinophils. IL-10 inhibits the early IFN-γ response. Created with Bio-Render.com.

3. Role of Cytokines in Chronic Heart Disease

In the acute phase of Chagas disease, the proinflammatory immune response should be regulated to control tissular and circulating parasitism. In some patients, this response is dysregulated, promoting the establishment of a persistent inflammatory process.

IFN-γ is considered the backbone for an effective Th1 response at the onset of infection; however, high levels of this cytokine during the chronic phase can affect the myocardium in some individuals. Its expression is associated with susceptibility to damage during this phase by inducing a cytotoxic and proinflammatory cellular response [41,42].

TNF-α is thought to induce the formation of reactive oxygen intermediates and promote the deposition of the extracellular matrix in the late phase of the disease, leading to cardiomyocyte necrosis and apoptosis [43,44]. In addition, it promotes the expression and high levels of IL-6, which contribute to the development of thrombotic processes by increasing endothelial platelet adhesion and stimulating platelet production [45].

IL-1β induces the production of peptides that deregulate osmosis in myocardial cells, producing the myocardial hypertrophy commonly described in these patients. Finally, IL-2 has been implicated in the abnormal activation of T lymphocytes in other dilated cardiomyopathies [46]; its involvement in chronic Chagas disease has been reported [47–49].

In the chronic phase, an increase in the circulating concentrations of several chemokines has been demonstrated, due to the increased expression of IL-8. This has implications in the induction of myocardial ischemia, since it is a platelet-attracting factor and favors the formation of platelet thrombi at the microvascular level, causing ischemia and tissue necrosis, important elements in the pathogenesis of lesions in Chagas disease [48,50]. In addition, there are reports of the involvement in this stage of the disease of some chemokine receptors, such as CXCL10 and CXCL9, which are expressed in the cardiac tissue and have been associated with a large infiltration of CD8+ T cells in the myocardium in a murine model [51]. The major cytokines involved in the pathogenesis of lesions in the chronic phase of Chagas disease are listed in Table 1.

Table 1. Major cytokines and chemokines involved in the chronic phase of Chagas disease (CHD). MO^a, monocytes; MA^b, macrophages; CE^c, endothelial cells; NE^d, neutrophils; CD4+ ^{e}T, CD4+ lymphocytes; NK^f, NK cells; CD^g, dendritic cells; Th2, Th2-type response; Th1, Th1 lymphocytes; O^h, other cell types.

Cytokine	Secreted by	Effect	Ref.
TNF-α	Th1, MA^b, MO^a, N^d, LT^f, NE^d and NK^f	Elevated levels of TNF-α, are associated with left ventricular dysfunction in patients with CHD.	[52] Ferreira et al., 2003
		Plasma concentrations of TNF-α are higher in chronic patients compared to healthy controls.	[53] Rocha et al., 2012
IL-12	MA^b, CD^g, Th1	Elevated levels of IL-12 have been implicated in the pathophysiology of CHD in humans.	[54] Pereira et al., 2018
IL-6	MA^b, CE^c, and Th2	IL-6 levels increase with the progression of heart failure in Chagas disease.	[55] López et al., 2006
IL-2	Th1, CD4+ ^{e}T	Patients with CHD have decreased peripheral IL-2 levels.	[49] Briceño et al., 1996
TGF-β	Th2, MA^b, O^h	Cardiac involvement is associated with increased activity of the TGF-β signaling pathway and expression of its key components.	[56] Ferreira et al., 2022
		Inhibition of TGF-β signaling reverses *T.-cruzi*-induced fibrosis and hypertrophy in chronic disease.	[57] Ferrão, et al., 2018
IFN-γ	Th1, NK^f	Patients with CHD have increased levels of IFN-γ upon infection.	[41] Abel et al., 2001
		The response to infection, along with IFN-γ production, can depress energy metabolism (myocardial ATP), affecting myocardial contractility, electrical conduction, and heart rate.	[58] Ferreira et al., 2014
		In patients with CHD, several lines of evidence point to the importance of IFN-γ in the pathogenesis of myocarditis and heart failure.	[59] Chevillard et al., 2018
IL-10	MA^b, MC^a, Th2, O^h	Low IL-10 expression is associated with cardiac dysfunction in terms of left ventricular ejection fraction.	[60] Costa et al., 1999
		In patients with cardiomyopathy, elevated serum IL-10 levels have been described in asymptomatic patients without cardiomegaly.	[61] Vasconcelos et al., 2015
IL-17	CD4+ ^{e}T	In the pathogenesis of myocarditis, IL-17 production is associated with decreased IL-12 expression.	[62] da Matta et al., 2010
		Patients with severe cardiac manifestations have a Th17 profile with significantly higher levels of IL-17 and IL-6 in addition to IFN-γ and IL-2.	[63] De Alba-Alvarado et al., 2018

Although studies investigating cytokine production following *T. cruzi* antigen presentation are scarce, some authors have specifically described surface, kinetoplast, and flagellar antigens. All of these antigens are inducers of the Th1 response, activating plasma cells via IL-6 and the effector mechanisms of CD8 T cells. It has been reported that the *r*TSA-1 and *r*Tc24 antigens are inducers of IFN-γ production in circulating T lymphocytes from patients with Chagas heart disease [64]; similarly, the Y-strain TcSA antigen leads to the production of IFN-γ, TNFα, and IL-2 by T CD4 lymphocytes, maintaining the Th1-type response described above (Figure 2). However, some authors agree that there is a loss of functionality of both CD4+ and CD8+ T lymphocytes. This may be important in the progression to chronicity as it maintains T lymphocytes in a senescent phenotype [65,66].

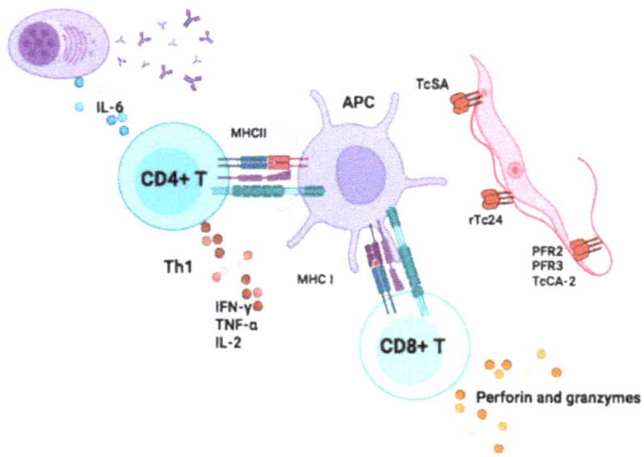

Figure 2. Cytokine induction by *T. cruzi* antigens. The TSA-1 and Tc24 antigens induce IFN-γ production by T lymphocytes. TcSA antigen induces IFN-γ, TNFα, and IL-2 production by CD4+ T and maintains the Th1 response of granzyme- and perforin-producing CD8+ T. Created with BioRender.com.

4. Cytokines with Profibrotic Effect

Cardiac fibrosis is an important process in the progression of the chronic phase of Chagas disease. TGF-β exerts master control in regulating the formation and degradation of the extracellular matrix, in addition to inducing the expression of essential components such as fibronectin, laminin, and collagen, and in the generation of fibrosis [67]. Other cytokines are also considered profibrotic in the myocardium, although they have not been fully identified in Chagas disease, including TNF-α, endothelin, platelet-derived growth factor (PDGF), and basic fibroblast growth factor (BFGF). In addition, cardiac fibroblasts have been described as acquiring a proinflammatory phenotype characterized by the secretion of cytokines and chemokines such as IL-8, IL-1β, CCL2, eotaxin, and TNF-α with extracellular matrix-degrading properties [68,69] (Figure 3)

In Chagas disease, myocardial fibrosis is involved in the genesis of tachyarrhythmias (ventricular tachycardia or fibrillation) because fibrosis can block the cardiac stimulus and reduce conduction velocity. In addition, collagen septa between the muscle bundles may interfere with the conduction of the electrical impulse, which is expressed on ECG tracing as a right bundle branch block or hemiblock of the left anterior fascicle of the bundle of His [70,71].

In a study of patients with Chagas heart disease, septal hypertrophy and left ventricular posterior wall hypertrophy were found. These lesions correlate with increased levels of proinflammatory and profibrotic cytokines [62,63]. Other authors have suggested that the persistent inflammation may be the consequence of a delayed hypersensitivity immune response, with activation of memory T cells and recruitment of mononuclear cells to the site of inflammation [72].

Several pathogenic mechanisms have been described as promoting tissue damage in the chronic phase of the disease, including inflammation, conduction and microvascular dysfunction, which collectively trigger the process of cardiac fibrosis [71]. Proinflammatory cytokines are also involved in these mechanisms; an important one, TNF-α, together with IFN-γ, induces the activation of the iNOS pathway in macrophages, which by producing nitric oxide, induce cardiac cell apoptosis and inhibit beta-adrenergic receptors [73]. In a cascade of events amplified by the continued production of IL-6 and TNF-α, the calcium pathway is inhibited, further impairing myocardial contractility and leading to cardiac conduction abnormalities [74] (Figure 3). In addition, the expression of IFN-γ and IL-12 can maintain the activation of CD8+ T lymphocytes, which produce granzyme b and perforin; both cellular factors contribute to necrosis and cytotoxic tissue damage, which can be repaired by the production of collagen [75]. On the other hand, cytokines such as IL-6 and IL-8 modify the vascular endothelium and contribute to the formation of platelet-derived microthrombi [76,77], which alter the perfusion of cardiac tissue, causing ischemia and tissue necrosis. Together with TGF-β, ischemia induces the production of metalloproteases for the degradation of the extracellular matrix, resulting in the formation of collagen and ultimately fibrosis [78] (Figure 3).

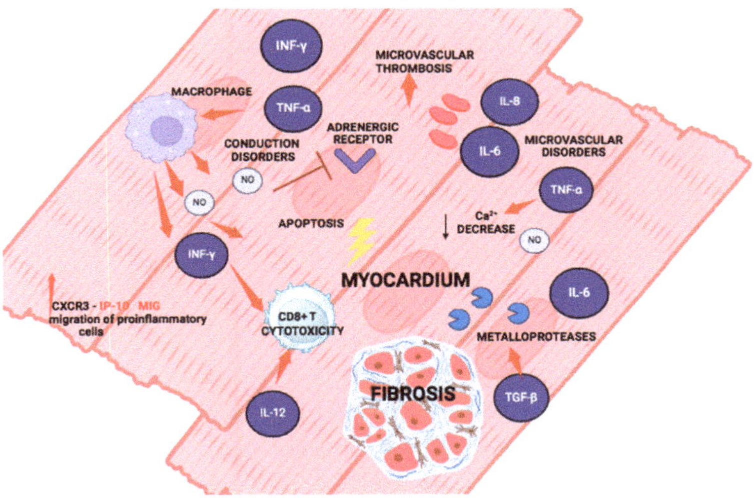

Figure 3. Fibrotic cytokines and chemokines in cardiac tissue during the chronic phase. Continuous Th1 cytokine production maintains NO in macrophages, promoting apoptosis. IL-12 maintains and IFN-γ stimulates CD8+ T cytotoxicity. CXCR3 and its ligands maintain the proinflammatory cell migration. IL-8 promotes thrombus formation and tissue ischemia. IL-6 and TNF-α inhibit the calcium pathway, affecting myocardial contractility. TGF-β induces metalloprotease production, resulting in extracellular matrix degradation, collagen production, and fibrosis. Created with BioRender.com.

5. Cytokines and Chemokines in Resistance and Susceptibility to *Trypanosoma Cruzi* Infection

Several authors emphasize the role of cytokines in Chagas disease because of their protective role during the acute phase (Table 2). A key function of cytokines and chemokines in controlling the parasite is to stimulate the functions of innate cells. Meanwhile, other authors suggest that these same cytokines, when deregulated or overexpressed during the chronic phase, may contribute to host susceptibility to the disease (Table 3).

Chemokine receptors (CXCR), such as CXCR3, promote the specific chemotaxis of CD8+ T cells to control infected cells [79]; however, ligands of these receptors, such as CXCL10 (IP-10), CXCL11, and CXCL9 (MIG), are produced by antigen-presenting cells in infected tissues, favoring the development of myocarditis with high levels of CD8+ T lymphocyte populations [26]; This chemokine receptor (CXCR3) continues to be studied and has been defined as an important mediator of the cytotoxic response [79]; on the other hand, CCR5 has been found to be overexpressed with a lower expression of CXCR4 in patients with mild heart disease compared to normal individuals, suggesting that these chemokines are involved in the development of early forms of Chagas cardiomyopathy [80,81]. Finally, resistance and susceptibility to infection or disease have been associated with cytokine and chemokine expression (Tables 2 and 3).

Table 2. Cytokines and chemokines reported to induce protection in Chagas disease.

Cytokine	Protection	Ref.
IL-12	High levels in asymptomatic patients infected with *T. cruzi*.	[38] Michailowsky et al., 2001
IFN-γ	IFN-γ production is induced in patients on drug treatment (nifurtimox).	[41] Bahia et al., 1998
IFN-γ and TNF-α	In vitro stimulation of macrophages with IFN-γ reduces the number of amastigotes in the acute phase.	[18] Cardoni et al., 1999
IL-17-A	Detected in an acute-phase mouse model. Involved in parasite control.	[62] da Matta Guedes et al., 2010 [80] Miyazaki et al., 2010
IL-6	It contributes to mounting the response to the parasite, as well as to host resistance to *T. cruzi* infection during the acute phase.	[82] Gao et al., 2002
IL-27	In patients with Chagas disease without severe cardiopathy, its production has shown a tolerogenic effect.	[83] Natale et al., 2021
CCL3 and CCL5	In a mouse model, they are involved in mucosal protection and activation of the B-cell defense.	[84] Sullivan et al., 2011

Table 3. Cytokines and chemokines reported as susceptibility inducers of CHD. CHD: chronic heart disease.

Cytokine	Susceptibility	Ref.
IL-1, IL-6, IL-15, IFN-γ, TNF-α	Elevated levels in patients with CHD.	[40] Gomes et al., 2003
IL-10	Low levels in patients with CHD. By inhibiting macrophage proliferation in the acute phase, it mediates susceptibility to *T. cruzi* infection.	[85] Reed et al., 1994
TNF-α	Elevated levels in patients with dilated cardiomyopathy and in chronic, asymptomatic, and symptomatic patients.	[52] Ferreira et al., 2003
IL-1β	Elevated levels in patients with CHD.	[46] Petersen and Burleigh, 2003
IL-17	High levels associated with a significant increase in myocardial destruction. Elevated circulating levels in younger patients with CHD.	[62] da Matta Guedes et al., 2010 [63] De alba et al., 2018
IL-13	In combination with arginase-1, it promotes susceptibility to *T. cruzi* infection.	[86] Abad et al., 2018
IL-15	Found in the clinical cardiac form of the disease.	[40] Gomes et al., 2003
TGF-β and IL-10	Stimulation of macrophage cultures with high levels of both cytokines inhibits the trypanocidal effect of IFN-γ.	[18] Cardoni et al., 1999
Surrogate proteins for TGF-β, MMP-2, MMP-9, and fibronectin	Responsible for tissue fibrosis in pathological cardiac remodeling in CHD.	[87] Medeiros et al., 2019 [88] Bautista-López et al., 2013 [89] Fares et al., 2013 [67] Baron et al., 2022
CXCL9 (MIG), CXCL10, CCL2 CCR5, MCP-1CCXCR3	Increased levels of these chemokines in the heart tissue of patients with CHD. They are associated with increased IFN-γ levels. Necessary for recruitment of inflammatory cells.	[84] Sullivan et al., 2011

6. Host Genetics and Cytokine Expression

Host genetics is another key factor in cytokine expression and may explain the variation in cytokine levels between patients as well as susceptibility to heart disease in the chronic phase [90].Some authors are studying cytokine and chemokine gene polymorphisms to determine the possible genetic basis of cytokine dysregulation associated with various diseasesThe search for repeated microsatellite polymorphisms and single-nucleotide polymorphisms has been proposed. The former are used in linkage studies in nuclear family samples, and the latter are usually used in case-control association studies in population samples. Several authors have proposed the following alleles as inducers of susceptibility and resistance in Chagas disease (Tables 4 and 5).

Table 4. Major cytokine and chemokine genes and alleles reported to be protective against the development of cardiopathy in Chagas disease. * Gene, ** single-nucleotide polymorphism, SNP.

Genetic Protection			
Cytokine	**Gene/Polymorphism**	**Function**	**Ref.**
IL-12	*IL12B* * *(rs2546893]* ** *(rs919766]* **	High expression in asymptomatic or acutely infected patients with *T. cruzi*. Significant in IFN-γ production and Th1 cell differentiation.	[35] Frade-Barros et al., 2020
IL-10	*IL10* * *(rs1800896]* **	It has been associated with reduced disease severity and upregulation of the anti-inflammatory immune response.	[24] Sathler-Avelar et al., 2003
CCR5	*−59029AG (rs1799987)* **	It has been associated with lower CCR5 expression, which may confer protection against the development of severe cardiac symptoms.	[91] Fernández-Mestre et al., 2004
CXCL10, CCL5, CXCL9	*CXCL9* * *(rs10336 CC* **), *CXCL10* * *(rs 3921 GG* **) *CCR5* * *(rs1799988CC* **)	These chemokines are considered master regulators of inflammatory cell migration in the myocardium. A reduction in their expression is associated with less-severe myocarditis.	[92] Nogueira et al., 2012

Table 5. Major cytokine and chemokine genes and alleles reported in susceptibility to the development of CHD. * Gene, ** single-nucleotide polymorphism, SNP. CHD: chronic heart disease.

Genetic Susceptibility			
Cytokine	**Gene/Polymorphism**	**Function**	**Ref.**
TNF family, sTNFR1 and sTNFR2	*sTNFR1 (rs767455)* ** *sTNFR2(rs1061624)* ** *TNFA−1031TT* * *TNFA−308GG* *, *TNFa2* * *TNFa7* *, *TNFa8* * *TNFb2* *, *TNFb4* * *TNFd5* *, *TNFd7** *TNFe2* *	They modulate TNF-α signaling. They are associated with systemic inflammation and progression of CHD.	[93] Alvarado-Arnez et al., 2018
IFN-γ	*IFNG +874T/A*	It has been associated with susceptibility to CHD in patients in Colombia.	[94] Torres et al.,2010
IL-12	*IL12B* *	It has been associated with susceptibility to CHD.	[95] Zafra et al., 2007
IL-1B	*IL1B +5810G* * *IL1B −31 +395 CT* ** *IL −1RN.4CC*	Its elevated expression has been associated with an excessive inflammatory response that contributes to tissue damage and progression in CHD.	[96] Flórez et al., 2011 [97] Cruz-Robles et al., 2009
IL-4	*IL4 −590C/T* *[rs2243250)* **	The polymorphism in the promoter region of the gene enhances the Th2 response, which limits the inflammatory process in the acute phase and favors parasite persistence.	[96] Flórez et al., 2011
IL-17	*IL17A* * *(rs2275913)* **	It has been significantly associated with an increased risk of developing CHD.	[98] Strauss et al., 2020
TGF-β	*TGFB1**	It confers susceptibility in patients with CHD in Peru.	[99] Calzada et al., 2009
CCR5	*CCR5 −2554T* * *(rs1799987)* ** *−2733G (rs1799988)* **	It affects CCR5 expression, which is associated with immune cell migration to the site of infection and the inflammatory response in CHD.	[91] Fernández-Mestre et al., 2004

Interestingly, several authors have expressed interest in identifying single-nucleotide polymorphisms (SNPs) in cytokine genes that could be used as prognostic markers and to identify groups of cases prone to developing severe heart disease with Chagas disease. However, there is a consensus that solid genetic biomarkers are lacking due to the plasticity of the immune response and the inflammatory profile of each patient. These are not only determined by genetic factors, but also depend on the biology of the parasite and its virulence, as well as the biology of the host itself.

7. Parasite Genetics

It has been hypothesized that genetic variability in discrete typing units (DTUs) within the *T. cruzi* taxon would promote different cytokine profiles in the host, as the genetic pattern of the parasite influences the clinical picture of the disease. However, Poveda, in 2014, did not find a differential cytokine profile, although they reported greater inflammation in patients infected with TcII strains compared to those infected with TcI, which could be due to pleiotropic and redundant functions of the circulating cytokines [100]. Finally, it is noteworthy that the cytokine profile is proinflammatory for all DTUs, which ultimately leads to the same clinical and pathological inflammatory signs in the heart, gastrointestinal tract, and nervous system, regardless of the DTU.

8. Balance of Proinflammatory and Anti-inflammatory Cytokines

Finally, it should be considered that a favorable outcome for the vertebrate is more likely if an adequate balance is established between the proinflammatory mediators involved in the antiparasitic Th1 response (IL-1, IL-6, IFN-γ, and TNF-α), which are expressed in the acute phase and subsequently regulated by anti-inflammatory cytokines (IL-10) [101]. In this regard, IL-10 not only inhibits the Th1 response, but also controls the responses of other cells (T and B lymphocytes); however, an excessive decrease in the levels of anti-inflammatory mediators (IL-10) and the deregulation of cytokines such as IFN-γ lead to a proinflammatory response with the continued presence of activated CD4+ and CD8+ T lymphocyte populations, which contributes to the progression of chronic disease [102] (see Figure 4).

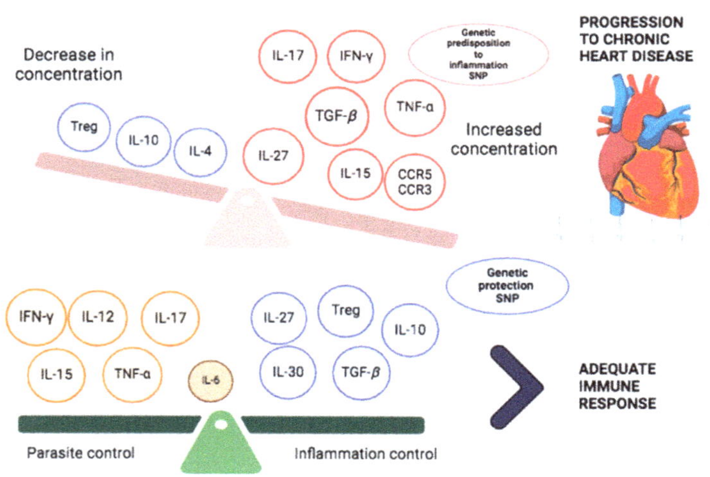

Figure 4. Balance between pro- and anti-inflammatory cytokines. At the top, there is sustained expression of proinflammatory/profibrotic cytokines (red) and their genes, predisposing to chronic disease. Initially, in the acute phase, proinflammatory cytokines are expressed to control parasitemia (yellow); subsequently, regulatory anti-inflammatory cytokines (blue) and their genes prevent and control inflammation. Created with BioRender.com.

Some cytokines may be at the center of the immune balance; among the most studied are IL-27, IL-17, and IL-6.

IL-27, generally produced by antigen-presenting cells, contains IL-30 as a subunit. Its main function is to regulate T and B lymphocytes, as both cells express a specific receptor for this cytokine. It has been shown experimentally that the administration of IL-27 improves the response of Th1 cells, both monofunctional and polyfunctional, in individuals without heart disease, along with an increase in the expression of IL-10 [103]. Its production has also been reported in other studies in patients with Chagas disease who do not have severe heart disease, suggesting that IL-27 production has a tolerogenic effect in these cases. As mentioned above, several authors have reported a marked exhaustion of the T-cell response in the chronic phase of Chagas disease [65,66]. In this regard, Natale (2021) suggests that this is due to the dysregulation between IL-7 and IL-27, and their receptors, which promotes the state of exhaustion of the T-cell response in the late phases of Chagas disease [83,103,104].

The function of IL-17 depends on the pro- or anti-inflammatory context in which it is expressed. There is evidence that it controls the Th1-type response or recruits IL-10-producing neutrophils, which favors host survival during acute infection by regulating an exacerbated immune response [22]. The levels of this cytokine have also been found to be elevated in patients with maternal–fetal infection associated with low parasitemia. Thus, IL-17 has been considered a protective cytokine during the acute phase [80]; however, like other proinflammatory cytokines, its uncontrolled overexpression is associated with a chronic inflammatory process in tissues [105,106]. Other studies suggest that IL-17 may contribute to pathogenicity when associated with Th1-type proinflammatory cytokines [62,107]; this response is even associated with chronic inflammatory autoimmune diseases [108]. Like IL-17, IL-6 can act with proinflammatory or anti-inflammatory function [109]. IL-6 signaling pathways are relevant in cardiomyocytes and macrophages involved in the anti-parasite response; however, most investigators support the hypothesis that this cytokine is important in inflammation and lesion pathogenesis in the chronic disease and has little effect on anti-inflammatory regulatory processes in Chagas disease [55,63,64,73,75].

The role of IL-10 and the tolerogenic dendritic cells that produce this cytokine, which plays a role in the regulation of inflammation by reducing proinflammatory Th1 profiles, which in turn modulates fibrosis, is currently under investigation. It has also been observed that IL-10, in combination with drugs such as fenofibrate, reduces the inflammatory cellular infiltrate and polarizes macrophages toward the M2 phenotype, favoring proper tissue remodeling [110].

There have been few studies on the involvement of TGF-β- and IL-10-producing regulatory T cells (Tregs), which may play an important role in regulating inflammation late in the acute phase and preventing clinical progression to the chronic phase [111] (Figure 4).

9. Conclusions

The immune response of the vertebrate host and the balance between proinflammatory and regulatory responses are key determinants of the clinical course of Chagas disease.

Th1- and Th17-type responses are required in the acute phase of the disease for parasite control by IFN-γ and IL-17. Later, their regulation by cytokines such as IL-10, IL-30, and IL-27 is equally important. However, the persistent expression of proinflammatory and profibrotic cytokines induces the progression of tissue lesions characteristic of cardiopathy in Chagas disease.

This work highlights the importance of studying these cytokines in situ, as shown in Figure 3, because most studies focus on the systemic response in the serum of infected patients. Thus, these cytokines can be considered as biomarkers of lesion development and clinical progression. On the other hand, they can be used to directly determine in tissues whether the inflammatory process in patients with Chagas disease is under control.

As an adjunct to specific antiparasitic therapies, cytokine levels may be useful as biomarkers of a cure or treatment failure, or to develop antiparasitic therapies in combination with drugs proven effective in modulating the inflammatory response to provide new alternatives in the treatment of Chagas-disease-related cardiopathies.

10. Perspective

Further therapeutic studies in patients at different clinical stages of the disease are needed to study the modulation of cytokine and chemokine activity that may favor the host response by controlling the excessive production of proinflammatory and profibrotic mediators involved in the chronic pathology of Chagas disease.

Author Contributions: Conceptualization and writing—original draft preparation M.C.d.A.-A. and M.I.B.-T.; supervision and revision P.M.S.-S. and E.Z.; visualization and editing image M.C.d.A.-A.; editing text M.I.B.-T., M.C.-B. and E.Z. All authors have read and agreed to the published version of the manuscript.

Funding: This study was partially financed by UNAM, grant numbers IN221121, IN212125, PAPIIT, and DGAPA; Facultad de Medicina, UNAM, project numbers FM/DI/004/2023, FM/DI/006/2024, and FM/DI/003/2022; and CONACHyT, CBF2023-2024-117.

Acknowledgments: We wish to thank Nayeli Torres Santiago for the image support, to Nadia A Fernandez Santos for the contribution to the original text edition and revison, and to Juan Francisco Rodríguez for their help with the final copyediting.

Conflicts of Interest: The authors declare no conflicts of interest.

References

1. National Autonomous University of Mexico. *Manual para el Diagnóstico de la Infección por Trypanosoma cruzi*; OPS/WHO: Mexico City, Mexico, 2006; 45p. Available online: https://www.researchgate.net/publication/277957592_Manual_para_el_Diagnostico_de_la_Infeccion_por_Trypanosoma_cruz (accessed on 1 September 2024).
2. Chagas, C. Nova tripanozomiaze humana: Estudos sobre a morfolojia e o ciclo evolutivo do Schizotrypanum cruzi n. gen., n. sp., ajente etiolojico de nova entidade morbida do homem. *Mem. Inst. Oswaldo Cruz* **1909**, *1*, 159–218. [CrossRef]
3. Ruiz-Piña, H.A.; Cruz-Reyes, A. The Opossum Didelphis Virginiana as a Synanthropic Reservoir of *Trypanosoma cruzi* in Dzidzilché, Yucatán, México. *Mem. Inst. Oswaldo Cruz* **2002**, *97*, 613–620. [CrossRef] [PubMed]
4. Del Valle Pugliese Uliarte, D.; Durante, L.A.T.; Rivas, D.M.; Villegas, V.G.; Maestre, K.S.; De Soto, A.J.B. Acute Miocarditis and Meningoencephalitis Caused by *Trypanosoma cruzi* in an Hiv-Seropositive Patient. *Rev. Cuba. Med. Trop.* **2014**, *66*, 458–464.
5. Quintal Avilés, R.; Zavala Velázquez, J.; Pinzón Cantarell, J. La enfermedad de Chagas en el estado de Yucatán México (informe de transmisores). *Salud Publica Mex.* **1976**, *18*, 999–1003.
6. Ferraresso, M.G.; Torre, A.C.; Piva, M.M.M.; Barcan, L. Chagas disease reactivation: Cutaneous manifestations in a transplanted patient. *An. Bras. Dermatol.* **2018**, *93*, 890–892. [CrossRef]
7. Teixeira, A.R.; Nascimento, R.J.; Sturm, N.R. Evolution and pathology in Chagas disease—A Review. *Mem. Inst. Oswaldo Cruz* **2006**, *101*, 463–491. [CrossRef]
8. Salazar-Schettino, P.M.; Bucio-Torres, M.I.; Cabrera-Bravo, M.; Citlalli De Alba-Alvarado, M.; Rocío Castillo-Saldaña, D.; Zenteno-Galindo, E.A.; Rojo-Medina, J.; Angélica Fernández-Santos, N.; Perera-Salazar, G. Enfermedad de Chagas en México. *Rev. Fac. Med.* **2016**, *59*, 6–16.
9. Sánchez, Y.S.; Talavera, R.V.; Bravo, C.B.; Huerta, L.V.; Benzaquen, E.C.; Díaz, F.D.; Echegaray, J.B.; Altamirano, J.G.; Oviedo, L.R. Estudio comparativo de alteraciones electrocardiográficas, frecuencia cardiaca y presión arterial entre seropositivos y seronegativos para *Trypanosoma cruzi* en el valle de Vítor, Arequipa-Perú. *Acta Médica Peru.* **2007**, *24*, 76–80.
10. Vinajá, D.; Aché, A. Alteraciones Electrocardiográficas en pacientes con Enfermedad de Chagas. Hospital José Rangel de Villa de Cura. 1998–2008. *Rev. Inst. Nac. Hig. Rafael Rangel* **2012**, *43*, 12–19.
11. Barbosa, M.M.; Nunes, M.C.P. Risk stratification in chagas disease. *Rev. Esp. Cardiol.* **2012**, *65*, 17–21. [CrossRef]
12. Köberle, F. Chagas' Disease and Chagas' Syndromes: The Pathology of American Trypanosomiasis. *Adv. Parasitol* **1968**, *6*, 63–116. [CrossRef] [PubMed]
13. Useche, Y.; Pérez, A.R.; de Meis, J.; Bonomo, A.; Savino, W. Central Nervous System Commitment in Chagas Disease. *Front. Immunol.* **2022**, *13*, 975106. [CrossRef]
14. Nunes, M.C.P.; Beaton, A.; Acquatella, H.; Bern, C.; Bolger, A.F.; Echeverría, L.E.; Dutra, W.O.; Gascon, J.; Morillo, C.A.; Oliveira-Filho, J.; et al. Chagas Cardiomyopathy: An Update of Current Clinical Knowledge and Management: A Scientific Statement From the American Heart Association. *Circulation* **2018**, *138*, e169–e209. [CrossRef] [PubMed]

15. De Fuentes-Vicente, J.A.; Vidal-López, D.G.; Flores-Villegas, A.L.; Moreno-Rodríguez, A.; De Alba-Alvarado, M.C.; Salazar-Schettino, P.M.; Rodríguez-López, M.H.; Gutiérrez-Cabrera, A.E. *Trypanosoma cruzi*: A review of biological and methodological factors in Mexican strains. *Acta Trop.* **2019**, *195*, 51–57. [CrossRef] [PubMed]
16. Moncayo, A. Progress towards interruption of transmission of Chagas disease. *Mem. Inst. Oswaldo Cruz* **1999**, *94*, 401–404. [CrossRef]
17. Abbas, A.K.; Lichtman, A.H.; Pillai, S. *Inmunología Celular y Molecular*; Elsevier: Amsterdam, The Netherlands, 2018; ISBN 8491132759.
18. Cardoni, R.L.; Antunez, M.I.; Abrami, A.A. Respuesta TH1 en la infeccion experimental con *Trypanosoma cruzi*. *Medicina* **1999**, *59*, 84–90.
19. Campos, M.A.S.; Almeida, I.C.; Takeuchi, O.; Akira, S.; Valente, E.P.; Procópio, D.O.; Travassos, L.R.; Smith, J.A.; Golenbock, D.T.; Gazzinelli, R.T. Activation of Toll-Like Receptor-2 by Glycosylphosphatidylinositol Anchors from a Protozoan Parasite. *J. Immunol.* **2001**, *167*, 416–423. [CrossRef]
20. Bartholomeu, D.C.; Ropert, C.; Melo, M.B.; Parroche, P.; Junqueira, C.F.; Teixeira, S.M.R.; Sirois, C.; Kasperkovitz, P.; Knetter, C.F.; Lien, E.; et al. Recruitment and Endo-Lysosomal Activation of TLR9 in Dendritic Cells Infected with *Trypanosoma cruzi*. *J. Immunol.* **2008**, *181*, 1333–1344. [CrossRef]
21. Campos, M.A.; Closel, M.; Valente, E.P.; Cardoso, J.E.; Akira, S.; Alvarez-Leite, J.I.; Ropert, C.; Gazzinelli, R.T. Impaired Production of Proinflammatory Cytokines and Host Resistance to Acute Infection with *Trypanosoma cruzi* in Mice Lacking Functional Myeloid Differentiation Factor 88. *J. Immunol.* **2004**, *172*, 1711–1718. [CrossRef]
22. Boari, T.T.; Vesely, A.; Bermejo, M.C.; Ramello, D.A.; Montes, M.C. IL-17RA Signaling Reduces Inflammation and Mortality during *Trypanosoma cruzi* Infection by Recruiting Suppressive IL-10-Producing Neutrophils. *PLoS Pathog.* **2012**, *8*, 1002658. [CrossRef]
23. Futosi, K.; Fodor, S.; Mócsai, A. Neutrophil cell surface receptors and their intracellular signal transduction pathways. *Int. Immunopharmacol.* **2013**, *17*, 638–650. [CrossRef]
24. Sathler-Avelar, R.; Lemos, E.M.; Reis, D.D.; Medrano-Mercado, N.; Araujo-Jorge, T.C.; Antas, P.R.Z.; Corrêa-Oliveira, R.; Teixeira-Carvalho, A.; Elói-Santos, S.M.; Favato, D.; et al. Phenotypic Features of Peripheral Blood Leucocytes during Early Stages of Human Infection with *Trypanosoma cruzi*. *Scand. J. Immunol.* **2003**, *58*, 655–663. [CrossRef] [PubMed]
25. Laucella, S.A.; Rottenberg, M.E.; Titto, E.H. Papel de las citoquininas en la resistencia y patología durante la infección con *Trypanosoma cruzi*. *Rev. Argent. Microbiol.* **1996**, *28*, 99–109. [PubMed]
26. Dos Santos, P.V.A.; Roffê, E.; Santiago, H.C.; Torres, R.A.; Marino, A.P.M.P.; Paiva, C.N.; Silva, A.A.; Gazzinelli, R.T.; Lannes-Vieira, J. Prevalence of $CD8^{+}\alpha\beta$ T cells in *Trypanosoma cruzi*-elicited myocarditis is associated with acquisition of CD62LLowLFA-1HighVLA-4High activation phenotype and expression of IFN-γ-inducible adhesion and chemoattractant molecules. *Microbes Infect.* **2001**, *3*, 971–984. [CrossRef] [PubMed]
27. Antúnez, M.I.; Cardoni, R.L. IL-12 and IFN-γ production, and NK cell activity, in acute and chronic experimental *Trypanosoma cruzi* infections. *Immunol. Lett.* **2000**, *71*, 103–109. [CrossRef] [PubMed]
28. Lawrence, T.; Natoli, G. Transcriptional regulation of macrophage polarization: Enabling diversity with identity. *Nat. Rev. Immunol.* **2011**, *11*, 750–761. [CrossRef]
29. Hovsepian, E.; Mirkin, G.A.; Penas, F.; Manzano, A.; Bartrons, R.; Goren, N.B. Modulation of inflammatory response and parasitism by 15-Deoxy-Δ12,14 prostaglandin J2 in *Trypanosoma cruzi*-infected cardiomyocytes. *Int. J. Parasitol.* **2011**, *41*, 553–562. [CrossRef]
30. Díaz-Gandarilla, J.A.; Osorio-Trujillo, C.; Hernández-Ramírez, V.I.; Talamás-Rohana, P. PPAR activation induces m1 macrophage polarization via cPLACOX-2 inhibition, activating ros production against Leishmania Mexicana. *Biomed. Res. Int.* **2013**, *2013*, 215283. [CrossRef]
31. Martins, R.M.; Alves, R.M.; Macedo, S.; Yoshida, N. Starvation and rapamycin differentially regulate host cell lysosome exocytosis and invasion by *Trypanosoma cruzi* metacyclic forms. *Cell. Microbiol.* **2011**, *13*, 943–954. [CrossRef]
32. Sag, D.; Carling, D.; Stout, R.D.; Suttles, J. Adenosine 5′-Monophosphate-Activated Protein Kinase Promotes Macrophage Polarization to an Anti-Inflammatory Functional Phenotype. *J. Immunol.* **2008**, *181*, 8633–8641. [CrossRef]
33. Basso, B. Modulation of immune response in experimental Chagas disease. *World J. Exp. Med.* **2013**, *3*, 1–10. [CrossRef] [PubMed]
34. Zanluqui, N.G. Macrophage Polarization in Chagas Disease. *J. Clin. Cell. Immunol.* **2015**, *6*, 317. [CrossRef]
35. Farage Frade-Barros, A.; Maria Ianni, B.; Cabantous, S.; Wide Pissetti, C.; Saba, B.; Tzu Lin-Wang, H.; Buck, P.; Antonio Marin-Neto, J.; Schmidt, A.; Dias, F. Polymorphisms in Genes Affecting Interferon-γ Production and Th1 T Cell Differentiation Are Associated with Progression to Chagas Disease Cardiomyopathy. *Front. Immunol.* **2020**, *11*, 1386. [CrossRef]
36. Teixeira, D.E.; Benchimol, M.; Crepaldi, P.H.; de Souza, W. Interactive Multimedia to Teach the Life Cycle of *Trypanosoma cruzi*, the Causative Agent of Chagas Disease. *PLoS Negl. Trop. Dis.* **2012**, *6*, e1749. [CrossRef] [PubMed]
37. Aliberti, J.C.S.; Cardoso, M.A.G.; Martins, G.A.; Gazzinelli, R.T.; Vieira, L.Q.; Silva, J.S. Interleukin-12 mediates resistance to *Trypanosoma cruzi* in mice and is produced by murine macrophages in response to live trypomastigotes. *Infect. Immun.* **1996**, *64*, 1961–1967. [CrossRef] [PubMed]
38. Michailowsky, V.; Silva, N.M.; Rocha, C.D.; Vieira, L.Q.; Lannes-Vieira, J.; Gazzinelli, R.T. Pivotal role of interleukin-12 and interferon-γ axis in controlling tissue parasitism and inflammation in the heart and central nervous system during *Trypanosoma cruzi* infection. *Am. J. Pathol.* **2001**, *159*, 1723–1733. [CrossRef]

39. Guiñazú, N.; Pellegrini, A.; Carrera-Silva, E.A.; Aoki, M.P.; Cabanillas, A.M.; Gìronés, N.; Fresno, M.; Cano, R.; Gea, S. Immunisation with a major *Trypanosoma cruzi* antigen promotes pro-inflammatory cytokines, nitric oxide production and increases TLR2 expression. *Int. J. Parasitol.* **2007**, *37*, 1243–1254. [CrossRef]
40. Gomes, J.A.S.; Bahia-Oliveira, L.M.G.; Rocha, M.O.C.; Martins-Filho, O.A.; Gazzinelli, G.; Correa-Oliveira, R. Evidence that development of severe cardiomyopathy in human Chagas' disease is due to a Th1-specific immune response. *Infect. Immun.* **2003**, *71*, 1185–1193. [CrossRef]
41. Bahia-Oliveira, L.M.G.; Gomes, J.A.S.; Rocha, M.O.C.; Moreira, M.C.V.; Lemos, E.M.; Luz, Z.M.P.; Pereira, M.E.S.; Coffman, R.L.; Dias, J.C.P.; Cançado, J.R.; et al. IFN-γ in human Chagas' disease: Protection or pathology? *Braz. J. Med. Biol. Res.* **1998**, *31*, 127–131. [CrossRef]
42. Abel, L.C.J.; Rizzo, L.V.; Ianni, B.; Albuquerque, F.; Bacal, F.; Carrara, D.; Bocchi, E.A.; Teixeira, H.C.; Mady, C.; Kalil, J.; et al. Chronic Chagas' disease cardiomyopathy patients display an increased IFN-γ response to *Trypanosoma cruzi* infection. *J. Autoimmun.* **2001**, *17*, 99–107. [CrossRef]
43. Lima, E.S.; Andrade, Z.A.; Andrade, S.G. TNF-α is expressed at sites of parasite and tissue destruction in the spleen of mice acutely infected with *Trypanosoma cruzi*. *Int. J. Exp. Pathol.* **2001**, *82*, 327–336. [CrossRef] [PubMed]
44. Engel, D.; Peshock, R.; Armstong, R.C.; Sivasubramanian, N.; Mann, D.L. Cardiac myocyte apoptosis provokes adverse cardiac remodeling in transgenic mice with targeted TNF overexpression. *Am. J. Physiol. Heart Circ. Physiol.* **2004**, *287*, H1303–H1311. [CrossRef] [PubMed]
45. Antunes, D.; Marins-Dos-Santos, A.; Ramos, M.T.; Mascarenhas, B.A.S.; De Carvalho Moreira, C.J.; Farias-De-Oliveira, D.A.; Savino, W.; Monteiro, R.Q.; De Meis, J. Oral Route Driven Acute *Trypanosoma cruzi* Infection Unravels an IL-6 Dependent Hemostatic Derangement. *Front. Immunol.* **2019**, *10*, 1073. [CrossRef] [PubMed]
46. Petersen, C.A.; Burleigh, B.A. Role for interleukin-1β in *Trypanosoma cruzi*-induced cardiomyocyte hypertrophy. *Infect. Immun.* **2003**, *71*, 4441–4447. [CrossRef] [PubMed]
47. Salazar-Schettino, P.M.; Cabrera-Bravo, M.; Vazquez-Antona, C.; Zenteno, E.; De Alba-Alvarado, M.; Torres Gutierrez, E.; Gomez, Y.G.; Perera-Salazar, M.G.; de la Torre, G.G.; Bucio-Torres, M.I. Chagas disease in Mexico: Report of 14 cases of chagasic cardiomyopathy in children. *Tohoku J. Exp. Med.* **2016**, *240*, 243–249. [CrossRef] [PubMed]
48. Adamopoulos, S.; Parissis, J.T.; Kremastinos, D.T. A glossary of circulating cytokines in chronic heart failure. *Eur. J. Heart Fail.* **2001**, *3*, 517–526. [CrossRef]
49. Briceño, L.; Mosca, W. Defective production of interleukin 2 in patients with Chagas' disease. Purified IL-2 augments in vitro response in patients with chagasic cardiomyopathy. *Mem. Inst. Oswaldo Cruz* **1996**, *91*, 601–607. [CrossRef]
50. Marin-Neto, J.A.; Cunha-Neto, E.; Maciel, B.C.; Simões, M.V. Pathogenesis of chronic Chagas heart disease. *Circulation* **2007**, *115*, 1109–1123. [CrossRef]
51. Talvani, A.; Ribeiro, C.S.; Aliberti, J.C.; Michailowsky, V.; Santos, P.V.; Murta, S.M.; Romanha, A.J.; Almeida, I.C.; Farber, J.; Lannes-Vieira, J.; et al. Kinetics of cytokine gene expression in experimental chagasic cardiomyopathy: Tissue parasitism and endogenous IFN-γ as important determinants of chemokine mRNA expression during infection with *Trypanosoma cruzi*. *Microbes Infect.* **2000**, *2*, 851–866. [CrossRef]
52. Ferreira, R.C.; Ianni, B.M.; Abel, L.C.; Buck, P.; Mady, C.; Kalil, J.; Cunha-Neto, E. Increased plasma levels of tumor necrosis factor-alpha in asymptomatic/"indeterminate" and Chagas disease cardiomyopathy patients. *Mem. Inst. Oswaldo Cruz* **2003**, *98*, 407–412. [CrossRef]
53. Rodrigues, D.B.R.; Dos Reis, M.A.; Romano, A.; Pereira, S.A.D.L.; Teixeira, V.D.P.A.; Tostes Junior, S.; Rodrigues, V. In situ expression of regulatory cytokines by heart inflammatory cells in chagas' disease patients with heart failure. *J. Immunol. Res.* **2012**, *2012*, 361730. [CrossRef]
54. de Sena Pereira, N.; Queiroga, T.B.D.; Nunes, D.F.; de Mesquita Andrade, C.; Nascimento, M.S.L.; Do-Valle-Matta, M.A.; da Câmara, A.C.J.; da Cunha Galvão, L.M.; Guedes, P.M.M.; Chiari, E. Innate immune receptors over expression correlate with chronic chagasic cardiomyopathy and digestive damage in patients. *PLoS Negl. Trop. Dis.* **2018**, *12*, e0006589. [CrossRef]
55. López, L.; Arai, K.; Giménez, E.; Jiménez, M.; Pascuzo, C.; Rodríguez-Bonfante, C.; Bonfante-Cabarcas, R. Las concentraciones séricas de interleucina-6 y proteína C reactiva se incrementan a medida que la enfermedad de Chagas evoluciona hacia el deterioro de la función cardíaca. *Rev. Española Cardiol.* **2006**, *59*, 50–56. [CrossRef]
56. Ferreira, R.R.; Waghabi, M.C.; Bailly, S.; Feige, J.J.; Hasslocher-Moreno, A.M.; Saraiva, R.M.; Araujo-Jorge, T.C. The Search for Biomarkers and Treatments in Chagas Disease: Insights From TGF-Beta Studies and Immunogenetics. *Front. Cell. Infect. Microbiol.* **2022**, *11*, 767576. [CrossRef] [PubMed]
57. Ferrão, P.M.; Nisimura, L.M.; Moreira, O.C.; Land, M.G.; Pereira, M.C.; de Mendonça-Lima, L.; Araujo-Jorge, T.C.; Waghabi, M.C.; Garzoni, L.R. Inhibition of TGF-β pathway reverts extracellular matrix remodeling in *T. cruzi*-infected cardiac spheroids. *Exp. Cell Res.* **2018**, *362*, 260–267. [CrossRef]
58. Ferreira, L.R.P. Interferon-γ and other inflammatory mediators in cardiomyocyte signaling during Chagas disease cardiomyopathy. *World J. Cardiol.* **2014**, *6*, 782. [CrossRef]
59. Chevillard, C.; Nunes, J.P.S.; Frade, A.F.; Almeida, R.R.; Pandey, R.P.; Nascimento, M.S.; Kalil, J.; Cunha-Neto, E. Disease Tolerance and Pathogen Resistance Genes May Underlie *Trypanosoma cruzi* Persistence and Differential Progression to Chagas Disease Cardiomyopathy. *Front. Immunol.* **2018**, *9*, 411639. [CrossRef]

60. Costa, G.C.; Rocha, M.O.D.C.; Moreira, P.R.; Menezes, C.A.S.; Silva, M.R.; Gollob, K.J.; Dutra, W.O. Functional IL-10 gene polymorphism is associated with Chagas disease cardiomyopathy. *J. Infect. Dis.* **2009**, *199*, 451–454. [CrossRef]
61. Vasconcelos, R.H.T.; Azevedo, E.A.N.; Diniz, G.T.N.; Cavalcanti, M.G.A.M.; de Oliveira, W.; de Morais, C.N.L.; Gomes, Y.M. Interleukin-10 and tumour necrosis factor-alpha serum levels in chronic Chagas disease patients. *Parasite Immunol.* **2015**, *37*, 376–379. [CrossRef]
62. Guedes, P.M.D.M.; Gutierrez, F.R.S.; Maia, F.L.; Milanezi, C.M.; Silva, G.K.; Pavanelli, W.R.; Silva, J.S. IL-17 produced during *Trypanosoma cruzi* infection plays a central role in regulating parasite-induced myocarditis. *PLoS Negl. Trop. Dis.* **2010**, *4*, e604. [CrossRef]
63. De Alba-Alvarado, M.; Salazar-Schettino, P.M.; Jiménez-Álvarez, L.; Cabrera-Bravo, M.; García-Sancho, C.; Zenteno, E.; Vazquez-Antona, C.; Cruz-Lagunas, A.; Zúñiga, J.; Bucio-Torres, M.I. Th-17 cytokines are associated with severity of *Trypanosoma cruzi* chronic infection in pediatric patients from endemic areas of Mexico. *Acta Trop.* **2018**, *178*, 134–141. [CrossRef] [PubMed]
64. Villanueva-Lizama, L.E.; Cruz-Chan, J.V.; Aguilar-Cetina, A.d.C.; Herrera-Sanchez, L.F.; Rodriguez-Perez, J.M.; Rosado-Vallado, M.E.; Ramirez-Sierra, M.J.; Ortega-Lopez, J.; Jones, K.; Hotez, P.; et al. *Trypanosoma cruzi* vaccine candidate antigens Tc24 and TSA-1 recall memory immune response associated with HLA-A and -B supertypes in Chagasic chronic patients from Mexico. *PLoS Negl. Trop. Dis.* **2018**, *12*, e0006240. [CrossRef] [PubMed]
65. Albareda, M.C.; Olivera, G.C.; Laucella, S.A.; Alvarez, M.G.; Fernandez, E.R.; Lococo, B.; Viotti, R.; Tarleton, R.L.; Postan, M. Chronic Human Infection with Trypanosoma Cruzi Drives CD4+ T Cells to Immune Senescence. *J Immunol* **2009**, *183*, 4103. [CrossRef] [PubMed]
66. Mateus, J.; Guerrero, P.; Lasso, P.; Cuervo, C.; González, J.M.; Puerta, C.J.; Cuéllar, A. An animal model of acute and chronic chagas disease with the reticulotropic Y strain of *Trypanosoma cruzi* that depicts the multifunctionality and dysfunctionality of T cells. *Front. Immunol.* **2019**, *10*, 450465. [CrossRef] [PubMed]
67. Araújo-Jorge, T.C.; Waghabi, M.C.; Soeiro, M.d.N.C.; Keramidas, M.; Bailly, S.; Feige, J.J. Pivotal role for TGF-beta in infectious heart disease: The case of *Trypanosoma cruzi* infection and consequent Chagasic myocardiopathy. *Cytokine Growth Factor Rev.* **2008**, *19*, 405–413. [CrossRef]
68. Baron, M.A.; Ferreira, L.R.P.; Teixeira, P.C.; Moretti, A.I.S.; Santos, R.H.B.; Frade, A.F.; Kuramoto, A.; Debbas, V.; Benvenuti, L.A.; Gaiotto, F.A.; et al. Matrix Metalloproteinase 2 and 9 Enzymatic Activities are Selectively Increased in the Myocardium of Chronic Chagas Disease Cardiomyopathy Patients: Role of TIMPs. *Front. Cell. Infect. Microbiol.* **2022**, *12*, 836242. [CrossRef]
69. Borthwick, L.A.; Wynn, T.A.; Fisher, A.J. Cytokine Mediated Tissue Fibrosis. *Biochim. Biophys. Acta* **2013**, *1832*, 1049. [CrossRef]
70. Dolber, P.C.; Spach, M.S. Thin Collagenous Septa in Cardiac Muscle. *Anat. Rec.* **1987**, *218*, 45–55. [CrossRef]
71. De Alba-Alvarado, M.C.; Torres-Gutiérrez, E.; Reynoso-Ducoing, O.A.; Zenteno-Galindo, E.; Cabrera-Bravo, M.; Guevara-Gómez, Y.; Salazar-Schettino, P.M.; Rivera-Fernández, N.; Bucio-Torres, M.I. Immunopathological Mechanisms Underlying Cardiac Damage in Chagas Disease. *Pathogens* **2023**, *12*, 335. [CrossRef]
72. Abrahamsohn, I.A.; Blotta, M.H.S.L.; Curotto, M.A. Enhancement of delayed-type hypersensitivity to *Trypanosoma cruzi* in mice treated with Mycobacterium bovis BCG and cyclophosphamide. *Infect. Immun.* **1981**, *31*, 1145–1151. [CrossRef]
73. Satoh, M.; Nakamura, M.; Tamura, G.; Makita, S.; Segawa, I.; Tashiro, A.; Satodate, R.; Hiramori, K. Inducible Nitric Oxide Synthase and Tumor Necrosis Factor-Alpha in Myocardium in Human Dilated Cardiomyopathy. *J. Am. Coll. Cardiol.* **1997**, *29*, 716–724. [CrossRef] [PubMed]
74. Kinugawa, K.I.; Takahashi, T.; Kohmoto, O.; Yao, A.; Aoyagi, T.; Momomura, S.I.; Hirata, Y.; Serizawa, T. Nitric oxide-mediated effects of interleukin-6 on [Ca2+]i and cell contraction in cultured chick ventricular myocytes. *Circ. Res.* **1994**, *75*, 285–295. [CrossRef] [PubMed]
75. Silverio, J.C.; De-Oliveira-Pinto, L.M.; Da Silva, A.A.; De Oliveira, G.M.; Lannes-Vieira, J. Perforin-expressing cytotoxic cells contribute to chronic cardiomyopathy in *Trypanosoma cruzi* infection. *Int. J. Exp. Pathol.* **2010**, *91*, 72–86. [CrossRef] [PubMed]
76. De Jonge, E.; Friederich, P.W.; Vlasuk, G.P.; Rote, W.E.; Vroom, M.B.; Levi, M.; Van der Poll, T. Activation of coagulation by administration of recombinant factor VIIa elicits interleukin 6 (IL-6) and IL-8 release in healthy human subjects. *Clin. Vaccine Immunol.* **2003**, *10*, 495–497. [CrossRef]
77. Pérez-Campos Mayoral, L.; Hernández-Huerta, M.T.; Papy-García, D.; Barritault, D.; Zenteno, E.; Sánchez Navarro, L.M.; Pérez-Campos Mayoral, E.; Matias Cervantes, C.A.; Martínez Cruz, M.; Mayoral Andrade, G.; et al. Immunothrombotic Dysregulation in Chagas Disease and COVID-19: A Comparative Study of Anticoagulation. *Mol. Cell. Biochem.* **2021**, *476*, 3815–3825. [CrossRef]
78. Kobayashi, T.; Kim, H.J.; Liu, X.; Sugiura, H.; Kohyama, T.; Fang, Q.; Wen, F.Q.; Abe, S.; Wang, X.; Atkinson, J.J.; et al. Matrix metalloproteinase-9 activates TGF-β and stimulates fibroblast contraction of collagen gels. *Am. J. Physiol.-Lung Cell. Mol. Physiol.* **2014**, *306*, L1006–L1015. [CrossRef]
79. Pontes Ferreira, C.; Cariste, L.M.; Ferri Moraschi, B.; Ferrarini Zanetti, B.; Won Han, S.; Araki Ribeiro, D.; Vieira Machado, A.; Lannes-Vieira, J.; Gazzinelli, R.T.; Vasconcelos, J.R.C. CXCR3 chemokine receptor guides *Trypanosoma cruzi*-specific T-cells triggered by DNA/adenovirus ASP2 vaccine to heart tissue after challenge. *PLoS Negl. Trop. Dis.* **2019**, *13*, e7597. [CrossRef]
80. Talvani, A.; Rocha, M.O.C.; Ribeiro, A.L.; Correa-Oliveira, R.; Teixeira, M.M. Chemokine Receptor Expression on the Surface of Peripheral Blood Mononuclear Cells in Chagas Disease. *J. Infect. Dis.* **2004**, *189*, 214–220. [CrossRef]
81. Gomes, J.A.S.; Bahia-Oliveira, L.M.G.; Rocha, M.O.C.; Busek, S.C.U.; Tekeira, M.M.; Silva, J.S.; Correa-Oliveira, R. Type 1 Chemokine Receptor Expression in Chagas' Disease Correlates with Morbidity in Cardiac Patients. *Infect Immun* **2005**, *73*, 7960–7966. [CrossRef]

82. Gao, W.; Pereira, M.A. Interleukin-6 is required for parasite specific response and host resistance to Trypanosomacruzi. *Int. J. Parasitol.* **2002**, *32*, 167–170. [CrossRef]
83. Natale, M.A.; Minning, T.; Albareda, M.C.; Eiro, M.D.C.; Álvarez, M.G.; Lococo, B.; Cesar, G.; Bertocchi, G.; Elias, M.J.; Caputo, M.B.; et al. Immune exhaustion in chronic Chagas disease: Pro-inflammatory and immunomodulatory action of IL-27 in vitro. *PLoS Negl. Trop. Dis.* **2021**, *15*, e0009473. [CrossRef] [PubMed]
84. Sullivan, N.L.; Eickhoff, C.S.; Zhang, X.; Giddings, O.K.; Lane, T.E.; Hoft, D.F. Importance of the CCR5-CCL5 axis for mucosal *Trypanosoma cruzi* protection and B cell activation. *J. Immunol.* **2011**, *187*, 1358–1368. [CrossRef] [PubMed]
85. Reed, S.G.; Brownell, C.E.; Russo, D.M.; Silva, J.S.; Grabstein, K.H.; Morrissey, P.J. IL-10 mediates susceptibility to *Trypanosoma cruzi* infection. *J. Immunol.* **1994**, *153*, 3135–3140. [CrossRef]
86. Dar, M.A.; Hölscher, C. Arginase-1 is responsible for IL-13-mediated susceptibility to *Trypanosoma cruzi* infection. *Front. Immunol.* **2018**, *9*, 2790. [CrossRef]
87. Medeiros, N.I.; Gomes, J.A.S.; Fiuza, J.A.; Sousa, G.R.; Almeida, E.F.; Novaes, R.O.; Rocha, V.L.S.; Chaves, A.T.; Dutra, W.O.; Rocha, M.O.C.; et al. MMP-2 and MMP-9 plasma levels are potential biomarkers for indeterminate and cardiac clinical forms progression in chronic Chagas disease. *Sci. Rep.* **2019**, *9*, 14170. [CrossRef] [PubMed]
88. Bautista-López, N.L.; Morillo, C.A.; López-Jaramillo, P.; Quiroz, R.; Luengas, C.; Silva, S.Y.; Galipeau, J.; Lalu, M.M.; Schulz, R. Matrix metalloproteinases 2 and 9 as diagnostic markers in the progression to Chagas cardiomyopathy. *Am. Heart J.* **2013**, *165*, 558–566. [CrossRef]
89. Fares, R.C.G.; Correa-Oliveira, R.; de Araújo, F.F.; Keesen, T.S.L.; Chaves, A.T.; Fiuza, J.A.; Ferreira, K.S.; Rocha, M.O.C.; Gomes, J.A.S. Identification of phenotypic markers of B cells from patients with Chagas disease. *Parasite Immunol.* **2013**, *35*, 214–223. [CrossRef] [PubMed]
90. Gómez, I.; Thomas, M.C.; Palacios, G.; Egui, A.; Carrilero, B.; Simón, M.; Valladares, B.; Segovia, M.; Carmelo, E.; López, M.C. Differential Expression of Immune Response Genes in Asymptomatic Chronic Chagas Disease Patients Versus Healthy Subjects. *Front. Cell Infect. Microbiol.* **2021**, *11*, 722984. [CrossRef]
91. Fernández-Mestre, M.T.; Montagnani, S.; Layrisse, Z. Is the CCR5-59029-G/G genotype a protective factor for cardiomyopathy in Chagas disease? *Hum. Immunol.* **2004**, *65*, 725–728. [CrossRef]
92. Nogueira, L.G.; Santos, R.H.B.; Ianni, B.M.; Fiorelli, A.I.; Mairena, E.C.; Benvenuti, L.A.; Frade, A.; Donadi, E.; Dias, F.; Saba, B.; et al. Myocardial Chemokine Expression and Intensity of Myocarditis in Chagas Cardiomyopathy Are Controlled by Polymorphisms in CXCL9 and CXCL10. *PLoS Negl. Trop. Dis.* **2012**, *6*, e1867. [CrossRef]
93. Alvarado-Arnez, L.E.; Batista, A.M.; Alves, S.M.; Melo, G.; de Lorena, V.M.B.; Cardoso, C.C.; Pereira, I.R.; Carrazzone, C.; Pacheco, A.G.; Oliveira, W.; et al. Single nucleotide polymorphisms of cytokine-related genes and association with clinical outcome in a Chagas disease case-control study from Brazil. *Mem. Inst. Oswaldo Cruz* **2018**, *113*, e170489. [CrossRef] [PubMed]
94. Torres, O.A.; Calzada, J.E.; Beraún, Y.; Morillo, C.A.; González, A.; González, C.I.; Martín, J. Role of the IFNG +874T/A polymorphism in Chagas disease in a Colombian population. *Infect. Genet. Evol.* **2010**, *10*, 682–685. [CrossRef] [PubMed]
95. Zafra, G.; Morillo, C.; Martín, J.; González, A.; González, C.I. Polymorphism in the 3′ UTR of the IL12B gene is associated with Chagas' disease cardiomyopathy. *Microbes Infect.* **2007**, *9*, 1049–1052. [CrossRef]
96. Flórez, O.; Martín, J.; González Rugeles, C.I. Interleukin 4, interleukin 4 receptor-α and interleukin 10 gene polymorphisms in Chagas disease. *Parasite Immunol.* **2011**, *33*, 506–511. [CrossRef] [PubMed]
97. Cruz-Robles, D.; Chvez-Gonzlez, J.P.; Cavazos-Quero, M.M.; Prez-Mndez, O.; Reyes, P.A.; Vargas-Alarcn, G. Association between IL-1B and IL-1RN Gene Polymorphisms and Chagas' Disease Development Susceptibility. *Immunol. Invest.* **2009**, *38*, 231–239. [CrossRef] [PubMed]
98. Strauss, M.; Palma-Vega, M.; Casares-Marfil, D.; Bosch-Nicolau, P.; Lo Presti, M.S.; Molina, I.; González, C.I.; Paglini, P.A.; Schijman, A.G.; Robello, C.; et al. Genetic polymorphisms of IL17A associated with Chagas disease: Results from a meta-analysis in Latin American populations. *Sci. Rep.* **2020**, *10*, 5015. [CrossRef]
99. Calzada, J.E.; Beraún, Y.; González, C.I.; Martín, J. Transforming growth factor beta 1 (TGFβ1) gene polymorphisms and Chagas disease susceptibility in Peruvian and Colombian patients. *Cytokine* **2009**, *45*, 149–153. [CrossRef]
100. Poveda, C.; Fresno, M.; Gironès, N.; Martins-Filho, O.A.; Ramírez, J.D.; Santi-Rocca, J.; Marin-Neto, J.A.; Morillo, C.A.; Rosas, F.; Guhl, F. Cytokine Profiling in Chagas Disease: Towards Understanding the Association with Infecting *Trypanosoma cruzi* Discrete Typing Units (A BENEFIT TRIAL Sub-Study). *PLoS ONE* **2014**, *9*, e91154. [CrossRef]
101. Hovsepian, E.; Penas, F.; Siffo, S.; Mirkin, G.A.; Goren, N.B. IL-10 Inhibits the NF-κB and ERK/MAPK-Mediated Production of Pro-Inflammatory Mediators by Up-Regulation of SOCS-3 in *Trypanosoma cruzi*-Infected Cardiomyocytes. *PLoS ONE* **2013**, *8*, e79445. [CrossRef]
102. de Souza Santos, E.; Silva, D.K.C.; Reis, B.P.Z.C.D.; Barreto, B.C.; Cardoso, C.M.A.; Santos, R.R.D.; Meira, C.S.; Soares, M.B.P. Immunomodulation for the Treatment of Chronic Chagas Disease Cardiomyopathy: A New Approach to an Old Enemy. *Front. Cell. Infect. Microbiol.* **2021**, *11*, 765879. [CrossRef]
103. Joseph, W.; Carl, J.; Bai, X.-F. IL27: Its Roles in the Induction and Inhibition of Inflammation. *Int. J. Clin. Exp. Pathol.* **2008**, *1*, 117–123.
104. Pérez-Antón, E.; Thomas, M.C.; Egui, A.; López, M.C. T-Cell Exhaustion Process during Chronic Infection Caused by Intracellular Trypanosomatids. *Ars. Pharmaceutica* **2019**, *60*, 65–78. [CrossRef]

105. Kitada, S.; Kayama, H.; Okuzaki, D.; Koga, R.; Kobayashi, M.; Arima, Y.; Kumanogoh, A.; Murakami, M.; Ikawa, M.; Takeda, K. BATF2 inhibits immunopathological Th17 responses by suppressing Il23a expression during *Trypanosoma cruzi* infection. *J. Exp. Med.* **2017**, *214*, 1313–1331. [CrossRef]
106. Albareda, M.C.; Perez-Mazliah, D.; Natale, M.A.; Castro-Eiro, M.; Alvarez, M.G.; Viotti, R.; Bertocchi, G.; Lococo, B.; Tarleton, R.L.; Laucella, S.A. Perturbed T Cell IL-7 Receptor Signaling in Chronic Chagas Disease. *J. Immunol.* **2015**, *194*, 3883–3889. [CrossRef] [PubMed]
107. Magalhães, L.M.D.; Villani, F.N.A.; Nunes, M.D.C.P.; Gollob, K.J.; Rocha, M.O.C.; Dutra, W.O. High Interleukin 17 Expression Is Correlated With Better Cardiac Function in Human Chagas Disease. *J. Infect. Dis.* **2013**, *207*, 661–665. [CrossRef]
108. Kim, B.S.; Park, Y.J.; Chung, Y. Targeting IL-17 in autoimmunity and inflammation. *Arch. Pharm. Res.* **2016**, *39*, 1537–1547. [CrossRef]
109. Luo, Y.; Zheng, S.G. Hall of fame among pro-inflammatory cytokines: Interleukin-6 gene and its transcriptional regulation mechanisms. *Front. Immunol.* **2016**, *7*, 236681. [CrossRef]
110. Rada, J.; Donato, M.; Penas, F.N.; Alba Soto, C.; Cevey, Á.C.; Pieralisi, A.V.; Gelpi, R.; Mirkin, G.A.; Goren, N.B. IL-10-Dependent and -Independent Mechanisms Are Involved in the Cardiac Pathology Modulation Mediated by Fenofibrate in an Experimental Model of Chagas Heart Disease. *Front. Immunol.* **2020**, *11*, 572178. [CrossRef]
111. de Araújo, F.F.; Corrêa-Oliveira, R.; Rocha, M.O.C.; Chaves, A.T.; Fiuza, J.A.; Fares, R.C.G.; Ferreira, K.S.; Nunes, M.C.P.; Keesen, T.S.; Damasio, M.P.S.; et al. Foxp3^{+}CD25^{high} CD4^{+} regulatory T cells from indeterminate patients with Chagas disease can suppress the effector cells and cytokines and reveal altered correlations with disease severity. *Immunobiology* **2012**, *217*, 768–777. [CrossRef]

pathogens

MDPI

Review

The Relevance of the Predominant Clonal Evolution (PCE) Model for the Molecular Epidemiology and Subspecific Taxonomy of *Trypanosoma cruzi*

Michel Tibayrenc

MIVEGEC, IRD, CNRS, University Montpellier, 34000 Montpellier, France; michel.tibayrenc@ird.fr

Abstract: The predominant clonal evolution (PCE) model is often misunderstood. Contrary to common belief, it is not restricted to strict mitotic clonality. Instead, it encompasses processes such as selfing, strong homogamy, and various forms of parthenogenesis, as widely acknowledged by researchers studying clonality. Moreover, the PCE model does not claim that genetic recombination is entirely absent or devoid of epidemiological and evolutionary significance. In this review, I will explore the reciprocal relationship between PCE and molecular epidemiology (strain typing) and discuss the implications of PCE for revising the subspecific nomenclature of *Trypanosoma cruzi*.

Keywords: parasite; fungus; bacterium; Chagas disease; strain typing; clonality; evolution; taxonomy; species concept

1. Introduction

Infection with *Trypanosoma cruzi* remains a critical health problem in Latin America and is becoming increasingly concerning in countries outside the traditional Chagas disease region, due to imported cases. Effective Chagas disease control necessitates socioeconomic measures alongside ongoing efforts in basic research, including the molecular evolution of *T. cruzi* [1].

Since the early 1960s, the advent of molecular tools such as Multilocus* Enzyme Electrophoresis* (MLEE*) and DNA-based markers—such as Microsatellites*, Multilocus Sequence Typing* (MLST*) [2], Pulse-Field Gel Electrophoresis* (PFGE*), Restriction Fragment Length Polymorphism (RFLP*), Randomly Amplified Polymorphic DNA (RAPD*), and Amplified Fragment Length Polymorphism (AFLP*)—have paved the way for the emergence of molecular epidemiology. Today, this field relies on highly discriminative markers such as metabarcoding* [3], low-stringency single-primer polymerase chain reaction (LSSP-PCR) [4], single nucleotide polymorphisms* (SNPs*), and culture-free genome-wide locus sequence typing (GLST) [5]. These molecular tools enable the tracking of epidemiological pathways by tagging pathogen strains with genetic markers.

For a long time, molecular epidemiology was based on an empirical interpretation of data: strains appearing identical based on a given marker were assumed to be identical, similar strains were considered similar, and highly divergent strains were deemed very different—without further inference. The integration of population genetic* and phylogenetic* concepts into molecular epidemiology—first applied to *Trypanosoma cruzi* [6,7] and later to other pathogens [8]—provided a deeper understanding of the evolutionary significance of pathogen strains. This advancement greatly enhanced the resolution and accuracy of molecular epidemiological studies.

The PCE model remains widely misunderstood by many researchers. Since it is highly relevant not only to molecular epidemiology but also to the subspecific taxonomy of pathogens, a fresh, updated clarification is warranted. This article does not aim to provide an exhaustive review of *T. cruzi* genetic diversity, but rather, to clarify the PCE model and its predictive power regarding *T. cruzi* subspecific variability.

2. Why Is Clonality vs. Sexuality Relevant for Both Molecular Epidemiology and Taxonomy?

In this paper, sexuality is broadly defined as any genetic exchange between different cells, whereas clonality encompasses all cases in which daughter cells are genetically identical or very similar to parental cells.

Both molecular epidemiology and subspecific taxonomy require stable "relevant units of analysis" (RUAs) [9]. If sexuality is frequent and leads to panmixia*, such stability is lost, as genetic lineages undergo constant recombination and quickly dissolve into the common gene pool.

This is why the long-standing clonality/sexuality debate [6,10,11] is crucial not only for understanding the fundamental biology of pathogens but also for designing molecular epidemiological tools and refining taxonomic classifications. To clarify this debate, it is essential to establish clear definitions of the terms "clone" and "clonality".

3. Clones and Clonality: Multiple Meanings

The primary task of science is to establish clear definitions for the concepts under study. These definitions may vary between authors and according to the purpose of a given study. A term can have multiple definitions, provided they are clear and coherent.

"Clone" is often understood in a cytological sense, referring to a set of daughter cells derived from a single progenitor cell through mitotic* division. This interpretation is commonly referred to as "strict clonality" [12,13] or "absolute clonality" [14]

In this article, as in many previous ones, "clone" and "clonality" are strictly defined in genetic terms. Clonality is confirmed in all cases where daughter cells are genetically identical or highly similar to the parental cell. This genetic definition of clonality is widely accepted among scientists studying clonal reproduction on clonal (parthenogenetic) vertebrates, including fish, amphibians, and reptiles (see, for example, [15]).

Genetic clonality thus encompasses not only "strict" (mitotic) clonality but also self-fertilization*, extreme homogamy* (inbreeding*), and various forms of parthenogenesis*. These processes all produce genetically identical clones and have similar impacts on pathogen population structures. This crucial point is often misunderstood by authors who adhere strictly to the definition of "strict" (mitotic) clonality.

4. Misconceptions Regarding the PCE Model Being "Challenged"

Since its initial formulation [6–8], the PCE model has undergone significant refinement and clarification [16]. However, it has been incorrectly criticized on the following points:

4.1. Total Absence of Genetic Recombination/Sexuality

A fundamental principle of the PCE model is that it does not claim genetic recombination or sexuality is completely absent. Rather, it asserts that such events are not frequent enough to disrupt the predominant clonal structure of the species under study. The mere detection of genetic exchange, whether experimentally or in natural populations [12,13,17–23], does not in itself challenge the PCE model. What matters is the actual impact of these exchanges on the species' population structure (see below).

Successful experimental recombination (e.g., [24]) provides valuable insights into the basic biology of *T. cruzi*. However, it does not indicate the frequency or significance of such events in natural populations of this parasite.

4.2. Genetic Recombination Has No or Little Evolutionary/Epidemiological Significance

The PCE model does not claim that genetic recombination or sexuality lacks evolutionary or epidemiological importance, as is sometimes mistakenly argued [21,25–29]. On the contrary, it is likely to be highly significant.

4.3. Use of Inadequate Genetic Tools

The PCE model is not an artifact resulting from outdated genetic approaches with insufficient resolution [22,30–32] or an insufficient number of markers [33]. For a long time [34], it has relied on the analysis of multiple genetic markers and the most recent high-resolution genomic* tools.

4.4. Improper Sampling

It has been suggested that the apparent lack or rarity of genetic exchange is not due to inherent biological properties of the species under study but rather to trivial geographical and/or temporal separation (the "Wahlund effect*" [33,35]; (see also "the starving sex hypothesis"). However, this criticism does not hold, as all studies addressing the PCE model have carefully accounted for this potential bias through rigorous sampling at both the sympatric and allopatric levels, across micro- and macro-geographical scales.

4.5. Inappropriate Evolutionary Scale

Another hypothesis [27,32] proposes that the apparent PCE in *T. cruzi* is due to genetic recombination/sexuality being inhibited between the species' main phylogenetic subdivisions (discrete typing units* [DTUs*] or near-clades*; see below) but occurring, to some extent, within each of them. While this hypothesis is testable, it has been challenged by microevolutionary analyses (see "the Russian doll* model" below).

5. Main Features of the PCE Model and the Case of *Trypanosoma cruzi*

After outlining what the PCE model is not, here are its main characteristics.

5.1. Mandatory Analysis of Multiple Genetic Loci

PCE results from long-term, widespread inhibition of sexuality, genetic exchange, and recombination*. Since genetic recombination involves the exchange of genetic material across two or more loci, identifying PCE necessarily requires analyzing at least two loci—ideally, as many as possible. This necessitates surveying multilocus genotypes (MLGs). In species undergoing PCE, these MLGs provide the most precise definition of the often vague and widely used term "strain". The more loci that are included in a study, the greater its resolution power.

This requirement invalidates studies based solely on the mini-exon gene, which has been widely used in *T. cruzi* research [36]. Since this gene family represents the variability of only a single locus, it cannot demonstrate a lack of recombination. However, once PCE has been confirmed through appropriate analyses, this marker can be used to characterize *T. cruzi* MLGs and genetic subdivisions [37]. Additionally, because genetic exchange can only be detected through variable loci, adding monomorphic loci does not enhance the analytical resolution.

Fortunately, since the pioneering studies on the genetic variability of the Chagas disease agent (*T. cruzi*) [38], numerous variable genetic loci have been analyzed, providing extensive data for research on genetic recombination.

5.2. Widespread Propagation of Unchanged MLGs Across Space and Time

This is the most intuitive consequence of clonal evolution. Under clonality, daughter cells are genetically identical—or nearly identical—to the parental cell, replicating as "genetic photocopies". When this process persists over evolutionary timescales, these unchanged MLGs can be detected across vast geographic and temporal ranges, at frequencies incompatible with panmictic expectations.

For example, in *T. cruzi*, the MLEE MLG 39 has been sampled in multiple locations across Bolivia, Chile, and Brazil, from human hosts and triatomine bugs, since the early 1980s [39]. It continues to be sampled today. Similarly, MLEE MLG 19 has been identified in triatomine vectors, wild mammals, and human hosts in Chile, Brazil, Venezuela, Colombia, and Bolivia. Additional cases of widely distributed MLEE MLGs are detailed in [39,40].

The ubiquity of these clonal multilocus genotypes cannot be explained by the Wahlund effect. If Wahlund's effect were responsible, these overrepresented MLGs would be confined to restricted geographical areas.

At higher resolution and shorter evolutionary timescales, widespread microsatellite MLGs have been identified in Bolivian *T. cruzi* strains (Messenger et al., 2015 [41]). Similarly, widespread MLST MLGs have been sampled across vast geographical distances in Brazil [2].

It is important to note that such widespread MLGs should not be considered "true" clones but rather families of closely related clones, or "clonets" (see below).

5.3. Linkage Disequilibrium (LD)

LD is a statistical measure indicating that genotypes at different loci do not recombine randomly, contrary to what would be expected under panmixia. In a panmictic system, knowing the genotype at one locus provides no predictive information about genotypes at other loci. This state is known as linkage equilibrium. LD, by contrast, describes the opposite scenario: knowing the genotype at one locus allows for high-probability predictions about genotypes at other loci.

This principle underpins the powerful technique of indirect typing, where a single genetic marker can be used to infer entire MLGs and DTUs. For example, PCR typing [42]; AFLP-PCR and PCR-RFLP [43]; amplification of a single gene, such as *TcSC5D* [44]; mini-exon gene analysis [37]; and fluorescent fragment length barcoding [45] are techniques that could be used.

Strong LD was detected early in *T. cruzi*, leading to the hypothesis that MLEE MLGs behave as genetic clones [6].

A particularly revealing case of LD occurs between genetic markers subject to different selective pressures, as this suggests that LD extends across the entire genome of the species. In *T. cruzi*, LD has been demonstrated between MLEE and RAPD [46,47], as well as between MLEE, RAPD, and PCR-RFLP [48]. LD has also been observed between the polymorphism of 12 antigen loci on one side, and MLEE/RAPD on the other [49].

LD in *T. cruzi* is also detectable on microevolutionary scales (see below: Russian Doll Evolution).

5.4. Multigene Bifurcating Trees (MGBTs)

MGBTs serve as a proof-by-contradiction for PCE. If recombination was frequent, it would erase any durable phylogenetic signal over time. The persistence of MGBTs, as well as their close similarities across ecosystems (wild vs. domestic), hosts (triatomine bugs vs. mammals), and geographic regions, provides the strongest evidence for clonal evolution. This has been noted by multiple authors working on the population genetics of pathogens [50–54].

MGBTs are ubiquitous in *T. cruzi*, both at the species subdivision level (see below: discrete typing units and near-clades) and at lower evolutionary scales (see below: Russian Doll Evolution).

6. Additional Key Concepts Related to the PCE Model

6.1. Clonets

The so-called "clones" identified through a given set of genetic markers are not true clones but rather families of closely related clones. Using markers with higher resolution power will further subdivide these "clones" into smaller units. The term "clonet" refers to MLGs* defined by a specific set of genetic markers in species undergoing PCE [55].

For instance, MLEE MLGs are clonets characterized by MLEE analysis. The widespread MLEE MLGs 19 and 39 [39,40] are such clonets. If microsatellites are used—offering greater resolution—additional genetic variability within each of these clonets becomes evident (Figure 1).

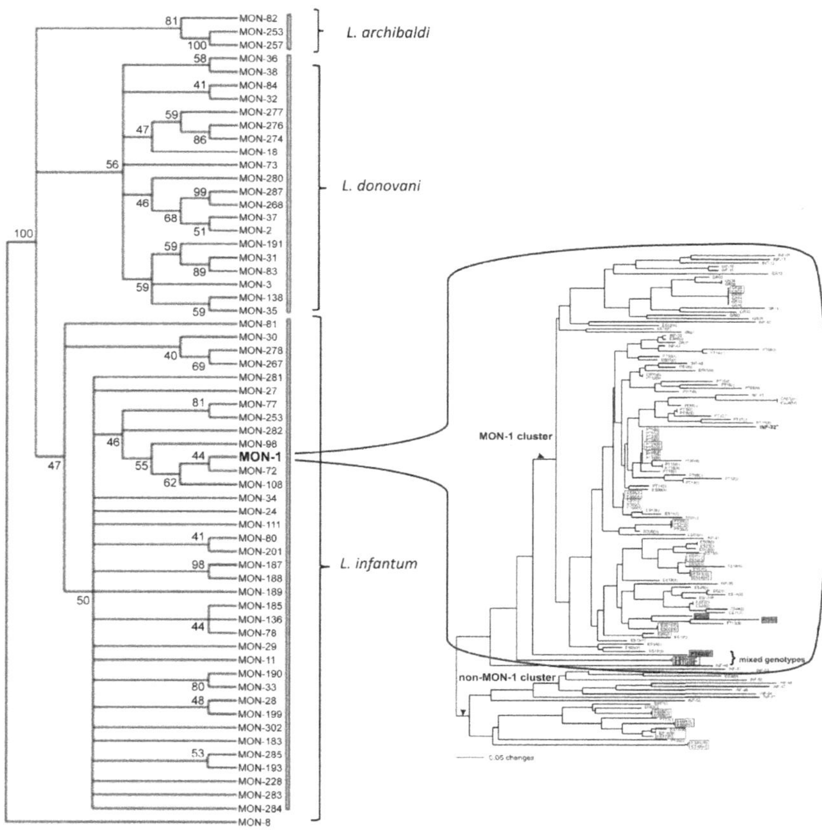

Figure 1. An illustration of the clonet concept and a typical Russian doll pattern* (RDP*) in *Leishmania infantum*. (**left**) A phylogenetic tree of the *Leishmania donovani/infantum* complex, based on MLEE analysis (adapted from Figure 1 in [56]). (**right**) The MLEE multilocus genotype "MON-1", which represents only a small fraction of the *L. infantum* species (**left**), is revealed to be highly heterogeneous and structured when analyzed using microsatellites, which offer greater discriminatory power than MLEE (adapted from Figure 2 in [57]). MON-1 is a typical MLEE clonet: rather than a single clone, it constitutes a family of closely related clones.

6.2. Discrete Typing Units (DTUs)

DTUs, a term coined to define molecular epidemiology units for pathogen species (Tibayrenc, 1998), have been widely used for *T. cruzi*. This concept is purely descriptive, without inferring evolutionary significance, though DTUs have been proposed as equivalent to clades [58].

Six primary DTUs have been identified within *T. cruzi* via both MLEE and RAPD: DTUs 1, 2a, 2b, 2c, 2e, and 2f [47]. These were later renamed Tc I to VI [59]. A seventh DTU, specifically isolated from bats, was identified [60].

6.3. Near-Clades

PCE does not imply an absolute absence of recombination. Actually, it is highly probable that strictly clonal pathogens do not exist. However, in many pathogen species, distinct phylogenetic lineages are clearly present [16]. Occasional genetic exchange makes the term "clade" inappropriate to describe them, as a clade is a genetic lineage that is strictly isolated from others. Now, genetic isolation is not absolute among these near-clades. Moreover, the near-clades Tc V and VI appear to have a hybrid origin [61,62]. The term "clade" is thus unsuitable in this case, since a clade is expected to have a unique ancestor. However, the concept of near-clade remains valid for these hybrid lineages. The term "near-clade" [34] was specifically coined to designate pathogen phylogenetic lineages whose distinctness is somewhat blurred by occasional genetic exchange. Nevertheless, near-clades can be clearly identified through phylogenetic analysis and remain stable over space and time. Near-clades thus correspond to DTUs whose evolutionary origins have been clarified. The near-clade concept offers a valuable means to simplify the otherwise confusing terminology found in the literature on pathogen genetic diversity (Table 1).

Table 1. The numerous imprecise terms used to describe genetic clusters within pathogen species. The near-clade concept provides a useful alternative to this confusing terminology. Clonets, DTUs, and near-clades serve as convenient and relevant units of analysis (RUAs) [9] for both basic and applied research.

Viruses	Bacteria	Parasitic Protozoa	Fungi
clades	clades	assemblages	AFLP groups
clusters	clonal complexes	clades	clades
genogroups	clonal lineages	clonal haplogroups	clonal groups
genotypes	clonal subgroups	clonal haplotypes	clonal lineages
groups	clusters	clonal lineages	clusters
lineages	eBurst groups	clonal types	clonal groups
major genotypes	family strains	clones	genetically distinct subgroups
major lineages	genetic groups	clonotypes	genotypes
phylogenetic groups	genoclouds	clusters	genotypic groups
phylogroups	genogroups	core subgroups	groups
subclades	genome groups	discrete typing units (DTUs)	lineages
subgenotypes	genomospecies	divergent entities	major clades
subgenotype clusters	genospecies	genetic clades	minor clades
subgroups	groups	genetic groups	molecular genotypes
sublineages	haplotypes	genetic types	molecular types

Table 1. *Cont.*

Viruses	Bacteria	Parasitic Protozoa	Fungi
substrains	lineages	genotypes	phylogenetic species
subtypes	major branches	groups	subclades
subvariants	major clusters	haplogroups	subclusters
types	main/major lineages	haplotypes	subgenotypes
variants	major phylogenetic groups	lesser subgroups	subgroups
	phylogenetic clades	lineages	subpopulations
	phylogenetic groups	main haplogroups	varieties
	phylogenetic groupings	major clades	
	phylogroups	major clonal lineages	
	populations	major groups	
	primary clusters	major monophyletic groups	
	principal genetic groups	phylogenetic lineages	
	pulsotypes	populations	
	secondary clusters	subassemblages	
	semi discrete lineages	subclades	
	subclades	subclusters	
	subclones	subgroups	
	subclusters	sublineages	
	subgroups	subpopulations	
	sublineages	subgenotypes	
	subpopulations	subgroups	
	subspecies	subspecies	
	subspecies groups	subtypes	
	subtypes	subtype groups	
		types	

6.4. Russian Doll Evolution; Russian Doll Patterns (RDPs)

In many pathogen species subdivided into major genetic units (DTUs; near-clades), the population structure and evolutionary traits observed within each near-clade resemble a miniature version of the overall species structure (Russian Doll Evolution [63]; Figure 1). This includes (i) widespread, stable MLGs; (ii) LD; and (iii) smaller near-clades identified by MGBTs. This indicates that PCE also operates within each near-clade at a microevolutionary level. This challenges the hypothesis [27,32] that posits that genetic isolation exists between the main genetic subdivisions (DTUs; near-clades) but is significantly weaker within them.

RDPs are widespread in *T. cruzi* and are observed across all six major DTUs/near-clades of this parasite. In Ecuadorian TcI stocks, Costales et al. [64] identified highly significant LD, with this DTU/near-clade clustering into two smaller near-clades statistically associated with sylvatic and domestic cycles. RDP was also observed within TcI [65] using microsatellite analysis. The distribution of smaller near-clades suggests partial geographic separation, though this factor alone does not fully explain the results (Figure 1). The authors found statistically significant LD within all populations, concluding that they predominantly evolved clonally. Llewellyn et al. [66] also detected strong LD within the formerly designated DTU TcIIc (now TcIII). RDP is clearly demonstrated in a recent study [67] that analyzed a concatenated phylogenetic tree* involving multiple DNA sequences (Figure 2).

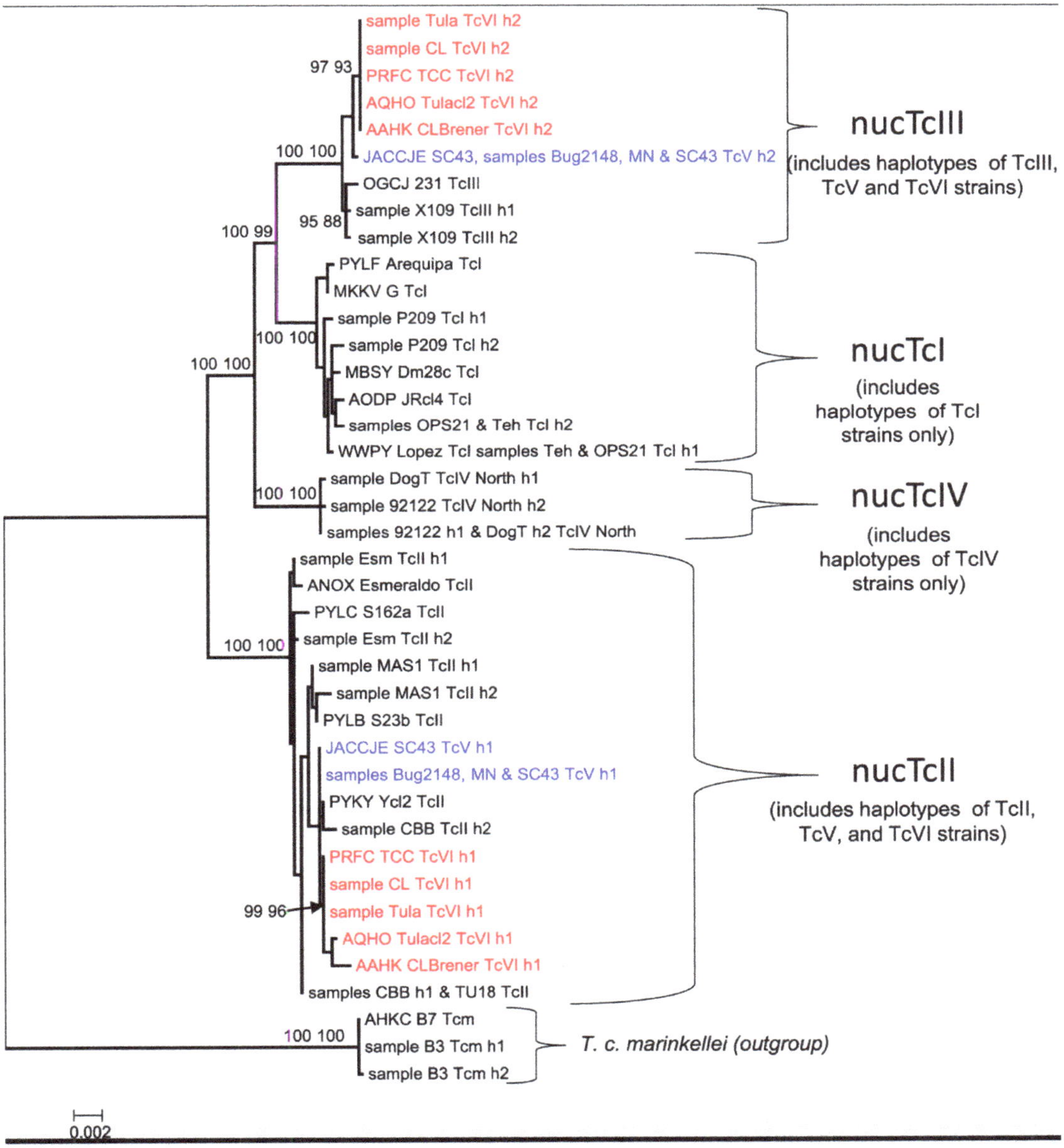

Figure 2. A concatenated phylogenetic tree of several nuclear genes, clearly showing smaller near-clades (RDPs) within each of the main near-clades of *Trypanosoma cruzi* [67].

Majeau et al. [68] constructed a concatenated phylogenetic tree based on the sequences of 30 nuclear genes. Their analysis confirmed a clear clustering into six main DTUs and revealed a strong RDP within TcI (Figure 3).

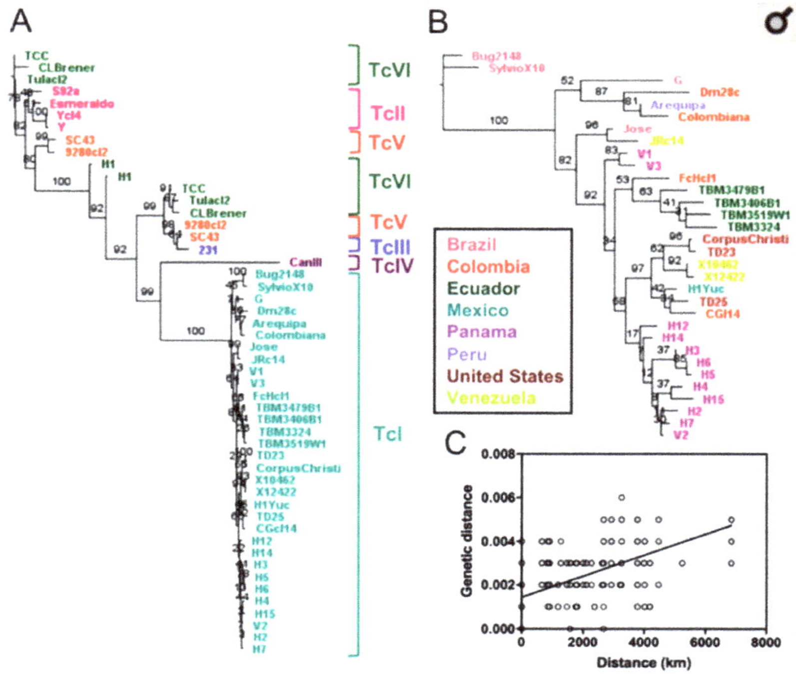

Figure 3. Two phylogenetic trees from concatenated sequences: (**A**) the whole *T. cruzi* taxon is subdivided into 6 near-clades or DTUs. (**B**) Additional subclustering within the TcI near-clade (Russian doll pattern). (**C**) Mantel test showing the correlation between geographical and genetic distances. (after Figure 5 of [68]).

Genetic clustering within TcI is further illustrated in Supplementary Figure S1 of [69].

Regarding North American strains, Flores-López et al. [70] produced a concatenated tree based on four loci, revealing a clear RDP within TcI, including group III-V-VI, TcIV, and its variant TcIV "USA" (their Supplementary Figure S5).

Messenger et al. [41], in a comprehensive survey of Bolivian *Trypanosoma cruzi* strains involving 199 clones from 68 isolates and utilizing 26 microsatellite loci plus 10 mitochondrial maxicircle genes, revealed clear substructuring (smaller, related clades; RDP) within the TcI lineage. Not only were smaller near-clades detected, but highly significant LD levels were also found, even within the smallest near-clades, along with repeated MLGs at frequencies incompatible with panmixia. Substructuring was partly explained by geographic separation. However, this Wahlund effect alone does not fully account for the observed patterns. Furthermore, 65 out of 68 isolates were multiclonal, providing ample opportunities for mating (see below: "starving sex hypothesis"). While microsatellite (nuclear) and mitochondrial gene phylogenies were generally congruent, some discrepancies were noted, possibly due to genetic exchange. Alternative explanations include differing selective pressures and evolutionary rates, which are not mutually exclusive.

6.5. Is Apparent Clonality in T. cruzi Attributable to a Lack of Mating Opportunity? The "Starving Sex" Hypothesis

The "starving sex" hypothesis [71] can be considered a specific case of the Wahlund effect, positing that mating is restricted because different genotypes are not present simultaneously in the same host. More broadly, it encompasses all cases where physical barriers (spatial or temporal) are the only obstacles to genetic exchange.

This hypothesis was proposed by Cibulskis [72] to explain apparent clonality in *T. cruzi*. However, multiple instances of multiclonal infections have been recorded in this parasite, both in hosts [73] and triatomine vectors [39]. Flores-López et al. [70] also reported mixed infections of TcI and TcIV in the USA.

Additionally, various multiclonal infections have been observed at a microevolutionary level among microclones characterized by microsatellites [41]. The "starving sex" hypothesis is therefore not a parsimonious explanation for *T. cruzi* clonality, as this parasite has ample opportunities for mating in its natural cycles.

7. Conclusions and Future Direction

T. cruzi exhibits all the hallmarks of the PCE model, which in turn provides excellent predictive power regarding its population genetic structure. Without exception, no *T. cruzi* population has been found to be panmictic, meaning that random genetic exchange does not occur. Meiotic processes have been detected in TcI [74], which is a highly relevant discovery in terms of basic evolution. However, this discovery does not support the claim [75] that these populations are "almost panmictic". These populations still display strong LD and smaller near-clades (Russian doll patterns) and thus conform to the PCE model. Meiosis is insufficiently frequent to disrupt the prevailing pattern of clonal evolution.

All surveyed *T. cruzi* populations exhibit widespread MLGs, LD, and MGBTs at all evolutionary scales.

From a molecular epidemiology perspective, clonets, DTUs, and near-clades provide practical analytical units for epidemiological surveillance due to their two key properties: (i) stability across space and time and (ii) ubiquity.

For the same reasons, DTUs/near-clades serve as a valuable framework for studying the medically relevant properties of the Chagas disease agent, including the epidemiology, pathogenicity, drug susceptibility, and vaccine potential. In particular, the major near-clades (TcI-VI) represent a promising resource for experimental evolution in *T. cruzi*, warranting further exploration.

Lastly, smaller DTUs/near-clades within TcI-VI could serve as the foundation for a subspecific taxonomy of *T. cruzi*, should experts deem it appropriate. This has already been proposed for the near-clade TcI [42,76]. However, it is advisable to adopt a flexible, informal nomenclature to avoid an excessive proliferation of Latin names.

The medical relevance of the PCE concept lies in its clear utility for molecular epidemiology (strain typing) and future research into the potential distinctive properties of the smaller near-clades (pathogenicity, treatment resistance).

Future development in this field should definitively rely on the latest progress in genomic studies, encompassing whole genome sequencing and Multi-SNP typing.

T. cruzi is not an isolated case; many pathogenic species, including bacteria, protozoa, fungi, and yeasts, exhibit similar PCE patterns, suggesting a convergent evolutionary trend among parasitic organisms [77].

Funding: This research received no external funding.

Data Availability Statement: Data are contained within the article.

Acknowledgments: We thank Jenny Telleria (IRD Montpellier, France) for designing Figure 1.

Conflicts of Interest: The author declares no conflict of interest.

Glossary of Specialized Terms

Amplification fragment length polymorphism (AFLP)	Selective amplification of genomic restriction fragments (obtained by RFLP) by PCR*.
Barcoding	The DNA barcode. Some DNA fragments are highly conserved within the same species and variable between species. These are genetic markers or barcodes. cf. metabarcoding.
Clade	An evolutionary lineage defined by cladistic analysis. A clade is monophyletic (it has a single common ancestor) and genetically isolated (it evolves independently) from other clades, with no genetic exchange. See cladistic analysis.
Cladistic analysis	A method of phylogenetic analysis that relies on the polarization of characters, distinguishing between ancestral (plesiomorphic) and derived (apomorphic) traits. Only apomorphic characters shared by all members of a given clade (synapomorphic characters) are considered phylogenetically informative.
Concatenated phylogenetic tree	Method of constructing a phylogenetic tree by concatenating different gene sequences that have been aligned into a supergene matrix.
Discrete typing unit (DTU)	A group of genetic stocks that are more closely related to each other than to any other stock, remain stable over time and space, and can be identified by specific genetic, molecular, or immunological markers called tags. DTUs serve as reliable analytical units in studies considering pathogen genetic diversity and are ideal targets for molecular epidemiology surveys.
Genetic recombination	The exchange of genetic material between different individuals at two or more genetic loci, resulting in offspring with genetic combinations distinct from those of either parent.
Genomics	While genetics focuses on the study of individual genes, genomics examines the entire genome as a whole.
Homogamy	The tendency of an organism to mate with individuals that are genetically very similar or identical to itself.
Inbreeding	cf. homogamy.
Isoenzymes	Different electrophoretic variants of a given enzyme reflecting the genetic variability of the gene encoding that enzyme within the surveyed population. Variations in migration patterns among isoenzymes of the same enzyme arise from differences in their overall electric charge, which is determined by the specific electric charges of the amino acids composing the enzyme. Consequently, electrophoretic differences correspond to variations in the amino acid sequence and, ultimately, differences in the upstream gene sequence.
Linkage disequilibrium	The nonrandom association of genotypes at different genetic loci. In a population with no linkage disequilibrium (i.e., random genetic recombination), knowing an individual's genotype at one locus does not provide information about their genotype at another locus. For example, in a randomly mating human population, knowing an individual's ABO blood group does not indicate their Rhesus blood group. Conversely, if such an association is observed, it suggests linkage disequilibrium, indicating limited genetic recombination, which can be quantified using various statistical methods.

Locus	The physical location of a gene on a chromosome. By extension, in genetic terminology, the term locus may also refer to the gene itself (plural: loci).
Metabarcoding	Use of Next Generation Sequencing. Metabarcoding is an extension of barcoding through the use of NGS technology, which allows blind and one-time identification of all species present in a sample. cf. barcoding.
Microsatellite	A short DNA sequence, typically 1–4 base pairs long, repeated in tandem along the DNA molecule. In many species, including pathogens, the number of repeats varies significantly between individuals and across populations and pathogen strains. The number of repeats at a specific locus defines microsatellite alleles. Microsatellites are present at hundreds of locations in most species. These markers evolve rapidly and offer high-resolution analysis.
Mitosis	Equational cell division that gives rise to genetically identical cells.
Multilocus	Referring to a trait involving multiple loci.
Multilocus enzyme electrophoresis (MLEE)	Isoenzyme* analysis involving a wide range of enzyme systems, each corresponding to one or more genetic loci. Multilocus enzyme electrophoresis (MLEE) has been extensively applied in population genetics across a diverse array of living organisms, including pathogens.
Multilocus genotype	A combined genotype determined by multiple genetic loci.
Multilocus sequence typing (MLST)	A method for characterizing pathogens based on sequencing several housekeeping genes.
Panmixia: panmictic, panmictic expectations	A genetic structure in which genetic exchange occurs randomly within a given population. Panmictic expectations are the confirmation of this state through various population genetics tests.
Parthenogenesis	A mode of reproduction observed in certain metazoans (e.g., insects, amphibians, fish, reptiles) that occurs without the genetic contribution of a mating partner.
Phylogenetics	A branch of genetics dedicated to reconstructing the evolutionary history and relationships of taxa or distinct evolutionary lineages.
Polymerase chain reaction (PCR)	A technique that amplifies the complementary strands of a target DNA sequence through a series of cycles until the desired amount of DNA is produced. PCR employs synthesized primers whose nucleotide sequences are complementary to the DNA flanking the target region. The DNA is heated to denature and separate the complementary strands and then cooled to allow the primers to bind to the flanking sequences. Taq DNA polymerase is added, and the reaction undergoes the necessary number of replication cycles to achieve amplification.
Population genetics	The study of genetic variation across space and time within and among populations. This field emphasizes the population or species as a whole rather than individual organisms (see also population genomics).
Population genomics	The study of genomic variation across space and time within and among populations.

Pulse field gel electrophoresis (PFGE)	The separation of large DNA fragments achieved through a specialized electrophoresis technique that uses alternately pulsed, perpendicularly oriented electrical fields. Strains sharing the same pulsed-field gel electrophoresis (PFGE) pattern are referred to as pulse types. In bacteria, the large DNA fragments are produced by the action of a low-frequency restriction enzyme (a bacterial endonuclease that cuts at a low frequency) on the bacterial chromosome. As such, PFGE is a specific form of restriction fragment length polymorphism (RFLP). In the case of parasitic protozoa (such as *Trypanosoma* and *Leishmania*) and yeasts, the large DNA fragments represent entire chromosomes, reflecting the organism's molecular karyotype.
Random primer amplified polymorphic DNA (RAPD)	In the classical polymerase chain reaction (PCR) method, the primers used are known DNA sequences, whereas the RAPD (random amplified polymorphic DNA) technique relies on primers with arbitrarily determined sequences. RAPD primers are typically 10 base pairs long, and the possible combinations are virtually unlimited. For a given genotype of an individual or strain, different primers will reveal different polymorphisms. RAPDs are an extremely powerful method for exploring the genetic variability of organisms. However, their use in routine strain identification is limited due to their lack of reproducibility.
Recombination (genetic)	cf. Genetic recombination.
Restriction fragment length polymorphism (RFLP)	DNA variability in a given organism can be detected using bacterial restriction endonucleases. These enzymes cut DNA at specific restriction sites defined by particular DNA sequences. The resulting polymorphism in DNA fragments can be visualized on gels, either directly through ethidium bromide staining or via Southern blot hybridization using specific probes.
Selfing: self-fertilization	Fertilization of an organism by itself, hence by a genotype that is identical to itself.
Sexuality	In this context, sexuality is used broadly to encompass all forms of genetic exchange between two distinct cells or individuals.
Single nucleotide polymorphism	Polymorphisms resulting from single nucleotide variations in the DNA sequence, known as single nucleotide polymorphisms (SNPs), contributing to differences among individuals, populations, and pathogen strains. SNPs are commonly used as high-resolution population markers.
Wahlund effect	In population genetics, the Wahlund effect traditionally refers to a heterozygote deficiency arising when two genetically distinct populations with different allele frequencies are mistakenly considered a single population, despite being separated by physical barriers (e.g., time or space). Here, we extend the term to refer to any apparent deviation from panmixia that results solely from physical obstacles (e.g., time or space) impeding genetic exchange.

References

1. de Sousa, A.S.; Vermeij, D.; Novaes Ramos, A., Jr.; Luquetti, A.O. Chagas Disease. *Lancet* **2024**, *403*, 203–218. [CrossRef] [PubMed]
2. Roman, F.; Iñiguez, A.M.; Yeo, M.; Jansen, A.M. Multilocus sequence typing: Genetic diversity in *Trypanosoma cruzi* I (TcI) isolates from Brazilian didelphids. *Parasit. Vectors* **2018**, *11*, 107. [CrossRef]
3. Barnabé, C.; Sempéré, G.; Manzanilla, V.; Millan, J.M.; Amblard-Rambert, A.; Waleckx, E. mbctools: A User-Friendly Metabarcoding and Cross-Platform Pipeline for Analyzing Multiple Amplicon Sequencing Data across a Large Diversity of Organisms. *Peer Community J.* **2024**, *4*, e114. [CrossRef]

4. Tavares de Oliveira, M.; Sulleiro, E.; da Silva, M.C.; Silgado, A.; de Lana, M.; da Silva, J.S.; Molina, I.; Marin-Neto, A. Intra-Discrete Typing Unit TcV Genetic Variability of *Trypanosoma cruzi* in Chronic Chagas' Disease Bolivian Immigrant Patients in Barcelona, Spain. *Front. Cardiovasc. Med.* **2021**, *8*, 665624.
5. Schwabl, P.; Sánchez, J.M.; Costales, J.A.; Sofiá Ocaña-Mayorga, S.; Segovia, M.; Carrasco, H.J.; Hernández, C.; Ramírez, J.D.; Lewis, M.D.; Grijalva, M.J.; et al. Culture-free genome-wide locus sequence typing (GLST) provides new perspectives on *Trypanosoma cruzi* dispersal and infection complexity. *PLoS Genet.* **2020**, *16*, e1009170. [CrossRef] [PubMed]
6. Tibayrenc, M.; Cariou, M.L.; Solignac, M.; Carlier, Y. Arguments génétiques contre l'existence d'une sexualité actuelle chez *Trypanosoma cruzi*; Implications taxinomiques. *C. R. Acad. Sci.* **1981**, *293*, 207–209.
7. Tibayrenc, M.; Ward, P.; Moya, A.; Ayala, F.J. Natural populations of *Trypanosoma cruzi*, the agent of Chagas'disease, have a complex multiclonal structure. *Proc. Nat. Acad. Sci. USA* **1986**, *83*, 115–119. [CrossRef]
8. Tibayrenc, M.; Kjellberg, F.; Ayala, F.J. A clonal theory of parasitic protozoa: The population structure of *Entamoeba, Giardia, Leishmania, Naegleria, Plasmodium, Trichomonas* and *Trypanosoma*, and its medical and taxonomical consequences. *Proc. Nat. Acad. Sci. USA* **1990**, *87*, 2414–2418. [CrossRef]
9. Tibayrenc, M.; Ayala, F.J. Relevant units of analysis for applied and basic research dealing with neglected transmissible diseases: The predominant clonal evolution model of pathogenic microorganisms. *PLoS Neglect. Trop. Dis.* **2017**, *11*, e0005293. [CrossRef]
10. Ochman, H.; Selander, R.K. Evidence for clonal population structure in *Escherichia coli*. *Proc. Nat. Acad. Sci. USA* **1984**, *81*, 198–201. [CrossRef]
11. Ørskov, F.; Ørskov, I. Summary of a workshop on the clone concept in the epidemiology, taxonomy, and evolution of the Enterobacteriaceae and other Bacteria. *J. Infect. Dis.* **1983**, *148*, 346–357. [CrossRef] [PubMed]
12. Rougeron, V.; De Meeûs, T.; Hide, M.; Waleckx, E.; Bermudez, H.; Arevalo, J.; Llanos-Cuentas, A.; Dujardin, J.C.; De Doncker, S.; Le Ray, D.; et al. Extreme inbreeding in *Leishmania braziliensis*. *Proc. Nat. Acad. Sci. USA* **2009**, *106*, 10224–10229. [CrossRef] [PubMed]
13. Rougeron, V.; De Meeûs, T.; Hide, M.; Le Falher, G.; Bucheton, B.; Dereure, J.; El-Safi, S.H.; Dessein, A.; Bañuls, A. Multifaceted Population Structure and Reproductive Strategy in *Leishmania donovani* Complex in One Sudanese Village. *PLoS Neglect. Trop. Dis.* **2011**, *5*, e1448. [CrossRef] [PubMed]
14. Sturm, N.R.; Campbell, D.A. Alternative lifestyles: The population structure of *Trypanosoma cruzi*. *Acta Trop.* **2010**, *115*, 35–43. [CrossRef]
15. Avise, J.C. Evolutionary perspectives on clonal reproduction in vertebrate animals. *Proc. Nat. Acad. Sci. USA* **2015**, *29*, 8867–8873. [CrossRef]
16. Tibayrenc, M. Molecular Epidemiology of Pathogenic Microorganisms and the Predominant Clonal Evolution (PCE) model. In *Genetics and Evolution of Infectious Diseases*, 3rd ed.; Tibayrenc, M., Ed.; Elsevier: Amsterdam, The Netherlands; Academic Press: Cambridge, MA, USA, 2024.
17. Calo, S.; Billmyre, B.B.; Heitman, J. Generators of Phenotypic Diversity in the Evolution of Pathogenic Microorganisms. *PLoS Pathog.* **2013**, *9*, e1003181. [CrossRef]
18. Feretzaki, F.; Heitman, J. Unisexual Reproduction Drives Evolution of Eukaryotic Microbial Pathogens. *PLoS Pathog.* **2013**, *9*, e1003674. [CrossRef]
19. Gibson, W.; Lewis, M.D.; Yeo, M.; Miles, M. Genetic Exchange in Trypanosomatids and its Relevance to Epidemiology. In *Genetics and Evolution of Infectious Diseases*, Tibayrenc, M., Ed.; Elsevier: Amsterdam, The Netherlands, 2024.
20. Gutiérrez-Corbo, C.; Domínguez-Asenjo, B.; Martínez-Valladares, M.; Pérez-Pertejo, Y.; García-Estrada, C.; Balaña-Fouce, R.; Reguera, R.M. Reproduction in Trypanosomatids. *Past Present. Biol.* **2021**, *10*, 471.
21. Miles, M.A.; Llewellyn, M.S.; Lewis, M.D.; Yeo, M.; Baleela, R.; Fitzpatrick, S.; Gaunt, M.W.; Mauricio, I.L. The molecular epidemiology and phylogeography of *Trypanosoma cruzi* and parallel research on *Leishmania*: Looking back and to the future. *J. Parasitol.* **2009**, *136*, 1509–1526. [CrossRef]
22. Rougeron, V.; De Meeûs, T.; Kako Ouraga, S.; Hide, M.; Bañuls, A. "Everything You Always Wanted to Know about Sex (but Were Afraid to Ask)" in Leishmania after Two Decades of Laboratory and Field Analyses. *PLoS Pathog.* **2010**, *6*, e1001004. [CrossRef]
23. Tomasini, N. Introgression of the Kinetoplast DNA: An Unusual Evolutionary Journey in *Trypanosoma cruzi*. *Curr. Genom.* **2018**, *19*, 133–139. [CrossRef] [PubMed]
24. Gaunt, M.W.; Yeo, M.; Frame, I.A.; Tothard, J.R.; Carrasco, H.J.; Taylor, M.C.; Mena, S.S.; Veazey, P.; Miles, G.A.; Acosta, N.; et al. Mechanism of genetic exchange in American trypanosomes. *Nature* **2003**, *421*, 936–939. [CrossRef] [PubMed]
25. Lewis, M.D.; Llewellyn, M.S.; Yeo, M.; Messenger, L.A.; Miles, M.A. Experimental and natural recombination in *Trypanosoma cruzi*. In *American Trypanosomiasis: Chagas Disease. One Hundred Years of Research*, 2nd ed.; Telleria, J., Tibayrenc, M., Eds.; Elsevier: Amsterdam, The Netherlands; Academic Press: Cambridge, MA, USA, 2017.
26. Messenger, L.A.; Miles, M. Evidence and importance of genetic exchange among field populations of *Trypanosoma cruzi*. *Acta Trop.* **2015**, *151*, 150–155. [CrossRef]

27. Ramírez, J.D.; Llewellyn, M.S. Reproductive clonality in protozoan pathogens—Truth or artefact? *Mol. Ecol.* **2014**, *23*, 4195–4202. [CrossRef] [PubMed]
28. Weedall, G.D.; Hall, N. Sexual reproduction and genetic exchange in parasitic protists. *J. Parasitol.* **2015**, *142*, S120–S127. [CrossRef]
29. Zingales, B.; Macedo, A.M. Fifteen Years after the Definition of *Trypanosoma cruzi* DTUs: What have we learned? *Life* **2023**, *13*, 2339. [CrossRef]
30. Gelanew, T.; Kuhls, K.; Hurissa, Z.; Weldegebreal, T.; Hailu, W.; Kassahun, A.; Abebe, T.; Hailu, A.; Schönian, G. Inference of Population Structure of *Leishmania donovani* Strains Isolated from Different Ethiopian Visceral Leishmaniasis Endemic Areas. *PLoS Negl. Trop. Dis.* **2010**, *4*, e889. [CrossRef]
31. Mauricio, I.L.; Yeo, M.; Baghaei, M.; Doto, D.; Pratlong, F.; Zemanova, E.; Dedet, J.P.; Lukes, J.; Miles, M.A. Towards multilocus sequence typing of the *Leishmania donovani* complex: Resolving genotypes and haplotypes for five polymorphic metabolic enzymes (ASAT, GPI, NH1, NH2, PGD). *Int. J. Parasitol.* **2006**, *36*, 757–769. [CrossRef]
32. Ramírez, J.D.; Llewellyn, M.S. Response to Tibayrenc and Ayala: Reproductive clonality in protozoan pathogens—Truth or artefact? *Mol. Ecol.* **2015**, *24*, 5782–5784. [CrossRef]
33. Shaik, J.S.; Dobson, D.E.; Sacks, D.L.; Beverley, S.M. *Leishmania* Sexual Reproductive Strategies as Resolved through Computational Methods Designed for Aneuploid Genomes. *Genes* **2021**, *12*, 167. [CrossRef]
34. Tibayrenc, M.; Ayala, F.J. Reproductive clonality of pathogens: A perspective on pathogenic viruses, bacteria, fungi, and parasitic protozoa. *Proc. Nat. Acad. Sci. USA* **2012**, *109*, E3305–E3313. [CrossRef] [PubMed]
35. Morrison, L.J.; Tweedie, A.; Black, A.; Pinchbeck, G.L.; Christley, R.M.; Schoenefeld, A.; Hertz-Fowler, C.; MacLeod, A.; Turner, C.M.R.; Tait, A. Discovery of Mating in the Major African Livestock Pathogen *Trypanos. Congolense*. *PLoS ONE* **2009**, *4*, e5564. [CrossRef]
36. Dumonteil, E.; Desale, H.; Tu, W.; Hernandez-Cuevas, N.; Shroyer, M.; Goff, K.; Marx, P.A.; Herrera, C. Intra-host *Trypanosoma cruzi* strain dynamics shape disease progression: The missing link in Chagas disease pathogenesis. *Microbiol. Spectr.* **2023**, *11*, e04236-22. [CrossRef]
37. Cura, C.I.; Mejía-Jaramillo, A.M.; Duffy, T.; Burgos, J.M.; Rodriguero, M.; Cardinal, M.V.; Kjos, S.; Gurgel-Gonçalves, R.; Blanchet, D.; De Pablos, L.M.; et al. *Trypanosoma* cruzi I genotypes in different geographical regions and transmission cycles based on a microsatellite motif of the intergenic spacer of spliced-leader genes. *Int. J. Parasitol.* **2010**, *40*, 1599–1607. [CrossRef] [PubMed]
38. Miles, M.A.; Souza, A.; Povoa, M.; Shaw, J.J.; Lainson, R.; Toyé, P.J. Isozymic heterogeneity of *Trypanosoma cruzi* in the first autochtonous patients with Chagas'disease in Amazonian Brazil. *Nature* **1978**, *272*, 819–821. [CrossRef]
39. Tibayrenc, M.; Ayala, F. Isozyme variability of *Trypanosoma cruzi*, the agent of Chagas' disease: Genetical, taxonomical and epidemiological significance. *Evolution* **1988**, *42*, 277–292. [CrossRef]
40. Barnabé, C.; Brisse, S.; Tibayrenc, M. Population structure and genetic typing of *Trypanosoma cruzi*, the agent of Chagas'disease: A multilocus enzyme electrophoresis approach. *Parasitology* **2000**, *150*, 513–526. [CrossRef] [PubMed]
41. Messenger, L.A.; Garcia, L.; Vanhove, M.; Huaranca, C.; Bustamante, M.; Torrico, M.; Torrico, F.; Miles, M.A.; Llewellyn, M.S. Ecological host fitting of *Trypanosoma cruzi* TcI in Bolivia: Mosaic population structure, hybridization and a role for humans in Andean parasite dispersal. *Mol. Ecol.* **2015**, *245*, 2406–2422. [CrossRef]
42. Ramírez, J.D.; Hernández, C. *Trypanosoma cruzi* I: Towards the need of genetic subdivision? Part II. *Acta Tropica* **2018**, *184*, 53–58. [CrossRef]
43. Rivas-Garcıá, L.; Carballo-Amador, M.A.; Flores-López, C.A. Design of a AFLP-PCR and PCR-RFLP test that identify the majority of discrete typing units of *Trypanosoma cruzi*. *PLoS ONE* **2020**, *15*, e0237180. [CrossRef]
44. Cosentino, R.O.; Agüero, F. A Simple Strain Typing Assay for *Trypanosoma cruzi*: Discrimination of Major Evolutionary Lineages from a Single Amplification Product. *PLoS Neglect. Trop. Dis.* **2012**, *6*, e1777. [CrossRef] [PubMed]
45. Hamilton, P.B.; Lewis, M.D.; Cruickshank, C.; Gaunt, M.W.; Yeo, M.; Llewellyn, M.S.; Valente, S.A.; Maia da Silva, F.; Stevens, J.R.; Miles, M.A.; et al. Identification and lineage genotyping of South American trypanosomes using fluorescent fragment length barcoding. *Infect. Genet. Evol.* **2011**, *11*, 44–51. [CrossRef]
46. Tibayrenc, M.; Neubauer, K.; Barnabé, C.; Guerrini, F.; Sarkeski, D.; Ayala, F.J. Genetic characterization of six parasitic protozoa: Parity of random-primer DNA typing and multilocus isoenzyme electrophoresis. *Proc. Natl. Acad. Sci. USA* **1993**, *90*, 1335–1339. [CrossRef] [PubMed]
47. Brisse, S.; Barnabé, C.; Tibayrenc, M. Identification of six *Trypanosoma cruzi* phylogenetic lineages by random amplified polymorphic DNA and multilocus enzyme electrophoresis. *Int. J. Parasitol.* **2000**, *30*, 35–44. [CrossRef]
48. Carrasco, H.J.; Segovia, M.; Llewellyn, M.S.; Morocoima, A.; Urdaneta-Morales, S.; Martínez, C.; Martínez, C.E.; Garcia, C.; Rodríguez, M.; Espinosa, R.; et al. Geographical Distribution of *Trypanosoma cruzi* Genotypes in Venezuela. *PLoS Neglect. Trop. Dis.* **2012**, *6*, e1707. [CrossRef]
49. Rozas, M.; De Doncker, S.; Adaui, V.; Coronado, X.; Barnabé, C.; Tibayrenc, M.; Solari, A.; Dujardin, J.C. Multilocus Polymerase Chain Reaction Restriction Fragment–Length Polymorphism Genotyping of *Trypanosoma cruzi* (Chagas Disease): Taxonomic and Clinical Applications. *J. Infect. Dis.* **2007**, *195*, 1381–1388. [CrossRef] [PubMed]

50. Hanage, W.P. Not So Simple After All: Bacteria, Their Population Genetics, and Recombination. *Cold Spring Harb. Perspect. Biol.* **2016**, *8*, a018069. [CrossRef]
51. Maiden, M.C.J. Multilocus Sequence Typing of Bacteria. *Ann. Rev. Microbiol.* **2006**, *60*, 561–588. [CrossRef]
52. Pearson, T.; Giffard, P.; Beckstrom-Sternberg, S.; Auerbach, R.; Hornstra, H.; Tuanyok, A.; Erin, P.; Price, E.P.; Glass, M.B.; Leadem, B.; et al. Phylogeographic reconstruction of a bacterial species with high levels of lateral gene transfer. *BMC Biol.* **2009**, *7*, 78. [CrossRef]
53. Pérez-Losada, M.; Browne, E.B.; Madsen, A.; Wirth, T.; Viscidi, R.P.; Crandall, K.A. Population genetics of microbial pathogens estimated from multilocus sequence typing (MLST) data. *Infect. Genet. Evol.* **2006**, *6*, 97–112. [CrossRef]
54. Posada, D.; Crandall, K.A.; Holmes, E.C. Recombination in evolutionary genomics. *Ann. Rev. Genet.* **2002**, *36*, 75–97. [CrossRef] [PubMed]
55. Tibayrenc, M.; Kjellberg, F.; Ayala, F.J. The clonal theory of parasitic protozoa: A taxonomic proposal applicable to other clonal organisms. *Bioscience* **1991**, *41*, 767–774. [CrossRef]
56. Pratlong, F.; Lami, P.; Ravel, C.; Balard, Y.; Dereure, J.; Serres, G.; El Baidouri, F.; Dedet, J.P. Geographical distribution and epidemiological features of Old World *Leishmania infantum* and *Leishmania donovani* foci, based on the isoenzyme analysis of 2277 strains. *J. Parasitol.* **2013**, *40*, 423–434. [CrossRef]
57. Kuhls, K.; Chicharro, C.; Cañavate, C.; Cortes, S.; Campino, L.; Haralambous, C.; Soteriadou, K.; Pratlong, F.; Dedet, J.P.; Mauricio, I.; et al. Differentiation and Gene Flow among European Populations of *Leishmania infantum* MON-1. *PLoS Negl. Trop. Dis.* **2008**, *2*, e261. [CrossRef]
58. Tibayrenc, M. Genetic epidemiology of parasitic protozoa and other infectious agents: The need for an integrated approach. *Int. J. Parasitol.* **1998**, *28*, 85–104. [CrossRef] [PubMed]
59. Zingales, B.; Miles, M.A.; Campbell, D.; Tibayrenc, M.; Macedo, A.M.; Teixeira, M.M.; Schijman, A.; Llewellyn, M.S.; Lages-Silva, E.; Machado, C.R.; et al. The revised *Trypanosoma cruzi* subspecific nomenclature: Rationale, epidemiological relevance and research applications. *Infect. Genet. Evol.* **2012**, *12*, 240–253. [CrossRef]
60. Marcili, A.; Lima, L.; Cavazzsana, M.; Junqueira, A.C.V.; Veludo, H.H.; Maia da Silva, F.; Campaner, M.; Paiva, F.; Nunes, V.L.B.; Teixeira, M.M.G. A new genotype of *Trypanosoma cruzi* associated with bats evidenced by phylogenetic analyses using SSU rDNA, cytochrome b and histone H2B genes and genotyping. *J. Parasitol.* **2009**, *136*, 641–655. [CrossRef]
61. Brisse, S.; Henriksson, J.; Barnabé, C.; Douzery, E.J.P.; Berkvens, D.; Serrano, M.; De Carvalho, M.R.C.; Buck, G.A.; Dujardin, J.C.; Tibayrenc, M. Evidence for genetic exchange and hybridization in *Trypanosoma cruzi* based on nucleotide sequences and molecular karyotype. *Infect. Genet. Evol.* **2003**, *2*, 173–183. [CrossRef]
62. Sturm, N.R.; Vargas, N.S.; Westenberger, S.J.; Zingales, B.; Campbell, D.A. Evidence for multiple hybrid groups in *Trypanosoma cruzi*. *Int. J. Parasitol.* **2003**, *33*, 269–279. [CrossRef]
63. Tibayrenc, M.; Ayala, F.J. How clonal are *Trypanosoma* and *Leishmania*? *Trends Parasitol.* **2013**, *29*, 264–269. [CrossRef]
64. Costales, J.A.; Jara-Palacio, M.A.; Llewellyn, M.S.; Messenger, L.A.; Ocaña-Mayorgaa, S.; Villacísa, A.G.; Tibayrenc, M.; Grijalva, M.J. *Trypanosoma cruzi* population dynamics in the Central Ecuadorian Coast. *Acta Tropica* **2015**, *151*, 88–93. [CrossRef] [PubMed]
65. Llewellyn, M.S.; Miles, M.A.; Carrasco, H.J.; Lewis, M.D.; Yeo, M.; Vargas, J.; Torrico, F.; Diosque, P.; Valente, V.; Valente, S.A.; et al. Genome-Scale Multilocus Microsatellite Typing of *Trypanosoma* cruzi Discrete Typing Unit I Reveals Phylogeographic Structure and Specific Genotypes Linked to Human Infection. *PLoS Pathog.* **2009**, *5*, e1000410. [CrossRef] [PubMed]
66. Llewellyn, M.S.; Lewis, M.D.; Acosta, N.; Yeo, M.; Carrasco, H.J.; Segovia, M.; Vargas, J.; Torrico, F.; Miles, M.A.; Gaunt, M. *Trypanosoma cruzi* IIc: Phylogenetic and phylogeographic insights from sequence and microsatellite analysis and potential impact on emergent Chagas disease. *PLoS Negl. Trop. Dis.* **2009**, *3*, e510. [CrossRef] [PubMed]
67. Barnabé, C.; Brenière, S.F.; Santillán-Guayasamín, S.; Douzery, E.J.P.; Waleckx, E. Revisiting gene typing and phylogeny of *Trypanosoma cruzi* reference strains: Comparison of the relevance of mitochondrial DNA, single-copy nuclear DNA, and the intergenic region of mini-exon gene. *Infect. Genet. Evol.* **2023**, *115*, 105504. [CrossRef]
68. Majeau, A.; Murphy, L.; Herrera, C.; Dumonteil, E. Assessing *Trypanosoma cruzi* Parasite Diversity through Comparative Genomics: Implications for Disease Epidemiology and Diagnostics. *Pathogens* **2021**, *10*, 212. [CrossRef]
69. Cruz-Saavedra, L.; Schwabl, P.; Vallejo, G.A.; Carranza, J.C.; Muñoz, M.; Patino, L.H.; Paniz-Mondolfi, A.; Llewellyn, M.S.; Ramírez, J.D. Genome plasticity driven by aneuploidy and loss of heterozygosity in *Trypanosoma cruzi*. *Microb. Genom.* **2022**, *8*, 843. [CrossRef]
70. Flores-López, C.; Mitchell, E.A.; Reisenman, C.E.; Sarkar, S.; Williamson, P.C.; Machado, C.A. Phylogenetic diversity of two common *Trypanosoma cruzi* lineages in the Southwestern United States. *Infect. Genet. Evol.* **2022**, *99*, 105251. [CrossRef] [PubMed]
71. Tibayrenc, M.; Ayala, F.J. New insights into clonality and panmixia in *Plasmodium* and *Toxoplasma*. *Adv. Parasitol.* **2014**, *84*, 253–268.
72. Cibulskis, R.E. The microdistribution of *Trypanosoma cruzi*. *Trans. R. Soc. Trop. Med. Hyg.* **1985**, *79*, 138–139. [CrossRef]

73. Brénière, S.F.; Tibayrenc, M.; Antezana, G.; Pabon, J.; Carrasco, R.; Selaès, H.; Desjeux, P. Résultats préliminaires en faveur d'une relation faible ou inexistante entre les formes cliniques de la maladie de Chagas et les souches isoenzymatiques de *Trypanosoma* cruzi. *C. R. Acad. Sci.* **1985**, *300*, 555–558.
74. Schwabl, P.; Imamura, H.; Van den Broeck, F.; Costales, J.A.; Maiguashca-Sánchez, J.; Miles, M.A.; Andersson, B.; Grijalva, M.J.; Llewellyn, M.S. Meiotic sex in Chagas disease parasite *Trypanosoma cruzi*. *Nat. Com.* **2019**, *10*, 3972. [CrossRef] [PubMed]
75. Rusman, F.; Floridia-Yapur, N.; Ragone, P.G.; Diosque, P.; Tomasini, N. Evidence of hybridization, mitochondrial introgression and biparental inheritance of the kDNA minicircles in *Trypanosoma* cruzi I. *PLoS Negl. Trop. Dis.* **2020**, *14*, e0007770. [CrossRef] [PubMed]
76. Guhl, F.; Ramírez, J.D. *Trypanosoma cruzi* I diversity: Towards the need of genetic subdivision? *Acta Tropica* **2011**, *119*, 1–4. [CrossRef]
77. Tibayrenc, M.; Ayala, F.J. Models in parasite and pathogen evolution: Genomic analysis reveals predominant clonality and progressive evolution at all evolutionary scales in parasitic protozoa, yeasts and bacteria. *Adv. Parasitol.* **2021**, *111*, 75–117. [PubMed]

pathogens

MDPI

Review

An Overview of *Trypanosoma cruzi* Biology Through the Lens of Proteomics: A Review

Jenny Telleria [1] and Jaime A. Costales [2,*]

1 Institut de Recherche Pour le Développement (IRD), UMR Intertryp IRD-CIRAD, 34398 Montpellier, Cedex 5, France; jenny.telleria@ird.fr

2 Centro de Investigación Para la Salud en América Latina, Pontificia Universidad Católica del Ecuador, Quito 170525, Ecuador

* Correspondence: jacostalesc@puce.edu.ec

Abstract: The protozoan parasite *Trypanosoma cruzi*, causative agent of Chagas disease, affects millions of people in endemic Latin American countries and beyond. In Latin America, Chagas disease is an important cause of death and disability, for which vaccines are lacking and improved treatment options are required. Additionally, the factors governing the development of a variety of clinical manifestations during Chagas disease, ranging from complete lack of symptoms to severe irreversible chronic organ damage (mainly cardiac or digestive), remain largely unknown. Much remains to be learned regarding the biology of *T. cruzi* in order to enhance our understanding of these lines of inquiry. In this context, proteomic methods have been leveraged to investigate different parasite strains, life-cycle forms, subcellular compartments, macromolecular complexes, signaling events and secreted molecules. The factors driving morphological transformation during the life cycle, the composition and functions of the parasite's organelles and secreted molecules as well as the determinants of pathogenicity have been explored via proteomic methods, yielding insights into the fundamental processes behind the parasite biology and informing drug design and vaccine development. Importantly, the correlation between the wide genetic and phenotypic variability displayed by *T. cruzi* has been examined through proteomic methods as well. Here, we review the literature on *T. cruzi* proteomics and discuss it in the light of its limitations and in the context of the parasite's genetic diversity.

Keywords: parasite; proteome; metabolism; biology; life cycle; gene expression; parasite–host interaction

1. Chagas Disease

Chagas disease is a chronic infection caused by the flagellated protozoan *Trypanosoma cruzi*, which is endemic to 21 countries in the Latin American continent. In recent years, due to human migration, Chagas disease has spread beyond the areas of endemic transmission as infected individuals have established residence in countries such as Canada and the United States, as well as European and Asian countries [1]. Currently, ~6 million people are infected and ~70 million live under risk of infection with *T. cruzi*. It is estimated that each year, the disease causes 12 thousand deaths, while 30 thousand new Chagas disease cases occur in the same period (https://www.paho.org/en/topics/chagas-disease, accessed on 19 March 2025).

The World Health Organization/Pan-American Health Organization (WHO/PAHO) has launched an initiative to implement comprehensive and sustainable actions which aimed to eliminate 30 communicable diseases and related conditions by 2030. Zoonotic diseases, such as Chagas disease, are included in the initiative (https://www.paho.org/en/elimination-initiative, accessed on 19 March 2025).

Chagas disease transmission takes place mainly through the activity hematophagous insects, commonly known as "kissing bugs", which harbor *T. cruzi* in the digestive tract and transmit it by the stercorarian route. However, the disease can also be transmitted, though less frequently, by the congenital or oral routes, as well as via transfusion of *T. cruzi*-infected blood and blood products, organ transplantation and sporadically, during laboratory accidents [2].

Chagas disease progresses through an initial symptomatic or asymptomatic acute phase (lasting ~2 months) and a chronic phase. The clinical manifestations of the latter phase are divided into indeterminate (asymptomatic) form and a symptomatic chronic form (which includes cardiomyopathy and/or digestive pathology).

Currently, no effective vaccines for Chagas disease have been developed and only two drugs are approved for parasitological treatment of *T. cruzi* infection: benznidazole (BNZ) and nifurtimox (NFX). Both drugs have considerable side effects, which can range from hypersensitivity reactions and/or anorexia, at the beginning of treatment, to peripheral neuropathies or even psychosis at the end of treatment. A mutagenic and teratogenic effect has been demonstrated in experimental animals, which has not been evidenced in humans. Both drugs are contraindicated in pregnant women [3,4]. NFX acts on the parasite's redox mechanism, while BNZ acts as a toxic agent [5]. Treatment is 100% successful in infants under age one; however, in older children and adults, effectiveness is only ~60% during the acute phase, while the effectiveness in the chronic phase is still a matter of debate [6,7]. Treatment failure may be related to the genetic diversity of the parasite and/or hosts [8].

2. Proteomics and *T. cruzi*

In 2005, as the product of the collaborative effort carried out by an international consortium, an initial version of the complete *T. cruzi* genome was published [9]. A year before, genomic databases for the parasite had been established and since, they have served as key resources in *T. cruzi* research [10]. Currently, this database is a bioinformatics resource center that houses 88 genomes of kinetoplastids.

As is the case for other trypanosomatids, transcription in *T. cruzi* is polycistronic, and precursor RNAs are subsequently processed into mRNAs via trans-splicing [11]. Therefore, gene expression regulation does not occur at the transcriptional level; instead, it takes place via controlling the stability and translation of specific mRNAs [12]. Indeed, lack of correlation between global transcriptomic and proteomic data has been reported [13], highlighting the importance of evaluating expression at the protein level in this parasite. Post-translational modifications also play an important role in protein function modulation by *T. cruzi*. [14]. As a result, proteomics has been recognized as a key tool to gain understanding of the global protein expression patterns in this parasite. A multitude of studies have leveraged proteomic methods to contribute to the understanding of various aspects of its biology, including developmental stage-specific protein expression, global protein content of subcellular fractions (membranes and organelles), post-translational modifications, virulence, infectivity and pathogenicity factors, metabolic pathway usage, drug resistance, as well as the parasite secretome and gene expression patterns associated with the parasite genotypes. These efforts have improved the understanding of the biology of *T. cruzi* and may help unveil mechanisms underlaying different Chagas clinical manifestations, allow the identification of novel antiparasitic drug targets and inform vaccine design.

3. Stage-Specific Global Protein Expression Patterns During the Parasite Life Cycle

T. cruzi possesses a digenetic life cycle, involving hematophagous triatomine bugs, commonly known as "kissing bugs", as well as a mammalian host (including humans). The parasite transitions through morphologically different stages during its life cycle, and alternates between replicative and non-replicative forms. In the triatomine bug's midgut, *T. cruzi* multiplies as epimastigotes, which later transform into metacyclic trypomastigotes (MTs), infectious to mammals, in the vector hindgut. Triatomines feed on the mammalian host's blood and defecate near the bite site. MTs are present in the triatomine bug's feces and contaminate the bite wound or mucous membranes, establishing intracellular infection in mammalian cells. The parasite multiplies in the host cell as amastigotes. After several replication rounds, amastigotes occupy most of the host cell cytoplasm and transform into trypomastigotes, which cause host cell rupture and escape into the blood stream. These free trypomastigotes infect additional mammalian host cells, disseminating the infection to other tissues. Finally, if an uninfected triatomine bug takes in blood form trypomastigotes (BFTs) up with a blood meal, the parasites will infect the bug's digestive tract, and transform into epimastigotes in the midgut, reinitiating the cycle (Figure 1).

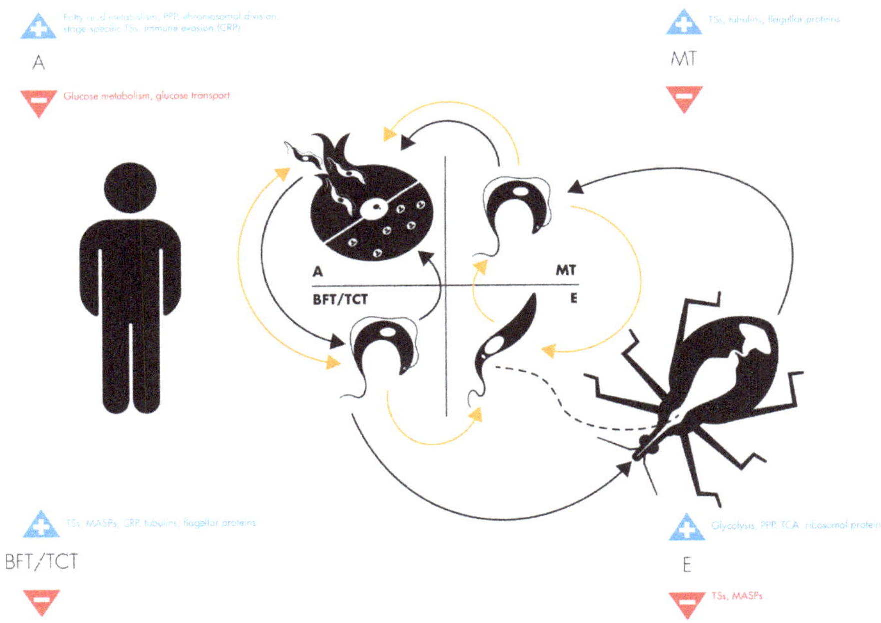

Figure 1. Life cycle of *T. cruzi*. The progression of *T. cruzi* through the different life cycle stages occurring in the mammalian host (**left**) and invertebrate host (**right**) is indicated by the black arrows. Orange arrows indicate the transitions that can be achieved in the laboratory using axenic or mammalian tissue culture (see text for details). Blue and red arrows indicate greater or lesser protein expression, respectively, in each life-cycle stage as reported in the reviewed proteomic studies. A = amastigote, BFT/TCT = blood form trypomastigote/tissue-culture-derived trypomastigote, E = epimastigote, MT = metacyclic trypomastigote.

In addition to displaying distinct morphologies, different life-cycle forms of *T. cruzi* possess specific characteristics. For instance, only epimastigotes and amastigotes are capable of replication; the former in the triatomine vector digestive tract and the later intracellularly, in the mammalian host. Epimastigotes are unable to infect mammalian cells;

however, they are infectious to triatomines. Meanwhile, trypomastigotes, both MT and BFT, are infectious to the mammalian host as well as to the triatomine vectors. Researchers have leveraged proteomic approaches to unveil protein expression patterns specific to each *T. cruzi* developmental stage, as well as during transition between stages.

A determining factor during these studies is the ability to generate *T. cruzi* populations comprised exclusively (or predominantly) of a given parasite stage, to be used as input for proteomic analysis. Different methods are available to obtain different life forms in the laboratory (Figure 1). Epimastigotes can be cultured axenically and, in stationary phase cultures, part of the parasite population spontaneously transforms into MTs. Such transformation can also be induced via nutritional stress, i.e., incubating axenic epimastigotes in low nutrient medium. MTs can then be affinity-purified from the cultures via anion-exchange chromatography on a diethylaminoethyl (DEAE)-cellulose column [15]. In contrast, the life-cycle stages present in the mammalian host must be obtained via intracellular infection. In experimentally infected animals, trypomastigotes (BFTs) egress from infected cells and reach the bloodstream, from where they can be collected. Cultured mammalian cells can be infected with MTs or BFTs, which establish intracellular infection, transform into amastigotes and multiply in the host cell cytoplasm. After ~five days of in vitro intracellular multiplication, amastigotes transform into trypomastigotes, rupture the host cell and are released into the culture medium. These are known as tissue-culture-derived trypomastigotes (TCTs) and, in general, they are thought to be equivalent to BFTs. Amastigotes can be obtained axenically, by exposing TCTs to low pH (pH 5–6), e.g., [16,17]. If MTs, TCTs, BFTs or amastigotes are cultured in axenic medium, they transform into epimastigotes and multiply. Therefore, the methods employed to obtain the parasite population to be studied via proteomics, as well as its purity, can significantly impact the experimental results.

3.1. Epimastigotes

Pioneering studies [14] delineated the global protein expression profile for epimastigotes. Proteins associated to metabolism, cytoskeleton assembly, protein targeting, cell cycle growth and protein synthesis were identified; tubulins, heat-shock proteins (HSPs) and prostaglandin F2-α were reported to be the most abundant proteins in this life stage. Subsequent studies reported the presence of multiple isoforms of tubulins and heat-shock proteins in this life stage [18]. Furthermore, Cordero and co-workers identified energy metabolism enzymes (glucose uptake and oxidation, i.e., glycolysis, pentose phosphate shunt or tricarboxylic acid cycle) in epimastigotes, in accordance with the high energy requirements of this proliferative stage of the parasite [19]. Additionally, epimastigotes express abundant enzymes to catalyze conversion of histidine (an abundant carbon source in the triatomine gut) into glutamate [20]. During the exponential growth phase, epimastigote cultures upregulate ribosomal proteins, consistent with the protein synthesis required for multiplication [21]. However, they have reduced expression of proteins associated with cell invasion and immune evasion (such as trans-sialidades, mucin-associated surface proteins (MASPs)) [20]. Altogether, these results indicate that epimastigotes are intensely active in terms of energetic metabolism and protein synthesis, and adapted to the use of glucose as a nutrient, along with amino acids available in the vector digestive tract. However, as expected, they display less virulence factors than mammalian-infective stages.

3.2. Metacyclic Trypomastigotes

Metacyclogenesis, the process by which epimastigotes transform into MTs, is of special interest, since it marks the evolution from non-infective trypomastigotes into the stage infective to mammals (including humans). Metacyclogenesis initiates in the hindgut of insect vectors, where the parasite is subjected to nutritional stress and confrontation with proteases. It has been suggested by proteomic analysis that glucose depletion leads to the elongation of the flagella, thereby increasing their area and facilitating nutrient uptake [22]. Fine characterization of protein expression patterns at different time-points of nutritional stress-induced metacyclogenesis has been attempted, identifying modulation of tubulins, some not meeting the expected molecular weights and isoelectric points in the 2D gels, suggesting varied post-translational modifications [23,24]. When reaching the stationary phase, axenically cultured epimastigotes upregulate proteins involved in autophagy (i.e., autophagy protein Apg6) [21]. Additionally, the universal minicircle sequence binding protein (UMSBP), which binds the origin of the kinetoplast replication, is acetylated in the stationary phase, despite its expression level being the same in the two growth phases, suggesting it could be a regulatory event. These changes have been suggested to mark the transition from exponential growth into metacyclogenesis [21].

Enzymes associated to the REDOX balance were found to display differential regulation in MT [24]. Additionally, cytoskeletal proteins, like actins and dyneins, were found to be upregulated in these infectious life stages, along with paraflagellar rod proteins [23]. Similar to epimastigotes, MTs express abundant enzymes to catalyze conversion of histidine into glutamate [20]. However, as opposed to epimastigotes, MTs must be equipped to face the mammalian immune system, and they express larger quantities of proteins involved in antioxidant defenses [20]. A large proportion of proteins identified in MTs are surface proteins (35 members of the trans-sialidase superfamily, one mucin, and one GP63 protease) involved in host–parasite interplay (evasion from the host immune response, invasion of mammalian cells and glycosylation of parasite surface molecules) [19].

It has also been shown that MTs have the ability to transform back into epimastigotes [25]. MTs incubated for 72 h liver infusion tryptose blood medium display a global protein expression pattern resembling that of epimastigotes, including downregulation of virulence factors, i.e., proteins related to parasite infectivity, adhesion and invasion, immune evasion and response to oxidative stress. These intermediary forms, termed "recently differentiated epimastigotes" (rdEpis) are infectious to cultured cells and experimental animals, and they are resistant to complement (a property regular epimastigotes do not have and which makes them not infectious in the human host).

3.3. Amastigotes

Amastigotes are intracellular replicative forms, which multiply inside mammalian host cells. Early proteomic studies showed that amastigotes display low expression of glucose transporters, and abundant expression of enzymes associated to fatty-acid oxidation into acetylCoA (AcoA), as well as subsequent AcoA oxidation [20]. As expected in a replicative form, proteins involved in chromosomal division are expressed by amastigotes along with enzymes involved in the oxidative phase of the pentose phosphate pathway, which generates precursors for nucleotide biosynthesis [26].

Proteomic studies have shown that amastigotes display stage-specific surface proteins, including trans-sialidases and other glycoproteins, likely stage-specific antigens [26]. A putative thymet oligopeptidase (TOP), which could potentially participate in degradation of proteasome-derived peptides by interfering with MHCI antigen presentation, was found to be overrepresented in amastigotes. Additionally, immune escape-related proteins, like complement regulatory protein and calreticulin, which promote parasite resistance to complement-mediated lysis, were found to be highly upregulated in amastigotes [16,25].

3.4. Blood Form Trypomastigotes/Tissue-Culture-Derived Trypomastigotes (BFT/TCTs)

In comparison to epimastigotes, TCTs were found to produce larger amounts of trans-sialidases and mucin-associated surface proteins (MASPs), surface proteins associated with cell invasion and immune evasion [20,27]. Additionally, trypomastigotes upregulate the production of complement regulatory proteins to prevent lysis by host complement [25].

The cell surface subproteome of TCTs has been reported to be largely similar to that of axenic amastigotes [16]. Stage-specific surface proteins identified corresponded mainly to trans-sialidases and glycoproteins, likely stage-specific antigens, though authors pointed out some of these proteins may not necessarily be associated with the plasma membrane and may also correspond to microvesicles/exosomes. However, greater expression of proteins related to flagellar motility, including paraflagellar rod components and tubulins, has been reported in TCTs compared to axenic (pH-induced) amastigotes and epimastigotes [26,28]. Additionally, metabolic enzymes and heat-shock proteins, among others, are differentially expressed between TCTs and amastigotes [28].

4. Subcellular Proteomes

Proteomics in combination with cell fractionation approaches have been employed to explore the protein components of different subcellular compartments in *T. cruzi* cells. It has been pointed out that an important limitation of such strategy is the inability to generate completely "pure" organelles or subcellular fractions and that, in conjunction with the high sensitivity of the proteomic methods, this creates the possibility of incorrect assignment of proteins' location [29]. Individual proteins identified as part of proteomes from purified organelles or subcellular fractions must, therefore, be considered only as candidates for a particular subcellular location, which must be experimentally confirmed. In some cases, researchers have confirmed predicted subcellular locations for selected subsets of the proteins of interest [30,31] via expression of recombinant proteins and microscopy.

Another limitation is that proteomic studies regarding subcellular localization of *T. cruzi* proteins have largely concentrated in epimastigotes. One important exception is Won and co-workers, who performed proteomic analysis of both epimastigotes and amastigotes and confirmed the location of a subset of proteins to the flagellum of both these parasite stages as well [31].

Early studies [29] analyzed the proteins in the organellar cell fraction of *T. cruzi* epimastigotes. A subset of these proteins was selected and their gene copy number, sequence variation, transmembrane domains and targeting signals were characterized. Importantly, the authors cloned the genes and expressed the corresponding proteins with a c-myc epitope tag in *T. cruzi* epimastigotes. Immunofluorescence microscopy revealed the localization of these proteins in different cellular compartments, such as endoplasmic reticulum, acidocalcisome, mitochondrion and putative cytoplasmic transport or delivery vesicles. Additional studies, which have concentrated on specific organelles or macromolecular complexes, are reviewed below.

4.1. Ribosomes

Ayub and co-workers purified *T. cruzi* ribosomes and analyzed them via mass spectrometry to characterize their proteome. *T. cruzi* ribosomal proteins were shown to have 50% sequence identity with yeast ribosomal proteins, although they are longer because of the presence of N- and C- terminal extensions unique to trypanosomatids. Importantly, *T. cruzi* 60S subunit proteins L22 and L44 were detected for the first time in this study [32].

4.2. Reservosomes

Epimastigotes obtain proteins and lipids from the culture medium by endocytosis and store them in bodies known as reservosomes, which contain secretory-system-derived proteolytic enzymes. Therefore, reservosomes are sites for both nutrient storage and degradation [33]. To characterize their molecular signature, Sant'Anna and co-workers purified epimastigote reservosomes by differential centrifugation, and analyzed their membrane-associated components via mass spectroscopy. The reservosome membrane fraction was found to be rich in proteins associated to hydrolase activity, carrier and ion-transport activities, nucleotide-binding, transferase activity, ion-binding and oxidoreductase activity. A large number of reservosomal proteins were related to metabolism. Proteins included enzymes and proton pumps, small GTPases of the Rab family, ABC transporters, lysosomal hydrolases and P-type H+-ATPase (TcHA3). Cruzipain, the major *T. cruzi* protease, was also detected. In this study, however, experimental confirmation of intracellular location was not pursued.

4.3. Contractile Vacuole

Contractile vacuoles are specialized organelles, which play a key role in allowing protists to maintain cell volume [34]. During its life cycle, *T. cruzi* develops in different environments, in some of which it encounters osmotic stress (vector digestive tract, mammalian kidney), and its contractile vacuole is critical to maintain cell-volume stability [35]. Ulrich and co-workers purified epimastigote contractile vacuole fractions using differential gradients and centrifugation. A large number (39) of members of the dispersed gene family (DFG-1) proteins, calpain-like cysteine peptidases, amastins (transmembrane glycoproteins characteristic of amastigotes), transport-related and intracellular proteins like small G-proteins (Rabs), soluble N-ethylmaleimide-sensitive factor (NSF), adaptor protein (SNAP) receptors (SNAREs), in addition to transporters and channels, among other proteins, were identified in these fractions [30]. A subset of these proteins was expressed with GFP tags and verified via fluorescence microscopy, including Rab11, Rab32, AP180, ATPase subunit B, VAMP1 and phosphate transporter, which predominantly localized to the vacuole bladder [30].

4.4. Plasma Membrane Fractions

Queiroz and co-workers studied plasma-membrane-associated proteins by two complementary methods: trypsinization (shave) of intact life cells and biotinylation of cell surface followed by affinity isolation with streptavidin columns. A control sample where no treatment was performed was included to account for secreted proteins. The resulting proteins were analyzed by LC-MS/MS. Authors concluded that both methods allowed characterization of plasma membrane proteins, in a complementary manner, and suggested they must be used in conjunction [36].

4.5. Nucleus

Proteomic approaches have been employed to explore the nuclear subproteome of *T. cruzi* by isolation of the nuclear fraction employing a sucrose gradient, centrifugation coupled with proteomic analysis [37]. LC-MS/MS identified 864 proteins; approximately one-third of them were previously uncharacterized and one-third had not previously reported via proteomic analysis. As expected, proteins identified were related to chromatin organization and DNA-binding functions, including centrins, histones, nucleolar proteins, nucleosome assembly proteins, ribosome proteins, transcription factors and structural nuclear proteins. Additionally, proteins involved in RNA processing were identified, including poly-A binding protein, RNA helicases, elongation factors and eukaryotic translation initiation factors. Additionally, 53 ribosomal proteins were identified, 25 of which were components of the 40S ribosomal subunit and the other 28 as components of the 60S subunit. Ribosome biosynthesis occurs primarily in the nucleoli and nucleoplasm and, therefore, detection of ribosomal proteins in the nuclear fraction is not surprising.

4.6. Glycosomes

Glycosomes are specialized *T. cruzi* organelles, related to peroxisomes [38]. As opposed to other eukaryotes, in which glycolysis occurs in the cytoplasm, the initial seven glycolytic reactions in *T. cruzi* occur in glycosomes, which contain enzymes associated with glycolysis and other central carbon metabolism pathways [38]. Several other metabolic reactions key for β-oxidation of fatty acids, purine salvage, pentose phosphate pathway, gluconeogenesis and biosynthesis of ether lipids, isoprenoids, sterols and pyrimidines take place in glycosomes [39]. The same authors found other enzymes for glycolysis, oxidative and non-oxidative branches of the pentose phosphate pathway (PPP) sugar nucleotide synthesis, ether lipid biosynthesis, lipid metabolism, purine and pyrimidine synthesis, sterol synthesis, response to oxidative stress, gluconeogenesis, membrane transporters. These data allowed the authors to confirm the presence of enzymes involved in these various metabolic pathways in the parasite's glycosomes. However, no experimental confirmation of protein localization was pursued in this study [39].

4.7. Spliceosomes

As opposed to most eukaryotes, trypanosomes transcribe protein-coding genes into polycistronic RNA, which subsequently is processed into individual RNAs via trans-splicing. During this process, 5′splice-leader (39-bp in length) RNAis added to each individual RNA in the polycistronic unit [40]. In order to characterize the proteins forming part of *T. cruzi* spliceosomes, Barbosa and co-workers expressed *T. cruzi* TAP-tagged spliceosome-associated proteins TcCwc21 and TcSmD1, and used them to affinity-purify other protein components of the spliceosomal macromolecular complex and subsequently identify them via mass spectrometry. A total of 115 proteins were identified, which have conserved orthologs of *T. brucei* and/or human spliceosomes, especially the U5-snRNP related, Sm/LSm proteins and helicases. A core set of protein components common to human and *T. brucei* spliceosomes was identified, namely SNRNP200, SNRNP116, PRPF8, HSPA8, SNRPD1, SNRPB and SNRPD2. Authors highlight the fact that these similarities between human and trypanosomatid spliceosomes indicate that molecular complexes carrying out cis- and trans-splicing are mostly similar.

4.8. Flagellum

The *T. cruzi* flagellum is composed of nine pairs of peripheral microtubules and two central microtubules [41]. Motor functions are attributed to this organelle; however, the distal end is a recognized domain of signaling of the environment [31]. In an elegant study, the same authors used the biotin ligase TurboID to attain proximity-dependent biotinylation of flagellar proteins in *T. cruzi* epimastigotes and amastigotes. These authors employed CRISPR/Cas-9 facilitated association of the TurboID ligase to the small myristoylated protein 1-1 (TcSMP1-1), which they had showed to localize to the amastigote flagellum in a previous study [42]. Using this approach, 218 candidate flagellar proteins in epimastigotes and 99 in amastigotes were identified, forty of which were common to the two developmental stages. The candidate protein list was shown to be enriched in orthologs to other trypanosomatid species for which a flagellar localization had been annotated. The authors then successfully validated the location of three of the candidate proteins into the parasite flagellum, namely calpain 1.3, cellular apoptosis susceptibility protein 3 (CARP3) and a hypothetical protein, for which they expressed epitope-tagged versions that could be visualized in fixed parasites by fluorescence microscopy.

5. *T. cruzi* Intra-Specific Diversity and Proteome

T. cruzi displays high genetic diversity with clearly structured genetic subpopulations. The parasite is believed to reproduce in primarily clonal fashion, with occasional episodes of hybridization/genetic exchange [43–46], although the possibility of more frequent genetic exchange has also been proposed. Based on genomic and mitochondrial DNA, six major genetic lineages have been identified, which are designated as "discrete typing units" (DTUs) TcI-TcVI [47,48]. Subsequently, an additional genetic lineage, Tcbat, was also described [49].

Most reports about *T. cruzi* proteomics in the literature analyze exclusively a single parasite strain or isolate. In the instances where more than one strain is analyzed, differences between strains become evident [50,51]. Among the small number of reports studying more than one parasite stock, Tavares de Oliveira, in 2018, reported differences across the proteomic patterns yielded by two parasite isolates from chagasic patients, one corresponding to DTU TcI and the other to TcII [52]. Similarly, distinct proteomic patterns were displayed by different isolates derived from triatomine bugs [53].

In an effort to examine the entire breath of *T. cruzi* diversity, Telleria and co-workers studied axenically cultured epimastigotes belonging to DTU TcI-TcVI, and found that the phylogenetic structure resulting from proteomic (2DE) analysis is highly correlated with that obtained by genetic marker analysis. Proteins like α-tubulin, β-tubulin, HSP85-like protein, elongation factor 2 and iron superoxide dismutase displayed DTU-specific expression patterns [54]. The authors observed various spots corresponding to tubulins and other proteins, which might result from post-translational modifications.

Interestingly, axenic cultures of *T. cruzi* epimastigotes containing a mix of parasites from different DTUs display global proteomic profiles different from those of the individual strains, indicating that expression of a subset of proteins can be potentiated or inhibited when strains are mixed [55]. The significance of mixed *T. cruzi* infections in vectors or human hosts is not understood. However, these experimental results suggest a possible communication equivalent to a molecular dialogue between genotypes. It is suggested that expression levels of certain proteins can be potentiated or inhibited when strains coexist [55].

Taken together, these results highlight the importance of strain and DTU choice during proteomic studies about *T. cruzi*. Care must be exercised when generalizing findings, because strain choice can significantly impact experimental results. Additionally, when analyzed together, parasites from different strains or DTUs can impact each other's global protein expression. The implications of the latter phenomenon in the infection of vectors and mammalian hosts remain unknown.

6. Parasite Pathogenicity Factors

As already mentioned, *T. cruzi* is a highly variable pathogen, and parasite strains differ greatly in terms of their infectivity and pathogenicity. Meanwhile, the molecular factors that determine pathogenicity and the clinical manifestations of Chagas disease are not totally clear. Therefore, a common strategy employed by researchers has been to characterize strains with different infectivity or pathogenicity via proteomics, in hopes to identify proteomic signatures or individual proteins expressed exclusively or primarily by pathogenic strains.

Some studies have compared strains with different pathogenicity within the same DTU [51]. Dias, in 2011, explored the expression pattern of two TcI strains, one isolated from acute infection and one from cardiac form chronic patients. They identified tryparedoxin peroxidase, lipoamide dehydrogenase, tyrosine amino transferase and HSP70 as overexpressed proteins in the acute Chagas isolate. Dias and co-workers and Zago and co-workers performed proteomic profiling of three TcI isolates (2DE-MALDI-TOF), two displaying higher parasitemia and larger chronic damages than the other. Proteins associated with cell invasion (e.g., cysteine peptidases, surface glycoprotein gp82 and cruzipain), cytoskeletal dynamics and motility (isoforms of actin and tubulin, cofilin-1, KMP11 and paraflagellar rod proteins) were enriched in epimastigotes and trypomastigote stages of the pathogenic strain [51,56]. Furthermore, proteins involved in redox homeostasis like tryparedoxin peroxidases, peroxiredoxins, thioredoxin-dependent peroxide reductase and iron superoxide dismutases were overexpressed in the trypomastigotes of the most pathogenic strains. The expression of several proteins related to redox homeostasis was confirmed experimentally. The authors suggested that the overexpression of these proteins constitutes a virulence factor in trypomastigotes of TcI strains.

Other studies have compared strains belonging to different DTUs [57]. These authors compared the VFRA strain (TcI), which produces chronic infection in mice, with the Y strain (TcII), which causes lethal acute infections in the mouse model, along with higher parasitemia and higher replication rates in mouse macrophages and in the heart tissue. Overall, despite belonging to different DTUs, the proteomes of both strains were found to be similar. The authors suggested that the differences in pathogenicity are correlated with higher production of antioxidant defenses in the VFRA strain and higher expression of proteins associated with parasite replication in the Y strain.

Finally, in what probably constitutes the best study of this kind, San Francisco and co-workers compared two clones derived from the same parental strain (H510 C8C3, TcI) with different pathogenicity. For 30 years, one of the clones (virulent) was passaged weekly in mice, while the other (low virulence) was passaged as axenic epimastigotes. This resulted in the selection of two polar phenotypes: the virulent clone produced higher parasitemia and parasite load in tissues (heart, liver, quadriceps and lung) in mice, and significantly higher infectivity over cultured cells. The proteomic analysis showed biological pathways enriched in the virulent strain were associated with bioenergetics and biosynthetic pathways (including the TCA cycle and citrate metabolism), along with virulence factors such as redoxins, cruzipain, trans-sialidases, complement resistance protein and Tc-85. Therefore, virulence would depend on increased capacity to generate energy output, upregulation of virulence factors and greater capacity to face oxidative stress [58].

7. *T. cruzi* Secretome

T. cruzi has been shown to secrete proteins in large (>100–200 nm) plasma-membrane-derived vesicles, as well as smaller vesicles (<100 nm), derived from microvesicular bodies in the flagellar pocket region. Additionally, free proteins without association with vesicles are released by the parasite [59].

Secretomes have been characterized for various developmental forms of *T. cruzi*, including epimastigotes [60], axenic amastigotes [61], trypomastigotes [62] and MTs [60]. Proteins released into the culture medium, either free or associated to vesicles, have been analyzed via mass spectrometry and qualitative differences have been identified between secretomes across parasite stages. For example, proteomic analysis of extracellular vesicles (EVs) indicates an expansion of trans-sialidases in trypomastigotes as compared to epimastigotes [62]. Mandacaru, in 2019, found an increased number of mucins and mucin-associated proteins during amastigogenesis, while the number of released trans-sialidases and surface gp63 proteases was reduced [60]. Egress of intracellular *T. cruzi* forms (amastigotes and trypomastigotes) upon host cell rupture at the end of the intracellular cycle is accompanied by the release of hundreds of parasite-derived proteins, most of which belong to multigenic families, such as trans-sialidase and trans-sialidase-like proteins, gp63 surface proteins, MASPs and DGF1 [50,61]. Together, these results suggest that protein secretion by *T. cruzi* takes place throughout the parasite's life cycle, tailored to the different environments the parasite must confront.

Additionally, differences in the number and composition of secretory vesicles have been reported to exist between parasite strains belonging to the TcI and TcII lineages [63]. These authors reported TCT from a TcI strain released a higher number of vesicles containing trans-sialidases and cruzipain (implicated in pathogenesis and immunopathology) than the TcII strain. The authors tested the hypothesis that the secreted proteins could modulate parasite–host interactions, by pre-treatment of tissue-cultured cells with vesicles prior to in vitro infection with the parasite. They determined that exposure to the vesicles modulated infection and parasite multiplication in vitro. Although, in our opinion, care should be exercised when extrapolating findings from a single strain to an entire DTU, these findings highlight that secretomes are not uniform among parasites from different strains and secreted proteins can impact cell infection and parasite multiplication. In the context of a highly variable parasite, such as *T. cruzi*, this adds a layer of complexity to host–parasite interactions during infection.

Characterization of the *T. cruzi* secretome has been largely performed using in vitro systems; however, characterization of parasite-derived extracellular vesicles has been attempted in vivo as well. Cortes-Serra, in 2020, studied EVs present in the serum of a chronic *T. cruzi*-infected patient, whose infection reactivated due to immunosuppression before heart transplant. They identified *T. cruzi* pyruvate phosphate dikinase (PPDK), a glycosomal enzyme, among the vesicular contents [64], although the role this protein may play during infection remains unknown. Studies involving larger patient cohorts are warranted.

Additionally, qualitative and quantitative variations of secreted proteins between strains might be related to infectiveness and evasion of the host mammalian immune response. The *T. cruzi* secretome represents a diverse assortment of proteins involved in parasite metabolism, signaling and nucleic acid binding, which might have a direct role in parasite survival and virulence [60].

8. Post-Translational Modifications

Post-translational modification is thought to play a key role in protein function regulation in *T. cruzi*. Phosphorylation, glycosylation, acetylation, nitrosylation, sumoylation in *T. cruzi* have been explored via proteomic approaches.

8.1. Chromatin Modifications

Chromatin is packed in nucleosomes, composed of two units of each core histone (H2A, H2B, H3 and H4) around which 146 bp of DNA are coiled around. Meanwhile, H1 links nucleosomes together [65]. Histone post-translational modifications are crucial because they impact chromatin structure, and, consequently, gene expression, modulating key cellular processes. Much has been learned through proteomic analysis about histone post-translational modification of histones. De Jesus and co-workers extracted chromatin from epimastigotes and MTs and used high-resolution mass spectrometry to identify post-translational modifications, detecting 44 novel modifications, including acetylations (18), monomethylations (7), dimethylations (8), trimethylation (7) and phosphorylations (4). These authors found that epimastigotes display more histone modifications than trypomastigotes, and reported several novel types of modifications [66].

Although the role of histone nitrosylation in *T. cruzi* remains to be determined, histone nitrosylation in H2B and H3 on Cys64 and Cys126 has also been described [67,68]. These authors employed chromatin immunoprecipitation (ChIP) with an anti-nitrotyrosine antibody to purify nitrated proteins, which were subsequently analyzed by mass spectrometry.

Histones H1, H2B, H2A and H3 were identified to be some of the most common nitrated DNA-binding proteins. In addition, they reported an increase in histones H1, H2B, H2A and H4 nitration levels when parasites were incubated with extracellular matrix, with which trypomastigotes interact in the contexts of host cell invasion. These results indicate that interaction with the extracellular matrix (ECM) influences chromatin structure. Additionally, [69] detected previously known acetylation sites, K10 and K14, in H4, as well as K7, K15 and K17 in H2B.v, reported a new H3 acetylation site at position K79.t. and [70] identified methylated sites in different histones: H4R90me2, H4R54me2, H3.VK95me3, H2BK97me, H1K10me2, H2BR61me, H2BR61me2 and H2AK102me2.

Histone deacetylases (HDACs) remove acetyl groups from histones, modulating chromatin compaction and, consequently, gene expression. De Oliveira de Santos et al. (2019) studied the effects of trichostatin A (TSA), an HDAC inhibitor, over histone acetylation in *T. cruzi* by quantitation of acetylated histones via LC-MS/MS [71]. Quantitative proteomic analyses indicated an increase in histone acetylation 72 h post-TSA treatment. The major effects on *T. cruzi* were not found to take place via histones, but rather via microtubule cytoskeleton dynamics and impaired kDNA segregation, resulting in polynucleated parasites with altered morphology.

8.2. Nitrosylation

As mentioned above, early in the process of cell invasion, *T. cruzi* trypomastigotes are known to interact with the ECM; a complex structure of more than 300 proteins and glycoproteins [72]. Using resin-assisted enrichment of thiols combined with mass spectrometry [68] mapped S-nitrosylated proteins from *T. cruzi*, and decrease in S-nitrosylation upon contact with the ECM was detected. S-nitrosylated proteins were found to be enriched in the ribosome. Additionally, proteins related to transport, carbohydrate (glycolysis/TCA cycle), lipid metabolism, gene transcription (histones), proteases and cytoskeleton remodeling, indicating that profound adjustments of the parasite physiology in response to ECM occur before host cell infection.

8.3. Phosphorylation

As in other organisms, reversible phosphorylation is believed to play a key role during signaling in *T. cruzi*. Two percent of this parasite's genome encodes kinases [73]. Proteomic approaches have been harnessed to gain insights into the role of phosphorylation in global cell signaling networks in *T. cruzi*. The authors of [74] attempted to generate a comprehensive phosphorylation map for *T. cruzi* epimastigotes via liquid LC-tandem MS, identifying 237 phospho-peptides in 199 individual proteins, along with mapping of 220 phosphorylation sites (including on serine, threonine and tyrosine). The existence of tyrosine-phosphorylated proteins was confirmed via Western blot. Phosphorylated proteins were involved in motility, transportation, metabolism (glycolytic enzymes), pathogenesis (trans-sialidases and dispersed gene family protein 1), nucleic acid dynamics and signaling. Importantly, phosphorylation of histone 2B was reported by these authors. Additionally, proteomic methods have been used to identify phosphorylation in serine 23 of histone 2B in *T. cruzi* [66]. Additional studies have applied super-SILAC (super-stable isotope labeling by amino acids in cell culture) and LC-MS/MS to characterize global protein phosphorylation patterns during nutritional stress-induced metacyclogenesis in *T. cruzi* [75], identifying 4205 protein groups and 3643 phosphopeptides, and locating 4846 phosphorylation sites. Phospho-sites modulated during metacyclogenesis were located in proteins associated with fatty acid synthesis, regulation of protein expression, structural components, stress response and intracellular signaling, suggesting that phosphorylation/dephosphorylation events trigger metayclogenesis.

8.4. Glycosylation

Mass-spectrometry-based glycoprofiling analysis of epimastigotes and trypomastigotes has been performed [76], identifying 1306 N-glycosylation sites in 690 *T. cruzi* proteins. This has allowed for detailed characterization of the N-linked and O-linked glycoproteome, and unveiled stage-specific glycoproteomic signatures, in terms of epimastigote- vs. trypomastigote-exclusive N-linked and O-linked glycopeptides. Greater glycosylation was identified in gp85/trans-sialidase, MASP, mucins and GP63 family proteins in trypomastigotes. Additionally, [77] employed proteomics to show that O-GlcNAcylation, a monosaccharide post-translational modification of serine and threonine residues, takes place in *T. cruzi*. Mass spectrometry allowed for the identification of 1271 putative O-GlcNAcylated proteins and six modification sequences. O-GlcNAcylation was found in clathrins, RNA helicases, DNA polymerases, trans-sialidases and uncharacterized proteins. This post-translational modification is believed to allow for protein function regulation in a similar fashion to phosphorylation.

8.5. Methylation

De Almeida, in 2023, applied mass-spectrometry-based proteomics (LC-MS/MS) to describe non-histone methylation in epimastigotes. This post-translational modification takes place in arginines and lysines. A total of 1253 methyl sites in 824 methylated proteins were identified in proteins involved in translation, RNA and DNA binding, amino acid and carbohydrate metabolism [70].

8.6. Non-Histone Acetylation

Moretti, in 2018, identified 89 ε-lysine-acetylated sites in 235 proteins in epimastigotes. The lysine-acetylated protein set is enriched in enzymes involved in oxidation/reduction balance, crucial for parasite survival. Furthermore, the eukaryotic translation initiation factor 5A (eIF5A) was determined to have an acetylation at lysine 50 (K50). It was proposed by the authors that different acetylation profiles could correlate with parasite environmental fitness. Therefore, the data they presented can be used to determine experimentally the role of each identified modification. Finally, they proposed that development of specific acetyltransferase and deacetylase inhibitors can potentially lead to the discovery of new treatment options for Chagas disease [69].

8.7. SUMOylation

The small ubiquitin-like modifier pathway (SUMO) is a post-translational modification, which modulates protein activity, stability and/or localization. [78] used bioinformatics to identify the existence of orthologous genes for the SUMOylation pathway in *T. cruzi* and confirmed this system is active in the four major life-cycle stages of the parasite via Western blot analysis of total parasite homogenates using anti-TcSUMO polyclonal antibodies. Additionally, these authors generated epimastigote lines expressing a double- tagged *T. cruzi* SUMO, and SUMOylated proteins were isolated by affinity chromatography. By two-dimensional liquid chromatography–mass spectrometry, a total of 236 proteins with diverse biological functions were identified as potential *T. cruzi* SUMO targets. Metacaspase-3 was experimentally validated as a SUMOylation substrate.

8.8. Pathway Crosstalk

Finally, it is important to consider that these post-translational modifications coexist in the parasite cells, and that they can potentially overlap and influence each other. For example, nitrosylation and phosphorylation were determined to occur in both kinases and phosphatases, suggesting crosstalk between them [68]. In addition, 171 *T. cruzi* proteins found to be methylated were previously reported to contain phosphorylation sites, including flagellar proteins and RNA-binding proteins, indicating that there may be an interplay between these different modifications in non-histone proteins [70].

9. Interactomes

Another strategy used by researchers to probe *T. cruzi* biology has been to employ mass spectrometry methods to identify the proteins specifically interacting with a given molecule, molecules, or macromolecular complex of interest. This has been of special utility when characterizing the REDOX system and post-translational mechanisms in the parasite, characterizing chromatin-associated proteins, as well as the composition of the macromolecular complexes associated with parasite mRNA translation.

9.1. Enzymes Related to Oxidative Stress Response

Throughout the life cycle, *T. cruzi* is exposed to oxidative stress imposed by reactive oxygen species (ROS), such as superoxide anions, hydrogen peroxide (H_2O_2) and hydroxyl radicals, derived from its own metabolism, host cells and external agents [79]. An imbalance between the production of ROS and the antioxidant defense system could be lethal to the parasite [79]. The parasite's antioxidant mechanism is based on the combined action of several molecules, such as tryparredoxin peroxidases (TcTXN1 y II), which tackle oxygen free radicals and hydrogen peroxide through peroxidase activity, and peroxynitrite reduction [80]. These molecules, in conjunction with trypanothione reductase, have functions similar to glutathione in mammalian cells, i.e., intervening in redox reactions dependent on NADPH [80]. Trypanothione is a thiol that gives up electrons for detoxification of hydroperoxides [81]. The ability of the parasite to manage ROS production and antioxidant capacity is crucial to its survival.

Interactomes of tryparedoxin 1 (TcTXN1) [82] and 2 (TcTXNII) [83], as well as the mitochondrial tryparedoxin reductase (TcMPx) [84], have been reported. Piñeyro and coworkers designed a 6-His-tagged TcTXN1 active site mutant lacking the resolving cysteine, which causes the enzyme to form heterodisulfide complexes with potential substrates, and expressed it in *T. cruzi* epimastigotes. The TcTXN1 complexes present in parasite lysates could then be affinity-purified thanks to the His-tag and subsequently analyzed by 2DE/MS. The strategy was validated, showing the interaction of the recombinant protein and cytosolic peroxiredoxin. This approach allowed the identification of 15 putative TcTXN1 interacting proteins, whose functions are related to oxidative metabolism and protein metabolism (synthesis, ubiquitination, degradation). A similar approach was employed by [83] to explore the interactome of TcTXNII. They expressed a recombinant TcTXNII lacking the resolving cysteine in the active site, expressed it, incubated it with parasite lysates and identified putative TcTXNII proteins separated in SDS-PAGE gels. Sixteen putative proteins were identified, which are involved in the antioxidant system, energy metabolism, cytoskeleton and protein translation. Finally, the interactome for mitochondrial tryparedoxin peroxidase (TcMPx) was explored under physiological and oxidative stress conditions (i.e., the latter involves exposing parasites to hydrogen peroxide) [84]. Tagged recombinant TcMPx was expressed in epimastigotes, parasites were lysed and the TcMPx interactions were explored via 2DE coupled to mass spectrometry. Under physiologic conditions, the identified proteins had functions related to oxidoreductase activity and stress response. Meanwhile, under oxidative stress conditions, identified proteins were involved in amino acid metabolism, processes related to microtubules, stress responses and protein polymerization.

9.2. Chromatin

Leandro de Jesus and co-workers purified chromatin from epimastigotes and trypomastigotes and studied associated proteins via high-resolution mass spectrometry. Micrococcal nuclease digestion showed both life stages have equivalent nucleosome contents. However, epimastigote chromatin contains a much larger diversity and number of associated proteins than that of trypomastigotes [85]. Additionally, proteins involved in DNA replication, such as proliferating cell nuclear antigen (PCNA), replication protein A (RPA) and DNA topoisomerases were detected only in epimastigote chromatin. Ribosomal proteins were found to be enriched in epimastigotes, which is also expected, because RNA maturation occurs mainly at the nuclear space in trypanosomes [86]. The number and diversity of proteins associated with trypomastigotes were much lower (1/3 of their chromatin was found to be histones in contrast to 1/20 in epimastigotes). Importantly, the differential expression of seven selected proteins was confirmed experimentally, via Western blot and chromatin immunoprecipitation (ChiP). Although proteins involved in the cytoskeleton, ribosome and mitochondria and metabolic processes were also detected, their role must be verified to rule out contamination.

9.3. Polysomes

In a study carried out by Alves and co-workers, epimastigote polysomes (groups of ribosomes joined to a single mRNA molecule) were separated into glucose gradient, either during the parasite growth phase or under nutritional stress. As expected, polysomes in growing parasites were much longer, containing up to several ribosomes, while in stressed parasites, polysomes were much shorter and less abundant. The larger percentage of proteins identified were ribosomal proteins (70% for growing and 59% for stressed parasites). Non-ribosomal proteins included RNA-binding proteins, cytoskeletal components, heat-shock response proteins and proteins related to metabolism and translation. Surprisingly, metabolic proteins were the largest group of non-ribosomal proteins. The authors suggest this might be due to multifunctional enzymes playing roles during translation [87].

Wippel and co-workers studied the interactome of the RNA-binding protein-9/ribonucleoprotein (RBP9-mRNP) complex. This complex is of interest because of the lack of transcriptional regulation in *T. cruzi* genes. RNA-binding proteins associate with mRNAs and others to form ribonucleoprotein complexes (mRNPs), which are crucial in post-transcriptional regulation. As part of the characterization of *T. cruzi* RBP9, an RNA-binding protein unique to trypanosomatids, the authors determined it is located in the cytoplasm in epimastigotes, that it is expressed during metacyclogenesis but not in MTs and that it is located in cytoplasmic translational complexes. They employed a proteomic shotgun approach to characterize the proteins interacting with RBP9. In growing epimastigotes, RBP9 interacts with proteins involved in RNA metabolism, namely ZC3H39, PABP1/2, ALBA3/4 and UBP1/2 and ribosomal proteins. Authors suggest RBP9 is part of mRNP complexes in the cytoplasm involved in mRNA translation regulation [88].

9.4. Inositol Phosphate Interactome

Mantilla and co-workers used an affinity-based approach to identify novel targets of inositol heptakis-phosphate (5-IP_7), a high-energy metabolite implicated in phosphorylation that regulates multiple cellular processes. Pull-down assays with immobilized 5-IP_7 coupled with mass spectrometry in *T. cruzi* epimastigotes and amastigotes were conducted [89]. The proteins identified suggest that 5-IP_7 participates in various metabolic and signaling pathways in epimastigotes and amastigotes, including ribosome biogenesis, and trafficking processes associated to the contractile vacuole.

10. Response to Compounds, Drug Resistance

Proteomic approaches have also been leveraged to identify characteristics that make the parasite resistant to the drugs available for treatment, including benznidazole [8,90], as well as the parasite response to other compounds, including quinone derivatives and naphthoimidazoles derived from β-lapachone piplartine, synthetic diamines and organometallic compounds [13,91–94]. A common theme in these studies is the modulation of proteins involved in antioxidative responses [8,13,91–94].

11. Perspectives and Conclusions

Proteomic approaches have played an important role in understanding key aspects of the biology of *T. cruzi*, and hold the potential to shed light over additional, not yet understood, fundamental features of parasite/host interplay during Chagas disease. The parasite's particularities regarding gene expression regulation (i.e., lack of transcriptional regulation) limit the applicability of transcriptomics and make proteomics even more valuable for studying the biology of *T. cruzi*.

Core protein expression features of the different life-cycle forms of *T. cruzi* have been described via proteomic methods. These studies have also shown that our understanding of the intermediary forms that may mediate stage transitions is still limited and needs to be further explored and more clearly defined.

Furthermore, subcellular proteome analysis has provided a window into organelle and macromolecular complex (e.g., polysome) composition and function in *T. cruzi*. However, complete, contamination-free, organelle or molecular complex purification is unattainable. This fact, coupled with the high sensitivity of mass spectrometry methods, creates the possibility of erroneous cellular location assignment for proteins. Therefore, experimental validation (e.g., immunofluorescent or electronic microscopy, chip, etc.) is essential to confirm putative location assignment obtained by proteomic methods. Although individual validation of each protein of interest is not possible, validation of at least some selected proteins is desirable. Such validation is missing in many proteomic studies dealing with *T. cruzi*.

The broad genetic diversity of *T. cruzi* is reflected in varying global protein expression patterns displayed by DTUs and strains. The choice of the strain to be employed in proteomic studies must, therefore, not be taken lightly. Additionally, most published proteomic studies are limited to a single *T. cruzi* strain. Although it might be impossible or impractical to include a large enough number of strains to represent the intra- and inter-DTU variability of the parasite, it must be emphasized that extrapolation of findings to other DTUs, and even to strains within the same DTU, must be carried out with caution. Ideally, experimental validation would be performed.

When comparing global proteomic patterns in pathogenic vs. non-pathogenic strains, the use of phenotypically different clones derived from the same parental strain is ideal. In our opinion, this constitutes the best experimental strategy reported in the literature so far for the purpose of contrasting proteomic patterns with the aim of identifying pathogenicity determinants, since it eliminates confounding factors such as phylogenetic differences between strains. However, widespread implementation of this strategy is prevented by the unavailability of such strain pairs. Importantly, it would be desirable to extend the findings from pathogenic vs. non-pathogenic strain pairs to a variety of different strains, validating the proteomic patterns across the entire breath of *T. cruzi* genetic diversity.

Finally, proteomic datasets offer new opportunities for analysis of protein localization and functionality, allowing the unraveling of poorly characterized parasite/host interactions. Proteomic data repositories open up the opportunity for in silico exploration, which might shed further light on fundamental parasite biology, and inform drug design and vaccine development.

Author Contributions: Conceptualization, J.A.C. and J.T.; writing—original draft preparation, J.A.C. and J.T.; writing—review and editing, J.A.C. and J.T. All authors have read and agreed to the published version of the manuscript.

Funding: The authors are grateful for the financial support provided by Pontificia Universidad Católica del Ecuador.

Data Availability Statement: No new data were created or analyzed in this study. Data sharing is not applicable to this article.

Acknowledgments: The authors would like to thank Mario Grijalva for critically reading the manuscript and Daniel Bustillos for assistance with the illustration.

Conflicts of Interest: The authors declare they do not have conflicts of interest.

References

1. Lidani, K.C.F.; Andrade, F.A.; Bavia, L.; Damasceno, F.S.; Beltrame, M.H.; Messias-Reason, I.J.; Sandri, T.L. Chagas Disease: From Discovery to a Worldwide Health Problem. *Front. Public Health* **2019**, *7*, 166. [CrossRef] [PubMed]
2. Coura, J.R. The main sceneries of Chagas disease transmission. The vectors, blood and oral transmissions—A comprehensive review. *Mem. Inst. Oswaldo Cruz* **2014**, *110*, 277–282. [CrossRef] [PubMed]
3. Apt, B.W.; Heitmann, G.I.; Jercic, L.M.I.; Jofré, M.L.; Muñoz, C.D.V.P.; Noemí, H.I.; San Martín, V.A.M.; Sapunar, P.J.; Torres, H.M.; Zulantay, A.I. Guías clínicas de la enfermedad de Chagas 2006: Parte VI. Tratamiento antiparasitario de la enfermedad de Chagas. *Rev. Chil. Infectol.* **2008**, *25*, 384–389. [CrossRef]
4. Rojo, G.; Castillo, C.; Duaso, J.; Liempi, A.; Droguett, D.; Galanti, N.; Maya, J.D.; López-Muñoz, R.; Kemmerling, U. Toxic and therapeutic effects of Nifurtimox and Benznidazol on *Trypanosoma cruzi* ex vivo infection of human placental chorionic villi explants. *Acta Trop.* **2014**, *132*, 112–118. [CrossRef]
5. Maya, J.D.; Cassels, B.K.; Iturriaga-Vásquez, P.; Ferreira, J.; Faúndez, M.; Galanti, N.; Ferreira, A.; Morello, A. Mode of action of natural and synthetic drugs against *Trypanosoma cruzi* and their interaction with the mammalian host. *Comp. Biochem. Physiol. A Mol. Integr. Physiol.* **2007**, *146*, 601–620. [CrossRef]
6. Pinazo, M.-J.; Muñoz, J.; Posada, E.; López-Chejade, P.; Gállego, M.; Ayala, E.; del Cacho, E.; Soy, D.; Gascon, J. Tolerance of Benznidazole in Treatment of Chagas' Disease in Adults. *Antimicrob. Agents Chemother.* **2010**, *54*, 4896–4899. [CrossRef]
7. Viotti, R.; Vigliano, C.A.; Álvarez, M.G.; Lococo, B.E.; Petti, M.A.; Bertocchi, G.L.; Armenti, A.H. The Impact of Socioeconomic Conditions on Chronic Chagas Disease Progression. *Rev. Esp. Cardiol. Engl. Ed.* **2009**, *62*, 1224–1232. [CrossRef]
8. Andrade, H.M.; Murta, S.M.F.; Chapeaurouge, A.; Perales, J.; Nirdé, P.; Romanha, A.J. Proteomic Analysis of *Trypanosoma cruzi* Resistance to Benznidazole. *J. Proteome Res.* **2008**, *7*, 2357–2367. [CrossRef]
9. El-Sayed, N.M. Comparative Genomics of Trypanosomatid Parasitic Protozoa. *Science* **2005**, *309*, 404–409. [CrossRef]
10. Luchtan, M. TcruziDB: An integrated *Trypanosoma cruzi* genome resource. *Nucleic Acids Res.* **2004**, *32*, D344–D346. [CrossRef]
11. Graham, S.V. Mechanisms of stage-regulated gene expression in kinetoplastida. *Parasitol. Today* **1995**, *11*, 217–223. [CrossRef] [PubMed]
12. Callejas-Hernández, F.; Gutierrez-Nogues, Á.; Rastrojo, A.; Gironès, N.; Fresno, M. Analysis of mRNA processing at whole transcriptome level, transcriptomic profile and genome sequence refinement of *Trypanosoma cruzi*. *Sci. Rep.* **2019**, *9*, 17376. [CrossRef]
13. Mosquillo, M.F.; Smircich, P.; Ciganda, M.; Lima, A.; Gambino, D.; Garat, B.; Pérez-Díaz, L. Comparative high-throughput analysis of the *Trypanosoma cruzi* response to organometallic compounds. *Metallomics* **2020**, *12*, 813–828. [CrossRef] [PubMed]
14. Parodi-Talice, A.; Durán, R.; Arrambide, N.; Prieto, V.; Dolores Piñeyro, M.; Pritsch, O.; Cayota, A.; Cerveñansky, C.; Robello, C. Proteome analysis of the causative agent of Chagas disease: *Trypanosoma cruzi*. *Int. J. Parasitol.* **2004**, *34*, 881–886. [CrossRef]
15. Teixeira, M.M.; Yoshida, N. Stage-specific surface antigens of metacyclic trypomastigotes of *Trypanosoma cruzi* identified by monoclonal antibodies. *Mol. Biochem. Parasitol.* **1986**, *18*, 271–282. [CrossRef]
16. Queiroz, R.M.L.; Charneau, S.; Bastos, I.M.D.; Santana, J.M.; Sousa, M.V.; Roepstorff, P.; Ricart, C.A.O. Cell Surface Proteome Analysis of Human-Hosted *Trypanosoma cruzi* Life Stages. *J. Proteome Res.* **2014**, *13*, 3530–3541. [CrossRef]
17. Tomlinson, S.; Vandekerckhove, F.; Frevert, U.; Nussenzweig, V. The induction of *Trypanosoma cruzi* trypomastigote to amastigote transformation by low pH. *Parasitology* **1995**, *110*, 547–554. [CrossRef]

18. Sodré, C.L.; Chapeaurouge, A.D.; Kalume, D.E.; De Mendonça Lima, L.; Perales, J.; Fernandes, O. Proteomic map of *Trypanosoma cruzi* CL Brener: The reference strain of the genome project. *Arch. Microbiol.* **2009**, *191*, 177–184. [CrossRef]
19. Cordero, E.M.; Nakayasu, E.S.; Gentil, L.G.; Yoshida, N.; Almeida, I.C.; Da Silveira, J.F. Proteomic Analysis of Detergent-Solubilized Membrane Proteins from Insect-Developmental Forms of *Trypanosoma cruzi*. *J. Proteome Res.* **2009**, *8*, 3642–3652. [CrossRef]
20. Atwood, J.A.; Weatherly, D.B.; Minning, T.A.; Bundy, B.; Cavola, C.; Opperdoes, F.R.; Orlando, R.; Tarleton, R.L. The *Trypanosoma cruzi* Proteome. *Science* **2005**, *309*, 473–476. [CrossRef]
21. Avila, C.C.; Mule, S.N.; Rosa-Fernandes, L.; Viner, R.; Barisón, M.J.; Costa-Martins, A.G.; De Oliveira, G.S.; Teixeira, M.M.G.; Marinho, C.R.F.; Silber, A.M.; et al. Proteome-Wide Analysis of *Trypanosoma cruzi* Exponential and Stationary Growth Phases Reveals a Subcellular Compartment-Specific Regulation. *Genes* **2018**, *9*, 413. [CrossRef] [PubMed]
22. Rodrigues, J.P.F.; Sant'ana, G.H.T.; Juliano, M.A.; Yoshida, N. Inhibition of Host Cell Lysosome Spreading by *Trypanosoma cruzi* Metacyclic Stage-Specific Surface Molecule gp90 Downregulates Parasite Invasion. *Infect. Immun.* **2017**, *85*, e00302-17. [CrossRef] [PubMed]
23. De Godoy, L.M.F.; Marchini, F.K.; Pavoni, D.P.; Rampazzo, R.D.C.P.; Probst, C.M.; Goldenberg, S.; Krieger, M.A. Quantitative proteomics of *Trypanosoma cruzi* during metacyclogenesis. *Proteomics* **2012**, *12*, 2694–2703. [CrossRef] [PubMed]
24. Parodi-Talice, A.; Monteiro-Goes, V.; Arrambide, N.; Avila, A.R.; Duran, R.; Correa, A.; Dallagiovanna, B.; Cayota, A.; Krieger, M.; Goldenberg, S.; et al. Proteomic analysis of metacyclic trypomastigotes undergoing *Trypanosoma cruzi* metacyclogenesis. *J. Mass Spectrom. JMS* **2007**, *42*, 1422–1432. [CrossRef]
25. Kessler, R.L.; Contreras, V.T.; Marliére, N.P.; Aparecida Guarneri, A.; Villamizar Silva, L.H.; Mazzarotto, G.A.C.A.; Batista, M.; Soccol, V.T.; Krieger, M.A.; Probst, C.M. Recently differentiated epimastigotes from *Trypanosoma cruzi* are infective to the mammalian host. *Mol. Microbiol.* **2017**, *104*, 712–736. [CrossRef]
26. Magalhães, A.; Queiroz, R.; Bastos, I.; Santana, J.; Sousa, M.; Ricart, C.; Charneau, S. Comparative Two-Dimensional Gel Electrophoresis of *Trypanosoma cruzi* Mammalian-Stage Forms in an Alkaline pH Range. *Protein Pept. Lett.* **2015**, *22*, 1066–1075. [CrossRef]
27. Magalhães, A.D.; Charneau, S.; Paba, J.; Guércio, R.A.; Teixeira, A.R.; Santana, J.M.; Sousa, M.V.; Ricart, C.A. *Trypanosoma cruzi* alkaline 2-DE: Optimization and application to comparative proteome analysis of flagellate life stages. *Proteome Sci.* **2008**, *6*, 24. [CrossRef]
28. Paba, J.; Santana, J.M.; Teixeira, A.R.L.; Fontes, W.; Sousa, M.V.; Ricart, C.A.O. Proteomic analysis of the human pathogen *Trypanosoma cruzi*. *Proteomics* **2004**, *4*, 1052–1059. [CrossRef]
29. Ferella, M.; Nilsson, D.; Darban, H.; Rodrigues, C.; Bontempi, E.J.; Docampo, R.; Andersson, B. Proteomics in *Trypanosoma cruzi*—Localization of novel proteins to various organelles. *Proteomics* **2008**, *8*, 2735–2749. [CrossRef]
30. Ulrich, P.N.; Jimenez, V.; Park, M.; Martins, V.P.; Atwood, J.; Moles, K.; Collins, D.; Rohloff, P.; Tarleton, R.; Moreno, S.N.J.; et al. Identification of Contractile Vacuole Proteins in *Trypanosoma cruzi*. *PLoS ONE* **2011**, *6*, e18013. [CrossRef]
31. Won, M.M.; Baublis, A.; Burleigh, B.A. Proximity-Dependent Biotinylation and Identification of Flagellar Proteins in *Trypanosoma cruzi*. *mSphere* **2023**, *8*, e00088-23. [CrossRef] [PubMed]
32. Ayub, M.J.; Atwood, J.; Nuccio, A.; Tarleton, R.; Levin, M.J. Proteomic analysis of the *Trypanosoma cruzi* ribosomal proteins. *Biochem. Biophys. Res. Commun.* **2009**, *382*, 30–34. [CrossRef] [PubMed]
33. Sant'Anna, C.; Nakayasu, E.S.; Pereira, M.G.; Lourenço, D.; De Souza, W.; Almeida, I.C.; Cunha-e-Silva, N.L. Subcellular proteomics of *Trypanosoma cruzi* reservosomes. *Proteomics* **2009**, *9*, 1782–1794. [CrossRef] [PubMed]
34. Rohloff, P.; Docampo, R. A contractile vacuole complex is involved in osmoregulation in *Trypanosoma cruzi*. *Exp. Parasitol.* **2008**, *118*, 17–24. [CrossRef]
35. Kollien, A.H.; Grospietsch, T.; Kleffmann, T.; Zerbst-Boroffka, I.; Schaub, G.A. Ionic composition of the rectal contents and excreta of the reduviid bug Triatoma infestans. *J. Insect Physiol.* **2001**, *47*, 739–747. [CrossRef]
36. Queiroz, R.M.L.; Charneau, S.; Motta, F.N.; Santana, J.M.; Roepstorff, P.; Ricart, C.A.O. Comprehensive Proteomic Analysis of *Trypanosoma cruzi* Epimastigote Cell Surface Proteins by Two Complementary Methods. *J. Proteome Res.* **2013**, *12*, 3255–3263. [CrossRef]
37. Santos Júnior, A.d.C.M.d.; Kalume, D.E.; Camargo, R.; Gómez-Mendoza, D.P.; Correa, J.R.; Charneau, S.; Sousa, M.V.d.; Lima, B.D.d.; Ricart, C.A.O. Unveiling the *Trypanosoma cruzi* Nuclear Proteome. *PLoS ONE* **2015**, *10*, e0138667. [CrossRef]
38. Quiñones, W.; Acosta, H.; Gonçalves, C.S.; Motta, M.C.M.; Gualdrón-López, M.; Michels, P.A.M. Structure, Properties, and Function of Glycosomes in *Trypanosoma cruzi*. *Front. Cell. Infect. Microbiol.* **2020**, *10*, 25. [CrossRef]
39. Acosta, H.; Burchmore, R.; Naula, C.; Gualdrón-López, M.; Quintero-Troconis, E.; Cáceres, A.J.; Michels, P.A.M.; Concepción, J.L.; Quiñones, W. Proteomic analysis of glycosomes from *Trypanosoma cruzi* epimastigotes. *Mol. Biochem. Parasitol.* **2019**, *229*, 62–74. [CrossRef]
40. Barbosa, R.L.; Da Cunha, J.P.C.; Menezes, A.T.; Melo, R.D.F.P.; Elias, M.C.; Silber, A.M.; Coltri, P.P. Proteomic analysis of *Trypanosoma cruzi* spliceosome complex. *J. Proteom.* **2020**, *223*, 103822. [CrossRef]

41. Ballesteros-Rodea, G.; Santillán, M.; Martínez-Calvillo, S.; Manning-Cela, R. Flagellar Motility of *Trypanosoma cruzi* Epimastigotes. *J. Biomed. Biotechnol.* **2012**, *2012*, 520380. [CrossRef]
42. Won, M.M.; Krüger, T.; Engstler, M.; Burleigh, B.A. The Intracellular Amastigote of *Trypanosoma cruzi* Maintains an Actively Beating Flagellum. *mBio* **2023**, *14*, e03556-22. [CrossRef] [PubMed]
43. Tibayrenc, M.; Ward, P.; Moya, A.; Ayala, F.J. Natural populations of *Trypanosoma cruzi*, the agent of Chagas disease, have a complex multiclonal structure. *Proc. Natl. Acad. Sci. USA* **1986**, *83*, 115–119. [CrossRef] [PubMed]
44. Tibayrenc, M.; Ayala, F.J. Genomics and High-Resolution Typing Confirm Predominant Clonal Evolution Down to a Microevolutionary Scale in *Trypanosoma cruzi*. *Pathogens* **2020**, *9*, 356. [CrossRef]
45. Tibayrenc, M.; Ayala, F.J. The population genetics of *Trypanosoma cruzi* revisited in the light of the predominant clonal evolution model. *Acta Trop.* **2015**, *151*, 156–165. [CrossRef]
46. Tibayrenc, M.; Ayala, F.J. Reproductive clonality of pathogens: A perspective on pathogenic viruses, bacteria, fungi, and parasitic protozoa. *Proc. Natl. Acad. Sci. USA* **2012**, *109*, E3305–E3313. [CrossRef]
47. Zingales, B.; Miles, M.A.; Campbell, D.A.; Tibayrenc, M.; Macedo, A.M.; Teixeira, M.M.G.; Schijman, A.G.; Llewellyn, M.S.; Lages-Silva, E.; Machado, C.R.; et al. The revised *Trypanosoma cruzi* subspecific nomenclature: Rationale, epidemiological relevance and research applications. *Infect. Genet. Evol.* **2012**, *12*, 240–253. [CrossRef]
48. Zingales, B.; Andrade, S.; Briones, M.; Campbell, D.; Chiari, E.; Fernandes, O.; Guhl, F.; Lages-Silva, E.; Macedo, A.; Machado, C.; et al. A new consensus for *Trypanosoma cruzi* intraspecific nomenclature: Second revision meeting recommends TcI to TcVI. *Mem. Inst. Oswaldo Cruz* **2009**, *104*, 1051–1054. [CrossRef]
49. Hamilton, P.B.; Teixeira, M.M.G.; Stevens, J.R. The evolution of *Trypanosoma cruzi*: The 'bat seeding' hypothesis. *Trends Parasitol.* **2012**, *28*, 136–141. [CrossRef]
50. Brossas, J.-Y.; Gulin, J.E.N.; Bisio, M.M.C.; Chapelle, M.; Marinach-Patrice, C.; Bordessoules, M.; Palazon Ruiz, G.; Vion, J.; Paris, L.; Altcheh, J.; et al. Secretome analysis of *Trypanosoma cruzi* by proteomics studies. *PLoS ONE* **2017**, *12*, e0185504. [CrossRef]
51. Zago, M.P.; Hosakote, Y.M.; Koo, S.; Dhiman, M.; Piñeyro, M.D.; Parodi-Talice, A.; Basombrio, M.A.; Robello, C.; Garg, N.J. TcI Isolates of *Trypanosoma cruzi* Exploit the Antioxidant Network for Enhanced Intracellular Survival in Macrophages and Virulence in Mice. *Infect. Immun.* **2016**, *84*, 1842–1856. [CrossRef] [PubMed]
52. Tavares De Oliveira, M.; Taciana Santos Silva, K.; Xavier Neves, L.; De Ornelas Toledo, M.J.; Castro-Borges, W.; De Lana, M. Differential expression of proteins in genetically distinct *Trypanosoma cruzi* samples (TcI and TcII DTUs) isolated from chronic Chagas disease cardiac patients. *Parasit. Vectors* **2018**, *11*, 611. [CrossRef] [PubMed]
53. Kikuchi, S.A.; Sodré, C.L.; Kalume, D.E.; Elias, C.G.R.; Santos, A.L.S.; De Nazaré Soeiro, M.; Meuser, M.; Chapeaurouge, A.; Perales, J.; Fernandes, O. Proteomic analysis of two *Trypanosoma cruzi* zymodeme 3 strains. *Exp. Parasitol.* **2010**, *126*, 540–551. [CrossRef] [PubMed]
54. Telleria, J.; Biron, D.G.; Brizard, J.-P.; Demettre, E.; Seveno, M.; Barnabe, C.; Ayala, F.J.; Tibayrenc, M. Phylogenetic character mapping of proteomic diversity shows high correlation with subspecific phylogenetic diversity in *Trypanosoma cruzi*. *Proc. Natl. Acad. Sci. USA* **2010**, *107*, 20411–20416. [CrossRef]
55. Machin, A.; Telleria, J.; Brizard, J.-P.; Demettre, E.; Séveno, M.; Ayala, F.J.; Tibayrenc, M. *Trypanosoma cruzi*: Gene Expression Surveyed by Proteomic Analysis Reveals Interaction between Different Genotypes in Mixed In Vitro Cultures. *PLoS ONE* **2014**, *9*, e95442. [CrossRef]
56. Díaz, M.L.; Solari, A.; González, C.I. Differential expression of *Trypanosoma cruzi* I associated with clinical forms of Chagas disease: Overexpression of oxidative stress proteins in acute patient isolate. *J. Proteom.* **2011**, *74*, 1673–1682. [CrossRef]
57. Herreros-Cabello, A.; Callejas-Hernández, F.; Fresno, M.; Gironès, N. Comparative proteomic analysis of trypomastigotes from *Trypanosoma cruzi* strains with different pathogenicity. *Infect. Genet. Evol.* **2019**, *76*, 104041. [CrossRef]
58. San Francisco, J.; Gutiérrez, B.; González, J. Role of Proteomics in the Study of *Trypanosoma cruzi* Biology. In *Biology of Trypanosoma cruzi*; De Souza, W., Ed.; IntechOpen: London, UK, 2019; ISBN 978-1-83968-203-2.
59. Bayer-Santos, E.; Aguilar-Bonavides, C.; Rodrigues, S.P.; Cordero, E.M.; Marques, A.F.; Varela-Ramirez, A.; Choi, H.; Yoshida, N.; Da Silveira, J.F.; Almeida, I.C. Proteomic Analysis of *Trypanosoma cruzi* Secretome: Characterization of Two Populations of Extracellular Vesicles and Soluble Proteins. *J. Proteome Res.* **2013**, *12*, 883–897. [CrossRef]
60. Mandacaru, S.C.; Queiroz, R.M.L.; Alborghetti, M.R.; De Oliveira, L.S.; De Lima, C.M.R.; Bastos, I.M.D.; Santana, J.M.; Roepstorff, P.; Ricart, C.A.O.; Charneau, S. Exoproteome profiling of *Trypanosoma cruzi* during amastigogenesis early stages. *PLoS ONE* **2019**, *14*, e0225386. [CrossRef]
61. Bautista-López, N.L.; Ndao, M.; Camargo, F.V.; Nara, T.; Annoura, T.; Hardie, D.B.; Borchers, C.H.; Jardim, A. Characterization and Diagnostic Application of *Trypanosoma cruzi* Trypomastigote Excreted-Secreted Antigens Shed in Extracellular Vesicles Released from Infected Mammalian Cells. *J. Clin. Microbiol.* **2017**, *55*, 744–758. [CrossRef]
62. Retana Moreira, L.; Prescilla-Ledezma, A.; Cornet-Gomez, A.; Linares, F.; Jódar-Reyes, A.B.; Fernandez, J.; Ibarrola Vannucci, A.K.; De Pablos, L.M.; Osuna, A. Biophysical and Biochemical Comparison of Extracellular Vesicles Produced by Infective and Non-Infective Stages of *Trypanosoma cruzi*. *Int. J. Mol. Sci.* **2021**, *22*, 5183. [CrossRef] [PubMed]

63. Ribeiro, K.S.; Vasconcellos, C.I.; Soares, R.P.; Mendes, M.T.; Ellis, C.C.; Aguilera-Flores, M.; De Almeida, I.C.; Schenkman, S.; Iwai, L.K.; Torrecilhas, A.C. Proteomic analysis reveals different composition of extracellular vesicles released by two *Trypanosoma cruzi* strains associated with their distinct interaction with host cells. *J. Extracell. Vesicles* **2018**, *7*, 1463779. [CrossRef]
64. Cortes-Serra, N.; Mendes, M.T.; Mazagatos, C.; Segui-Barber, J.; Ellis, C.C.; Ballart, C.; Garcia-Alvarez, A.; Gállego, M.; Gascon, J.; Almeida, I.C.; et al. Plasma-Derived Extracellular Vesicles as Potential Biomarkers in Heart Transplant Patient with Chronic Chagas Disease. *Emerg. Infect. Dis.* **2020**, *26*, 1846–1851. [CrossRef] [PubMed]
65. Bannister, A.J.; Kouzarides, T. Regulation of chromatin by histone modifications. *Cell Res.* **2011**, *21*, 381–395. [CrossRef]
66. de Jesus, T.C.L.; Nunes, V.S.; Lopes, M.d.C.; Martil, D.E.; Iwai, L.K.; Moretti, N.S.; Machado, F.C.; de Lima-Stein, M.L.; Thiemann, O.H.; Elias, M.C.; et al. Chromatin Proteomics Reveals Variable Histone Modifications during the Life Cycle of *Trypanosoma cruzi*. *J. Proteome Res.* **2016**, *15*, 2039–2051. [CrossRef]
67. Magalhães, R.D.M.; Mattos, E.C.; Rozanski, A.; Galante, P.A.F.; Palmisano, G.; Cruz, A.K.; Colli, W.; Camargo, A.A.; Alves, M.J.M. Global changes in nitration levels and DNA binding profile of *Trypanosoma cruzi* histones induced by incubation with host extracellular matrix. *PLoS Negl. Trop. Dis.* **2020**, *14*, e0008262. [CrossRef]
68. Mule, S.N.; Manchola, N.C.; De Oliveira, G.S.; Pereira, M.; Magalhães, R.D.M.; Teixeira, A.A.; Colli, W.; Alves, M.J.M.; Palmisano, G. Proteome-wide modulation of S-nitrosylation in *Trypanosoma cruzi* trypomastigotes upon interaction with the host extracellular matrix. *J. Proteom.* **2021**, *231*, 104020. [CrossRef]
69. Moretti, N.S.; Cestari, I.; Anupama, A.; Stuart, K.; Schenkman, S. Comparative Proteomic Analysis of Lysine Acetylation in Trypanosomes. *J. Proteome Res.* **2018**, *17*, 374–385. [CrossRef]
70. De Almeida, R.F.; Lucena, A.C.R.; Batista, M.; Marchini, F.K.; De Godoy, L.M.F. Non-histone protein methylation in *Trypanosoma cruzi* epimastigotes. *Proteomics* **2023**, *23*, 2200230. [CrossRef]
71. De Oliveira Santos, J.; Zuma, A.A.; De Luna Vitorino, F.N.; Da Cunha, J.P.C.; De Souza, W.; Motta, M.C.M. Trichostatin A induces *Trypanosoma cruzi* histone and tubulin acetylation: Effects on cell division and microtubule cytoskeleton remodelling. *Parasitology* **2019**, *146*, 543–552. [CrossRef]
72. Pereira, M.; Soares, C.; Canuto, G.A.B.; Tavares, M.F.M.; Colli, W.; Alves, M.J.M. Down Regulation of NO Signaling in *Trypanosoma cruzi* upon Parasite-Extracellular Matrix Interaction: Changes in Protein Modification by Nitrosylation and Nitration. *PLoS Negl. Trop. Dis.* **2015**, *9*, e0003683. [CrossRef] [PubMed]
73. Parsons, M.; Worthey, E.A.; Ward, P.N.; Mottram, J.C. Comparative analysis of the kinomes of three pathogenic trypanosomatids: *Leishmania major*, *Trypanosoma brucei* and *Trypanosoma cruzi*. *BMC Genom.* **2005**, *6*, 127. [CrossRef]
74. Nakayasu, E.S.; Gaynor, M.R.; Sobreira, T.J.P.; Ross, J.A.; Almeida, I.C. Phosphoproteomic analysis of the human pathogen *Trypanosoma cruzi* at the epimastigote stage. *Proteomics* **2009**, *9*, 3489–3506. [CrossRef] [PubMed]
75. Lucena, A.C.R.; Amorim, J.C.; Van De Paula Lima, C.E.; Batista, M.; Krieger, M.A.; De Godoy, L.M.F.; Marchini, F.K. Quantitative phosphoproteome and proteome analyses emphasize the influence of phosphorylation events during the nutritional stress of *Trypanosoma cruzi*: The initial moments of in vitro metacyclogenesis. *Cell Stress Chaperones* **2019**, *24*, 927–936. [CrossRef]
76. Alves, M.J.M.; Kawahara, R.; Viner, R.; Colli, W.; Mattos, E.C.; Thaysen-Andersen, M.; Larsen, M.R.; Palmisano, G. Comprehensive glycoprofiling of the epimastigote and trypomastigote stages of *Trypanosoma cruzi*. *J. Proteom.* **2017**, *151*, 182–192. [CrossRef]
77. Torres-Gutiérrez, E.; Pérez-Cervera, Y.; Camoin, L.; Zenteno, E.; Aquino-Gil, M.O.; Lefebvre, T.; Cabrera-Bravo, M.; Reynoso-Ducoing, O.; Bucio-Torres, M.I.; Salazar-Schettino, P.M. Identification of O-Glcnacylated Proteins in *Trypanosoma cruzi*. *Front. Endocrinol.* **2019**, *10*, 199. [CrossRef]
78. Bayona, J.C.; Nakayasu, E.S.; Laverrière, M.; Aguilar, C.; Sobreira, T.J.P.; Choi, H.; Nesvizhskii, A.I.; Almeida, I.C.; Cazzulo, J.J.; Alvarez, V.E. SUMOylation Pathway in *Trypanosoma cruzi*: Functional Characterization and Proteomic Analysis of Target Proteins. *Mol. Cell. Proteom.* **2011**, *10*, M110.007369. [CrossRef]
79. Igoillo-Esteve, M.; Maugeri, D.; Stern, A.L.; Beluardi, P.; Cazzulo, J.J. The pentose phosphate pathway in *Trypanosoma cruzi*: A potential target for the chemotherapy of Chagas disease. *An. Acad. Bras. Ciênc.* **2007**, *79*, 649–663. [CrossRef]
80. Pineyro, M.; Pizarro, J.; Lema, F.; Pritsch, O.; Cayota, A.; Bentley, G.; Robello, C. Crystal structure of the tryparedoxin peroxidase from the human parasite. *J. Struct. Biol.* **2005**, *150*, 11–22. [CrossRef]
81. Ilari, A.; Fiorillo, A.; Genovese, I.; Colotti, G. Polyamine-Trypanothione Pathway: An Update. *Future Med. Chem.* **2017**, *9*, 61–77. [CrossRef]
82. Piñeyro, M.D.; Parodi-Talice, A.; Portela, M.; Arias, D.G.; Guerrero, S.A.; Robello, C. Molecular characterization and interactome analysis of *Trypanosoma cruzi* Tryparedoxin 1. *J. Proteom.* **2011**, *74*, 1683–1692. [CrossRef] [PubMed]
83. Arias, D.G.; Piñeyro, M.D.; Iglesias, A.A.; Guerrero, S.A.; Robello, C. Molecular characterization and interactome analysis of *Trypanosoma cruzi* tryparedoxin II. *J. Proteom.* **2015**, *120*, 95–104. [CrossRef]
84. Peloso, E.F.; Dias, L.; Queiroz, R.M.L.; Leme, A.F.P.P.; Pereira, C.N.; Carnielli, C.M.; Werneck, C.C.; Sousa, M.V.; Ricart, C.A.O.; Gadelha, F.R. *Trypanosoma cruzi* mitochondrial tryparedoxin peroxidase is located throughout the cell and its pull down provides one step towards the understanding of its mechanism of action. *Biochim. Biophys. Acta BBA-Proteins Proteom.* **2016**, *1864*, 1–10. [CrossRef]

85. Leandro de Jesus, T.C.; Calderano, S.G.; Vitorino, F.N.d.L.; Llanos, R.P.; Lopes, M.d.C.; de Araújo, C.B.; Thiemann, O.H.; Reis, M.d.S.; Elias, M.C.; Chagas da Cunha, J.P. Quantitative Proteomic Analysis of Replicative and Nonreplicative Forms Reveals Important Insights into Chromatin Biology of *Trypanosoma cruzi*. *Mol. Cell. Proteom.* **2017**, *16*, 23–38. [CrossRef] [PubMed]
86. Günzl, A. The Pre-mRNA Splicing Machinery of Trypanosomes: Complex or Simplified? *Eukaryot. Cell* **2010**, *9*, 1159–1170. [CrossRef]
87. Alves, L.R.; Ávila, A.R.; Correa, A.; Holetz, F.B.; Mansur, F.C.B.; Manque, P.A.; De Menezes, J.P.B.; Buck, G.A.; Krieger, M.A.; Goldenberg, S. Proteomic analysis reveals the dynamic association of proteins with translated mRNAs in *Trypanosoma cruzi*. *Gene* **2010**, *452*, 72–78. [CrossRef]
88. Wippel, H.H.; Inoue, A.H.; Vidal, N.M.; Costa, J.F.D.; Marcon, B.H.; Romagnoli, B.A.A.; Santos, M.D.M.; Carvalho, P.C.; Goldenberg, S.; Alves, L.R. Assessing the partners of the RBP9-mRNP complex in *Trypanosoma cruzi* using shotgun proteomics and RNA-seq. *RNA Biol.* **2018**, *15*, 1106–1118. [CrossRef]
89. Mantilla, B.S.; Kalesh, K.; Brown, N.W.; Fiedler, D.; Docampo, R. Affinity-based proteomics reveals novel targets of inositol pyrophosphate (5-IP_7)-dependent phosphorylation and binding in *Trypanosoma cruzi* replicative stages. *Mol. Microbiol.* **2021**, *115*, 986–1004. [CrossRef]
90. Trochine, A.; Alvarez, G.; Corre, S.; Faral-Tello, P.; Durán, R.; Batthyany, C.I.; Cerecetto, H.; González, M.; Robello, C. *Trypanosoma cruzi* chemical proteomics using immobilized benznidazole. *Exp. Parasitol.* **2014**, *140*, 33–38. [CrossRef]
91. Azeredo, C.M.; Saraiva, M.F.; De Oliveira, M.R.; Barbosa, G.; De Almeida, M.V.; De Souza, M.V.N.; Soares, M.J. The terpenic diamine GIB24 inhibits the growth of *Trypanosoma cruzi* epimastigotes and intracellular amastigotes, with proteomic analysis of drug-resistant epimastigotes. *Chem. Biol. Interact.* **2020**, *330*, 109165. [CrossRef]
92. Brunoro, G.V.F.; Faça, V.M.; Caminha, M.A.; Ferreira, A.T.d.S.; Trugilho, M.; de Moura, K.C.G.; Perales, J.; Valente, R.H.; Menna-Barreto, R.F.S. Differential Gel Electrophoresis (DIGE) Evaluation of Naphthoimidazoles Mode of Action: A Study in *Trypanosoma cruzi* Bloodstream Trypomastigotes. *PLoS Negl. Trop. Dis.* **2016**, *10*, e0004951. [CrossRef] [PubMed]
93. Menna-Barreto, R.F.S.; Beghini, D.G.; Ferreira, A.T.S.; Pinto, A.V.; De Castro, S.L.; Perales, J. A proteomic analysis of the mechanism of action of naphthoimidazoles in *Trypanosoma cruzi* epimastigotes in vitro. *J. Proteom.* **2010**, *73*, 2306–2315. [CrossRef] [PubMed]
94. Vieira, G.A.L.; Silva, M.T.A.D.; Regasini, L.O.; Cotinguiba, F.; Laure, H.J.; Rosa, J.C.; Furlan, M.; Cicarelli, R.M.B. *Trypanosoma cruzi*: Analysis of two different strains after piplartine treatment. *Braz. J. Infect. Dis.* **2018**, *22*, 208–218. [CrossRef] [PubMed]

pathogens

MDPI

Review

Trypanosoma cruzi Transmission Through Blood Samples and Derivatives: Main Routes, Control Strategies, and Recent Advancements in Blood Banks

Aline Nefertiti Silva da Gama and Maria de Nazaré Correia Soeiro *

Cell Biology Laboratory, Oswaldo Cruz Institute, Oswaldo Cruz Foundation, Rio de Janeiro 21040-900, RJ, Brazil; alinenefertitigama@gmail.com
* Correspondence: soeiro@ioc.fiocruz.br

Abstract: Neglected Tropical Diseases are a group of 25 conditions caused by diverse agents. They mostly affect people with poorer health outcomes, particularly preventable diseases. The social determinants of health influence the development and progression of these poverty diseases, with inadequate sanitation presenting chronicity, high morbidity, and economic impacts. Chagas disease, a prominent Neglected Tropical Disease caused by the intracellular pathogen *Trypanosoma cruzi*, is endemic in Latin America but is increasing as a global concern due to population migration. It is transmitted through insect vectors, congenitally, orally via contaminated food and beverage, via transfusions and organ donation, and due to laboratory accidents, among other minor relevant routes. As a silent illness, with many infected individuals remaining asymptomatic, it contributes to underdiagnosis, and delayed treatment that involves nitro derivatives is often discontinued due to side effects. Chagas disease spreads in non-endemic areas like the United States of America and Europe. Blood screening practices vary, with endemic regions implementing universal testing, while non-endemic areas rely on selective methods. Recent innovations, such as riboflavin–ultraviolet light treatment and arylimidamide compounds, represent promising alternatives to reduce transfusion transmission. This review presents an analysis of *Trypanosoma cruzi* transmission through blood and derivatives, addressing the main routes, globally implemented control strategies, and recent advancements in blood bank safety.

Keywords: Chagas disease; blood banks; neglected tropical diseases

1. Introduction

Neglected Tropical Diseases comprise a group of 25 conditions caused by various agents, including viruses, bacteria, fungi, protozoa, helminths, and toxins [1]. These diseases share several common characteristics: (i) a strong association with inadequate or absent basic sanitation; (ii) the potential for transmission control through prophylaxis and treatment, often hindered by the lack of adequate public policies; (iii) high prevalence among impoverished populations, affecting the most vulnerable groups; (iv) a tendency toward chronicity due to late diagnosis and inadequate treatment; and (v) significant economic impact due to high morbidity rates, including healthcare costs and disability-related pensions [2,3]. These diseases affect over one billion people worldwide, causing approximately 200,000 deaths annually and leading to the loss of about 25 million disability-adjusted life years [4].

Chagas disease belongs to the group of Neglected Tropical Diseases and was discovered 115 years ago by Brazilian physician Carlos Chagas. Chagas disease is caused by the

https://doi.org/10.3390/pathogens14020133

protozoan *Trypanosoma cruzi* and is endemic in 21 countries in Latin America, occurring from the southern United States to northern Argentina and Chile [5,6]. Currently, about 6–7 million people globally are infected, with 75 million people in the Americas at risk of contracting the disease [7]. As described by Chagas [8], transmission can occur through various routes: (i) vector borne, when the triatomine insect defecates during its blood meal, and the fecal material containing the parasite is introduced through the bite wound in the skin or mucous membranes; (ii) congenital (from mother to child) during pregnancy or delivery; (iii) orally, involving the consumption of food or beverages contaminated with infected triatomines or their feces; (iv) laboratory accidents; and (v) organ transplantation and blood transfusion [9–11].

Most acute infections are asymptomatic or exhibit nonspecific manifestations, characterized by a short duration, often 4–8 weeks, of displaying circulating parasites in the blood [12]. This clinical diversity contributes to many cases remaining undiagnosed or unnotified. Without appropriate treatment during the acute phase, the symptoms and parasitaemia are naturally controlled by an effective host immune response, and the individuals remain in the indeterminate form of the chronic phase, without presenting any symptoms for decades or even for their entire lives [13,14]. Nevertheless, 30 to 40% of infected people progress from the indeterminate form to the symptomatic chronic phase, presenting severe clinical manifestations, including digestive, cardiac, mixed (cardio-digestive), and neurological forms [14].

Treatment relies on two nitro derivatives, nifurtimox and benznidazole. Both cause severe side effects and require prolonged treatments, which often result in therapy discontinuation, in addition to the existence of parasites naturally resistant to both drugs [15,16]. Previously limited to the Americas, Chagas disease has become a global concern due to population mobility and urbanization. Today, cases are reported in countries such as Canada, the United States of America, and even parts of Europe, Africa, and the Western Pacific [17,18]. Over the past 40 years, the migration of *Trypanosoma cruzi*-infected individuals from endemic countries to non-endemic regions, particularly in North America and Europe, has raised concerns about blood safety across vast geographic areas [19]. Although initiatives have been implemented to mitigate the risk of transfusion-transmitted infections in non-endemic regions, there is still no standardized or universal approach, which hinders effective control of this transmission route [20]. This paper presents a review of the transmission of Chagas disease through blood and its derivatives, addressing the main infection routes, globally implemented control strategies, and recent advancements in technologies and medications to enhance blood bank safety. Finally, the impact of these initiatives on combating transfusion-transmitted infections and promoting public health in different regions of the world is discussed.

2. Blood and Blood Derivatives Transmission

Approximately 4.6 million Latin American migrants live in Europe, with an estimated 68,000 to 122,000 undiagnosed cases of Chagas disease. Determining the exact number of undocumented migrants is extremely challenging, although they are likely to have the highest prevalence rates of the disease [21]. The transfusional transmission of Chagas disease is widely recognized and has been the subject of studies since the first reports in Brazil in 1949 and 1952 [22,23]. While blood transmission, like the classic vectorial transmission, is under control in endemic regions, with serological testing being mandatory in some countries such as Brazil [24], this is not universally observed in non-endemic regions such as the United States of America, Canada, Asia, Australia, and Japan. Unlike Latin America, where control policies are more established, the process of globalization and increasing migration have contributed to the spread of Chagas disease to non-endemic areas, making vertical (mother to child) and transfusional transmission a global concern [20,25]. Many individuals affected by Chagas disease remain asymptomatic and, therefore, are unaware of their infection, which increases the risk of transmission, especially considering that, in some countries, blood donation is remunerated as an incentive. This situation represents a significant threat to the safety of the blood supply in non-endemic regions [26].

The first case of transfusional transmission of *Trypanosoma cruzi* in the United States of America was verified in 1987, and then in 1989 in Canada, and in the early 2000s in Europe, highlighting the urgent need to implement mitigation strategies in these areas [27]. The transmission of *Trypanosoma cruzi* through blood and its derivatives occurs when contaminated components are transfused to recipients [28,29]. All blood components, except for frozen fresh plasma derivatives, are considered potentially infectious. The parasite can sustain its viability for up to 18 days at 4 °C and up to 250 days at room temperature [30,31]. Although cryopreserved stocks may remain viable in frozen vials, there is no clear evidence of transmission through cellular components that were frozen and thawed [32].

In endemic countries, the likelihood of infection following the transfusion of a single unit of blood from an infected donor has been estimated at around 10–20% [33]. The infectivity of different blood components varies, with platelets being the most frequently associated with transfusional transmission of *Trypanosoma cruzi*. According to evidence from non-endemic regions, the transmission risk ranges from approximately 6–26% for platelets and only about 0–4% for packed red blood cells, plasma, and cryoprecipitates, since the bloodstream parasites concentrate in the same region as leucocytes and platelets after differential centrifugation [34]. Platelets are frequently involved in *Trypanosoma cruzi* transmission, particularly in immunocompromised patients or those undergoing oncological treatment [26]. The probability of transfusional transmission depends on multiple factors, such as the amount of blood transfused, the parasite strain, and the specific discrete type units, among others [35–37].

3. Control and Prevention Strategies in Blood Banks

Since the establishment of the practice of blood transfusion in 1940 in Latin America, Chagas disease has become a major risk factor, with the first case of transmission reported in 1952 in Brazil, but it has only been mandatory to perform serological testing in blood banks since 1969 [24]. Currently, screening is mandatory in all countries where the parasite circulates naturally, including the United States of America [38]. According to the Pan American Health Organization, in 2010, the prevalence of infection varied greatly between countries, being highest in Argentina, Paraguay, and Bolivia and relatively low in places such as Costa Rica, Chile, and Brazil. In the United States of America, although the prevalence is low, there are significant regional differences, with higher numbers in Florida and Texas [33]. While universal screening may not be economically feasible in settings with very low prevalence, it is ethically unacceptable for those recipients who may receive infected blood. Thus, even with advances in control, sporadic infections may still occur, highlighting the need to improve test sensitivity and continuously monitor mitigation strategies.

Thus, screening and control strategies to mitigate transfusion transmission of *Trypanosoma cruzi* vary widely among countries, reflecting their epidemiological realities and available resources. While endemic countries adopt universal screening and mandatory testing, non-endemic regions rely on targeted questionnaires and selective testing. Despite advances, challenges remain, highlighting the need for continued improvements, such as improving screening technologies and implementing more comprehensive global policies to ensure the safety of the blood supply.

Control and Safety Strategies in Blood Donations in Endemic and Non-Endemic Countries

As described above, Brazil has been a pioneer in implementing mandatory universal screening. Serological testing in blood banks uses advanced methods such as chemiluminescent and enzyme-linked immunosorbent assay, with confirmations through indirect immunofluorescence and Western blot. Brazil has one of the most robust legislations, ensuring that all donors are tested for *Trypanosoma cruzi*. In positive cases, the products are discarded, and the donors are permanently excluded. The prevalence of infection in Brazil is relatively low compared to other Latin American countries, but population mobility requires continuous surveillance [39,40]. In Argentina, Bolivia, and Paraguay, there is still a high prevalence of *Trypanosoma cruzi* infection, which makes mandatory universal blood screening a crucial measure. Similarly to Brazil, these countries use highly sensitive serological methods. Local legislation ensures strict control of donations, mitigating the risks of transfusion transmission [20].

In the United States of America, blood donor screening for antibodies to Trypanosoma cruzi was revived in 2007 as part of measures to ensure the safety of the blood supply. The process combines the application of questionnaires to assess risk factors, such as travel history or residence in endemic areas, with external serological testing for high-risk donors [41]. Methods such as chemiluminescent and enzyme-linked immunosorbent assay are widely used, complemented by advanced technologies such as indirect immunofluorescence and nucleic acid testing. Donors with positive results are duly notified, permanently excluded from the donation system, and have their products discarded to avoid transmission risks [20,27]. In cases of positive screening for antibodies to *Trypanosoma cruzi*, mandatory additional diagnostic tests are recommended. The diagnosis of chronic Chagas disease is based on positive results in at least two different serological tests, using specific antigenic techniques and guidelines, since no single test has sufficient sensitivity and specificity to confirm infection [10]. Among the most common methods are enzyme-linked immunosorbent assay with recombinant antigens, immunoblot using secreted–excreted trypomastigote antigens, and indirect immunofluorescence [42,43]. In cases of discrepant results, such as patients with a positive result in the Wiener recombinant antigen enzyme immunoassay test and a negative result in secreted–excreted trypomastigote antigens, the diagnostic algorithm adopted by the Centers for Disease Control and Prevention includes performing a third test to resolve the discordance. This procedure ensures greater diagnostic accuracy, minimizing the occurrence of false positives or false negatives, and guarantees appropriate patient management and blood supply safety [43].

In Canada, the United Kingdom, and Australia, screening for Trypanosoma cruzi infection in blood donors is based on epidemiological questionnaires and, in some cases, educational strategies [27]. In Canada, screening involves epidemiological questionnaires and selected serological tests for high-risk donors, complemented by polymerase chain reaction assays with Trypanosoma cruzi-specific primers [44]. In the United Kingdom, strategies focus on epidemiological questionnaires to identify donors with possible exposure to the parasite. The country also invests in educational campaigns, using printed materials and social media to inform about the risks associated with the disease. Although universal testing is not adopted, donors considered high risk are permanently excluded from the donation system. In Australia, screening is based solely on risk factors such as travel history and residence in endemic areas. Universal serological testing is not performed, which makes the use of epidemiological questionnaires an essential practice. In all these countries, educational actions play an important role in raising donor awareness about the risks of Chagas disease and safety measures, with an emphasis on education through printed materials, videos, and social media campaigns [27].

In France, Germany, and Spain, governments implement distinct strategies to mitigate the risk of transfusion-transmitted Trypanosoma cruzi. In France, the approach incorporates serological testing for Trypanosoma cruzi antibodies, employing highly sensitive methodologies such as enzyme-linked immunosorbent and chemiluminescent assay. In addition, pathogen inactivation/pathogen reduction methodologies, including the Intercept system, are used but cover less than 25% of blood products [27]. Germany adopts a risk-based approach centered on pre-donation screening by targeted questionnaires, which assess donors' exposure to potential infectious agents, and also consider factors such as birth or residence in endemic regions and/or potential contact with vectors. Also, systematic testing of collected samples is conducted using serological methods to detect antibodies and nucleic acid amplification techniques [45,46]. These strategies are complemented by the continuous monitoring of epidemiological trends, allowing adjustments to screening practices as necessary. Simultaneously, donor education is emphasized to raise awareness about the importance of honesty during the screening process. In Spain, serological screening is performed mainly on donors with risk factors identified in the previously completed questionnaires [27].

4. Chemoprophylaxis in Blood Banks

Due to the nature of Chagas disease pathology, in asymptomatic individuals during the chronic phase, parasitemia is characteristically low and intermittent in peripheral blood, which often allows the infection to remain unblocked, even decades after initial exposure [14,47]. This condition poses a significant risk of parasite transmission through blood transfusions [48]. Given the large number of asymptomatic, non-expressing, and untreated infected individuals, transfusion transmission still stands out as one of the main routes of dissemination of *Trypanosoma cruzi*, especially in non-endemic regions. To mitigate this issue, the implementation of methods for the detection and inactivation of the parasite in blood banks became an essential component of transfusion safety protocols. In this context, Nussenszweig and colleagues were pioneers in elucidating the effects of gentian violet—a mixture of triarylmethane dyes (crystal violet, methyl violet, and brilliant green)—on *Trypanosoma cruzi* trypomastigotes, advocating its application in blood banks as a prophylactic measure against transfusion-transmitted Chagas disease [49]. This dye was introduced into blood banks across Latin America during the 1960s, following recommendations from the World Health Organization [50,51]. However, a significant limitation associated with its use was the intense purple staining it caused on the skin and mucous membranes of transfusion recipients, raising concerns regarding its acceptability [52,53].

Chemoprophylaxis against transfusion-transmitted Chagas disease was initially performed by adding gentian violet at a concentration of 1:4000 to blood suspected of being contaminated, for a period of 24 h, the necessary duration for the eradication of *Trypanosoma cruzi* [54]. This method was useful in controlling transmission in regions highly endemic for Chagas disease in the past. In this context, innovative initiatives for the implementation of new diagnostic tests in blood banks and the investigation of chemoprophylaxis strategies have become essential.

Advanced Methodologies and Novel Compounds for Blood Bank Safety

The parasites can survive the storage conditions of blood components, becoming a challenge for transfusion safety. In this context, the efficacy of the combination of leukoreduction by filtration and treatment with riboflavin and ultraviolet light to reduce the *Trypanosoma cruzi* load in artificially contaminated whole blood units was evaluated [55]. The leukoreductive filtration was able to reduce the number of live parasites in the blood by more than 99%. However, the combination of filtration with treatment with riboflavin and ultraviolet light was even more effective, eliminating viable parasites in the samples tested. The number of circulating parasites in whole blood in chronic patients is generally low, approximately 1 parasite equivalent/mL or less [47], which makes parasite detection even more difficult, especially after blood fractionation. Furthermore, the riboflavin–ultraviolet light method has been shown to irreversibly damage the nucleic acids of pathogens, preventing their replication, without compromising blood components essential for transfusion [56]. In this context, leukoreduction emerges as an effective strategy to prevent transfusion-transmitted *Trypanosoma cruzi* infection, especially in endemic and non-endemic countries with high rates of immigration from Latin America.

The highest concentration of parasites was detected in plasma, followed by red blood cells before leukoreduction, evidencing a significant risk of transmission of *Trypanosoma cruzi* associated with non-leukoreduced red blood cells. Thus, the combined implementation of different approaches such as ultraviolet light and riboflavin use along with leukoreduction may represent a significant advance in reducing the risk of *Trypanosoma cruzi* transmission in blood banks. This approach has shown promise as a complementary alternative to measures already implemented in blood banks, since it overcomes the limitations of serological screening and other methods that may fail to detect recently infected donors or those with low parasitaemia.

In a recent study, arylimidamides demonstrated high trypanocidal activity under blood bank storage conditions, such as at 4 °C. In in vitro studies, the arylimidamides DB745 and DB766 presented a safe profile on blood cells, not causing hemolysis at concentrations of up to 96 μM [57]. Arylimidamides are a class of chemical compounds derived from classical aromatic amidines, but with minor structural modifications. They are called "reverse amidines" and differ from diamidines in that the amidine is linked to the aromatic nucleus of the molecule through a nitrogen atom, instead of being mediated by a carbon atom. This structural change gives arylimidamides differentiated physicochemical properties, improving their efficacy and safety in combating pathogens, such as protozoan parasites including *Trypanosoma cruzi* [58]. Furthermore, arylimidamides showed rapid action, almost eliminating bloodstream trypomastigotes, the form that circulates in the blood of mammalian hosts, after only two hours of in vitro incubation. Among the compounds tested, the arylimidamide DB1831 was the most promising, preventing acute infections in experimental mouse models due to arylimidamide-pre-treated trypomastigotes inoculum. Thus, arylimidamides have the potential to replace old methods, such as the use of gentian violet, which has undesirable side effects such as purple coloration of the skin and mucous membranes. Arylimidamides may represent an effective alternative for chemoprophylaxis of blood samples, contributing significantly to transfusion safety and the control of iatrogenic transmission of Chagas disease.

5. Conclusions and Perspectives

Transfusion transmission of *Trypanosoma cruzi* remains a significant concern, especially in non-endemic regions, where population mobility and the absence of universal serological screening make it difficult to control this route of dissemination of *Trypanosoma cruzi* [20]. Strategic advances, such as the combination of leukoreduction with riboflavin-based treatments and ultraviolet light, have demonstrated efficacy in eliminating parasites in blood components, representing a promising solution to complement traditional serological screening practices. Leukoreduction, a process that removes leukocytes from blood components, plays an important role in reducing the transmission of intracellular pathogens such as cytomegalovirus, human T-lymphotropic virus type I/II, and trypanosomatids, including *Leishmania* and *Trypanosoma cruzi* [55,59–61]. Furthermore, innovative compounds, such as arylimidamides, have emerged as a promising and innovative approach in non-clinical studies and deserve further analysis. These compounds exhibit high trypanocidal activity under blood bank storage conditions, with proven safety for blood cells and the ability to eliminate parasitaemia in the short term. Among the compounds tested, DB1831 showed promising results, preventing acute infections in experimental mouse models [57].

Regarding new avenues for future research, it is important to state that although blood safety interventions have proven effective at avoiding transmission of human immune deficient virus, human B and C hepatitis, and West Nile virus, the transmission of parasites by transfusion still represents a relevant challenge [62], posing significant risks to blood recipients. Even though parasitic infections such as Malaria and Chagas disease, among others, are often regionally spread, due to human mobility, they represent a current and growing challenge to other non-endemic areas. Also, infections due to *Babesia microti*, a tick-borne protozoan parasite, are increasing, with over 200 cases of transfusion-transmitted infection recorded in North America and infections steadily increasing worldwide, including in Australia, Asia, and Europe [62].

Then, technological and therapeutic advances represent a milestone in improving transfusion safety and controlling the iatrogenic transmission of parasites such as *Trypanosoma cruzi*. The implementation of innovative approaches combined with education and dissemination of information about this parasitosis can also contribute significantly to global public health, especially in a scenario of increasing population mobility and urbanization. Continued investment in research and global standardization policies are essential to ensure blood safety in all affected regions, promoting advances in the fight against blood transmission of the etiological agent of Chagas disease.

Author Contributions: Conceptualization, M.d.N.C.S. and A.N.S.d.G.; methodology, M.d.N.C.S. and A.N.S.d.G.; resources, M.d.N.C.S.; writing—original draft preparation, A.N.S.d.G.; writing—review and editing, M.d.N.C.S.; supervision, M.d.N.C.S.; project administration, M.d.N.C.S.; funding acquisition, M.d.N.C.S. All authors have read and agreed to the published version of the manuscript.

Funding: This research was funded by the E_20/2023—Programa Cientista do Nosso Estado—2023 (grant number E-26/203.969/2024) from the Fundação Carlos Chagas Filho de Amparo à Pesquisa do Estado do Rio de Janeiro (FAPERJ), FAPERJ grant number E-26/206.026/2022, the UNIVERSAL Project (CNPq grant number 404166/2021-0), and the Research Productivity Project (CNPq grant number 312170/2023-8). This study was also supported by Fiocruz.

Conflicts of Interest: The authors declare no conflicts of interest.

References

1. WHO Neglected Tropical Diseases—GLOBAL. Available online: https://www.who.int/health-topics/neglected-tropical-diseases (accessed on 3 October 2023).
2. Engels, D.; Zhou, X.-N. Neglected tropical diseases: An effective global response to local poverty-related disease priorities. *Infect. Dis. Poverty* **2020**, *9*, 10. [CrossRef]
3. Hotez, P.J.; Aksoy, S.; Brindley, P.J.; Kamhawi, S. What constitutes a neglected tropical disease? *PLoS Neglected Trop. Dis.* **2020**, *14*, e0008001. [CrossRef]
4. Weld, E.D.; Waitt, C.; Barnes, K.; Garcia Bournissen, F. Twice neglected? Neglected diseases in neglected populations. *Br. J. Clin. Pharmacol.* **2022**, *88*, 367–373. [CrossRef] [PubMed]
5. Kropf, S.P.; Lima, N.T. The history of Chagas disease: Reflections on science in action. *Memórias Inst. Oswaldo Cruz* **2022**, *117*, e200372. [CrossRef]
6. Lynn, M.K.; Bossak, B.H.; Sandifer, P.A.; Watson, A.; Nolan, M.S. Contemporary autochthonous human Chagas disease in the USA. *Acta Trop.* **2020**, *205*, 105361. [CrossRef]
7. Chagas Disease—PAHO/WHO | Pan American Health Organization. Available online: https://www.paho.org/en/topics/chagas-disease (accessed on 17 December 2023).
8. Chagas, C. Nova tripanozomiaze humana: Estudos sobre a morfolojia e o ciclo evolutivo do *Schizotrypanum cruzi* n. gen., n. sp., ajente etiolojico de nova entidade morbida do homem. *Memórias Inst. Oswaldo Cruz* **1909**, *1*, 159–218. [CrossRef]
9. de Sousa, A.S.; Vermeij, D.; Ramos, A.N.; Luquetti, A.O. Chagas disease. *Lancet* **2024**, *403*, 203–218. [CrossRef]

10. Swett, M.C.; Rayes, D.L.; Campos, S.V.; Kumar, R.N. Chagas disease: Epidemiology, diagnosis, and treatment. *Curr. Cardiovasc. Rep.* **2024**, *26*, 1105–1112. [CrossRef] [PubMed]
11. Rassi, A.; Rassi, A.; Marin-Neto, J.A. Chagas disease. *Lancet* **2010**, *375*, 1388–1402. [CrossRef]
12. Bern, C.; Messenger, L.A.; Whitman, J.D.; Maguire, J.H. Chagas disease in the United States: A public health approach. *Clin. Microbiol. Rev.* **2019**, *33*, e00023-19. [CrossRef] [PubMed]
13. Ribeiro, A.L.P.; RAISE Study Collaborators; Machado, Í.; Cousin, E.; Perel, P.; Demacq, C.; Geissbühler, Y.; de Souza, A.; Liprandi, A.S.; Nascimento, B.R.; et al. The burden of Chagas disease in the contemporary world: The RAISE study. *Glob. Heart* **2024**, *19*, 2. [CrossRef] [PubMed]
14. Echavarría, N.G.; Echeverría, L.E.; Stewart, M.; Gallego, C.; Saldarriaga, C. Chagas disease: Chronic Chagas cardiomyopathy. *Curr. Probl. Cardiol.* **2021**, *46*, 100507. [CrossRef] [PubMed]
15. Zuma, A.A.; de Souza, W. Chagas disease chemotherapy: What do we know so far? *Curr. Pharm. Des.* **2021**, *27*, 3963–3995. [CrossRef] [PubMed]
16. Pérez-Molina, J.A.; Crespillo-Andújar, C.; Bosch-Nicolau, P.; Molina, I. Trypanocidal treatment of Chagas disease. *Enfermedades Infecc. Microbiol. Clin. (Engl. Ed.)* **2021**, *39*, 458–470. [CrossRef] [PubMed]
17. Coura, J.R.; Viñas, P.A. Chagas Disease: A new worldwide challenge. *Nature* **2010**, *465*, S6–S7. [CrossRef] [PubMed]
18. Symptoms, Transmission, and Current Treatments for Chagas Disease | DNDi. Available online: https://dndi.org/diseases/chagas/facts (accessed on 3 December 2024).
19. Elkheir, N.; Carter, J.; García-Mingo, A.; Chiodini, P. Chagas disease in non-endemic settings. *Br. Med. J.* **2021**, *373*, n901. [CrossRef] [PubMed]
20. Shikanai Yasuda, M.A. Emerging and reemerging forms of *Trypanosoma cruzi* transmission. *Memórias Instuto Oswaldo Cruz* **2022**, *117*, e210033. [CrossRef]
21. Ullrich, J.; Guertler, L.; Quenzel, E.; Weinauer, F.; Rößler, D.; Kalus, U.; Pruß, A.; Albajar-Viñas, P.; Pritsch, M. Adding a piece to the puzzle of Latin American blood donors and the potential risk of *Trypanosoma cruzi* transmission in Germany. *Front. Cell. Infect. Microbiol.* **2022**, *12*, 1014134. [CrossRef]
22. Wendel, S. Transfusion transmitted Chagas disease: Is it really under control? *Acta Trop.* **2010**, *115*, 28–34. [CrossRef]
23. Pedreira De Freitas, J.L.; Amato Neto, V.; Sonntag, R.; Biancalana, A.; Nussenzveig, V.; Barreto, J.G. First tests on the accidental transmission of Chagas disease to man by blood transfusion. *Rev. Paul. Med.* **1952**, *40*, 36–40.
24. da Costa, A.C.; Rocha, E.A.; da Silva Filho, J.D.; Fidalgo, A.S.O.d.B.V.; Nunes, F.M.M.; Viana, C.E.M.; Gomes, V.B.A.F.; Oliveira, M.d.F. Prevalence of *Trypanosoma cruzi* infection in blood donors. *Arq. Bras. Cardiol.* **2020**, *115*, 1082–1091. [CrossRef]
25. Lidani, K.C.F.; Andrade, F.A.; Bavia, L.; Damasceno, F.S.; Beltrame, M.H.; Messias-Reason, I.J.; Sandri, T.L. Chagas disease: From discovery to a worldwide health problem. *Front. Public Health* **2019**, *7*, 166. [CrossRef]
26. Angheben, A.; Boix, L.; Buonfrate, D.; Gobbi, F.; Bisoffi, Z.; Pupella, S.; Gandini, G.; Aprili, G. Chagas disease and transfusion medicine: A perspective from non-endemic countries. *Blood Transfus.* **2015**, *13*, 540–550. [CrossRef] [PubMed]
27. Crowder, L.A.; Wendel, S.; Bloch, E.M.; O'Brien, S.F.; Delage, G.; Sauleda, S.; Leiby, D.A.; WP-TTID Subgroup on Parasites. International survey of strategies to mitigate transfusion-transmitted *Trypanosoma cruzi* in non-endemic countries, 2016–2018. *Vox Sang.* **2022**, *117*, 58–63. [CrossRef]
28. Wendel, S. A quimioprofilaxia de doenças transmissíveis por transfusão em componentes lábeis hemoterápicos. *Rev. Soc. Bras. Med. Trop.* **2002**, *35*, 275–281. [CrossRef] [PubMed]
29. Bonametti, A.M.; Castelo Filho, A.; Ramos, L.R.; Baldy, J.L.d.S.; Matsuo, T. Infecção por *Trypanosoma cruzi* em candidatos a doador de sangue. *Rev. Saúde Pública* **1998**, *32*, 566–571. [CrossRef] [PubMed]
30. González-Guzmán, S.; Paredes-Cervantes, V.; Bagu, E.T.; Crescencio-Trujillo, J.A.; Guerra-Marquez, Á.; Rivas, N.; Alejandre-Aguilar, R.; Bermúdez-Torres, E.; González-Cano, P. Seroprevalence and geographical distribution of sero-positive blood donors to *Trypanosoma cruzi* at the central blood bank of the national medical center "La Raza". *Transfusion* **2019**, *59*, 639–647. [CrossRef]
31. Wendel, S.; Gonzaga, A.L. Chagas' disease and blood transfusion: A new world problem? *Vox Sang.* **1993**, *64*, 1–12. [CrossRef] [PubMed]
32. Martin, D.L.; Goodhew, B.; Czaicki, N.; Foster, K.; Rajbhandary, S.; Hunter, S.; Brubaker, S.A. *Trypanosoma cruzi* survival following cold storage: Possible implications for tissue banking. *PLoS ONE* **2014**, *9*, e95398. [CrossRef]
33. Cucunubá, Z.M.; Gutiérrez-Romero, S.A.; Ramírez, J.-D.; Velásquez-Ortiz, N.; Ceccarelli, S.; Parra-Henao, G.; Henao-Martínez, A.F.; Rabinovich, J.; Basáñez, M.-G.; Nouvellet, P.; et al. The epidemiology of Chagas disease in the Americas. *Lancet Reg. Health Am.* **2024**, *37*, 100881. [CrossRef] [PubMed]
34. Fernandes, C.D.; Tiecher, F.M.; Balbinot, M.M.; Liarte, D.B.; Scholl, D.; Steindel, M.; Romanha, A. Efficacy of benznidazol treatment for asymptomatic chagasic patients from State of Rio Grande Do Sul evaluated during a three years follow-up. *Memórias Inst. Oswaldo Cruz* **2009**, *104*, 27–32. [CrossRef] [PubMed]

35. Dias, J.C.P.; Ramos, A.N., Jr.; Gontijo, E.D.; Luquetti, A.; Shikanai-Yasuda, M.A.; Coura, J.R.; Torres, R.M.; Melo, J.R.d.C.; de Almeida, E.A.; de Oliveira, W., Jr.; et al. II consenso brasileiro em doença de Chagas, 2015. *Epidemiol. Serviços Saúde* **2016**, *25*, 7–86. [CrossRef]
36. Carrazzone, C.F.V.; de Brito, A.M.; Gomes, Y.M. Importância da avaliação sorológica pré-transfusional em receptores de sangue. *Rev. Bras. Hematol. Hemoter.* **2004**, *26*, 93–98. [CrossRef]
37. Moraes-Souza, H.; Bordin, J.O. Strategies for prevention of transfusion-associated Chagas' disease. *Transfus. Med. Rev.* **1996**, *10*, 161–170. [CrossRef]
38. Lynn, M.K.; Dye-Braumuller, K.C.; Beatty, N.L.; Dorn, P.L.; Klotz, S.A.; Stramer, S.L.; Townsend, R.L.; Kamel, H.; Vannoy, J.M.; Sadler, P.; et al. Evidence of likely autochthonous Chagas disease in the Southwestern United States: A case series of *Trypanosoma cruzi* seropositive blood donors. *Transfusion* **2022**, *62*, 1808–1817. [CrossRef]
39. de Andrade, G.N.; Bosch-Nicolau, P.; Nascimento, B.R.; Martins-Melo, F.R.; Perel, P.; Geissbühler, Y.; Demacq, C.; Quijano, M.; Mosser, J.F.; Cousin, E.; et al. Prevalence of Chagas disease among Latin American immigrants in non-endemic countries: An updated systematic review and meta-analysis. *Lancet Reg. Health Eur.* **2024**, *46*, 101040. [CrossRef] [PubMed]
40. Martins-Melo, F.R.; Carneiro, M.; Ribeiro, A.L.P.; Bezerra, J.M.T.; Werneck, G.L. Burden of Chagas disease in Brazil, 1990–2016: Findings from the global burden of disease study 2016. *Int. J. Parasitol.* **2019**, *49*, 301–310. [CrossRef] [PubMed]
41. Bern, C.; Montgomery, S.P.; Katz, L.; Caglioti, S.; Stramer, S.L. Chagas disease and the US blood supply. *Curr. Opin. Infect. Dis.* **2008**, *21*, 476–482. [CrossRef] [PubMed]
42. Forsyth, C.J.; Manne-Goehler, J.; Bern, C.; Whitman, J.; Hochberg, N.S.; Edwards, M.; Marcus, R.; Beatty, N.L.; Castro-Sesquen, Y.E.; Coyle, C.; et al. Recommendations for screening and diagnosis of Chagas disease in the United States. *J. Infect. Dis.* **2021**, *225*, 1601–1610. [CrossRef] [PubMed]
43. Turabelidze, G.; Vasudevan, A.; Rojas-Moreno, C.; Montgomery, S.P.; Baker, M.; Pratt, D.; Enyeart, S. Autochthonous Chagas disease—Missouri, 2018. *Morb. Mortal. Wkly. Rep.* **2020**, *69*, 193–195. [CrossRef]
44. Fearon, M.A.; Scalia, V.; Huang, M.; Dines, I.; Ndao, M.; Lagacé-Wiens, P. A case of vertical transmission of Chagas disease contracted via blood transfusion in Canada. *Can. J. Infect. Dis. Med. Microbiol.* **2013**, *24*, 32–34. [CrossRef] [PubMed]
45. Guggenbühl Noller, J.M.; Froeschl, G.; Eisermann, P.; Jochum, J.; Theuring, S.; Reiter-Owona, I.; Bissinger, A.L.; Hoelscher, M.; Bakuli, A.; von Sonnenburg, F.-J.F.; et al. Describing nearly two decades of Chagas disease in Germany and the lessons learned: A retrospective study on screening, detection, diagnosis, and treatment of *Trypanosoma cruzi* infection from 2000–2018. *BMC Infect. Dis.* **2020**, *20*, 919. [CrossRef] [PubMed]
46. Wirth, M.; Gálvez, R.I.; Jochum, J.; Strauss, R.; Kristensen, K.; Stich, A.; Stegemann, M.; Stahl, P.; Puchner, K.P.; Strasen, J.; et al. Knowledge, attitudes, behaviors, and serological status related to Chagas disease among Latin American migrants in Germany: A cross-sectional study in six German cities. *Front. Cell. Infect. Microbiol.* **2023**, *12*, 1047281. [CrossRef] [PubMed]
47. Cancino-Faure, B.; Fisa, R.; Riera, C.; Bula, I.; Girona-Llobera, E.; Jimenez-Marco, T. Evidence of meaningful levels of *Trypanosoma cruzi* in platelet concentrates from seropositive blood donors. *Transfusion* **2015**, *55*, 1249–1255. [CrossRef] [PubMed]
48. Benjamin, R.J.; Stramer, S.L.; Leiby, D.A.; Dodd, R.Y.; Fearon, M.; Castro, E. *Trypanosoma cruzi* infection in North America and Spain: Evidence in support of transfusion transmission (CME). *Transfusion* **2012**, *52*, 1913–1921. [CrossRef] [PubMed]
49. Nussenzweig, V.; Sonntag, R.; Biancalana, A.; De Freitas, J.L.P.; Amato Neto, V.; Kloetzel, J. Effect of triphenylmethane dyes on *Trypanosoma cruzi* in vitro; use of gentian violet in prevention of transmission of Chagas disease by blood transfusion. *Hospital* **1953**, *44*, 731–744. [PubMed]
50. Souza, H.M. De Quimioprofilaxia da doença de Chagas transfusional. Realidade atual. *Rev. Soc. Bras. Med. Trop.* **1989**, *22*, 1–3. [CrossRef] [PubMed]
51. Docampo, R.; Moreno, S.N.; Muniz, R.P.; Cruz, F.S.; Mason, R.P. Light-enhanced free radical formation and trypanocidal action of gentian violet (crystal violet). *Science* **1983**, *220*, 1292–1295. [CrossRef] [PubMed]
52. Rezende, J.; Zupello, W.; Bafutto, M. O problema da transmissão da doença de Chagas por transfusão de sangue. Emprego da violeta de genciana como medida profilática. *Rev. Goiana Medicina.* **1965**, *11*, 35–47.
53. Moraes-Souza, H.; Kerbauy, J.; Barretto, O.C.; Pühler, G.M.; Nonoyama, K.; Juliano, Y. Metabolism and preservation of fresh and stored erythrocytes in blood treated with gentian violet. *Braz. J. Med. Biol. Res.* **1988**, *21*, 241–246. [PubMed]
54. Andrade, J.P.; Marin-Neto, J.A.; Paola, A.a.V.; Vilas-Boas, F.; Oliveira, G.M.M.; Bacal, F.; Bocchi, E.A.; Almeida, D.R.; Fragata Filho, A.A.; Moreira, M.C.V.; et al. I diretriz Latino-Americana para o diagnóstico e tratamento da cardiopatia chagásica. *Arq. Bras. Cardiologia.* **2011**, *97*, 01–48. [CrossRef]
55. Jimenez-Marco, T.; Cancino-Faure, B.; Girona-Llobera, E.; Alcover, M.M.; Riera, C.; Fisa, R. The effectiveness of riboflavin and ultraviolet light pathogen reduction technology in eliminating *Trypanosoma cruzi* from leukoreduced whole blood. *Transfusion* **2017**, *57*, 1440–1447. [CrossRef] [PubMed]
56. Mundt, J.M.; Rouse, L.; Van den Bossche, J.; Goodrich, R.P. Chemical and biological mechanisms of pathogen reduction technologies. *Photochem. Photobiol.* **2014**, *90*, 957–964. [CrossRef] [PubMed]

57. Timm, B.L.; da Gama, A.N.S.; Batista, M.M.; Batista, D.d.G.J.; Boykin, D.W.; De Koning, H.P.; Correia Soeiro, M.d.N. Arylimidamides have potential for chemoprophylaxis against blood-transmitted Chagas disease. *Pathogens* **2023**, *12*, 701. [CrossRef]
58. Guedes-da-Silva, F.; Batista, D.; Meuser, M.; Demarque, K.; Fulco, T.; Araújo, J.; Da Silva, P.; Da Silva, C.F.; Patrick, D.A.; Bakunova, S.M.; et al. In vitro and in vivo trypanosomicidal action of novel arylimidamides against *Trypanosoma cruzi*. *Antimicrob. Agents Chemother.* **2016**, *60*, 2425–2434. [CrossRef] [PubMed]
59. Pereira, L.Q.; Tanaka, S.C.S.V.; Ferreira-Silva, M.M.; Gomes, F.V.B.A.F.; Santana, M.P.; Aguiar, P.R.; de Araújo Pereira, G.; Gómez-Hernández, C.; Junior, V.R.; De Vito, F.B.; et al. Leukoreduction as a control measure in transfusion transmission of visceral leishmaniasis. *Transfusion* **2023**, *63*, 1044–1049. [CrossRef] [PubMed]
60. Murphy, E.L. Infection with Hhuman T-lymphotropic virus types-1 and -2 (HTLV-1 and -2): Implications for blood transfusion safety. *Transfus. Clin. Biol.* **2016**, *23*, 13–19. [CrossRef]
61. Ziemann, M.; Hennig, H. Prevention of transfusion-transmitted cytomegalovirus infections: Which is the optimal strategy? *Transfus. Med. Hemotherapy* **2013**, *41*, 40–44. [CrossRef] [PubMed]
62. Leiby, D.A.; O'Brien, S.F.; Wendel, S.; Nguyen, M.L.; Delage, G.; Devare, S.G.; Hardiman, A.; Nakhasi, H.L.; Sauleda, S.; Bloch, E.M.; et al. International survey on the impact of parasitic infections: Frequency of transmission and current mitigation strategies. *Vox Sang.* **2019**, *114*, 17–27. [CrossRef] [PubMed]

pathogens

MDPI

Review

Molecular Markers for the Phylogenetic Reconstruction of *Trypanosoma cruzi*: A Quantitative Review

David Ramírez-Delgado [1] and Carlos Alberto Flores-López [2,*]

1 Independent Researcher, Ensenada 22860, BC, Mexico; linfocitoth1@gmail.com
2 Facultad de Ciencias, Universidad Autónoma de Baja California, Ensenada 22860, BC, Mexico
* Correspondence: cflores2@uabc.edu.mx

Abstract: *Trypanosoma cruzi* is the parasite responsible for Chagas disease, which has a significant amount of genetic diversification among the species complex. Many efforts are routinely made to characterize the genetic lineages of *T. cruzi* circulating in a particular geographic area. However, the genetic loci used to typify the genetic lineages of *T. cruzi* have not been consistent between studies. We report a quantitative analysis of the phylogenetic power that is acquired from the commonly used genetic loci that are employed for the typification of *T. cruzi* into its current taxonomic nomenclature. Based on three quantitative criteria (the number of phylogenetic informative characters, number of available reference sequences in public repositories, and accessibility to DNA sequences for their use as outgroup sequences), we examine and discuss the most appropriate genetic loci for the genetic typification of *T. cruzi*. Although the mini-exon gene is by far the locus that has been most widely used, it is not the most appropriate marker for the typification of *T. cruzi* based on the construction of a resolved phylogenetic tree. Overall, the mitochondrial COII-NDI locus stands out as the best molecular marker for this purpose, followed by the Cytochrome b and the Lathosterol oxidase genes.

Keywords: *Trypanosoma cruzi* phylogenetics; discrete typing units (DTUs); phylogenetic resolution; *Trypanosoma cruzi* molecular markers; *Trypanosoma cruzi* genetic diversity; phylogenetically informative characters

1. Biological Rationale for Genetically Characterizing *Trypanosoma cruzi*

Trypanosoma cruzi is a kinetoplastid parasite that requires two species to complete its life cycle [1]. It has been described in at least nine orders of mammals in the American continent [2]. *T. cruzi* is a heteroxenous parasite, requiring two different hosts to complete its life cycle. It requires a vertebrate host (primarily mammals of the orders Marsupialia, Rodentia, and Carnivora) and a vector invertebrate host (species within the Triatominae subfamily). The hematophagous triatomines become infected when they feed on an infected vertebrate host. Inside the triatomines, the parasite multiplies within the hindgut. Subsequently, the parasite is transmitted to a vertebrate host by parasites present in the feces of the vector if they enter either via contact with the bite site or through a mucosa (e.g., through the conjunctiva or mucous). Alternatively, a mammal can become infected if the mammal ingests the infected vector [3]. Inside the mammal host, the parasite develops intracellularly, where it has the potential to invade a wide a variety of tissues, eventually rupturing these cells and releasing parasites into the bloodstream. In the bloodstream, the parasite can infect new tissues within the mammal or be picked up by a triatomine vector feeding on that host to continue its life cycle [4].

 https://doi.org/10.3390/pathogens14010072

Molecular clock analyses have estimated that *T. cruzi* basal genetic lineages started diverging approximately two million years ago [5], which is in accordance with the significant amount of genetic divergence observed between strains [6–13].

This large genetic diversity observed within *T. cruzi* is potentially a result of the parasite's broad distribution of hosts and habitats, leading to particular genomic adaptations. Although a clear picture of the actual evolutionary history of how this adaptation took place has not been revealed, there is a likelihood that if different evolutionary lineages of the parasite evolved within different niches (i.e., hosts and vectors), this would have led to specific genomic adaptations unique to each lineage. Alongside the neutral genetic divergence expected of predominantly clonal populations, these factors most likely contributed to the significant current genetic divergence observed within *T. cruzi* lineages.

Concerning the genetic differences that have been identified in *T. cruzi*, these appear to have certain biogeographical patterns [8,14–19]. Due to these patterns, it has been hypothesized that there could be a correlation between the genetic background of the parasite and the diverse pathological outcomes of Chagas disease. However, this association remains to be determined since only a handful of studies have found a potential correlation [19–23], while others have not [14,17]. Larger studies are needed to evaluate this. Nonetheless, the different pathological outcomes that can be developed from an infected patient have been well studied.

Within a human host, the parasite can infect multiple tissues and cause various pathologies. The disease classically presents itself with a short incubation period (1 to 2 weeks). After this, the acute phase begins, lasting 8 to 12 weeks. Although parasitemia is present, in most cases, it remains asymptomatic or mild, and non-specific symptoms are observed (fever, malaise, anorexia, headache, diarrhea, myalgia, and lymphadenopathy) [24–26]. The Romaña's sign could also be presented if the conjunctiva was the entry point [24]. In total, the severe acute phase occurs in less than 5% of cases and resolves spontaneously [25]. Nevertheless, 30% of untreated patients progress to a chronic phase, which can be categorized into indeterminate, cardiac, digestive, or neurological forms, the latter occurring in sporadic cases [27,28]. Patients in the indeterminate form remain asymptomatic for a prolonged period (e.g., decades) [24]. Regarding the genetic diversity of *T. cruzi*, this pathological outcome of Chagas disease is the most common form observed in Brazil. However, the cardiac and cardio-digestive forms are less frequent [15]. On the other hand, in the chronic forms, patients exhibit cardiovascular and digestive symptoms driven by an exacerbated inflammatory immune response. The digestive form involves the megaesophagus or megacolon observed in severe cases. As mentioned above, these presentations of Chagas disease are observed in Argentina, Brazil, Bolivia, Chile, Paraguay, and Uruguay but are virtually absent in Venezuela and Central America [15,19,27,29]. However, chronic cardiac manifestations are commonly observed in patients born in Argentina, Bolivia, Colombia, Venezuela, and Brazil [21].

As previously mentioned, there is no consensus on a possible correlation between the genetic background of *T. cruzi* and the diverse array of pathologies linked with Chagas disease. Furthermore, recent approaches to explain the complexity of the outcomes of Chagas disease involve the host–vector–pathogen tripartite interaction complex [13,15,27]. However, the genetic description of a biogeographical correlation has been thoroughly described [13,18,19,23,30], which gives grounds for the typification of the genetic lineages of this important human parasite.

2. Taxonomic Classification of *Trypanosoma cruzi*

Trypanosoma cruzi was described more than one hundred years ago by Carlos Chagas [1], and advances in molecular technologies have enabled its classification with greater precision. Since morphological differences are not apparent between different *T. cruzi* lineages, a molecular description is needed to identify the different genetic lineages of the parasite. The first attempt to classify *T. cruzi* with a molecular approach was based on the enzyme profile obtained by comparing the electrophoresis pattern retrieved from multiple enzymes of particular *T. cruzi* strains, given the fact that different strains have different alleles for the same enzymes and hence produce different protein bands when separated in electrophoresis by size and charge. This technique, called isoenzyme analysis, resulted in *T. cruzi* being classified into zymodemes (Z1, Z2 and Z3) [31]. Morel et al. [32] was the first to use the term schizodemes to classify *T. cruzi*. This classification was based on the restriction fragments that are obtained from the amplification of the *T. cruzi* minicircle gene. Following the schizodeme characterization, a more sensitive way to study kDNA polymorphism was the low-stringency single specific primer–PCR (LSSP-PCR) method, developed by Vago et al. [33]. In parallel, additional techniques were employed, such as the random amplified polymorphic DNA (RAPD) technique [10,11,34–36]. This approach was consistent with the description of two major lineages previously identified through the zymodeme analysis [31,37]. Moreover, with the development of molecular biology technologies, different approaches were applied to solve the elusive classification of this infectious agent. Eventually, the polymorphism of a specific messenger RNA (mRNA) sequence, known as mini-exon, was employed. The mini-exon gene is a trypanosomatid unique mRNA transcript [7,11,15,38]. Initially, the use of the mini-exon allowed the clustering of *T. cruzi* strains into two groups: Group 1, which yielded 300 bp PCR amplicons, and Group 2, which yielded 350 bp amplicons [11]. The subsequent analysis of Z3 strains identified a sub-clade inside group 2 [7]. Notably, trypanosomatids possess a unique organization of their ribosomal RNA (rRNA). Both the small and large subunits of their rRNA are significantly larger than those found in typical eukaryotic species. Within this organization, the small and large subunits are separated by two internal transcribed spacers (ITS1 and ITS2), which have been used to differentiate *T. cruzi* strains, as well as the products of 24SDNA rRNA [11,15]. The effort to demonstrate the phylogenetic relationships between the various strains of *T. cruzi* has employed other strategies, such as karyotypic comparisons [39,40]. However, the discovery of microsatellites provided another method for analyzing the *T. cruzi* population structure and phylogenetic relationship strains [9,41].

Overall, even though a diverse array of techniques have been employed to classify *T. cruzi* strains (MLEE, rRNA, mini-exon, and karyotype), the outcome has not been consistent. Subsequently, once DNA sequencing technologies were developed, the window was opened for researchers to start analyzing DNA sequence data. At the molecular level, DNA sequence analysis always provides the highest resolution and, in most cases, offers insights into resolving classification issues. So, it is not surprising that the DNA sequence of *T. cruzi* was also sought, and as a result, the first DNA sequences for *T. cruzi* were analyzed, and thus, data with a higher resolution to unravel the genetic background of *T. cruzi* strains were adopted. These first studies focused on reconstructing the evolutionary history of *T. cruzi* based on constructing a phylogenetic tree from ribosomal RNA sequences and the mini-exon gene [11,42]. This effort initially classified *T. cruzi* into two major groups designated as *T. cruzi* I and *T. cruzi* II [6,11,43]. Subsequent analyses subdivided TcII into five subgroups (IIa, IIb, IIc, IId, and IIe) [11,44]. However, more refined phylogenetic analysis revealed TcII to be a paraphyletic group [45], and thus, the division of *T. cruzi* into two major lineages was eventually modified. As a result of these efforts to classify *T. cruzi*, a novel nomenclature was introduced [17]. The updated classification scheme was based on a multilocus genotyping approach, which divided the genetic diversity into six groups named "Discrete Typing Units" (DTUs), which include TcI, TcII, TcIII, TcIV, TcV, and TcVI [46]. Furthermore, based on thorough phylogenetic analyses, the current classification status based on six DTUs has been challenged, where an alternative classification based on three mitochondrial clusters appears to be more consistent with DNA sequence data [47]. Nonetheless, since the six-DTU classification system was established, additional lineages have been discovered (TcBat and TcIV-USA), suggesting the true genetic diversity of the parasite will become greater once more exhaustive sampling is conducted [48,49]. Consequently, a new upgraded classification system should not be discarded.

In view of the current classification system, the methodology applied to identify the DTU of a particular *T. cruzi* sample usually requires the PCR amplification of a genetic locus (locus), followed by the comparison of the pattern obtained from the particular locus with reference sequences. Some methods exclude the need to determine the DNA sequence of the particular locus used for the typification of the parasite [50,51]. For example, the commonly used mini-exon locus consists of typifying *T. cruzi* based on the distinct amplicon sizes that are recovered from the PCR amplification and subsequent visualization in an agarose gel electrophoresis [7,11,42,52]. However, the highest resolution is obtained by the construction of a phylogenetic tree based on the DNA sequences of the target samples [12]. This phylogenetic analysis involves the additional procedure of sequencing the PCR amplicons. Therefore, using loci with appropriate features for the typification of *T. cruzi* DTUs is an important step in the procedure adopted.

3. Traits of a Suitable Genetic Locus for Typifying *T. cruzi*

The molecular markers used in reconstructing phylogenetic trees require them to have particular traits. One of the most important traits is having sufficient synapomorphies (shared and derived DNA polymorphisms), also known as phylogenetic informative characters (PICs) [53]. PICs are DNA polymorphisms shared by at least two individuals within the DNA dataset and are thus considered vital for reconstructing the respective sequences' evolutionary history and allowing a fully resolved phylogenetic tree to be constructed from the sequences of interest. In other words, they are the number of positions within the DNA sequence alignment that are useful for typifying the genetic lineages of *T. cruzi* (DTUs). The number of PICs needs to be balanced in the genetic locus being considered since there has to be enough DNA sequence conservation between the sequences being compared to ensure homologous sites are appropriately being compared (i.e., a correctly constructed DNA sequence alignment), as well as being conserved enough in the regions where the primers will attach to guarantee the effectivity of the primers across the taxonomic clade intended to be amplified [54,55]. In parallel, the genetic locus should not be too conserved within the amplicon region, or it will not have sufficient PICs to construct a resolved phylogenetic tree.

The phylogenetic tree construction is a powerful taxonomic classification tool; however, it needs to meet some assumptions. In this context, the identification of DNA polymorphisms as derived (apomorphies) or ancestral characters (plesiomorphies) is usually accomplished when an outgroup sequence is included in the analysis, which should preferentially consist of a species that shares relatively recent common ancestry with the ingroup (sequences whose phylogenetic reconstruction is the objective of the study) to qualify as an appropriate outgroup sequence [56–59]. Thus, in the case of *T. cruzi*, these could be homologous DNA sequences of *T. cruzi marinkellei*, *T. vespertilionis*, *T. rangeli*, or other closely related taxa that have not suffered too much sequence divergence compared to the ingroup. In contrast, *T. brucei*, *Leishmania* spp., or other kinetoplastids with a significant divergence from *T. cruzi* would not be appropriate outgroups.

Furthermore, it would be desirable for the molecular marker of interest to have reference sequences of all *T. cruzi* DTUs available in a public database (e.g., GenBank or TritrypDB), allowing their inclusion in the phylogenetic analysis. Otherwise, a study would have to sequence reference strains to compare the target sequences with the reference DTUs, thus enabling the genetic typification of the target *T. cruzi* sample. Considering that the objective of constructing the phylogenetic tree relies on including reference sequences in the DNA sequence dataset, working with a molecular marker (MM) with several reference sequences a priori is desirable. Therefore, the number and diversity of *T. cruzi* sequences available for a particular MM are advantageous attributes.

An additional key factor to consider in the reconstruction of the evolutionary history of a group of species based on a phylogenetic tree is the use of orthologous sequences (homologous sequences whose divergence dates back to a speciation event) in the analysis [60–62]. Failure to do so (i.e., including paralogous sequences) could result in erroneous conclusions since a phylogenetic tree constructed from paralogous sequences (homologous sequences whose divergence dates to a duplication event) will reflect the evolutionary history of the gene and not the species, resulting in the erroneous clustering of sequences [63]. In other words, using MMs that are not part of a gene family with genetic divergence among its gene members is preferable. Fortunately, none of the MMs commonly used to typify *T. cruzi* have relied on gene copies that are suspected to be a part of a gene family. It should be clarified that genetic loci that are found in more than a single copy are not useless; on the contrary, they can be beneficial for diagnostic purposes, as is the case with the minicircle or mini-exon genes, given their high copy number in the kinetoplastid genomes, resulting in their efficacy as molecular markers for the diagnosis of a particular species. Despite this, since DNA polymorphisms can be found within the various copies of the paralogous sequences, their use in phylogenetic inferences is not ideal. In short, the use of orthologous sequences is favored for the reconstruction of the evolutionary history of a particular group of species [61], including the typification of *T. cruzi*.

Finally, a further consideration for an effective molecular marker used to typify a particular species is the size of the molecular marker. Given that the PCR has been the method of choice for amplifying any genetic locus in the typification of *T. cruzi*, it is desirable for the locus to be amplified in a single PCR. In other words, it is desirable for the amplicon size to not be larger than the capacity of the commonly used DNA polymerases [64]. Overall, except for the maxicircle that consists of a concatenation of various mitochondrial genes, all other MMs commonly used to typify *T. cruzi* can be amplified in a single PCR.

In summary, the most suitable loci for typifying *T. cruzi* into DTUs should be analyzed based on having numerous phylogenetically informative characters (synapomorphies), the availability of a sequence that can be used as an outgroup, and an appropriate amount of reference sequences available in public repositories (e.g., GenBank). Therefore, the objective of this review was to provide a quantitative evaluation of the most used genetic locus that has been consistently used for typifying *T. cruzi* into DTUs. Given that the most commonly used loci for the typification of *T. cruzi* are amplified in a single PCR and are not paralogs of diverse gene families, our criteria to evaluate the most appropriate MMs in this review are based on (1) the number of PICs found within each MM; (2) the number of available reference *T. cruzi* sequences in GenBank; and (3) the availability of suitable outgroups.

4. Quantitative Approach for the Evaluation of MMs

A bibliography search was performed for published articles that included the keywords "cruzi", "DTU" and "GenBank" to identify the most common sequenced genetic loci that have been used to typify *T. cruzi* DTUs and whose sequences have been made available through their repository in GenBank. When genetic loci were identified that had been used in multiple studies, the corresponding GenBank accession of one of the reference sequences reported was used as a query for a search of homologous *T. cruzi* sequences (the GenBank accession code used in each dataset is reported in Table 1). The homologous DNA sequence search was conducted with BLAST [65], with default parameters. Based on these search criteria, all homologous sequences that covered a minimum threshold of 90% of the query sequence were downloaded from GenBank. This threshold was set to limit the construction of each dataset to sequences that could be confidently aligned. In the case of the maxicircle dataset, given that this dataset consists of the concatenation of several mitochondrial genes, the query cover threshold was lowered to 85% since most homologous hits were just below the 90% query cover.

Table 1. Summary of traits for the molecular markers most utilized for typifying *Trypanosoma cruzi*.

MM	bp	n	nG	PICs	% PICs	Acc. No.	OS	GL
COII-NDI	1119	179	645	176	15.72	HQ604893.1	T. c. m.	MT
Maxicircle	3681	120	916	110	2.98	KP136828.1	sev.	MT
Cyt b	533	252	286	99	18.57	HQ713679.1	T. c. m.	MT
Lathosterol oxidase	837	310	339	98	11.70	JN050587.1	T. c. m.	NUC
18S rRNA	641	348	870	86	13.41	LT220268.1	T. c. m. *	NUC
mini-exon	278	1138	1643	63	22.66	AM259467.1	NA	NUC
5S rRNA	550	16	190	47	8.54	M59503.1	NA *	NUC
DHFR-TS	1375	98	164	42	3.05	MF784874.1	T. c. m.	NUC
MK	615	47	71	25	4.06	MH671655.1	NA	NUC
MSH2	737	116	113	21	2.84	MF074182.1	T. r.	NUC
GPI	612	61	291	21	3.43	KT737478.1	T. c. m.	NUC
CatL-like *	478	71	96	19	3.97	JF421288.1	T. c. m.	NUC
Superoxide dismutase (sup. dism.)	570	11	79	18	3.15	XM_808937.1	T. r.	NUC
gGAPDH	853	23	79	17	1.99	OQ095397.1	T. c. m.	NUC
ITS1	1244	6	112	15	1.20	AF362831.1	T. r.	NUC
Tc24	675	49	51	13	1.92	OL781152.1	T. con.	NUC
HSP70	696	27	32	12	1.72	ON653039.1	T. c. m.	NUC
Minicircle *	71	48	50	10	14.08	AJ747914.1	Vert.	MT
5S rRNA (M5)	115	168	189	2	1.73	XR_008255252.1	T. r.	NUC

MM: molecular marker. bp: number of nucleotides included in DNA sequence alignment dataset. n: number of *T. cruzi* sequences included in dataset. nG: number of homologous sequences available of *T. cruzi* in GenBank as of 6 September 2024. PICs: number of SNPs that are phylogenetically informative characters. %PICs: percentage of dataset with phylogenetic informative characters. Acc. No.: GenBank accession code of sequence used as a query to create dataset. OS: closest related outgroup species. GL: genomic location. MT: located in mitochondrial genome. NUC: located in nuclear genome. T. c. m.: *Trypanosoma cruzi marinkellei*. T. r: *Trypanosoma rangeli*. T. con: *Trypanosoma conorhini*. sev: several different closely related species, but varies given the gene section analyzed since the maxicircle MM is concatenation of different loci. Vert: only found vertebrate DNA. NA: not available. NA *: not available that cover >90% of query. Minicircle *: dataset had the divergent regions removed with gBlocks (see methods for justification). T. c. m. *: The outgroup sequence clusters within the *T. cruzi* ingroup sequences, potentially reflecting that either some *T. cruzi* labeled sequences in GenBank are misidentified, or the outgroup sequence is misidentified. This could be overestimating the number of PICs. CatL-like *: several sequences labeled as *T. cruzi* were removed from the dataset since they appear to be misidentified (sequences removed: MZ087227.1, MZ087228.1, M90067.1, MZ087241.1, MZ087239.1, MZ087242.1, MZ087240.1, MZ087238.1, MZ087237.1, XM_797951.1, MZ087236.1, JF421346.1).

All constructed datasets were aligned with MUSCLE in SeaView version 4.0 [66,67]. The minicircle dataset had to be purged of the most divergent sequence regions from the alignment with the use of gBlocks version 0.91b [68] since the sequences were too divergent to construct an alignment confidently. Excluding the minicircle dataset, the homologous *T. cruzi* DNA sequences retrieved made up the DNA alignments of each genetic locus (all constructed alignments were manually checked and included in the Supplementary Material).

A verification of the authenticity of all sequences belonging to *T. cruzi* was performed by constructing a neighbor-joining tree with the inclusion of an outgroup sequence. In most cases, a *T. cruzi marinkellei* homologous sequence was used, with the exception being cases where a *T. c. marinkellei* homologous sequence was not available, meaning that another closely related taxon was used (e.g., *T. conorhini* with Tc24). Within the phylogenetic tree, any sequences that appeared more distantly related (i.e., outside of the monophyletic clade formed by *T. cruzi* sequences) to the ingroup and outgroup were discarded from the dataset since they could have represented misidentified sequences that were erroneously labeled as *T. cruzi* and therefore would have artificially increased the number of PICs estimated from each locus. This verification step could not be carried out for the mini-exon, 5S rRNA, MK or minicircle loci, which have the potential of having slightly higher PICs within their datasets.

To assess the availability of outgroup sequences for each MM, a BLAST search was performed on the NCBI using the representative query DNA sequence (labeled 'Ac.' in Table 1), excluding *T. cruzi* sequences. Based on these searches, the closest related species to *T. cruzi* identified was recorded in the column labeled "Out" in Table 1.

The PAUP version 4.0 software package (Swofford, 1991) was used to estimate the number of phylogenetically informative characters (alternatively called parsimony informative sites) within each MM. Table 1 summarizes the measured traits for each genetic locus.

5. Highest-Ranking Genetic Loci from the Quantitative Review

According to the bibliographic review performed, 19 genetic loci were identified as the most frequent molecular markers used to typify the genetic background of *T. cruzi*. These genetic loci were quite variable in the range of traits measured: (1) amplicon sizes between 115 and 3681 bp; (2) the number of *T. cruzi* reference sequences publicly available between 32 and 1643; and (3) the number of phylogenetic informative characters constituting 2–176 (Table 1). In total, 4 of the 19 markers are found within the mitochondrial genome, while the remaining loci are found in the nuclear genome, with the top three ranking molecular markers being mitochondrial.

Based on the quantitatively reviewed criteria, the COII-NDI gene segment is the best locus for *T. cruzi* typing (Table 1 and Figure 1A). This mitochondrial locus codes for the cytochrome c oxidase subunit II and NADH dehydrogenase subunit I proteins, which play a part in the mitochondrial electron transport chain and function in ATP synthesis. Given these enzymes' vital role, both proteins are ubiquitous in eukaryotes. Nevertheless, a particular trait of kinetoplastids is the ligation of the two genes within the mitochondrial genome. This particular trait is favorable for its use as a molecular marker, given the high abundance of PICs among both genes, resulting in a well-resolved phylogenetic tree (Figure 2A). In *T. cruzi*, the analyzed COII-NDI dataset had a total of 176 PICs. This high number of PICs resulted in COII-NDI having 60% more PICs than the second-ranking locus with the most number of PICs (Table 1), making COII-NDI the top molecular marker in terms of the phylogenetic resolution that can be obtained for the genetic typification of *T. cruzi* based on a single locus. Furthermore, it has a significant amount of available reference *T. cruzi* sequences, ranking number #4 (Figure 1B) as the locus with the most number of reference sequences available in GenBank at the time of this study (645 *T. cruzi* homologous DNA sequences), of which 179 sequences covered the minimum 90% query cover threshold that was set for the construction of each locus dataset (Table 1). Thus, constructing a *T. cruzi* phylogenetic tree with a substantial number of *T. cruzi* reference sequences is achievable with the COII-NDI locus.

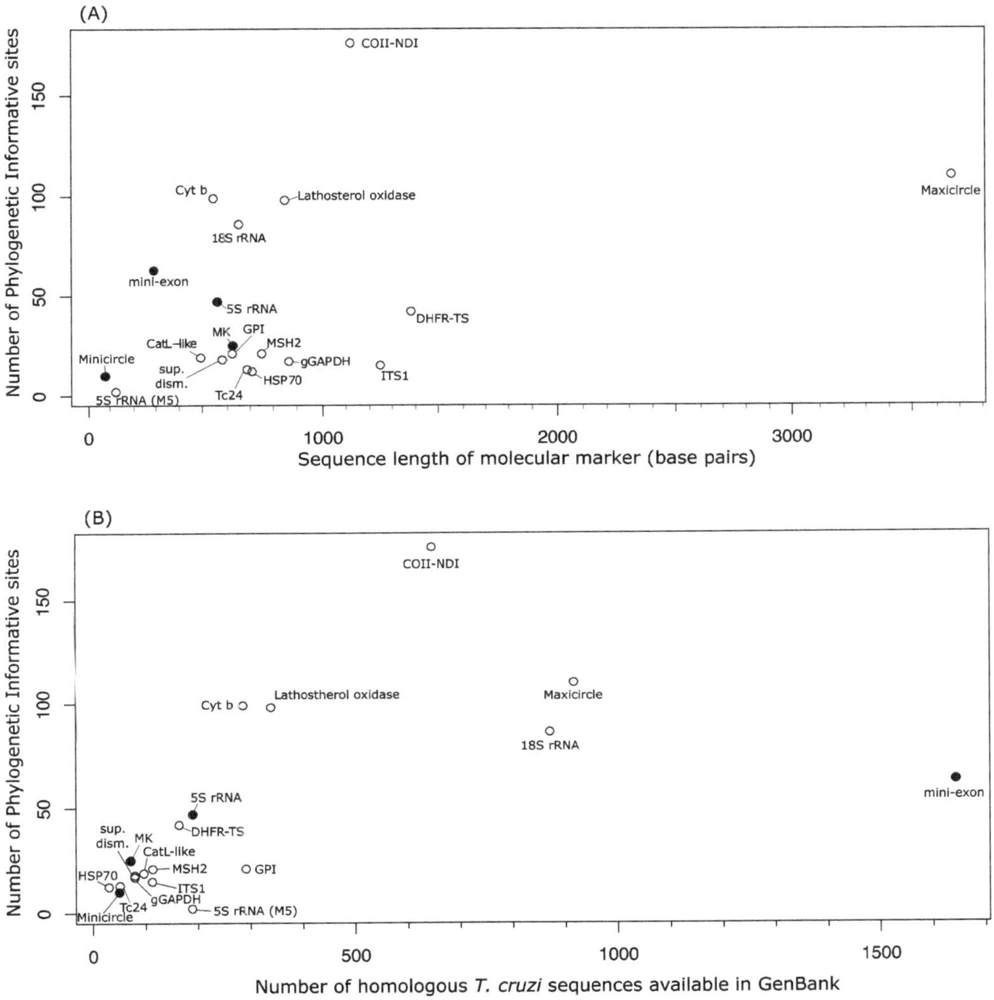

Figure 1. A quantitative evaluation of molecular markers for typifying *Trypanosoma cruzi*. (**A**) The number of phylogenetic informative characters (PICs) identified for the most utilized molecular markers used to typify the genetic background of *Trypanosoma cruzi* lineages. (**B**) The number of *T. cruzi* reference DNA sequences available for each MM as of June 2024. The metrics underscore the variability in phylogenetic informative traits between the most common MMs used to typify *T. cruzi*. Closed circles represent MMs that lack an appropriate outgroup. Open circles represent MMs with appropriate outgroup sequences in public repositories.

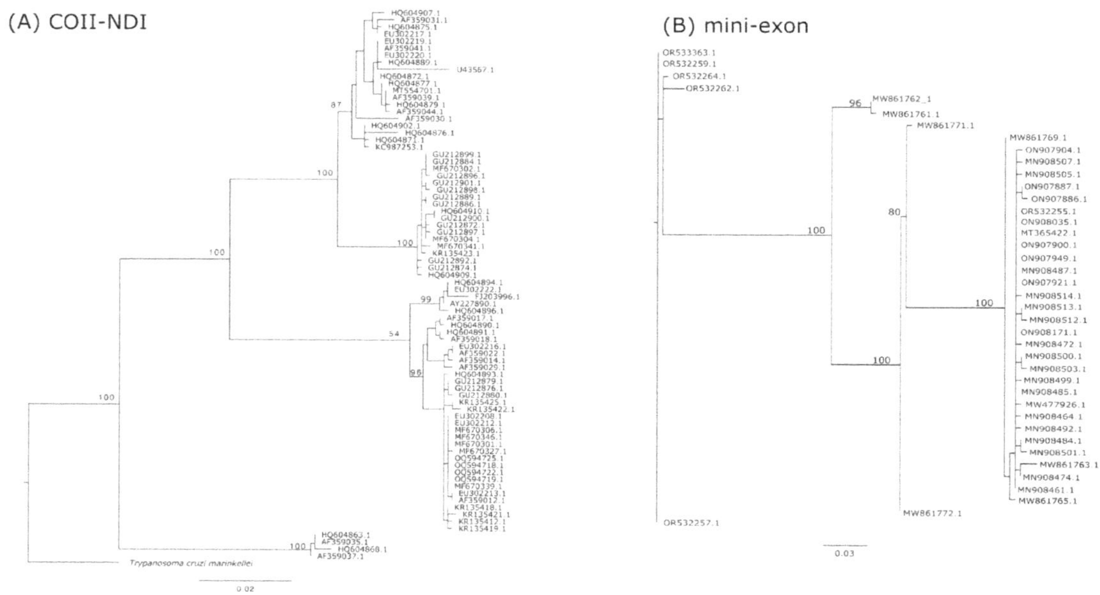

Figure 2. A phylogenetic resolution of the COII-NDI and mini-exon molecular markers to characterize the genetic diversity of *Trypanosoma cruzi.* (**A**) A maximum likelihood tree constructed from the COII-NDI locus. The tree was rooted with *T. cruzi marinkellei.* (**B**) A maximum likelihood tree constructed with the mini-exon locus. Given the absence of an appropriate outgroup for the mini-exon marker, the tree was not rooted. The comparison presents the differences in the phylogenetic resolution obtained from both MMs, underscoring the phylogenetic limitation of an MM with a lower amount of PICs and the lack of having an appropriate outgroup sequence to root the phylogenetic tree. The values above branches represent bootstrap values based on 1000 replicates. Both trees were constructed in W-IQ-TREE (http://www.iqtree.org), using IQ-TREE efficient tree reconstruction [69] and the UFBoot—ultrafast bootstrap approximation [70]. Phylogenetic trees were constructed from a limited number of the available DNA sequences in GenBank. To see the phylogenetic trees that include the totality of DNA sequences available in GenBank, see Supplementary Figure S1.

Moreover, COII-NDI has a large variety of closely related *Trypanosoma* spp. DNA sequences that can be used as outgroups to root the phylogenetic tree, which is not surprising given the widespread use of the maxicircle in kinetoplastids studies [45,71,72]. Thus, based on the criteria evaluated, the COII-NDI genetic locus is the best option for the genetic typification of *T. cruzi* DTUs.

Following the rank in terms of the number of PICs found within the genetic loci is the maxicircle with 110 PICs (Table 1 and Figure 1A). This genetic locus is unique to kinetoplastids since it is found in the genome that gives kinetoplastids their name (the "kinetoplast"). As a molecular marker, only a fragment of the whole maxicircle has usually been used to typify *T. cruzi* (approximately 3600 bp out of the roughly 20–40 kb that codify the maxicircle DNA sequence). The maxicircle and minicircle are good options for kinetoplastid diagnostic purposes since they have many PICs within the locus typically sequenced [73–75]. However, a potential obstacle is the large size of the PCR amplicon, which is obtained by the concatenation of multiple PCRs that target the genes from the maxicircle [8], which might be difficult when funds are limited. However, this molecular marker has a large amplicon size (the largest of all genetic loci typically used to typify *T. cruzi* DTUs), which is part of why it has so many PICs. However, the percentage of PICs along the sequence is quite low, resulting in a percentage of PICs (%PICs; defined as the percentage of the ratio between the number of PICs in an MM and the size of the MM) of only 3% (Table 1). By comparison, most other loci have %PICs much larger than the maxicircle (Table 1). Concerning the number of reference sequences available for this genetic locus and the availability of outgroup sequences, a total of 916 *T. cruzi* homologous sequences were found at the time of the BLAST search (Table 1 and Figure 1B), of which 120 covered the 90% query cover of the concatenated maxicircle sequence that was used as a query (Table 1). This number of references indicates that the maxicircle has an extensive number of available *T. cruzi* reference sequences. As for the accessibility of outgroups, they are mostly available depending on the genetic locus of the maxicircle that is used. But overall, appropriate outgroup sequences are available.

Next on the ranking list are the Cytochrome b (Cyt b) gene segment and the lathosterol oxidase gene with 99 and 98 PICs, respectively (Table 1 and Figure 1A). The Cyt b locus is found within the mitochondrial genome and encodes a protein with the same name that functions within the electron transport chain to produce ATPs. Given its vital function within a cell, the gene can be found in a broad array of eukaryotes and prokaryotes, which has resulted in its popular use as a molecular marker for a wide range of phylogenetic studies within a broad diversity of eukaryotic taxa. The lathosterol oxidase protein is encoded by the SC5D gene (located in the nuclear genome), encoding an enzyme used in the biosynthesis of cholesterol, a metabolic function most commonly found within animals and some protozoans. Both genes practically have the same amount of PICs (99 and 98, respectively) but have slightly different amplicon sizes (approximately 530 vs. 840, respectively), which means Cyt b has a larger %PIC per locus than the Lathostherol oxidase gene (18.5% vs. 11.7%, respectively) (Table 1). Regarding the number of reference sequences available for both loci, Cyt b had 286 homologous *T. cruzi* sequences, of which 252 passed the threshold to be used in our dataset. The Lathostherol oxidase gene had 339 homologous sequences, with 310 covering our threshold (Table 1 and Figure 1B). Furthermore, both molecular markers have DNA sequences of closely related Trypanosomatid species available for their use as outgroups in the corresponding phylogenetic analyses (Table 1), making these loci overall good candidates for their use in the genetic typification of *T. cruzi*.

The gene 18S rRNA trails the previous molecular markers, with 86 PICs from a DNA section of 640 bp in length. The genetic segment that codifies this rRNA is part of the ribosome's eukaryotic small subunit (40S) and is thus found in all eukaryotes. Due to its ubiquity in eukaryotes and the number of PICs, the 18S rRNA gene has been very commonly used in phylogenetic studies from a wide array of eukaryotic taxa. It therefore has a wide availability of closely related species that can be used as outgroups, as well as an important number of *T. cruzi* sequences in GenBank (Table 1 and Figure 1B). Concerning its utility based on the number of available reference *T. cruzi* sequences, it ranks #3, with a total of 870 homologous *T. cruzi* sequences found in GenBank at the time of the analysis, subtly trailing behind the maxicircle (Figure 1B). However, a potential issue is that upon inspection, the outgroup DNA sequence used for this MM (*T. cruzi marinkellei*) clusters within the ingroup of *T. cruzi* sequences. This potential issue could imply that either the particular outgroup sequences used (ON687560.1 and FJ001664.2) were misidentified and were actually *T. cruzi*, or most likely, some of the sequences currently labeled in GenBank as *T. cruzi* for this MM were the erroneously labeled sequences (i.e., are not *T. cruzi*). This underscores the importance of validating the accuracy of species identification in GenBank sequences, especially when they are used for phylogenetic studies. We recommend performing this validation through a preliminary phylogenetic analysis to ensure sequences cluster within the intended species once a proper outgroup sequence is included.

With respect to this issue, this hypothetical inclusion of non-*T. cruzi* sequences into the 18S rRNA dataset would cause the number of PICs to artificially increase and hence give a falsely overestimated genetic diversity. Nonetheless, the number of *T. cruzi* references available for the 18S rRNA gene makes this genetic locus another suitable candidate for its use in the genetic typification of *T. cruzi*. With 870 *T. cruzi* sequences, 18S rRNA ranks as one the most extensively used molecular markers for *T. cruzi*. Additionally, given the high number of gene copies within the nuclear genome, this genetic locus is easily amplified when working with low-quantity DNA templates.

Subsequently, after the 18S rRNA gene, the SL mini-exon ranks #6 in the list with 63 phylogenetic informative characters (Table 1 and Figure 1A). The biological nature of this MM is quite interesting, given that the mini-exon gene is a trypanosomatid-unique mRNA transcript that is 35 bp long and is spliced to the 5′ end of all mRNAs in most, if not all, trypanosomatid species [7,11,15,76–80]. The functionality of the leader sequence is hypothesized to stabilize the mRNA and provide a recognition site for ribosomes, contributing to ensuring a proper translation for the polycistronic mRNA used by kinetoplastids. As a result of the significant amount of leader sequences that are needed for kinetoplastids to edit all of their mRNAs, the parasite has adapted to have a large number of mini-exon genes composed of tandemly repeated coding blocks [15], within its genome, which in *T. cruzi* is estimated to be approximately 200 copies [81]. This abundance in copies is ideal for any attempts to use it as a diagnostic method for the presence of *T. cruzi*. For instance, targeting a molecular marker that has many copies within the genome of *T. cruzi* is a good option when it is required to typify the lineages of *T. cruzi* from field-collected specimens that do not go through a culture enrichment process of *T. cruzi*. Given that the starting amount of the DNA template used might be low, targeting a high-copy genetic locus increases the likelihood of effectively amplifying the PCR-targeted sequence. Ricardo P. Souto and colleagues proposed using this unique DNA sequence of kinetoplastids to typify *T. cruzi* strains in 1996 [11]. Since then, it has been widely used, and it is therefore not surprising for it to have become the most commonly used MM for the typification of *T. cruzi*, which is why it is the MM with the largest number of *T. cruzi* sequences currently available in GenBank, with a total of 1643 at the time of this study (Table 1 and Figure 1B).

Additionally, as mentioned previously, the number of PICs found within the mini-exon is not insignificant (63 sites) (Table 1), and the small size of this PCR amplicon explains how this MM has the largest percentage of PICs per locus (%PICs) of all the genetic loci analyzed (22.6%) (Table 1). Nonetheless, one of its major disadvantages is the inability to include proper outgroup sequences for this marker in the corresponding phylogenetic analyses. The reason appears to be the high divergence within the DNA mini-exon sequence among kinetoplastid species and within *T. cruzi* lineages, which is why it was proposed as a molecular marker in the first place. Since it appears to have significant differences in the amplicon length between lineages, it has been used to typify *T. cruzi* strains based solely on the size amplified via the PCR [11,82]. Henceforth, this extreme polymorphism hinders the ability to properly identify the homologous sequences in closely related taxa, resulting in the failure to include proper outgroups for the rooting of a phylogenetic analysis and thus interfering with the ability to reconstruct a well-resolved phylogenetic tree (Figure 2B). These excessive polymorphisms are evidenced by the fact that exhaustive homologous sequence searches with BLAST fail to identify homologous sequences that cover the entirety of the query and only find homologous sequences that cover <25% of the query coverage. The limitation of including outgroup DNA sequences is not trivial since using outgroup sequences is vital for adequately rooting a phylogenetic tree. Even though these molecular markers are useful to typify the major genetic lineages of *T. cruzi* by analyzing the amplicon sizes, their use for phylogenetic analyses (i.e., a higher resolution to determine the diversity within a DTU) is not recommended.

Another potential outcome of the inability to include an outgroup with the mini-exon dataset was an overestimated number of PICs for this MM due to the failure to complete the second checkpoint to guarantee the exclusion of not-*T. cruzi* DNA sequences. This second checkpoint was not accomplished because it required the use of an outgroup.

In positions #7 and #8, trailing behind the mini-exon, are the 5S rRNA and dihydrofolate reductase–thymidylate synthase (DHFR-TS) loci, with 47 and 42 phylogenetic informative characters, respectively. While 5S rRNA codifies a ribosomal RNA that forms part of the small ribosomal subunit (which means this gene is pretty much universal to all eukaryotes), the DHFR-TS gene codifies a bifunctional protein that plays a vital role in the replication of DNA and is unique to Trypanosomatids and other protozoa and has thus been proposed as a potential drug target. This bifunctional enzyme plays a role in reducing dihydrofolate to tetrahydrofolate within the folate cycle, which is essential for synthesizing nucleotides. Meanwhile, the thymidylate synthase enzyme function is needed to catalyze the conversion of dUMP to dTMP within the cycle of thymine synthesis, and thus, it is vital for DNA replication. The first phylogenetic tree published with the 5S rRNA locus in the reconstruction of the evolutionary history of *T. cruzi* was conducted by Westenberger et al. [83], while Machado and Ayala reported the first phylogenetic analysis for the DHFR-TS locus [45]. Both genetic loci have practically the same number of reference sequences available in GenBank (190 and 164, respectively) (Figure 1B), although the respective datasets were significantly different between each molecular marker, given the scarcity of homolog *T. cruzi* sequences that covered the minimum 90% query cover threshold that was set for the construction of each locus (Table 1). Conversely, appropriate outgroup sequences were only available for the DHFR-TS. The 5S rRNA locus does not have suitable outgroup sequences that cover more than 30% of the *T. cruzi* %S rRNA sequences and thus has this critical limitation as an MM.

Following the DHFR-TS gene are the remaining eleven loci, forming a cluster of genes with a significantly lower number of PICs (Figure 1A) than the rest. Within this cluster are the following genes: mevalonate kinase (MK), MutS Homolog 2 (MSH2), Glucose-6-phosphate isomerase (GPI), cathepsin L-like (CatL-like), superoxide dismutase, glycosomal glyceraldehyde phosphate dehydrogenase (gGAPDH), internal transcribed spacer 1 (ITS1), the trypomastigote excretory–secretory protein (Tc24), heat-shock protein 70 (HSP70), minicircle and the 5S rRNA (M5) loci. With PICs ranging between 2 and 25, respectively (Table 1), the number of reference sequences available in GenBank for each locus was also significantly smaller than for the previously discussed loci analyzed (Figure 1B), perhaps as a result of the low resolution that their use can accomplish. Although the majority of these loci have outgroup sequences available in GenBank, the low amount of reference sequences and, most importantly, the low number of phylogenetically informative characters justify the exclusion of their use as individual molecular markers used to typify the genetic diversity of *T. cruzi*.

A particular reference to the minicircle is needed since the extreme sequence divergence within this marker does not allow reliable DNA sequence alignments. Consequently, divergent DNA alignment sections were removed with gBlocks, resulting in a smaller dataset. While this reduces the number of PICs, retaining unreliable DNA alignments would unmistakably result in an erroneous phylogeny [68]. Additionally, the lack of an outgroup sequence (Table 1) further supports the exclusion of the minicircle as a suitable molecular marker for typifying *T. cruzi* based on a phylogenetic reconstruction.

Therefore, as a whole, the phylogenetic robustness that can be obtained from molecular markers that fulfill all three major traits evaluated in this review (a high number of phylogenetic informative characters, a significant amount of *T. cruzi* reference sequences and the availability of appropriate outgroup sequences) can be clearly observed in Figure 2. This figure juxtaposes the maximum likelihood phylogenetic trees of the best-ranked genetic locus versus one that does not have an outgroup (COII-NDI and the mini-exon, respectively). The divergence in the branch lengths and clustering positions of sequences identified with the COII-NDI phylogeny, as well as the importance of rooting the tree with an appropriate outgroup, underscores the need to carefully consider the parameters used in a quantitative review for the proper selection of MMs to typify *T. cruzi*.

6. Concluding Remarks

Although sequencing technologies are becoming more accessible, PCR amplicons remain essential for the genetic typing of *T. cruzi*, especially in understanding crucial aspects of its biology. Selecting an appropriate molecular marker (MM) depends heavily on the research question. For instance, detecting intra-DTU differences often requires loci with sufficient DNA polymorphisms that will allow the reconstruction of a detailed phylogenetic tree. To address these considerations, we provide a quantitative review of the most frequently employed genetic loci, evaluating them on three key criteria: (1) the number of phylogenetic informative characters; (2) the availability of *T. cruzi* reference sequences in public repositories; and (3) the accessibility of DNA sequences from related taxa for an outgroup comparison. This effort identified nineteen genetic loci as the most commonly applied for the genetic typification of *T. cruzi* strains. Of the nineteen loci analyzed, COII-NDI ranked highest due to its extensive outgroup sequence availability, abundant reference sequences for *T. cruzi*, and high number of phylogenetically informative characters—60% more than the next marker. The utility of fulfilling these traits can be clearly observed in Figure 2, where the phylogenetic resolution obtained from COII-NDI clearly is much higher than the mini-exon tree, which has a significantly lower number of PICs and does not have an appropriate outgroup sequence.

As a whole, mitochondrial markers, including the maxicircle and the Cytochrome b loci, are also highly effective markers to typify *T. cruzi*. Among the nuclear loci, lathosterol oxidase and 18S rRNA ranked best. When feasible, we recommend using a combination of these top-ranking loci for comprehensive genetic typing. If limited to a single locus, COII-NDI is the most suitable option.

Supplementary Materials: The following supporting information can be downloaded at: https://www.mdpi.com/article/10.3390/pathogens14010072/s1, Figure S1: Comparison of phylogenetic trees (COII-NDI & mini-exon) constructed with all *T. cruzi* available sequences in GenBank as of July 2024.

Author Contributions: Conceptualization: C.A.F.-L.; Data curation and validation: D.R.-D. and C.A.F.-L.; Analyses: D.R.-D. and C.A.F.-L.; Writing: D.R.-D. and C.A.F.-L. All authors have read and agreed to the published version of the manuscript.

Funding: This research received no external funding. The APC was funded by the "Convocatoria Especial de Apoyo a Publicación 2025" by Universidad Autónoma de Baja California, México.

Institutional Review Board Statement: Not applicable.

Informed Consent Statement: Not applicable.

Data Availability Statement: No new data were created or analyzed in this study. The data used for this review was publicly available in GenBank, it was actively analyzed as part of our study.

Acknowledgments: We wish to thank two anonymous reviewers for providing feedback on the manuscript and Universidad Autónoma de Baja California.

Conflicts of Interest: The authors declare no conflict of interest.

References

1. Chagas, C. Nova Tripanozomiaze Humana: Estudos Sobre a Morfolojia e o Ciclo Evolutivo Do *Schizotrypanum cruzi* n. Gen., n. sp., Ajente Etiolojico de Nova Entidade Morbida Do Homem. *Mem. Inst. Oswaldo Cruz* **1909**, *1*, 159–218. [CrossRef]
2. Brenière, S.F.; Waleckx, E.; Barnabé, C. Over Six Thousand *Trypanosoma cruzi* Strains Classified into Discrete Typing Units (DTUs): Attempt at an Inventory. *PLoS Negl. Trop. Dis.* **2016**, *10*, e0004792. [CrossRef]
3. Velásquez-Ortiz, N.; Ramírez, J.D. Understanding the Oral Transmission of *Trypanosoma cruzi* as a Veterinary and Medical Foodborne Zoonosis. *Res. Vet. Sci.* **2020**, *132*, 448–461. [CrossRef]
4. Martín-Escolano, J.; Marín, C.; Rosales, M.J.; Tsaousis, A.D.; Medina-Carmona, E.; Martín-Escolano, R. An Updated View of the *Trypanosoma cruzi* Life Cycle: Intervention Points for an Effective Treatment. *ACS Infect. Dis.* **2022**, *8*, 1107–1115. [CrossRef]
5. Flores-López, C.A.; Machado, C.A. Analyses of 32 Loci Clarify Phylogenetic Relationships among *Trypanosoma cruzi* Lineages and Support a Single Hybridization Prior to Human Contact. *PLoS Negl. Trop. Dis.* **2011**, *5*, e1272. [CrossRef] [PubMed]
6. Brisse, S.; Barnabé, C.; Bañuls, A.L.; Sidibé, I.; Noël, S.; Tibayrenc, M. A Phylogenetic Analysis of the *Trypanosoma cruzi* genome project CL Brener reference strain by multilocus enzyme electrophoresis and multiprimer random amplified polymorphic DNA Fingerprinting. *Mol. Biochem. Parasitol.* **1998**, *92*, 253–256. [CrossRef]
7. Fernandes, O.; Sturm, N.R.; Derré, R.; Campbell, D.A. The Mini-Exon Gene: A Genetic Marker for Zymodeme III of *Trypanosoma cruzi*. *Mol. Biochem. Parasitol.* **1998**, *95*, 129–133. [CrossRef] [PubMed]
8. Messenger, L.A.; Yeo, M.; Lewis, M.D.; Llewellyn, M.S.; Miles, M.A. Molecular Genotyping of *Trypanosoma cruzi* for Lineage Assigment and Population Genetics. In *Parasite Genomics Protocls. Methods in Molecular Biology*; Peacock, C., Ed.; Human Press: New York, NY, USA, 2015; pp. 1–365. ISBN 9781493914388.
9. Oliveira, R.P.; Broude, N.E.; Macedo, A.M.; Cantor, C.R.; Smith, C.L.; Pena, S.D.J. Probing the Genetic Population Structure of *Trypanosoma cruzi* with Polymorphic Microsatellites. *Proc. Natl. Acad. Sci. USA* **1998**, *95*, 3776–3780. [CrossRef] [PubMed]
10. Revollo, S.; Oury, B.; Laurent, J.P.; Barnabé, C.; Quesney, V.; Carrière, V.; Noël, S.; Tibayrenc, M. *Trypanosoma cruzi*: Impact of Clonal Evolution of the Parasite on Its Biological and Medical Properties. *Exp. Parasitol.* **1998**, *89*, 30–39. [CrossRef]
11. Souto, R.P.; Fernandes, O.; Macedo, A.M.; Campbell, D.A.; Zingales, B. DNA Markers Define Two Major Phylogenetic Lineages of *Trypanosoma cruzi*. *Mol. Biochem. Parasitol.* **1996**, *83*, 141–152. [CrossRef]
12. Wang, W.; Peng, D.; Baptista, R.P.; Li, Y.; Kissinger, J.C.; Tarleton, R.L. Strain-Specific Genome Evolution in *Trypanosoma cruzi*, the Agent of Chagas Disease. *PLoS Pathog.* **2021**, *17*, e1009254. [CrossRef] [PubMed]
13. Zingales, B.; Macedo, A.M. Fifteen Years after the Definition of *Trypanosoma cruzi* DTUs: What Have We Learned? *Life* **2023**, *13*, 2339. [CrossRef] [PubMed]
14. Dorn, P.L.; McClure, A.G.; Gallaspy, M.D.; Waleckx, E.; Woods, A.S.; Monroy, M.C.; Stevens, L. The Diversity of the Chagas Parasite, *Trypanosoma cruzi*, Infecting the Main Central American Vector, Triatoma Dimidiata, from Mexico to Colombia. *PLoS Negl. Trop. Dis.* **2017**, *11*, e0005878. [CrossRef]
15. Macedo, A.M.; Machado, C.R.; Oliveira, R.P.; Pena, S.D.J. *Trypanosoma cruzi*: Genetic Structure of Populations and Relevance of Genetic Variability to the Pathogenesis of Chagas Disease. *Mem. Inst. Oswaldo Cruz* **2004**, *99*, 1–12. [CrossRef] [PubMed]
16. Tavares de Oliveira, M.; Sulleiro, E.; Silgado Gimenez, A.; de Lana, M.; Zingales, B.; Santana da Silva, J.; Marin-Neto, J.A.; Molina, I. Quantification of Parasite Burden of *Trypanosoma cruzi* and Identification of Discrete Typing Units (Dtus) in Blood Samples of Latin American Immigrants Residing in Barcelona, Spain. *PLoS Negl. Trop. Dis.* **2020**, *14*, e0008311. [CrossRef]
17. Zingales, B.; Andrade, S.G.; Briones, M.R.S.; Campbell, D.A.; Chiari, E.; Fernandes, O.; Guhl, F.; Lages-Silva, E.; Macedo, A.M.; Machado, C.R.; et al. A New Consensus for *Trypanosoma cruzi* Intraspecific Nomenclature: Second Revision Meeting Recommends TcI to TcVI. *Mem. Inst. Oswaldo Cruz* **2009**, *104*, 1051–1054. [CrossRef]
18. Zingales, B.; Miles, M.A.; Campbell, D.A.; Tibayrenc, M.; Macedo, A.M.; Teixeira, M.M.G.; Schijman, A.G.; Llewellyn, M.S.; Lages-Silva, E.; Machado, C.R.; et al. The Revised *Trypanosoma cruzi* Subspecific Nomenclature: Rationale, Epidemiological Relevance and Research Applications. *Infect. Genet. Evol.* **2012**, *12*, 240–253. [CrossRef] [PubMed]
19. Zingales, B.; Bartholomeu, D.C. *Trypanosoma cruzi* Genetic Diversity: Impact on Transmission Cycles and Chagas Disease. *Mem. Inst. Oswaldo Cruz* **2022**, *117*, e210193. [CrossRef] [PubMed]
20. Santi-Rocca, J.; Fernandez-Cortes, F.; Chillón-Marinas, C.; González-Rubio, M.L.; Martin, D.; Gironès, N.; Fresno, M. A Multi-Parametric Analysis of *Trypanosoma cruzi* Infection: Common Pathophysiologic Patterns beyond Extreme Heterogeneity of Host Responses. *Sci. Rep.* **2017**, *7*, 8893. [CrossRef]
21. Nielebock, M.A.P.; Moreira, O.C.; das Chagas Xavier, S.C.; de Freitas Campos Miranda, L.; de Lima, A.C.B.; de Jesus Sales Pereira, T.O.; Hasslocher-Moreno, A.M.; Britto, C.; Sangenis, L.H.C.; Saraiva, R.M. Association between *Trypanosoma cruzi* DTU TcII and Chronic Chagas Disease Clinical Presentation and Outcome in an Urban Cohort in Brazil. *PLoS ONE* **2020**, *15*, e0243008. [CrossRef] [PubMed]
22. Rodrigues-dos-Santos, Í.; Melo, M.F.; de Castro, L.; Hasslocher-Moreno, A.M.; do Brasil, P.E.A.A.; Silvestre de Sousa, A.; Britto, C.; Moreira, O.C. Exploring the Parasite Load and Molecular Diversity of *Trypanosoma cruzi* in Patients with Chronic Chagas Disease from Different Regions of Brazil. *PLoS Negl. Trop. Dis.* **2018**, *12*, e0006939. [CrossRef] [PubMed]

23. Silvestrini, M.M.A.; Alessio, G.D.; Frias, B.E.D.; Sales Júnior, P.A.; Araújo, M.S.S.; Silvestrini, C.M.A.; Brito Alvim de Melo, G.E.; Martins-Filho, O.A.; Teixeira-Carvalho, A.; Martins, H.R. New Insights into *Trypanosoma cruzi* Genetic Diversity, and Its Influence on Parasite Biology and Clinical Outcomes. *Front. Immunol.* **2024**, *15*, 1342431. [CrossRef] [PubMed]
24. Malik, L.H.; Singh, G.D.; Amsterdam, E.A. The Epidemiology, Clinical Manifestations, and Management of Chagas Heart Disease. *Clin. Cardiol.* **2015**, *38*, 565–569. [CrossRef]
25. Pérez-Molina, J.A.; Molina, I. Chagas Disease. *Lancet* **2018**, *391*, 82–94. [CrossRef]
26. Rodríguez-Bejarano, O.H.; Avendaño, C.; Patarroyo, M.A. Mechanisms Associated with *Trypanosoma cruzi* Host Target Cell Adhesion, Recognition and Internalization. *Life* **2021**, *11*, 534. [CrossRef] [PubMed]
27. Medina-Rincón, G.J.; Gallo-Bernal, S.; Jiménez, P.A.; Cruz-Saavedra, L.; Ramírez, J.D.; Rodríguez, M.J.; Medina-Mur, R.; Díaz-Nassif, G.; Valderrama-Achury, M.D.; Medina, H.M. Molecular and Clinical Aspects of Chronic Manifestations in Chagas Disease: A State-of-the-Art Review. *Pathogens* **2021**, *10*, 1493. [CrossRef] [PubMed]
28. Nunes, M.C.P.; Dones, W.; Morillo, C.A.; Encina, J.J.; Ribeiro, A.L. Chagas Disease: An Overview of Clinical and Epidemiological Aspects. *J. Am. Coll. Cardiol.* **2013**, *62*, 767–776. [CrossRef] [PubMed]
29. Luquetti, A.O.; Miles, M.A.; Rassi, A.; De Rezende, J.M.; De Souza, A.A.; Póvoa, M.M.; Rodrigues, I. *Trypanosoma cruzi*: Zymodemes Associated with Acute and Chronic Chagas' Disease in Central Brazil. *Trans. R. Soc. Trop. Med. Hyg.* **1986**, *80*, 462–470. [CrossRef]
30. Miles, M.A.; Llewellyn, M.S.; Lewis, M.D.; Yeo, M.; Baleela, R.; Fitzpatrick, S.; Gaunt, M.W.; Mauricio, I.L. The Molecular Epidemiology and Phylogeography of *Trypanosoma cruzi* and Parallel Research on Leishmania: Looking Back and to the Future. *Parasitology* **2009**, *136*, 1509–1528. [CrossRef]
31. Miles, M.A.; Souza, A.; Povoa, M.; Shaw, J.J.; Laison, R.; Toye, P.J. Isozymic Heterogeneity of *Trypanosoma cruzi*. *Nature* **1978**, *272*, 819–821. [CrossRef] [PubMed]
32. Morel, C.; Chiari, E.; Camargo, E.P.; Mattei, D.M.; Romanha, A.J.; Simpson, L. Strains and Clones of *Trypanosoma cruzi* Can Be Characterized by Pattern of Restriction Endonuclease Products of Kinetoplast DNA Minicircles. *Proc. Natl. Acad. Sci. USA* **1980**, *77*, 6810–6814. [CrossRef] [PubMed]
33. Vago, A.R.; Macedo, A.M.; Adad, S.J.; D'Avila Reis, D.; Correa-Oliveira, R. PCR Detection of *Trypanosoma cruzi* DNA in Oesophageal Tissues of Patients with Chronic Digestive Chagas' Disease. *Lancet* **1996**, *348*, 891–892. [CrossRef]
34. Brisse, S.; Barnabé, C.; Tibayrenc, M. Identification of Six *Trypanosoma cruzi* Phylogenetic Lineages by Random Amplified Polymorphic DNA and Multilocus Enzyme Electrophoresis. *Int. J. Parasitol.* **2000**, *30*, 35–44. [CrossRef]
35. Steindel, M.; Neto, E.D.; de Menezes, C.L.P.; Romanha, A.J.; Simpson, A.J.G. Random Amplified Polymorphic DNA Analysis of *Trypanosoma cruzi* Strains. *Mol. Biochem. Parasitol.* **1993**, *60*, 71–79. [CrossRef]
36. Zingales, B.; Elizabeth, M.; Pereira, S.; Oliveira, R.P.; Almeida, K.A.; Umezawa, E.S.; Souto, R.P.; Vargas, N.; Isabel, M.; Franco, J.; et al. *Trypanosoma cruzi* Genome Project: Biological Characteristics and Molecular Typing of Clone CL Brener. *Acta Trop.* **1997**, *68*, 159–173. [CrossRef]
37. Tibayrenc, M.; Hoffmann, A.; Poch, O.; Echalar, L.; Lepont, F.; Lemesre, J.L.; Desjeux, P.; Ayala, F.J. Additional Data on *Trypanosoma cruzi* Isozymic Strains Encountered in Bolivian Domestic Transmission Cycles. *Trans. R. Soc. Trop. Med. Hyg.* **1986**, *80*, 442–447. [CrossRef]
38. Borst, P.; Fairlamb, A.H. Surface Receptors and Transporters of Trypanosoma Brucei. *Annu. Rev. Microbiol.* **1998**, *52*, 745–778. [CrossRef] [PubMed]
39. Gibson, W.C.; Miles, M.A. The Karyotype and Ploidy of *Trypanosoma cruzi*. *EMBO J.* **1986**, *5*, 1299–1305. [CrossRef]
40. Henriksson, J.; Åslund, L.; Pettersson, U. Karyotype Variability in *Trypanosoma cruzi*. *Parasitol. Today* **1996**, *12*, 108–114. [CrossRef] [PubMed]
41. Macedo, A.M.; Pimenta, J.R.; De Aguiar, R.S.; Melo, A.I.R.; Chiari, E.; Zingales, B.; Pena, S.D.J.; Oliveira, R.P. Usefulness of Microsatellite Typing in Population Genetic Studies of *Trypanosoma cruzi*. *Mem. Inst. Oswaldo Cruz* **2001**, *96*, 407–413. [CrossRef]
42. Brisse, S.; Verhoef, J.; Tibayrenc, M. Characterisation of Large and Small Subunit RRNA and Mini-Exon Genes Further Supports the Distinction of Six *Trypanosoma cruzi* Lineages. *Int. J. Parasitol.* **2001**, *31*, 1218–1226. [CrossRef] [PubMed]
43. Zingales, B.; Stolf, B.S.; Souto, R.P.; Fernandes, O.; Briones, M.R.S. Epidemiology, Biochemistry and Evolution of *Trypanosoma cruzi* Lineages Based on Ribosomal RNA Sequences. *Mem. Inst. Oswaldo Cruz* **1999**, *94*, 159–164. [CrossRef] [PubMed]
44. Stothard, J.R.; Frame, I.A.; Carrasco, H.J.; Miles, M.A. On the Molecular Taxonomy of *Trypanosoma cruzi* Using Riboprinting. *Parasitology* **1998**, *117*, 243–247. [CrossRef] [PubMed]
45. Machado, C.A.; Ayala, F.J. Nucleotide Sequences Provide Evidence of Genetic Exchange among Distantly Related Lineages of *Trypanosoma cruzi*. *Proc. Natl. Acad. Sci. USA* **2001**, *98*, 7396–7401. [CrossRef]
46. Tibayrenc, M. Genetic Epidemiology of Parasitic Protozoa and Other Infectious Agents: The Need for an Integrated Approach. *Int. J. Parasitol.* **1998**, *28*, 85–104. [CrossRef]

47. Barnabé, C.; Mobarec, H.I.; Jurado, M.R.; Cortez, J.A.; Brenière, S.F. Reconsideration of the Seven Discrete Typing Units within the Species *Trypanosoma cruzi*, a New Proposal of Three Reliable Mitochondrial Clades. *Infect. Genet. Evol.* **2016**, *39*, 176–186. [CrossRef] [PubMed]
48. Flores-López, C.A.; Mitchell, E.A.; Reisenman, C.E.; Sarkar, S.; Williamson, P.C.; Machado, C.A. Phylogenetic Diversity of Two Common *Trypanosoma cruzi* Lineages in the Southwestern United States. *Infect. Genet. Evol.* **2022**, *99*, 105251. [CrossRef] [PubMed]
49. Pinto, C.M.; Kalko, E.K.V.; Cottontail. I.; Wellinghausen, N.; Cottontail, V.M. TcBat a Bat-Exclusive Lineage of *Trypanosoma cruzi* in the Panama Canal Zone, with Comments on Its Classification and the Use of the 18S RRNA Gene for Lineage Identification. *Infect. Genet. Evol.* **2012**, *12*, 1328–1332. [CrossRef] [PubMed]
50. Dassanayake, R.S.; Samaranayake, L.P. Randomly Amplified Polymorphic DNA Fingerprinting: The Basics. In *Methods in Molecular Biology (Clifton, N.J.)*; Bartlett, J.M.S., Stirling, D., Eds.; Human Press: San Diego, CA, USA, 2003; Volume 226, pp. 117–122, ISBN 978-0-89603-642-0.
51. Rivas-Garciá, L.; Carballo-Amador, M.A.; Flores-Lópezid, C.A. Design of a AFLP-PCR and PCR-RFLP Test That Identify the Majority of Discrete Typing Units of *Trypanosoma cruzi*. *PLoS ONE* **2020**, *15*, e0237180. [CrossRef] [PubMed]
52. Murthy, V.K.; Dibbern, K.M.; Campbell, D.A. PCR Amplification of Mini-Exon Genes Differentiates *Trypanosoma cruzi* from Trypanosoma Rangeli. *Mol. Cell. Probes* **1992**, *6*, 237–243. [CrossRef] [PubMed]
53. Felsenstein, J. Parsimony in Systematics: Biological and Statistical Issues. *Annu. Rev. Ecol. Syst.* **1983**, *14*, 313–333. [CrossRef]
54. Tezuka, A.; Matsushima, N.; Nemoto, Y.; Akashi, H.D.; Kawata, M.; Makino, T. Comprehensive Primer Design for Analysis of Population Genetics in Non-Sequenced Organisms. *PLoS ONE* **2012**, *7*, e32314. [CrossRef]
55. You, F.M.; Huo, N.; Gu, Y.Q.; Lazo, G.R.; Dvorak, J.; Anderson, O.D. ConservedPrimers 2.0: A High-Throughput Pipeline for Comparative Genome Referenced Intron-Flanking PCR Primer Design and Its Application in Wheat SNP Discovery. *BMC Bioinform.* **2009**, *10*, 331. [CrossRef] [PubMed]
56. Lyons-Weiler, J.; Hoelzer, G.A.; Tausch, R.J. Optimal Outgroup Analysis. *Biol. J. Linn. Soc.* **1998**, *64*, 493–511. [CrossRef]
57. Muse, S.V.; Weir, B.S. Testing for Equality of Evolutionary Rates. *Genetics* **1992**, *132*, 269–276. [CrossRef] [PubMed]
58. Schneider, M.C.; Jancloes, M.; Buss, D.F.; Aldighieri, S.; Bertherat, E.; Najera, P.; Galan, D.I.; Durski, K.; Espinal, M.A. Leptospirosis: A Silent Epidemic Disease. *Int. J. Environ. Res. Public Health* **2013**, *10*, 7229–7234. [CrossRef] [PubMed]
59. Smith, A.B. Rooting Molecular Trees: Problems and Strategies. *Biol. J. Linn. Soc.* **1994**, *51*, 279–292. [CrossRef]
60. Dufayard, J.F.; Duret, L.; Penel, S.; Gouy, M.; Rechenmann, F.; Perrière, G. Tree Pattern Matching in Phylogenetic Trees: Automatic Search for Orthologs or Paralogs in Homologous Gene Sequence Databases. *Bioinformatics* **2005**, *21*, 2596–2603. [CrossRef] [PubMed]
61. Koonin, E.V. Orthologs, Paralogs, and Evolutionary Genomics. *Annu. Rev. Genet.* **2005**, *39*, 309–338. [CrossRef]
62. Wu, F.; Mueller, L.A.; Crouzillat, D.; Pétiard, V.; Tanksley, S.D. Combining Bioinformatics and Phylogenetics to Identify Large Sets of Single-Copy Orthologous Genes (COSII) for Comparative, Evolutionary and Systematic Studies: A Test Case in the Euasterid Plant Clade. *Genetics* **2006**, *174*, 1407–1420. [CrossRef]
63. Pett, W.; Adamski, M.; Adamska, M.; Francis, W.R.; Eitel, M.; Pisani, D.; Wörheide, G. The Role of Homology and Orthology in the Phylogenomic Analysis of Metazoan Gene Content. *Mol. Biol. Evol.* **2019**, *36*, 643–649. [CrossRef] [PubMed]
64. Grunenwald, H. Optimization of Polymerase Chain Reactions. In *Methods in Molecular Biology (Clifton, N.J.)*; Bartlett, J.M.S., Stirling, D., Eds.; Human Press: Clifton, NJ, USA, 2003; Volume 226, pp. 89–100.
65. Altschul, S.F.; Gish, W.; Miller, W.; Myers, E.W.; Lipman, D.J. Basic Local Alignment Search Tool. *J. Mol. Biol.* **1990**, *215*, 403–410. [CrossRef] [PubMed]
66. Edgar, R.C. MUSCLE: Multiple Sequence Alignment with High Accuracy and High Throughput. *Nucleic Acids Res.* **2004**, *32*, 1792–1797. [CrossRef]
67. Galtier, N.; Gouy, M.; Gautier, C. Seaview and Phylo_ Win: Two Graphic Tools for Sequence Alignment and Molecular Phylogeny. *Bioinformatics* **1996**, *12*, 543–548. [CrossRef] [PubMed]
68. Talavera, G.; Castresana, J. Improvement of Phylogenies after Removing Divergent and Ambiguously Aligned Blocks from Protein Sequence Alignments. *Syst. Biol.* **2007**, *56*, 564–577. [CrossRef] [PubMed]
69. Nguyen, L.T.; Schmidt, H.A.; Von Haeseler, A.; Minh, B.Q. IQ-TREE: A Fast and Effective Stochastic Algorithm for Estimating Maximum-Likelihood Phylogenies. *Mol. Biol. Evol.* **2015**, *32*, 268–274. [CrossRef] [PubMed]
70. Minh, B.Q.; Nguyen, M.A.T.; Von Haeseler, A. Ultrafast Approximation for Phylogenetic Bootstrap. *Mol. Biol. Evol.* **2013**, *30*, 1188–1195. [CrossRef] [PubMed]
71. Botero, A.; Kapeller, I.; Cooper, C.; Clode, P.L.; Shlomai, J.; Thompson, R.C.A. The Kinetoplast DNA of the Australian Trypanosome, *Trypanosoma copemani*, Shares Features with *Trypanosoma cruzi* and *Trypanosoma lewisi*. *Int. J. Parasitol.* **2018**, *48*, 691–700. [CrossRef] [PubMed]
72. Wang, J.F.; Lin, R.H.; Zhang, X.; Hide, G.; Lun, Z.R.; Lai, D.H. Novel Insertions in the Mitochondrial Maxicircle of *Trypanosoma musculi*, a Mouse Trypanosome. *Parasitology* **2022**, *149*, 1546–1555. [CrossRef]

73. Callejas-Hernández, F.; Herreros-Cabello, A.; del Moral-Salmoral, J.; Fresno, M.; Gironès, N. The Complete Mitochondrial DNA of *Trypanosoma cruzi*: Maxicircles and Minicircles. *Front. Cell. Infect. Microbiol.* **2021**, *11*, 672448. [CrossRef] [PubMed]
74. Berna, L.; Greif, G.; Pita, S.; Faral-Tello, P.; Dıaz-Viraque, F.; De Cassia Moreira De Souza, R.; Vallejo, G.A.; Alvarez-Valin, F.; Robello, C. Maxicircle Architecture and Evolutionary Insights into *Trypanosoma cruzi* Complex. *PLoS Negl. Trop. Dis.* **2021**, *15*, e0009719. [CrossRef] [PubMed]
75. Gómez-Palacio, A.; Cruz-Saavedra, L.; Van den Broeck, F.; Geerts, M.; Pita, S.; Vallejo, G.A.; Carranza, J.C.; Ramírez, J.D. High-Throughput Analysis of the *Trypanosoma cruzi* Minicirculome (McDNA) Unveils Structural Variation and Functional Diversity. *Sci. Rep.* **2024**, *14*, 5578. [CrossRef] [PubMed]
76. Borst, P. Discontinuous Transcription and Antigenic Variation in Trypanosomes. *Annu. Rev. Biochem.* **1986**, *55*, 701–732. [CrossRef] [PubMed]
77. Clayton, C. Regulation of Gene Expression in Trypanosomatids: Living with Polycistronic Transcription. *Open Biol.* **2019**, *9*, 190072. [CrossRef]
78. Nascimento, J.d.F.; Kelly, S.; Sunter, J.; Carrington, M. Codon Choice Directs Constitutive MRNA Levels in Trypanosomes. *eLife* **2018**, *7*, e32467. [CrossRef] [PubMed]
79. De Lange, T.; Liu, A.Y.C.; Van der Ploeg, L.H.T.; Borst, P.; Tromp, M.C.; Van Boom, J.H. Tandem Repetition of the 5′ Mini-Exon of Variant Surface Glycoprotein Genes: A Multiple Promoter for VSG Gene Transcription? *Cell* **1983**, *34*, 891–900. [CrossRef] [PubMed]
80. Fadda, A.; Ryten, M.; Droll, D.; Rojas, F.; Färber, V.; Haanstra, J.R.; Merce, C.; Bakker, B.M.; Matthews, K.; Clayton, C. Transcriptome-Wide Analysis of Trypanosome MRNA Decay Reveals Complex Degradation Kinetics and Suggests a Role for Co-Transcriptional Degradation in Determining MRNA Levels. *Mol. Microbiol.* **2014**, *94*, 307–326. [CrossRef]
81. El-Sayed, N.M.; Myler, P.J.; Bartholomeu, D.; Nilsson, D.; Aggarwal, G.; Tran, A.-N.; Ghedin, E.; Worthey, E.A.; Delcher, A.L.; Gaëlle, B.; et al. The Genome Sequence of *Trypanosoma cruzi*, Etiologic Agent of Chagas Disease. *Science* **2005**, *309*, 409–415. [CrossRef] [PubMed]
82. Herrera, C.; Bargues, M.D.; Fajardo, A.; Montilla, M.; Triana, O.; Vallejo, G.A.; Guhl, F. Identifying Four *Trypanosoma cruzi* I Isolate Haplotypes from Different Geographic Regions in Colombia. *Infect. Genet. Evol.* **2007**, *7*, 535–539. [CrossRef]
83. Westenberger, S.J.; Sturm, N.R.; Campbell, D.A. *Trypanosoma cruzi* 5S RRNA Arrays Define Five Groups and Indicate the Geographic Origins of an Ancestor of the Heterozygous Hybrids. *Int. J. Parasitol.* **2006**, *36*, 337–346. [CrossRef] [PubMed]

pathogens

MDPI

Review

Trypanosoma cruzi: Genomic Diversity and Structure

Alfonso Herreros-Cabello [1], Francisco Callejas-Hernández [2], Núria Gironès [1,3,*] and Manuel Fresno [1,3,*]

[1] Centro de Biología Molecular Severo Ochoa, Consejo Superior de Investigaciones Científicas, Universidad Autónoma de Madrid, Cantoblanco, 28049 Madrid, Spain
[2] Bloomberg School of Public Health, Johns Hopkins University, Baltimore, MD 21205, USA; bio.fcallejas@gmail.com
[3] Instituto Sanitario de Investigación Princesa, 28006 Madrid, Spain
* Correspondence: ngirones@cbm.csic.es (N.G.); mfresno@cbm.csic.es (M.F.)

Abstract: *Trypanosoma cruzi* is the causative agent of Chagas disease, a neglected tropical disease, and one of the most important parasitic diseases worldwide. The first genome of *T. cruzi* was sequenced in 2005, and its complexity made assembly and annotation challenging. Nowadays, new sequencing methods have improved some strains' genome sequence and annotation, revealing this parasite's extensive genetic diversity and complexity. In this review, we examine the genetic diversity, the genomic structure, and the principal multi-gene families involved in the pathogenicity of *T. cruzi*. The *T. cruzi* genome sequence is divided into two compartments: the core (conserved) and the disruptive (variable in length and multicopy gene families among strains). The disruptive region has also been described as genome plasticity and plays a key role in the parasite survival and infection process. This region comprises several multi-gene families, including trans-sialidases, mucins, and mucin-associated surface proteins (MASPs). Trans-sialidases are the most prevalent genes in the genome with a key role in the infection process, while mucins and MASPs are also significant glycosylated proteins expressed on the parasite surface, essential for its biological functions, as host–parasite interaction, host cell invasion or protection against the host immune system, in both insect and mammalian stages. Collectively, in this review, some of the most recent advances in the structure and composition of the *T. cruzi* genome are reviewed.

Keywords: *Trypanosoma cruzi*; strains; genomic structure; multi-gene families

1. Introduction to the Biology and Genetics of *Trypanosoma cruzi*

Recent advancements in the Trypanosomatidae family encompass a range of areas, including new understandings of trypanosomatid genetics and sexual processes, biodiversity, population structure, and host–parasite interactions [1–5]. Their ability to adapt to various environmental conditions and their high biological diversity allow these protists to significantly impact all biotic communities. The parasite *Trypanosoma cruzi (T. cruzi)* is responsible for Chagas disease, also known as American Trypanosomiasis, a chronic endemic, neglected tropical disease of Latin America.

The life cycle of *T. cruzi* is highly complex, involving four distinct stages in the parasite, an invertebrate hematophagous triatomine vector, and a broad range of mammalian hosts. In the midgut of the triatomine vector, the non-infective epimastigote stage replicates and then differentiates into the infective metacyclic trypomastigote in the hindgut. This infective form is then released in the feces or urine after a blood meal, allowing it to invade the host through mucosal surfaces or breaks in the skin. Once inside the host cells, *T. cruzi* escapes from the parasitophorous vacuole and transforms into the replicative amastigote. After several rounds of intracellular replication through binary fission, the amastigotes differentiate into trypomastigotes, which then lyse the host cell. The parasites enter the bloodstream, spreading the infection by invading other cells and tissues. Finally, when a triatomine takes a blood meal from an infected host, it ingests the bloodstream trypomastigotes, which then transform back into epimastigotes in the midgut of the vector, completing the life cycle [6].

T. cruzi is predominantly diploid and typically reproduces asexually through binary division, but there is evidence of natural hybridization, genetic exchange between strains, and sexual reproduction [7–12]. The population genetics of *T. cruzi* have sparked significant interest, leading to two contrasting views. One hypothesis, known as the clonal theory, suggests that *T. cruzi* exemplifies the predominant clonal evolution (PCE) model of pathogens, sharing many characteristics with other parasitic protozoa, fungi, and bacteria [13,14]. However, some researchers have shown that *T. cruzi* can reproduce sexually through a mechanism like classic meiosis, challenging the PCE model and suggesting it might not fully represent the biological complexity of this parasite [7,15], although it is important to highlight that the main form of reproduction of *T. cruzi* is clonal. Moreover, according to this clonal model, new genotypes have been assumed to develop through the gradual accumulation of distinct mutations, with minimal impact from rare genetic recombination events. Nevertheless, the observation of genetic exchange in vitro, the occurrence of sexual reproduction in nature, and the potential coexistence of clonal and sexually panmictic populations supports the existence of a sexual cycle in *T. cruzi* [9,16,17].

During mitosis, the genome of *T. cruzi* does not undergo condensation to form chromosomes, which hinders its visualization using conventional techniques [18,19]. Instead, the parasite karyotype has been determined using molecular biology techniques, such as pulsed-field gel electrophoresis (PFGE) combined with Southern blot analysis. These studies have unveiled significant molecular variability in both the size (0.45–4 Mb) and number (19–40) of chromosomes among strains and even within clones of the same strain [20].

Apart from the nuclear genome, trypanosomatids have an extra genome with the same levels of complexity as nuclear, defined by highly conserved and variable regions among strains. It is the mitochondrial or kinetoplast genome. Kinetoplastids possess a single large mitochondrion per cell [1] and their mitochondrial DNA forms a network of concatenated circular molecules known as the kinetoplast, consisting of maxicircles (~20–60 kb) and thousands of minicircles (0.5–10 kb) of varying sizes depending on the strains [21,22]. Maxicircles contain the mitochondrial genes found in other eukaryotes, and a divergent/variable region that can be more challenging to sequence due to its repetitive nature [23]. On the other hand, minicircles are unique to kinetoplastids and play a direct role in a U-insertion/deletion editing system by encoding guide RNAs (gRNAs) that are used for the maturation of the maxicircle transcripts [24].

2. Genomic Variability of *T. cruzi*

T. cruzi displays a very large genetic diversity. Nonetheless, there is no consensus on the heterogeneity of these populations. More than 6000 *T. cruzi* strains have been reported [25], and differences of up to 40% in the DNA content of clones of distinct strains have been detected, a variability that is primarily attributed to the nuclear DNA, with the highest recorded difference being 47.5% [26,27]. This wide variety of *T. cruzi* strains and the multitude of their genotypes have long been acknowledged due to their complex impact on the transmission cycles, epidemiology, and pathogenesis [28–30].

2.1. Genetic Classification of T. cruzi

Researchers have sought ways to categorize the *T. cruzi* strains primarily based on their biological and genomic differences, which gave rise to several classifications with different nomenclatures. We summarize here the most relevant classifications. In the 1990s, two main clades of *T. cruzi* (I and II) were identified based on biological, biochemical, and molecular criteria [31,32]. Subsequent analyses at the beginning of the XXI century, using gene and intergenic region sequences, suggested that this parasite can be classified into three major lineages, designated as clades A, B, and C [33], what was confirmed by another research that also described a fourth clade (D) and the presence of B–C hybrids [17]. Altogether, this data raised the possibility that *T. cruzi* could be a group of sub-species rather than just one. Furthermore, Brisse et al. proposed an alternative classification consisting of six phylogenetic lineages and proposed a change in the nomenclature of the initial *T. cruzi* groups, from I and II to I and IIa–e, based on phylogenetic information from multi-locus enzyme electrophoresis (MLEE) and random amplified polymorphic DNA (RAPD) markers [34].

Finally, after a Second Satellite Meeting in 2009, the six groups were numbered TcI to TcVI and termed as discrete typing units (DTUs) [35,36]. DTUs are defined as "sets of stocks that are genetically more related to each other than to any other stock and that are identifiable by common genetic, molecular or immunological markers" [37]. In this classification, TcV and TcVI represent hybrid lineages derived from haplotypes TcII and TcIII. Additionally, a seventh DTU initially restricted to bats (TcBat) was also described, closely related to TcI [38], and Flores-López et al. described by phylogenetic analysis a TcIV-USA lineage as a monophyletic group different to the South American TcIV and the other DTUs [39], emphasizing the fact that as additional vectors, hosts, and geographical areas are explored, we may discover and describe additional lineages.

However, some simplifications of these models have been proposed. First, in 2016, through the analysis of mitochondrial sequence genealogies, a new simplification of the DTU model was described. This analysis revealed three distinct mitochondrial clades showing a strong correlation with the previous DTUs: mtTcI corresponded to TcI, mtTcII to TcII, and mtTcIII included TcIII, TcIV, TcV, and TcVI [40]. Second, in 2021, another study performed a phylogenetic analysis using coding regions of maxicircles from up to 29 strains and 1108 single copy nuclear genes from all the DTUs, suggesting that *T. cruzi* is a complex of species composed of 2 major groups [41]. Group 1 would contain the previous lineages TcI, TcIII, and TcIV, and group 2 would contain the previous TcII, while hybrid strains (TcV and TcVI) would remain as a product of TcII and TcIII hybridization.

2.2. Strain Diversity

The first version of a *T. cruzi* genome, derived from the CL Brener strain, was published in 2005 [42], constituting a huge achievement for the molecular biology of the parasite. This strain was chosen to produce the *T. cruzi* genome reference given its high reproducible in vitro models of the acute phase, and a well characterized susceptibility to Benznidazole [43]. However, its hybrid nature was confirmed, and two haplotypes were identified to constitute the nuclear genome. These haplotypes were named according to their closest ancestor: Esmeraldo-like, derived from a TcII, and non-Esmeraldo-like, derived from a TcIII ancestor. Nonetheless, this assembly described for the first time the complex gene repertoire of *T. cruzi*. Unlike its closest counterparts *Leishmania major* and *T. brucei*, which have approximately 20–25% repetitive sequences in their genomes, *T. cruzi* exhibits around 50% repetition, complicating genome analysis and assembly [44].

Currently, multiple *T. cruzi* genomes are available in the National Center for Biotechnology Information (NCBI) database, facilitating the study of phenotypic, pathogenic, and complex variations among strains. Table 1 summarizes the available genomes for the different strains of *T. cruzi* with an associated publication, while Table 2 summarizes other available genomes that are not related to a research publication.

After the CL Brener assembly, the following sequenced strains were assembled above all with short-read sequencing methods (e.g., Illumina/Roche 454/Ion Torrent). This was the case for Sylvio X10/1 [45], Dm28c [46], Y [47], 231 [48], G, CL [49], and SC43 [50] strains of *T. cruzi* and the B7 strain of *T. cruzi marinkellei* [51]. Also, Gómez et al. sequenced six strains belonging to the different DTUs (except DTU TcIV) with Ion Torrent [52], and Reis-Cunha et al. assembled eleven TcII strains and two TcI strains with Illumina or Roche 454 [18,53]. Although these methods generate a high number of reads with low error rates, they are limited in their ability to produce complete chromosome reconstructions from short reads, leading to highly fragmented assemblies. This fragmentation can result in over-, under-, or misrepresentation of genes or entire chromosomal regions. Therefore, long-read sequencing methods (e.g., PacBio/PacBio HiFi/Nanopore) are considered more suitable for trypanosomatid genomes [54]. This technology enables the sequencing of long genetic fragments, circumventing the complex and repetitive nature of these genomes.

Long-read sequencing methods have the potential to produce genomes with a better resolution. Examples of strains assembled by these methods are Bug2148 [47], Dm28c, TCC [55], Tulahuen [56], and Dm25 [57]. However, these long-read methods exhibit a slightly higher error rate than short-read methods, being necessary to increase sequencing coverage (increasing the total cost per nucleotide) or combine technologies to minimize errors. Hybrid approaches have enhanced the assembly process, as demonstrated with the assembly of Berenice [58], Brazil A4, Y clone C6 [59], and STIB980 [60] strains.

Given the knowledge generated about genome assembly and genomic variability, we can conclude that more efforts are needed to complete as many genomes as possible to capture population dynamics and their evolution over time, especially for organisms, such as *T. cruzi*, whose complexity makes this process particularly difficult (and, therefore, expensive). More importantly, given the highly heterozygous nature and variable genomic repertoire between strains, we need as many genomes as strains described for *T. cruzi* for accurate transcriptomic and genomic analyses. Given the vast genomic variability and lack of synteny between strains (even classified into the same or close DTUs) [56], it is not possible to conclude that the latest genomes generated from hybrid assemblies and lower fragmentation levels can be used as universal genome references.

Considering the origin of the strains, the DTUs with more studied members are TcI and TcII, while to date there are no genomic assemblies for any TcIV strain. Notably, the origin of Bug2148 is under debate, since originally it was assigned to TcV, but some genomic and phylogenetic research has associated it with TcI [41,59–61], pointing out the possible mistake about the first assignation of this strain. Interestingly, Y and Dm28c strains have been assembled by different research groups and methods with very distinct results. While Y strain assemblies performed by short-read methods displayed sizes between 15 and 39 Mb, the combination of Illumina and Nanopore techniques gave 47 Mb. In the case of Dm28c, the difference is even more significant; short-read methods displayed sizes between 17 and 27 Mb, and the use of PacBio showed 53 Mb.

Transcriptomic data has also proven valuable for correcting and re-annotating previously assembled genomes. For instance, RNA-seq data improved the genome annotation of the Sylvio X10/1 strain, revealing that 79.95% of the genome corresponds to coding sequences, compared to the previous estimate of 37.73%. These findings suggest that the total genome size for Sylvio X10/1 is larger than previously reported, at least 51 Mb [62].

Table 1. Genome data of strains of *T. cruzi* and the B7 strain of *T. cruzi* marinkellei. Contig N50: is a statistic median such that 50% of the whole assembly is contained in contigs equal to or larger than this value.

<table>
<tr><th>Strain</th><th>DTU</th><th>Size (Mb)</th><th>Contigs</th><th>Contig N50 (kb)</th><th>%GC</th><th>Date of Version</th><th>Sequencing Method</th><th>Reference</th></tr>
<tr><td>G</td><td>I</td><td>25.17</td><td>5531</td><td>6.70</td><td>50</td><td>11-2018</td><td rowspan="2">Roche 454</td><td>[49]</td></tr>
<tr><td>Dm28c</td><td>I</td><td>27.35</td><td>1210</td><td>78.39</td><td>50.5</td><td>11-2013</td><td>[46]</td></tr>
<tr><td>Dm28c</td><td>I</td><td>53.27</td><td>636</td><td>317.64</td><td>51.5</td><td>05-2018</td><td>PacBio</td><td>[55]</td></tr>
<tr><td>Dm28</td><td>I</td><td>17.23</td><td>6541</td><td>3.66</td><td>48.5</td><td>07-2021</td><td rowspan="2">Ion Torrent</td><td rowspan="2">[52]</td></tr>
<tr><td>B.M. López</td><td>I</td><td>18.51</td><td>5923</td><td>5.13</td><td>48.5</td><td>02-2020</td></tr>
<tr><td>Sylvio X10/1</td><td>I</td><td>38.59</td><td>27,019</td><td>2.31</td><td>51</td><td>10-2012</td><td>Roche 454 + Illumina</td><td>[45]</td></tr>
<tr><td>STIB980</td><td>I</td><td>27.90</td><td>400</td><td>165.58</td><td>50.5</td><td>11-2023</td><td>Illumina + Nanopore</td><td>[60]</td></tr>
<tr><td>Brazil clone A4</td><td>I</td><td>45.56</td><td>697</td><td>191.35</td><td>51.5</td><td>11-2020</td><td>Illumina + PacBio</td><td>[59]</td></tr>
<tr><td>Dm25</td><td>I</td><td>45.40</td><td>179</td><td>496.17</td><td>51.5</td><td>02-2024</td><td>PacBio HiFi</td><td>[57]</td></tr>
<tr><td>Arequipa</td><td>I</td><td>19.05</td><td>10,332</td><td>1.91</td><td>51</td><td rowspan="2"></td><td rowspan="2">Roche 454</td><td rowspan="2">[18]</td></tr>
<tr><td>Colombiana</td><td>I</td><td>30.85</td><td>9547</td><td>4.90</td><td>51</td></tr>
<tr><td>S11</td><td>II</td><td>28.48</td><td>32,451</td><td>1.75</td><td>49</td><td rowspan="11">09-2018</td><td rowspan="10">Illumina</td><td rowspan="10">[53]</td></tr>
<tr><td>S154a</td><td>II</td><td>19.27</td><td>17,529</td><td>1.72</td><td>49</td></tr>
<tr><td>S15</td><td>II</td><td>27.51</td><td>31,694</td><td>2.00</td><td>49</td></tr>
<tr><td>S162</td><td>II</td><td>27.30</td><td>30,605</td><td>1.85</td><td>49</td></tr>
<tr><td>S23b</td><td>II</td><td>28.13</td><td>32,315</td><td>1.87</td><td>49</td></tr>
<tr><td>S44a</td><td>II</td><td>17.19</td><td>16,687</td><td>2.16</td><td>49</td></tr>
<tr><td>S92a</td><td>II</td><td>27.08</td><td>31,256</td><td>1.91</td><td>49</td></tr>
<tr><td>Y cl2</td><td>II</td><td>25.91</td><td>26,074</td><td>2.03</td><td>49</td></tr>
<tr><td>Y cl4</td><td>II</td><td>26.14</td><td>26,957</td><td>2.06</td><td>49</td></tr>
<tr><td>Y cl6</td><td>II</td><td>25.78</td><td>26,253</td><td>2.05</td><td>49</td></tr>
<tr><td>Y nc</td><td>II</td><td>29.99</td><td>9164</td><td>5.13</td><td>50.5</td><td>Roche 454</td><td>[18]</td></tr>
<tr><td>Y</td><td>II</td><td>39.04</td><td>9821</td><td>11.96</td><td>50</td><td>10-2017</td><td>Illumina</td><td>[47]</td></tr>
<tr><td>Y</td><td>II</td><td>15.55</td><td>6942</td><td>2.89</td><td>50</td><td>07-2021</td><td>Ion Torrent</td><td>[52]</td></tr>
<tr><td>Y clone C6</td><td>II</td><td>47.22</td><td>477</td><td>396.94</td><td>51.5</td><td>11-2020</td><td>Illumina + PacBio</td><td>[59]</td></tr>
<tr><td>Berenice</td><td>II</td><td>40.80</td><td>934</td><td>148.96</td><td>51</td><td>06-2020</td><td>Illumina + Nanopore</td><td>[58]</td></tr>
<tr><td>Ikiakarora</td><td>III</td><td>18.49</td><td>11,096</td><td>2.19</td><td>48.5</td><td>02-2020</td><td>Ion Torrent</td><td>[52]</td></tr>
<tr><td>231</td><td>III</td><td>35.36</td><td>8469</td><td>5.30</td><td>48.6</td><td>01-2018</td><td>Illumina</td><td>[48]</td></tr>
<tr><td>SOL</td><td>V</td><td>20.06</td><td>11,944</td><td>2.17</td><td>49.5</td><td>07-2021</td><td>Ion Torrent</td><td>[52]</td></tr>
<tr><td>Bug2148</td><td>I/V</td><td>55.16</td><td>929</td><td>200.36</td><td>51.5</td><td>10-2017</td><td>PacBio</td><td>[47]</td></tr>
<tr><td>SC43</td><td>V</td><td>79.9</td><td>1318</td><td>238.74</td><td>51.5</td><td>11-2020</td><td>Illumina</td><td>[50]</td></tr>
<tr><td>CL</td><td>VI</td><td>26.77</td><td>6344</td><td>4.07</td><td>50.5</td><td>11-2018</td><td>Roche 454</td><td>[49]</td></tr>
<tr><td>TCC</td><td>VI</td><td>87.06</td><td>1236</td><td>264.20</td><td>51.5</td><td>05-2018</td><td>PacBio</td><td>[55]</td></tr>
<tr><td>Tulahuen</td><td>VI</td><td>48.46</td><td>75</td><td>872.48</td><td>52</td><td>12-2023</td><td>Nanopore</td><td>[56]</td></tr>
<tr><td>CL Brener</td><td>VI</td><td>19.53</td><td>11,101</td><td>2.3</td><td>49.5</td><td>07-2021</td><td>Ion Torrent</td><td>[52]</td></tr>
<tr><td>CL Brener</td><td>VI</td><td>89.94</td><td>32,746</td><td>14.67</td><td>51.5</td><td>08-2005</td><td>Sanger</td><td>[42]</td></tr>
<tr><td>B7</td><td>---</td><td>34.23</td><td>23,154</td><td>2.85</td><td>51</td><td>10-2012</td><td>Roche 454 + Illumina</td><td>[51]</td></tr>
</table>

Table 2. Genome data of different *T. cruzi* strains without an associated publication. Contig N50: is a statistic median such that 50% of the whole assembly is contained in contigs equal to or larger than this value.

Strain	DTU	Size (Mb)	Contigs	Contig N50 (kb)	%GC	Date of Version	Sequencing Method
JR cl4	I	41.48	18,103	7.41	51.5	01-2013	Roche 454
Tula cl2	I	83.51	53,083	2.19	51.5	04-2013	Roche 454
Dm28c	I	50.93	1028	110.59	51.5	09-2017	PacBio
H1	I	27.34	11,257	16.44	49.5	02-2023	Illumina + PacBio + Nanopore
Esmeraldo cl3	II	38.08	20,187	5.35	51	01-2013	Roche 454

3. Genomic Structure

3.1. Chromosomes and Ploidy

Mitosis in *T. cruzi* occurs without a complete disassembly of the nuclear envelope. Although nucleosomes are present, chromatin does not condense into visible chromosomes, rendering classic cytogenetic studies unsuitable. Consequently, the karyotype of *T. cruzi* has been determined using molecular biology techniques, primarily PFGE in combination with Southern blotting [18,19,63,64]. Chromosomal band sizes range from 0.45 to 4 Mb, and the number of chromosomes has mainly been estimated by using probes as genetic markers. The number of chromosomes per haploid genome varies from 19 to 40, indicating that *T. cruzi* is predominantly diploid, although the sizes of homologous chromosomes as well as the chromosome lengths among clones of the same strain, strains from different DTUs, and even strains within the same DTU can differ significantly [18,20,65,66]. The analysis of ploidy or chromosomal copy number variation (CCNV) in this parasite was not feasible until the advent of the NGS methods.

The repetitive nature of the *T. cruzi* genome has led to difficulties in the genome assembly, making it challenging to determine the chromosome number and structure. Nevertheless, efforts to achieve full-length chromosome sequencing employed a combined strategy using synteny maps with *T. brucei* chromosomes and BAC end sequencing, resulting in the identification of 41 virtual chromosomes for the CL Brener strain [67]. Although the long-read methods overcame the high fragmentation limitations imposed by the short-read methods, producing contigs bigger than 1 Mb, likely covering entire chromosomes, some regions of the genomes remain fragmented. Therefore, the exact number and organization of chromosomes will ultimately be determined through the integration of long-read sequencing methods, optical mapping techniques, and polymer-based modeling, a field with huge advancements in the past decade [68].

T. cruzi is generally considered as a diploid organism. However, aneuploidies have been extensively studied in trypanosomatids, for example, in *Leishmania* spp., where 'mosaic aneuploidies' represent ploidy variations between isolates of the same strain and even among individual cells within the same population. These aneuploidies are associated with drug resistance, gene expression regulation, and host adaptation [69–71]. In contrast, *T. brucei* exhibits ploidy stability [72]. For *T. cruzi*, CCNV analysis depends on the quality of the assembled reference genome. Strains from different DTUs have shown that, like *Leishmania* spp., aneuploidy patterns vary both among and within strains and DTUs [18]. Despite this limitation, it has been observed that strains from DTU TcI appear to be more stable, whereas strains from DTUs TcII and TcIII exhibit a high degree of aneuploidies, including monosomies, trisomies, and tetrasomies [53].

These findings suggest that aneuploidy events in *T. cruzi* may facilitate gene expansion and alterations in gene expression, which is particularly crucial for parasites that rely on post-transcriptional mechanisms for gene regulation. While aneuploidies are generally associated with deleterious phenotypes in many eukaryotes, they may play a role in species-specific adaptations during trypanosomatid evolution. This could impact multi-gene families that are essential for establishing productive infections in mammalian hosts [73]. Regarding gene expansion processes, just a few mechanisms have been proposed. The main mechanism consists of homologous recombination at the disruptive compartment of the genome, mainly at the telomeric and sub-telomeric regions. This mechanism includes the participation of transposable elements' nucleases, which introduce chromosome breakages, and the subsequent repair (by homologous recombination) introduces new gene copies, variants, and pseudogenes. This mechanism also considers (in a lesser proportion) the possibility of recombination between non-homologous chromosomes (ectopic recombination) [74–77].

3.2. Genome Organization

T. cruzi displays significant genomic plasticity and an unusual gene organization with tandemly repeated sequences, multi-gene families, and retrotransposons constituting over 50% of its genome. This genomic plasticity is linked to its genetic composition and is compartmentalized into two principal regions of protein-coding genes. The first is the core compartment, which contains highly conserved genes with known, putative, and constitutive functions, typically annotated as hypothetical conserved genes that exhibit synteny with other species, such as *Leishmania major* and *T. brucei*. The second is the non-syntenic disruptive strains-specific compartment, primarily formed by genes that evolve rapidly, including the surface multi-gene families [78,79].

Furthermore, the *T. cruzi* genome comprises three main types of DNA: first, repetitive sequences, including retrotransposons, tandem repeats, and short repeat elements; second, coding sequences of multi-copy gene families, such as those encoding surface proteins and virulence factors; third, coding sequences of single-copy genes that are conserved across strains and species, although the total number estimation of these single-copy genes differs between hybrid and non-hybrid strains [79]. Notably, approximately 50% of the genetic content has unknown functions [47], which aligns with proteomic studies of strains of different DTUs, where 40–50% of the total proteins are of unknown function [80–83]. This highlights the urgent need for studies to unravel these significant gaps in our understanding of *T. cruzi* genetics and biology.

The core and disruptive compartments exhibit variable %G+C content. Indeed, genes with high recombination rates and constant evolution tend to have elevated levels of G+C [84,85], which is correlated with the fact that the *T. cruzi* core compartment has a G+C content of ~48%, while the disruptive compartment has ~53%. Other studies also confirmed that variations in the G+C content are correlated with specific telomeric repeats in *T. cruzi*, such as the hexameric repeat TTAGGG and poly-T structures [75,86]. Precisely, the telomeric and sub-telomeric regions in this parasite are hotspots for DNA recombination, leading to extensive genetic variations and continuous evolutionary processes, affecting the relative abundance and gene organization [74,87].

3.3. Replication Origin

In eukaryotic organisms, chromosomes are replicated from numerous DNA replication origins (ORIs), ranging from hundreds to thousands, which are identified by the binding of the origin recognition complex (ORC). In *T. cruzi*, the ORIs of the CL Brener strain were mapped using marker frequency analysis sequencing (MFA-seq) [88], identifying 103 and 110 putative ORIs in each haplotype of this hybrid strain. The analysis revealed that some replication initiation sites are located at the boundaries of transcription units, as in *Leishmania major* and *T. brucei*. Notably, most of the predicted ORIs were abundant within coding DNA sequences and showed a high G+C content (averaging 65%), compared to the average of 54% along the genome. Additionally, DNA combing analysis of the same strain showed a median inter-origin distance of 171.1 kb (mean 208.3 ± 29.24 kb) [89].

While some ORIs in *T. cruzi* are found in non-transcribed regions, like those in *T. brucei* and *Leishmania major*, many are strategically positioned in sub-telomeric regions, particularly near disperse gene family 1 (DGF-1) genes, where they could generate genetic variability in multi-gene families [88]. The transcription orientation towards telomeres suggests that the high density of ORIs in sub-telomeric regions would lead to head-on collisions between transcription and replication processes, as replisomes move towards the chromosome centers. These findings indicate that such collisions are frequent, contributing to genetic variability, as evidenced by increased single nucleotide polymorphism (SNP) levels in sub-telomeric regions and DGF-1 genes containing putative ORIs.

3.4. Chromatin, Transcription and Gene Regulation of T. cruzi

Nucleus and chromatin are different depending on the *T. cruzi* stage. In epimastigotes and amastigotes, the nucleus is spherical with a large nucleolus and high quantities of euchromatin (transcriptionally active), while the trypomastigotes display an elongated nucleus, absence of nucleolus, and abundant heterochromatin (transcriptionally inactive), which is dispersed in the nucleoplasm [90–92]. Nuclear morphology alterations start at the onset of metacyclogenesis, with heterochromatin progressively spreading throughout the nucleus in intermediate forms. The precise mechanisms governing and regulating these ultrastructural changes during metacyclogenesis remain largely unknown [90]. Moreover, in a recent study evaluating the open chromatin status genome-wide, the core and disruptive genomic compartments of the *T. cruzi* genome exhibited huge differences in open chromatin enrichment. The core genomic region was highly enriched in open chromatin compared to the disruptive compartment [93].

In trypanosomatids, the organization of genes, gene expression regulation, and RNA metabolism display distinct characteristics compared to other eukaryotes. Genes are arranged into directional gene clusters (DGCs), which are separated by strand-switch regions (SSRs) where transcription directions either converge or diverge [94]. These DGCs are then transcribed as extensive polycistronic transcription units (PTUs), and the mRNA maturation involves co-transcriptional trans-splicing of a capped spliced leader (SL) RNA and polyadenylation [95]. Figure 1 shows this polycistronic transcription of *T. cruzi*.

Furthermore, it is widely recognized that post-transcriptional regulation plays a predominant role in gene expression regulation [96]. By modifying RNA transcripts after they are produced, *T. cruzi* can quickly respond to changes in its environment favoring its survival, and it can do it in a stage-specific manner, which is essential for its ability to adapt to different environments and hosts. Also, this process is particularly important because *T. cruzi* lacks the typical transcriptional control mechanisms found in other organisms [97]. Nevertheless, recent studies have underscored the significant influence of histone variants, histone post-translational modifications, base J (a hypermodified nucleobase of kinetoplastids), nucleosome positioning, and chromatin organization on gene expression regulation, cell cycle control, and differentiation [93,98–100].

Additionally, only a limited number of transcriptional regulators have been identified in trypanosomes, and the influence of genome spatial organization on gene expression remains poorly understood. A recent study mapped genome-wide chromatin interactions in *T. cruzi* using chromosome conformation capture (Hi-C), displaying that the core and disruptive regions create two 3D-chromatin compartments, named C and D. Both chromatin compartments presented distinctions in levels of nucleosome positioning, DNA methylation, and chromatin interactions, which produce a change in genome expression dynamics. The authors concluded that the trypanosome genome is structured into chromatin-folding domains, with transcription being influenced by the local chromatin architecture. They proposed a model suggesting that epigenetic mechanisms play a role in regulating gene expression in trypanosomes [94].

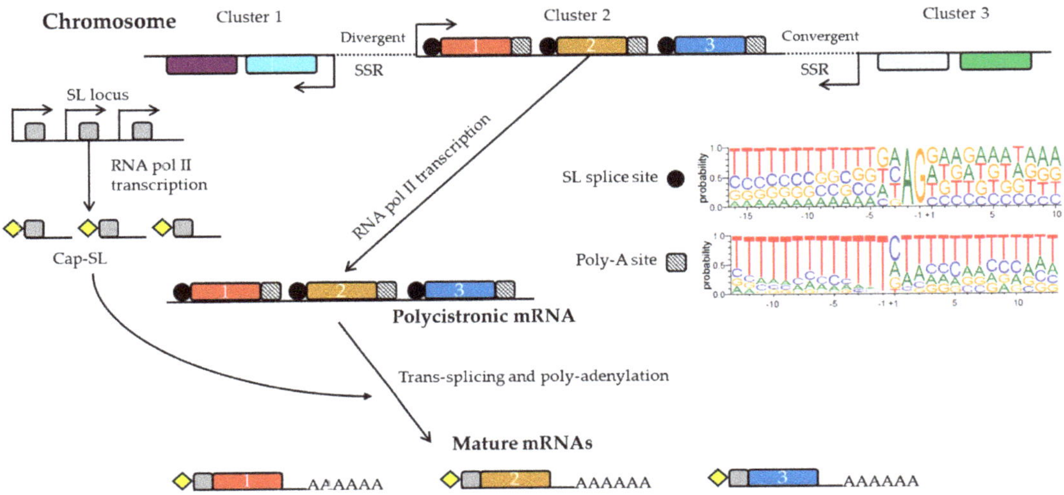

Figure 1. Polycistronic transcription of *T. cruzi* and processing. Gene clusters in the genome are transcribed to polycistronic mRNAs by the RNA polymerase II, as displayed by Cluster 2. Divergent and convergent SSRs, as well as the spliced leader (SL) and the poly-A insertion sites with the nucleotide distribution in *T. cruzi* according to Callejas-Hernández et al. [62], are shown. Then, this polycistronic transcript is processed by trans-splicing and polyadenylation to produce the mature and individual mRNAs. For the trans-splicing, a capped SL RNA (grey box) is added to the 5′ end of every mRNA. The SL locus is in a different chromosome and is also transcribed by the RNA polymerase II. Cap modification is shown by a yellow diamond.

As previously noted, genes in *T. cruzi* are organized into non-overlapping clusters on the same DNA strand, often with unrelated predicted functions. These genes are transcribed as polycistronic units and subsequently undergo trans-splicing and polyadenylation. Although no shared consensus motifs or patterns have been identified among the SSRs, these are functionally active. For example, transcription initiation and termination occur within SSRs, and they are also implicated in DNA replication origin and centromeric functions [101–103]. Also, SSRs exhibit a distinct composition compared to the rest of the genome and different higher intrinsic curvatures [104], related to the transcriptional regulation. Notably, SSRs from the disruptive compartment are longer than those from the core compartment, with mean lengths of approximately 4.5 kb and 1.5 kb, respectively [78].

The mRNA maturation by trans-splicing is a unique RNA processing mechanism in which two exons from different genomic locations combine to form a single transcript [105]. In the particular case of *T. cruzi*, this involves inserting a 39-nucleotide sequence at the 5′ end of each transcript, known as the mini-exon or spliced leader (SL), which is transcribed from a tandem array as a precursor of approximately 140 nucleotides and undergoes capping modification. The insertion of this Cap-SL stabilizes the mRNA and facilitates the excision of each mRNA from the PTU, allowing for final polyadenylation [106]. Furthermore, the AG dinucleotide has been identified as the consensus sequence for SL trans-splicing in *T. cruzi* [62], *Leishmania major* [107], and *T. brucei* [108]. Nevertheless, slight variations in the nucleotide composition surrounding the AG dinucleotide suggest species-specific mechanisms for mRNA maturation.

Finally, the other well-known process for the mRNA maturation is polyadenylation. The typical AAUAAA polyadenylation signal found in eukaryotes is absent in trypanosomatids. A recent study showed that *T. cruzi* utilizes a single nucleotide as the most likely polyadenylation signal, with cytosine being the most frequent (45.3%) and thymine the least frequent (6.79%) [62]. This is distinct from other trypanosomatid species, such as *Leishmania major* and *T. brucei*, which use an AA dinucleotide as the most probable polyadenylation signal [107,108].

4. Multi-Gene Families of *T. cruzi*

T. cruzi harbors many multi-gene families, some comprising hundreds of members, which contribute to the repetitive nature of its genome. Extensive research has been conducted to elucidate the structure, distribution, and functions of these multi-gene families, which predominantly encode surface proteins that play crucial roles throughout the *T. cruzi* life cycle, from facilitating effective host–cell interactions and invasion to providing protection against the host immune system. Additionally, they exhibit significant expansion and continuous evolution, resulting in considerable diversity among strains [109].

Berná et al. identified multi-gene families, such as TSs, mucins, and MASPs, within the disruptive compartment of the *T. cruzi* genome, while retrotransposon hot spot proteins (RHS), GP63, and DGF-1 families were found in both disruptive and core compartments [55]. The specific reasons for this difference remain unknown. Copy numbers of these multi-gene families in the genomes of various *T. cruzi* strains are illustrated in Figure 2. Data indicates that the TS family is the most expanded multi-gene family, followed by MASPs and RHS, although this pattern is not consistent across all strains. For example, SylvioX10 presents a huge number of DGF-1 members in CL Brener but just a few RHSs. The high variability among strains may be attributed to strain-specific genetic profiles, the accuracy of the genome assemblies, and genomic plasticity. This diversity is supported by the observed differences in infection kinetics, virulence, and immune responses among *T. cruzi* strains [110,111].

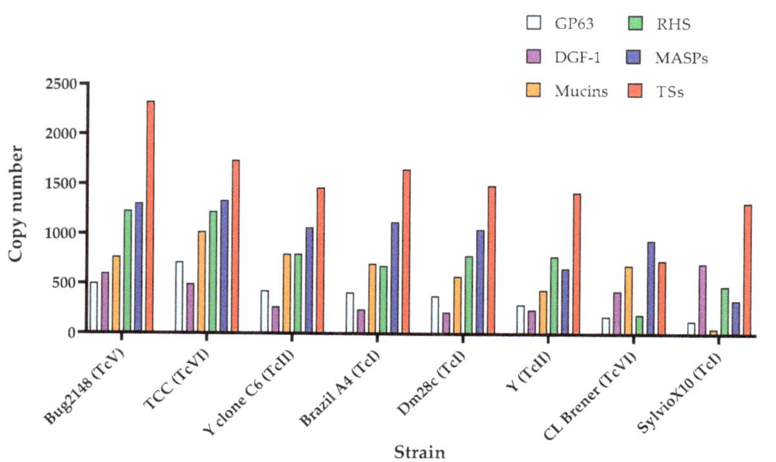

Figure 2. Copy number of the six principal multi-gene families in different strains of *T. cruzi*. TSs: trans-sialidases. RHS: retrotransposon hot spot proteins. MASPs: mucin-associated surface proteins. DGF-1: disperse gene family 1 members. Based on the data of Berná et al. [55], Callejas-Hernández et al. [47,62], Wang et al. [59], and El-Sayed et al. [42].

4.1. Trans-Sialidases (TSs)

This multi-gene family is the largest of *T. cruzi*, playing a vital role in host–parasite interactions [112,113]. Their members are located on the membrane surfaces of metacyclic and bloodstream trypomastigotes, as well as intracellular amastigotes, and they are mainly distributed along the flagellum, cell body, and flagellar pocket [114]. Also, TSs have been found differently expressed among strains in extracellular vesicles [115]. Interestingly, the TS copy number is significantly smaller in *T. brucei* and is absent in *Leishmania major* [44].

These proteins display 2 structural motifs, a lectin-like motif plus a beta barrel motif, and they can present a glycosylphosphatidylinositol (GPI) anchor. However, it can be removed by the action of a phosphatidylinositol phospholipase C, allowing TSs to be released into the bloodstream. Also, some researchers have established that all members share a VTVxNVxLYNR canonical motif [116], although some TSs exhibit remarkable motif degeneration in compliance with recent analysis [79].

According to the genome annotation of the distinct strain assemblies of Figure 2, independently of the DTU of origin, all strains possess more than 1000 TS proteins. Notably, Bug2148, TCC, and Brazil A4 are the strains with more members, strains assembled by long-read sequencing methods or the combination of both long- and short-read methods, while CL Brener is the strain with least members, likely because it was not sequenced by these methods. Many TS genes are located near telomeric and sub-telomeric regions, which can cause assembly collapses and lead to under- or over-representation of these genes [75]. This supports the use of both long- and short-read methods for the assembly of *T. cruzi* genomes and the fact that there may still be errors in estimates of the real number of TS members in some parasite genomes. Also, this suggests that TS expansion is partly due to their chromosomal location and the selective pressure exerted by the host immune system, as TS proteins are targets of both humoral and cell-mediated immune responses [117].

Regarding the biological function of this family, the best-characterized function is the trans-sialidase catalytic activity, first described in 1980 [118]. Subsequent studies have demonstrated that *T. cruzi* is unable to synthesize its own sialic acids and utilizes TS enzymes to incorporate sialic acids from host cell sialoglycoconjugates into acceptor molecules on their membranes [79]. This sialylation process confers a negatively charged coat that protects trypomastigotes from being targeted by human anti-α-galactosyl antibodies and the complement system [119,120]. Further research suggests that TS activity is crucial for *T. cruzi* survival and the establishment of an effective infection [120]. Notably, although neuraminidase activity was identified in TS enzymes, it is only active in the presence of suitable Gal acceptors [121].

Nevertheless, the critical residues necessary for catalytic activity have been identified in only a few genes, and additional roles related to host–ligand interactions and immune regulation have been proposed [116,122]. Indeed, TS enzymes interact with various mammalian host cells, including neurons, Schwann cells, cardiac fibroblasts, B cells, thymocytes, CD4+ and CD8+ T cells, platelets, and endothelial cells [123], although many of their ligands remain unknown. Therefore, given that not all members of this protein family seem to exhibit TS activity, it may be advisable to consider renaming the family to better reflect its diverse functional characteristics, and more studies are mandatory to unravel the exact function of each member in the family.

Regarding the classification of the TSs, the last aggrupation was proposed in 2011 [122]. This classification was established by a sequence cluster analysis with the CL Brener strain separating the TSs in 8 groups. Groups II and V displayed the largest number of members, accounting for approximately 70% of the TSs analyzed. However, this study had two very important limitations considering the genome complexity of this parasite: it was done with the annotated proteins from just one strain, CL Brener, and its genome proceeds from a non-NGS assembly, which is not advisable for genomes with highly repetitive sequences, like *T. cruzi*. Therefore, this classification may be very far from reality.

Even so, different research groups have classified the trans-sialidases detected in their assemblies according to this 8-group classification. Figure 3 represents the distribution of the groups among these strains, considering only these annotated trans-sialidases. Group V is the predominant TS group, constituting approximately 50% of the TSs in each strain, except for Brazil A4, in which it accounts for 55%. In contrast, groups I, III, IV, and VII are significantly smaller. The overall profile is consistent across strains, although some exhibit notable variations. For instance, the Y and Bug2148 strains have over 10% of group I TSs, whereas other strains have less than 4%. Also, certain strains lack specific groups almost entirely, such as Brazil A4 with group III, Y with group IV, and Dm28c with group VII. Regarding group VI TSs, the TCC, Dm28c, and Brazil A4 strains display more than 12%, while other strains have less than 8%.

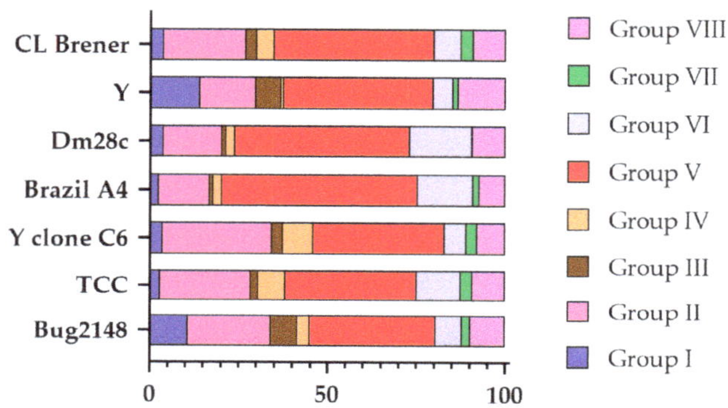

Figure 3. TS group distribution according to the annotated trans-sialidases in each strain of *T. cruzi* based on the data of Berná et al. [55], Callejas-Hernández et al. [47], Freitas et al. [122], and Wang et al. [59]. The percentage of each group is shown with different colors.

4.2. Mucin-Associated Surface Proteins (MASPs)

MASPs form the second largest multi-gene family of *T. cruzi*. They receive that name from their cluster position among mucin gene groups. Although both proteins differ in sequence, they are structurally similar [124]. MASPs display conserved N- and C-terminal domains encoding a signal peptide and a GPI-anchor addition site, respectively. These proteins are GPI-anchored to the membrane and predominantly expressed during the bloodstream trypomastigote stage [109], although they are also expressed in amastigotes and epimastigotes [80,125]. Moreover, MASPs are differentially expressed in clones from the same original population [126], and previous studies have demonstrated that their expression varies based on the type of host cell infected and across successive passages in mouse models [127]. The central region of MASPs varies in both sequence and length (176–645 amino acids), containing numerous repetitive motifs. Single amino acid repeats, particularly those with glutamic acid, are common, comprising about 27% of the identified motifs, while at least four potential O-glycosylation sites per sequence have been identified, with 70% of these sites being threonines [124].

Regarding the functions of the MASPs members, the specific roles in the parasite life cycle remain largely unclear. However, some researchers have begun to shed light on certain functions. The overexpression of MASPs in intracellular parasites, before amastigote division, suggests that these proteins play a crucial role in the survival and multiplication of intracellular amastigotes [124,128]. Additionally, MASP49 has been proposed as a virulence factor of *T. cruzi* [129], and the high degree of polymorphisms in the MASPs along with its location at the surface of trypomastigotes strongly suggests participation in host–parasite interactions [130]. In fact, MASP proteins have been demonstrated to stimulate antibody production during the acute infection stage [127]. Consequently, *T. cruzi* could express these surface proteins to produce unspecific T cell and antibody responses, facilitating parasite persistence within the mammalian hosts due to the variations in the extensive range of antigenic peptides within the MASP family. Furthermore, the N-terminal motif of MASPs has been shown to partially inhibit complement-mediated lysis [131], and immature MASPs, as well as the C-terminal portion, may be included in trypomastigote exosomes and exhibit immunogenic properties [131,132].

Considering the possible classifications of this multi-gene family, MASP members were divided into 7 subgroups (M1–M7) in recent research using hierarchical clustering analysis based on protein sequence similarity [130]. However, as is the case for the TS family, this classification was proposed just with the sequences of the CL Brener strain, hence a revision is needed. Figure 4 shows the abundance of each subgroup. Subgroups M1–M4 are the largest, while M5–M7 represent the smallest ones. Interestingly, in this study, the authors examined the antigenicity of these MASP subgroups, revealing that the recognition profiles are distinct between them. These profiles tend to vary during the acute infection stage and after multiple consecutive passages in murine models.

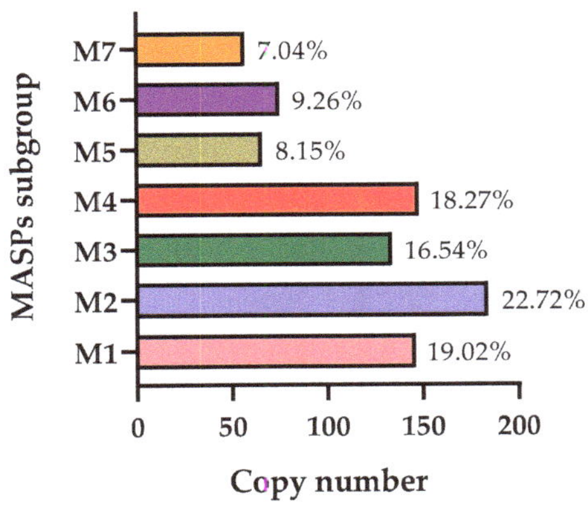

Figure 4. Copy number of each MASP subgroup in the CL Brener strain. The percentage of each subgroup is displayed.

4.3. Mucins

This multi-gene family is the most abundantly expressed in the membrane of *T. cruzi* and ranks as the fourth largest multi-gene family [133]. Mucins display a GPI-anchor, a dense array of O-linked oligosaccharides attached to serine and/or threonine residues, and they are the main acceptors of sialic acid in the parasite membrane [134]. Moreover, they are widely distributed over the cell body, flagellum, and flagellar pocket of the different *T. cruzi* stages [135].

Regarding their role in the infection process, two primary functions have been associated with mucins. First, they protect the parasite from host defense mechanisms, supporting the persistence of *T. cruzi* [109]. For example, epitopes in mucins generate strong but delayed and non-protective CD8+ T cell responses [136]. Also, mucins members are expressed in large amounts in all the parasite stages conferring high resistance against proteases and glycosidases, and their sialylation by TSs protects against the complement-independent lysis induced by human anti-α-galactosyl antibodies [133]. Second, mucins facilitate the attachment and invasion of specific host cells [137].

Mucins are categorized into two subfamilies, TcMUC and TcSMUG, based on structural and biological criteria (Figure 5). TcMUC proteins are exclusively expressed during the mammalian stages of the parasite (amastigotes and bloodstream trypomastigotes), whereas TcSMUG proteins are found in the insect-dwelling forms (epimastigotes and metacyclic trypomastigotes) [138]. The greater diversity observed in TcMUC proteins compared to TcSMUG proteins is linked to the immune system pressures they encounter in mammalian hosts [137] and their chromosomal localization near telomeric regions [47]. Furthermore, TcMUC proteins feature a signal peptide, a GPI anchor, and a central region rich in threonines that are targets for O-glycosylation and subsequent sialylation, which may explain the higher glycosylation levels observed in mucins of amastigotes and bloodstream trypomastigotes compared to those in epimastigotes [137].

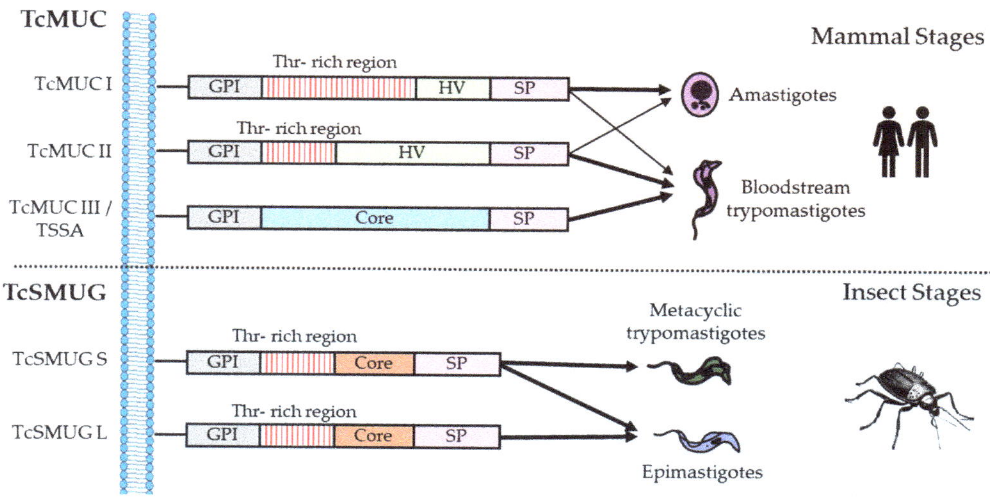

Figure 5. Mucin classification. Schematic representation of each type of mucin is displayed, as well as the life cycle stages in which they are expressed. The width of the arrow corresponds to the expression levels in each stage. SP: signal peptide. HV: hypervariable region.

TcMUC genes are subclassified into TcMUC I, II, and III. TcMUC I genes are more abundant in amastigotes, while TcMUC II genes are predominantly found in the membrane lipid rafts of bloodstream trypomastigotes and they are known as tGPI mucins [114]. Structurally, the principal difference between the TcMUC I and II members lies in the central region of the proteins. TcMUC I genes feature a short hypervariable (HV) section and numerous tandem repeats of the Thr_8-Lys-Pro_2 sequence, although different variations in this consensus sequence have been observed in the *T. cruzi* genome assemblies with NGS methods. In contrast, TcMUC II genes possess a central region with a longer HV section and fewer tandem repeats, which are still rich in threonine and proline. Some studies have suggested that TcMUC II genes may have evolved from TcMUC I genes or vice versa [139]. Also, TcMUCII members exhibit differential expressions among strains of O-linked α-galactosyl residues, which are crucial for inducing anti–α-Gal antibodies. Variations in their structures likely have significant implications for the diagnosis and immunopathology of Chagas disease, depending on the strain [140].

Besides, TcMUC group III is constituted by a single-copy gene of a mucin-like protein known as TSSA (Trypomastigote Small Surface Antigen), which is only present in the bloodstream of trypomastigote membranes and is not a sialic acid acceptor. TSSA appears to play a role in host cell invasion as an adhesion molecule [141]. Unlike TcMUC I and II genes, it does not display a specific Thr-rich region, and there is a debate about their sequence composition. While a sequence analysis has shown a high content of serine and threonine residues for O-glycosylation [142], another study described TSSA as a rather lowly glycosylated molecule [143].

TcSMUGs (*T. cruzi* small mucin-like genes) consist of two groups of genes, designated as L (large) and S (small), which exhibit over 80% sequence identity, but differ in their genomic structure. TcSMUG S genes form the backbone of GP35/50 mucins, which are expressed during the insect-dwelling stages [144]. In metacyclic trypomastigotes, some researchers have shown that GP35/50 mucins bind to target cells, inducing a bidirectional Ca^{2+} response that facilitates cell invasion [145], while others have suggested that mucin-mediated invasion is accomplished through interaction with host cell annexin A2 and clathrin-dependent endocytosis [146]. Conversely, in epimastigotes, GP35/50 mucins play a protective role against proteases in the insect intestinal tract [147]. Furthermore, TcSMUG S genes, unlike TcSMUG L genes, accept sialic acid residues transferred to the parasite membrane by TSs. Hence, TcSMUG L products are likely involved in attachment to the luminal midgut surface of the insects and are exclusive to the epimastigote stage [148,149].

4.4. DGF-1

The disperse gene family 1 (DGF-1) is the fifth largest multi-gene family of *T. cruzi*, and it was placed among the highest expressed proteins in this parasite [150]. Its composition includes the presence of eight to nine transmembrane hydrophobic helices at their C-terminal end, potentially serving as membrane anchors. These genes also exhibit six epidermal growth factor-1-like (EGF-1) domains and one EGF-2-like domain, spaced approximately 400 amino acids apart, along with lectin-binding motifs. Furthermore, integrin-like sequences suggest a cell surface localization, implying roles in cell–cell interactions or signal transduction. Finally, they lack a GPI anchor and around half of the members display a canonical signal peptide [151,152].

Based on a neighbor-joining (NJ) tree and the distribution of functional domains in the CL Brener strain, DFG-1 members were categorized into two principal groups (A and B), which included most of the proteins. However, the rest of the members were clustered in other small groups (C–E) [152]. Therefore, as we explained before for the other multi-gene families, more studies are mandatory in different strains to confirm this classification. Also, the NJ method for phylogeny analysis is less accurate than maximum likelihood or maximum parsimony methods for the calculation of phylogenies [153]. Additionally, phylogenetic analysis has suggested that in the DGF-1 family, the high sequence similarity among members indicates that these events are either recent or subjected to homogenization processes [152]. Interestingly, DGF-1 copies have been identified in the sub-telomeric regions of BAC-telomere recombinants of the CL Brener strain, and these DGF-1 copies are flanked by trans-sialidases and RHS elements [75]. These sub-telomeric localization of DGF-1 has been confirmed across various *T. cruzi* strains, with significant variation in copy numbers among strains [154].

Regarding the possible roles of DGF-1 proteins in the life cycle of *T. cruzi*, different homology, glycoproteomics, and structural analyses suggest that they function as receptors or bidirectional signal transducers, potentially regulated by cyclic AMP (cAMP), nucleoside analogs, and non-cAMP kinases [154–156]. These findings, along with in silico predictions, underscore the potential of DGF-1 genes to interact with other proteins, with these interactions being subject to regulatory mechanisms. However, despite the extensive data yielded by new omics technologies and advancements in immunology, the specific functions of this prominent and ubiquitous protein family remain elusive.

5. Conclusions

T. cruzi exhibits significant genomic complexity and extensive genetic diversity among strains, as demonstrated by the analyses of multiple strains and the differences observed among them at both the sequence and infection levels. Its genome is specifically organized into two distinct compartments, the core and the disruptive, which display different gene composition. Research into this genetic diversity has yielded vital insights into the parasite pathogenicity, and the complex array of genetic variations within *T. cruzi* populations poses key challenges in developing effective vaccines and therapies for Chagas disease. Notably, advances in NGS technologies, with the integration of long- and short-read sequencing methods, have significantly enhanced the quality of the *T. cruzi* genomes, shedding light on this extensive genetic diversity and complexity.

Furthermore, the information described about the different multi-gene families of *T. cruzi* confirms their critical role for the infection processes and parasite survival. These families have undergone significant growth and continuous evolution, enhancing the *T. cruzi* capacity to adapt, survive, and infect both insect and mammalian hosts. Most of the multi-gene families are exclusive to this type of parasites, which makes them conducive to the design of specific vaccines. Functional genomics, omics approaches, new structural and sequence analysis with the most recent assembled genomes and in vivo research may provide comprehensive insights into their biological functions. Altogether, these advances may allow the design of effective new treatments against Chagas disease.

Author Contributions: Conceptualization, A.H.-C., N.G. and M.F.; methodology, A.H.-C. and F.C.-H.; formal analysis, A.H.-C. and F.C.-H.; investigation, A.H.-C.; writing—original draft preparation, A.H.-C.; writing—review and editing, A.H.-C., F.C.-H., N.G. and M.F.; supervision, N.G. and M.F.; project administration, N.G. and M.F.; funding acquisition, N.G. and M.F. All authors have read and agreed to the published version of the manuscript.

Funding: This research was funded by "Ministerio de Ciencia e Innovación" (Spain), grant number PID2021-123389OB-I00 and PID2022-137487OB-I00.

Acknowledgments: We acknowledge the Genomics and NGS Core Facility (GENGS) at the Centro de Biología Molecular Severo Ochoa (CBMSO, CSIC-UAM) Madrid (Spain), which obtains general funding from both institutions, for their advice in the database searches.

Conflicts of Interest: The authors declare no conflicts of interest.

References

1. Maslov, D.A.; Opperdoes, F.R.; Kostygov, A.Y.; Hashimi, H.; Lukeš, J.; Yurchenko, V. Recent Advances in Trypanosomatid Research: Genome Organization, Expression, Metabolism, Taxonomy and Evolution. *Parasitology* **2019**, *146*, 1–27. [CrossRef] [PubMed]
2. Lukeš, J.; Butenko, A.; Hashimi, H.; Maslov, D.A.; Votýpka, J.; Yurchenko, V. Trypanosomatids Are Much More than Just Trypanosomes: Clues from the Expanded Family Tree. *Trends Parasitol.* **2018**, *34*, 466–480. [CrossRef] [PubMed]
3. Hajduk, S.; Ochsenreiter, T. RNA Editing in Kinetoplastids. *RNA Biol.* **2010**, *7*, 229–236. [CrossRef] [PubMed]
4. Jackson, A.P. Genome Evolution in Trypanosomatid Parasites. *Parasitology* **2015**, *142* (Suppl. 1), S40–S56. [CrossRef]
5. Kaufer, A.; Ellis, J.; Stark, D.; Barratt, J. The Evolution of Trypanosomatid Taxonomy. *Parasit. Vectors* **2017**, *10*, 287. [CrossRef] [PubMed]
6. Rodrigues, J.C.F.; Godinho, J.L.P.; de Souza, W. Biology of Human Pathogenic Trypanosomatids: Epidemiology, Lifecycle and Ultrastructure. *Subcell. Biochem.* **2014**, *74*, 1–42. [CrossRef] [PubMed]
7. Schwabl, P.; Imamura, H.; Van den Broeck, F.; Costales, J.A.; Maiguashca-Sánchez, J.; Miles, M.A.; Andersson, B.; Grijalva, M.J.; Llewellyn, M.S. Meiotic Sex in Chagas Disease Parasite *Trypanosoma cruzi*. *Nat. Commun.* **2019**, *10*, 3972. [CrossRef]
8. Berry, A.S.F.; Salazar-Sánchez, R.; Castillo-Neyra, R.; Borrini-Mayorí, K.; Chipana-Ramos, C.; Vargas-Maquera, M.; Ancca-Juarez, J.; Náquira-Velarde, C.; Levy, M.Z.; Brisson, D.; et al. Sexual Reproduction in a Natural *Trypanosoma cruzi* Population. *PLoS Negl. Trop. Dis.* **2019**, *13*, e0007392. [CrossRef] [PubMed]

9. Gaunt, M.W.; Yeo, M.; Frame, I.A.; Stothard, J.R.; Carrasco, H.J.; Taylor, M.C.; Mena, S.S.; Veazey, P.; Miles, G.A.J.; Acosta, N.; et al. Mechanism of Genetic Exchange in American Trypanosomes. *Nature* **2003**, *421*, 936–939. [CrossRef]
10. de Paula Baptista, R.; D'Ávila, D.A.; Segatto, M.; do Valle, Í.F.; Franco, G.R.; Valadares, H.M.; Gontijo, E.D.; da Cunha Galvão, L.M.; Pena, S.D.; Chiari, E.; et al. Evidence of Substantial Recombination among *Trypanosoma cruzi* II Strains from Minas Gerais. *Infect. Genet. Evol. J. Mol. Epidemiol. Evol. Genet. Infect. Dis.* **2014**, *22*, 183–191. [CrossRef]
11. Minning, T.A.; Weatherly, D.B.; Flibotte, S.; Tarleton, R.L. Widespread, Focal Copy Number Variations (CNV) and Whole Chromosome Aneuploidies in *Trypanosoma cruzi* Strains Revealed by Array Comparative Genomic Hybridization. *BMC Genom.* **2011**, *12*, 139. [CrossRef] [PubMed]
12. Messenger, L.A.; Miles, M.A. Evidence and Importance of Genetic Exchange among Field Populations of *Trypanosoma cruzi*. *Acta Trop.* **2015**, *151*, 150–155. [CrossRef] [PubMed]
13. Tibayrenc, M.; Ayala, F.J. Reproductive Clonality of Pathogens: A Perspective on Pathogenic Viruses, Bacteria, Fungi, and Parasitic Protozoa. *Proc. Natl. Acad. Sci. USA* **2012**, *109*, E3305–E3313. [CrossRef]
14. Tibayrenc, M.; Ayala, F.J. The Population Genetics of *Trypanosoma cruzi* Revisited in the Light of the Predominant Clonal Evolution Model. *Acta Trop.* **2015**, *151*, 156. [CrossRef] [PubMed]
15. Ramírez, J.D.; Llewellyn, M.S. Reproductive Clonality in Protozoan Pathogens—Truth or Artefact? *Mol. Ecol.* **2014**, *23*, 4195–4202. [CrossRef] [PubMed]
16. Zingales, B.; Macedo, A.M. Fifteen Years after the Definition of *Trypanosoma cruzi* DTUs: What Have We Learned? *Life* **2023**, *13*, 2339. [CrossRef]
17. Machado, C.A.; Ayala, F.J. Nucleotide Sequences Provide Evidence of Genetic Exchange among Distantly Related Lineages of *Trypanosoma cruzi*. *Proc. Natl. Acad. Sci. USA* **2001**, *98*, 7396. [CrossRef] [PubMed]
18. Reis-Cunha, J.L.; Rodrigues-Luiz, G.F.; Valdivia, H.O.; Baptista, R.P.; Mendes, T.A.O.; de Morais, G.L.; Guedes, R.; Macedo, A.M.; Bern, C.; Gilman, R.H.; et al. Chromosomal Copy Number Variation Reveals Differential Levels of Genomic Plasticity in Distinct *Trypanosoma cruzi* Strains. *BMC Genom.* **2015**, *16*, 499. [CrossRef]
19. Souza, R.T.; Lima, F.M.; Barros, R.M.; Cortez, D.R.; Santos, M.F.; Cordero, E.M.; Ruiz, J.C.; Goldenberg, S.; Teixeira, M.M.G.; da Silveira, J.F. Genome Size, Karyotype Polymorphism and Chromosomal Evolution in *Trypanosoma cruzi*. *PLoS ONE* **2011**, *6*, e23042. [CrossRef] [PubMed]
20. Henriksson, J.; Dujardin, J.C.; Barnabé, C.; Brisse, S.; Timperman, G.; Venegas, J.; Pettersson, U.; Tibayrenc, M.; Solari, A. Chromosomal Size Variation in *Trypanosoma cruzi* Is Mainly Progressive and Is Evolutionarily Informative. *Parasitology* **2002**, *124*, 277–286. [CrossRef] [PubMed]
21. Lukes, J.; Guilbride, D.L.; Votýpka, J.; Zíková, A.; Benne, R.; Englund, P.T. Kinetoplast DNA Network: Evolution of an Improbable Structure. *Eukaryot. Cell* **2002**, *1*, 495–502. [CrossRef]
22. Thomas, S.; Martinez, L.I.T.; Westenberger, S.J.; Sturm, N.R. A Population Study of the Minicircles in *Trypanosoma cruzi*: Predicting Guide RNAs in the Absence of Empirical RNA Editing. *BMC Genom.* **2007**, *8*, 133. [CrossRef]
23. Gerasimov, E.S.; Zamyatnina, K.A.; Matveeva, N.S.; Rudenskaya, Y.A.; Kraeva, N.; Kolesnikov, A.A.; Yurchenko, V. Common Structural Patterns in the Maxicircle Divergent Region of Trypanosomatidae. *Pathogens* **2020**, *9*, 100. [CrossRef] [PubMed]
24. Aphasizhev, R.; Aphasizheva, I. Mitochondrial RNA Editing in Trypanosomes: Small RNAs in Control. *Biochimie* **2014**, *100*, 125–131. [CrossRef] [PubMed]
25. Brenière, S.F.; Waleckx, E.; Barnabé, C. Over Six Thousand *Trypanosoma cruzi* Strains Classified into Discrete Typing Units (DTUs): Attempt at an Inventory. *PLoS Negl. Trop. Dis.* **2016**, *10*, e0004792. [CrossRef] [PubMed]
26. Lewis, M.D.; Llewellyn, M.S.; Gaunt, M.W.; Yeo, M.; Carrasco, H.J.; Miles, M.A. Flow Cytometric Analysis and Microsatellite Genotyping Reveal Extensive DNA Content Variation in *Trypanosoma cruzi* Populations and Expose Contrasts between Natural and Experimental Hybrids. *Int. J. Parasitol.* **2009**, *39*, 1305–1317. [CrossRef] [PubMed]
27. Dvorak, J.A.; Hall, T.E.; Crane, M.S.; Engel, J.C.; McDaniel, J.P.; Uriegas, R. *Trypanosoma cruzi*: Flow Cytometric Analysis. I. Analysis of Total DNA/Organism by Means of Mithramycin-Induced Fluorescence. *J. Protozool.* **1982**, *29*, 430–437. [CrossRef] [PubMed]
28. Zingales, B.; Bartholomeu, D.C. *Trypanosoma cruzi* Genetic Diversity: Impact on Transmission Cycles and Chagas Disease. *Mem. Inst. Oswaldo Cruz* **2022**, *117*, e210193. [CrossRef] [PubMed]
29. Zingales, B. *Trypanosoma cruzi* Genetic Diversity: Something New for Something Known about Chagas Disease Manifestations, Serodiagnosis and Drug Sensitivity. *Acta Trop.* **2018**, *184*, 38–52. [CrossRef]
30. Miles, M.A.; Llewellyn, M.S.; Lewis, M.D.; Yeo, M.; Baleela, R.; Fitzpatrick, S.; Gaunt, M.W.; Mauricio, I.L. The Molecular Epidemiology and Phylogeography of *Trypanosoma cruzi* and Parallel Research on Leishmania: Looking Back and to the Future. *Parasitology* **2009**, *136*, 1509–1528. [CrossRef] [PubMed]
31. Souto, R.P.; Fernandes, O.; Macedo, A.M.; Campbell, D.A.; Zingales, B. DNA Markers Define Two Major Phylogenetic Lineages of *Trypanosoma cruzi*. *Mol. Biochem. Parasitol.* **1996**, *83*, 141–152. [CrossRef]

32. Tibayrenc, M.; Neubauer, K.; Barnabé, C.; Guerrini, F.; Skarecky, D.; Ayala, F.J. Genetic Characterization of Six Parasitic Protozoa: Parity between Random-Primer DNA Typing and Multilocus Enzyme Electrophoresis. *Proc. Natl. Acad. Sci. USA* **1993**, *90*, 1335–1339. [CrossRef]
33. Robello, C.; Gamarro, F.; Castanys, S.; Alvarez-Valin, F. Evolutionary Relationships in *Trypanosoma cruzi*: Molecular Phylogenetics Supports the Existence of a New Major Lineage of Strains. *Gene* **2000**, *246*, 331–338. [CrossRef] [PubMed]
34. Brisse, S.; Barnabé, C.; Tibayrenc, M. Identification of Six *Trypanosoma cruzi* Phylogenetic Lineages by Random Amplified Polymorphic DNA and Multilocus Enzyme Electrophoresis. *Int. J. Parasitol.* **2000**, *30*, 35–44. [CrossRef] [PubMed]
35. Zingales, B.; Miles, M.A.; Campbell, D.A.; Tibayrenc, M.; Macedo, A.M.; Teixeira, M.M.G.; Schijman, A.G.; Llewellyn, M.S.; Lages-Silva, E.; Machado, C.R.; et al. The Revised *Trypanosoma cruzi* Subspecific Nomenclature: Rationale, Epidemiological Relevance and Research Applications. *Infect. Genet. Evol. J. Mol. Epidemiol. Evol. Genet. Infect. Dis.* **2012**, *12*, 240–253. [CrossRef] [PubMed]
36. Zingales, B.; Andrade, S.G.; Briones, M.R.S.; Campbell, D.A.; Chiari, E.; Fernandes, O.; Guhl, F.; Lages-Silva, E.; Macedo, A.M.; Machado, C.R.; et al. A New Consensus for *Trypanosoma cruzi* Intraspecific Nomenclature: Second Revision Meeting Recommends TcI to TcVI. *Mem. Inst. Oswaldo Cruz* **2009**, *104*, 1051–1054. [CrossRef]
37. Tibayrenc, M. Genetic Epidemiology of Parasitic Protozoa and Other Infectious Agents: The Need for an Integrated Approach. *Int. J. Parasitol.* **1998**, *28*, 85–104. [CrossRef]
38. Lima, L.; Espinosa-Álvarez, O.; Ortiz, P.A.; Trejo-Varón, J.A.; Carranza, J.C.; Pinto, C.M.; Serrano, M.G.; Buck, G.A.; Camargo, E.P.; Teixeira, M.M.G. Genetic Diversity of *Trypanosoma cruzi* in Bats, and Multilocus Phylogenetic and Phylogeographical Analyses Supporting Tcbat as an Independent DTU (Discrete Typing Unit). *Acta Trop.* **2015**, *151*, 166–177. [CrossRef]
39. Flores-López, C.A.; Mitchell, E.A.; Reisenman, C.E.; Sarkar, S.; Williamson, P.C.; Machado, C.A. Phylogenetic Diversity of Two Common *Trypanosoma cruzi* Lineages in the Southwestern United States. *Infect. Genet. Evol. J. Mol. Epidemiol. Evol. Genet. Infect. Dis.* **2022**, *99*, 105251. [CrossRef]
40. Barnabé, C.; Mobarec, H.I.; Jurado, M.R.; Cortez, J.A.; Brenière, S.F. Reconsideration of the Seven Discrete Typing Units within the Species *Trypanosoma cruzi*, a New Proposal of Three Reliable Mitochondrial Clades. *Infect. Genet. Evol. J. Mol. Epidemiol. Evol. Genet. Infect. Dis.* **2016**, *39*, 176–186. [CrossRef]
41. Berná, L.; Greif, G.; Pita, S.; Faral-Tello, P.; Díaz-Viraqué, F.; Souza, R.D.C.M.D.; Vallejo, G.A.; Alvarez-Valin, F.; Robello, C. Maxicircle Architecture and Evolutionary Insights into *Trypanosoma cruzi* Complex. *PLoS Negl. Trop. Dis.* **2021**, *15*, e0009719. [CrossRef] [PubMed]
42. El-Sayed, N.M.; Myler, P.J.; Bartholomeu, D.C.; Nilsson, D.; Aggarwal, G.; Tran, A.-N.; Ghedin, E.; Worthey, E.A.; Delcher, A.L.; Blandin, G.; et al. The Genome Sequence of *Trypanosoma cruzi*, Etiologic Agent of Chagas Disease. *Science* **2005**, *309*, 409–415. [CrossRef]
43. Reis-Cunha, J.L.; Bartholomeu, D.C. *Trypanosoma cruzi* Genome Assemblies: Challenges and Milestones of Assembling a Highly Repetitive and Complex Genome. *Methods Mol. Biol.* **2019**, *1955*, 1–22. [CrossRef] [PubMed]
44. El-Sayed, N.M.; Myler, P.J.; Blandin, G.; Berriman, M.; Crabtree, J.; Aggarwal, G.; Caler, E.; Renauld, H.; Worthey, E.A.; Hertz-Fowler, C.; et al. Comparative Genomics of Trypanosomatid Parasitic Protozoa. *Science* **2005**, *309*, 404–409. [CrossRef] [PubMed]
45. Franzén, O.; Ochaya, S.; Sherwood, E.; Lewis, M.D.; Llewellyn, M.S.; Miles, M.A.; Andersson, B. Shotgun Sequencing Analysis of *Trypanosoma cruzi* I Sylvio X10/1 and Comparison with T. Cruzi VI CL Brener. *PLoS Negl. Trop. Dis.* **2011**, *5*, e984. [CrossRef] [PubMed]
46. Grisard, E.C.; Teixeira, S.M.R.; de Almeida, L.G.P.; Stoco, P.H.; Gerber, A.L.; Talavera-López, C.; Lima, O.C.; Andersson, B.; de Vasconcelos, A.T.R. *Trypanosoma cruzi* Clone Dm28c Draft Genome Sequence. *Genome Announc.* **2014**, *2*, e01114. [CrossRef]
47. Callejas-Hernández, F.; Rastrojo, A.; Poveda, C.; Gironès, N.; Fresno, M. Genomic Assemblies of Newly Sequenced *Trypanosoma cruzi* Strains Reveal New Genomic Expansion and Greater Complexity. *Sci. Rep.* **2018**, *8*, 14631. [CrossRef]
48. Baptista, R.P.; Reis-Cunha, J.L.; DeBarry, J.D.; Chiari, E.; Kissinger, J.C.; Bartholomeu, D.C.; Macedo, A.M. Assembly of Highly Repetitive Genomes Using Short Reads: The Genome of Discrete Typing Unit III *Trypanosoma cruzi* Strain 231. *Microb. Genom.* **2018**, *4*, e000156. [CrossRef] [PubMed]
49. Bradwell, K.R.; Koparde, V.N.; Matveyev, A.V.; Serrano, M.G.; Alves, J.M.P.; Parikh, H.; Huang, B.; Lee, V.; Espinosa-Alvarez, O.; Ortiz, P.A.; et al. Genomic Comparison of *Trypanosoma conorhini* and *Trypanosoma rangeli* to *Trypanosoma cruzi* Strains of High and Low Virulence. *BMC Genom.* **2018**, *19*, 770. [CrossRef] [PubMed]
50. DeCuir, J.; Tu, W.; Dumonteil, E.; Herrera, C. Sequence of *Trypanosoma cruzi* Reference Strain SC43 Nuclear Genome and Kinetoplast Maxicircle Confirms a Strong Genetic Structure among Closely Related Parasite Discrete Typing Units. *Genome* **2020**, *64*, 525. [CrossRef]
51. Franzén, O.; Talavera-López, C.; Ochaya, S.; Butler, C.E.; Messenger, L.A.; Lewis, M.D.; Llewellyn, M.S.; Marinkelle, C.J.; Tyler, K.M.; Miles, M.A.; et al. Comparative Genomic Analysis of Human Infective *Trypanosoma cruzi* Lineages with the Bat-Restricted Subspecies *T. cruzi marinkellei*. *BMC Genom.* **2012**, *13*, 531. [CrossRef] [PubMed]

52. Gómez, I.; López, M.C.; Rastrojo, A.; Lorenzo-Díaz, F.; Requena, J.M.; Aguado, B.; Valladares, B.; Thomas, M.C. Variability of the Pr77 Sequence of L1Tc Retrotransposon among Six *T. cruzi* Strains Belonging to Different Discrete Typing Units (DTUs). *Acta Trop.* **2021**, *222*, 106053. [CrossRef] [PubMed]
53. Reis-Cunha, J.L.; Baptista, R.P.; Rodrigues-Luiz, G.F.; Coqueiro-dos-Santos, A.; Valdivia, H.O.; de Almeida, L.V.; Cardoso, M.S.; D'Ávila, D.A.; Dias, F.H.C.; Fujiwara, R.T.; et al. Whole Genome Sequencing of *Trypanosoma cruzi* Field Isolates Reveals Extensive Genomic Variability and Complex Aneuploidy Patterns within TcII DTU. *BMC Genom.* **2018**, *19*, 816. [CrossRef] [PubMed]
54. Camacho, E.; González-de la Fuente, S.; Rastrojo, A.; Peiró-Pastor, R.; Solana, J.C.; Tabera, L.; Gamarro, F.; Carrasco-Ramiro, F.; Requena, J.M.; Aguado, B. Complete Assembly of the Leishmania Donovani (HU3 Strain) Genome and Transcriptome Annotation. *Sci. Rep.* **2019**, *9*, 6127. [CrossRef]
55. Berná, L.; Rodriguez, M.; Chiribao, M.L.; Parodi-Talice, A.; Pita, S.; Rijo, G.; Alvarez-Valin, F.; Robello, C. Expanding an Expanded Genome: Long-Read Sequencing of *Trypanosoma cruzi*. *Microb. Genom.* **2018**, *4*, e000177. [CrossRef] [PubMed]
56. Hakim, J.M.C.; Gutierrez Guarnizo, S.A.; Málaga Machaca, E.; Gilman, R.H.; Mugnier, M.R. Whole-Genome Assembly of a Hybrid *Trypanosoma cruzi* Strain Assembled with Nanopore Sequencing Alone. *G3 Genes Genomes Genet.* **2024**, *14*, jkae076. [CrossRef]
57. Hoyos Sanchez, M.C.; Ospina Zapata, H.S.; Suarez, B.D.; Ospina, C.; Barbosa, H.J.; Carranza Martinez, J.C.; Vallejo, G.A.; Urrea Montes, D.; Duitama, J. A Phased Genome Assembly of a Colombian *Trypanosoma cruzi* TcI Strain and the Evolution of Gene Families. *Sci. Rep.* **2024**, *14*, 2054. [CrossRef] [PubMed]
58. Díaz-Viraqué, F.; Pita, S.; Greif, G.; de Souza, R.d.C.M.; Iraola, G.; Robello, C. Nanopore Sequencing Significantly Improves Genome Assembly of the Protozoan Parasite *Trypanosoma cruzi*. *Genome Biol. Evol.* **2019**, *11*, 1952–1957. [CrossRef] [PubMed]
59. Wang, W.; Peng, D.; Baptista, R.P.; Li, Y.; Kissinger, J.C.; Tarleton, R.L. Strain-Specific Genome Evolution in *Trypanosoma cruzi*, the Agent of Chagas Disease. *PLoS Pathog.* **2021**, *17*, e1009254. [CrossRef] [PubMed]
60. Fesser, A.; Beilstein, S.; Kaiser, M.; Schmidt, R.S.; Mäser, P. *Trypanosoma cruzi* STIB980: A TcI Strain for Drug Discovery and Reverse Genetics. *Pathogens* **2023**, *12*, 1217. [CrossRef] [PubMed]
61. Majeau, A.; Murphy, L.; Herrera, C.; Dumonteil, E. Assessing *Trypanosoma cruzi* Parasite Diversity through Comparative Genomics: Implications for Disease Epidemiology and Diagnostics. *Pathogens* **2021**, *10*, 212. [CrossRef] [PubMed]
62. Callejas-Hernández, F.; Gutierrez-Nogues, Á.; Rastrojo, A.; Gironès, N.; Fresno, M. Analysis of mRNA Processing at Whole Transcriptome Level, Transcriptomic Profile and Genome Sequence Refinement of *Trypanosoma cruzi*. *Sci. Rep.* **2019**, *9*, 17376. [CrossRef] [PubMed]
63. Henriksson, J.; Aslund, L.; Pettersson, U. Karyotype Variability in *Trypanosoma cruzi*. *Parasitol. Today* **1996**, *12*, 108–114. [CrossRef]
64. Santos, M.R.; Cano, M.I.; Schijman, A.; Lorenzi, H.; Vázquez, M.; Levin, M.J.; Ramirez, J.L.; Brandão, A.; Degrave, W.M.; da Silveira, J.F. The *Trypanosoma cruzi* Genome Project: Nuclear Karyotype and Gene Mapping of Clone CL Brener. *Mem. Inst. Oswaldo Cruz* **1997**, *92*, 821–828. [CrossRef] [PubMed]
65. Henriksson, J.; Porcel, B.; Rydåker, M.; Ruiz, A.; Sabaj, V.; Galanti, N.; Cazzulo, J.J.; Frasch, A.C.; Pettersson, U. Chromosome Specific Markers Reveal Conserved Linkage Groups in Spite of Extensive Chromosomal Size Variation in *Trypanosoma cruzi*. *Mol. Biochem. Parasitol.* **1995**, *73*, 63–74. [CrossRef]
66. Vargas, N.; Pedroso, A.; Zingales, B. Chromosomal Polymorphism, Gene Synteny and Genome Size in *T. cruzi* I and *T. cruzi* II Groups. *Mol. Biochem. Parasitol.* **2004**, *138*, 131–141. [CrossRef] [PubMed]
67. Weatherly, D.B.; Boehlke, C.; Tarleton, R.L. Chromosome Level Assembly of the Hybrid *Trypanosoma cruzi* Genome. *BMC Genom.* **2009**, *10*, 255. [CrossRef]
68. Parmar, J.J.; Woringer, M.; Zimmer, C. How the Genome Folds: The Biophysics of Four-Dimensional Chromatin Organization. *Annu. Rev. Biophys.* **2019**, *48*, 231–253. [CrossRef] [PubMed]
69. Downing, T.; Imamura, H.; Decuypere, S.; Clark, T.G.; Coombs, G.H.; Cotton, J.A.; Hilley, J.D.; de Doncker, S.; Maes, I.; Mottram, J.C.; et al. Whole Genome Sequencing of Multiple Leishmania Donovani Clinical Isolates Provides Insights into Population Structure and Mechanisms of Drug Resistance. *Genome Res.* **2011**, *21*, 2143–2156. [CrossRef]
70. Dujardin, J.-C.; Mannaert, A.; Durrant, C.; Cotton, J.A. Mosaic Aneuploidy in Leishmania: The Perspective of Whole Genome Sequencing. *Trends Parasitol.* **2014**, *30*, 554–555. [CrossRef] [PubMed]
71. Mannaert, A.; Downing, T.; Imamura, H.; Dujardin, J.-C. Adaptive Mechanisms in Pathogens: Universal Aneuploidy in Leishmania. *Trends Parasitol.* **2012**, *28*, 370–376. [CrossRef]
72. Almeida, L.V.; Coqueiro-Dos-Santos, A.; Rodriguez-Luiz, G.F.; McCulloch, R.; Bartholomeu, D.C.; Reis-Cunha, J.L. Chromosomal Copy Number Variation Analysis by next Generation Sequencing Confirms Ploidy Stability in *Trypanosoma brucei* Subspecies. *Microb. Genom.* **2018**, *4*, e000223. [CrossRef] [PubMed]
73. Reis-Cunha, J.L.; Valdivia, H.O.; Bartholomeu, D.C. Gene and Chromosomal Copy Number Variations as an Adaptive Mechanism Towards a Parasitic Lifestyle in Trypanosomatids. *Curr. Genom.* **2018**, *19*, 87–97. [CrossRef] [PubMed]
74. Ramirez, J.L. An Evolutionary View of *Trypanosoma cruzi* Telomeres. *Front. Cell. Infect. Microbiol.* **2019**, *9*, 439. [CrossRef]

75. Kim, D.; Chiurillo, M.A.; El-Sayed, N.; Jones, K.; Santos, M.R.M.; Porcile, P.E.; Andersson, B.; Myler, P.; da Silveira, J.F.; Ramírez, J.L. Telomere and Subtelomere of *Trypanosoma cruzi* Chromosomes Are Enriched in (Pseudo) Genes of Retrotransposon Hot Spot and Trans-Sialidase-like Gene Families: The Origins of *T. cruzi* Telomeres. *Gene* **2005**, *346*, 153–161. [CrossRef] [PubMed]
76. Azuaje, F.J.; Ramirez, J.L.; Da Silveira, J.F. In Silico, Biologically-Inspired Modelling of Genomic Variation Generation in Surface Proteins of *Trypanosoma cruzi*. *Kinetoplastid Biol. Dis.* **2007**, *6*, 6. [CrossRef]
77. Dreesen, O.; Li, B.; Cross, G.A.M. Telomere Structure and Function in Trypanosomes: A Proposal. *Nat. Rev. Microbiol.* **2007**, *5*, 70–75. [CrossRef] [PubMed]
78. Berná, L.; Pita, S.; Chiribao, M.L.; Parodi-Talice, A.; Alvarez-Valin, F.; Robello, C. Biology of the *Trypanosoma cruzi* Genome. In *Biology of Trypanosoma cruzi*; IntechOpen: London, UK, 2019; ISBN 978-1-83968-204-9.
79. Herreros-Cabello, A.; Callejas-Hernández, F.; Gironès, N.; Fresno, M. *Trypanosoma cruzi* Genome: Organization, Multi-Gene Families, Transcription, and Biological Implications. *Genes* **2020**, *11*, 1196. [CrossRef] [PubMed]
80. Atwood, J.A.; Weatherly, D.B.; Minning, T.A.; Bundy, B.; Cavola, C.; Opperdoes, F.R.; Orlando, R.; Tarleton, R.L. The *Trypanosoma cruzi* Proteome. *Science* **2005**, *309*, 473–476. [CrossRef]
81. Herreros-Cabello, A.; Callejas-Hernández, F.; Fresno, M.; Gironès, N. Comparative Proteomic Analysis of Trypomastigotes from *Trypanosoma cruzi* Strains with Different Pathogenicity. *Infect. Genet. Evol. J. Mol. Epidemiol. Evol. Genet. Infect. Dis.* **2019**, *76*, 104041. [CrossRef] [PubMed]
82. Avila, C.C.; Mule, S.N.; Rosa-Fernandes, L.; Viner, R.; Barisón, M.J.; Costa-Martins, A.G.; de Oliveira, G.S.; Teixeira, M.M.G.; Marinho, C.R.F.; Silber, A.M.; et al. Proteome-Wide Analysis of *Trypanosoma cruzi* Exponential and Stationary Growth Phases Reveals a Subcellular Compartment-Specific Regulation. *Genes* **2018**, *9*, 413. [CrossRef] [PubMed]
83. de Godoy, L.M.F.; Marchini, F.K.; Pavoni, D.P.; Rampazzo, R.d.C.P.; Probst, C.M.; Goldenberg, S.; Krieger, M.A. Quantitative Proteomics of *Trypanosoma cruzi* during Metacyclogenesis. *Proteomics* **2012**, *12*, 2694–2703. [CrossRef] [PubMed]
84. Kudla, G.; Helwak, A.; Lipinski, L. Gene Conversion and GC-Content Evolution in Mammalian Hsp70. *Mol. Biol. Evol.* **2004**, *21*, 1438–1444. [CrossRef] [PubMed]
85. Galtier, N. Gene Conversion Drives GC Content Evolution in Mammalian Histones. *Trends Genet. TIG* **2003**, *19*, 65–68. [CrossRef]
86. Chiurillo, M.A.; Cano, I.; Da Silveira, J.F.; Ramirez, J.L. Organization of Telomeric and Sub-Telomeric Regions of Chromosomes from the Protozoan Parasite *Trypanosoma cruzi*. *Mol. Biochem. Parasitol.* **1999**, *100*, 173–183. [CrossRef]
87. Moraes Barros, R.R.; Marini, M.M.; Antônio, C.R.; Cortez, D.R.; Miyake, A.M.; Lima, F.M.; Ruiz, J.C.; Bartholomeu, D.C.; Chiurillo, M.A.; Ramirez, J.L.; et al. Anatomy and Evolution of Telomeric and Subtelomeric Regions in the Human Protozoan Parasite *Trypanosoma cruzi*. *BMC Genom.* **2012**, *13*, 229. [CrossRef]
88. de Araujo, C.B.; da Cunha, J.P.C.; Inada, D.T.; Damasceno, J.; Lima, A.R.J.; Hiraiwa, P.; Marques, C.; Gonçalves, E.; Nishiyama-Junior, M.Y.; McCulloch, R.; et al. Replication Origin Location Might Contribute to Genetic Variability in *Trypanosoma cruzi*. *BMC Genom.* **2020**, *21*, 414. [CrossRef]
89. de Araujo, C.B.; Calderano, S.G.; Elias, M.C. The Dynamics of Replication in *Trypanosoma cruzi* Parasites by Single-Molecule Analysis. *J. Eukaryot. Microbiol.* **2019**, *66*, 514–518. [CrossRef]
90. Gonçalves, C.S.; Ávila, A.R.; de Souza, W.; Motta, M.C.M.; Cavalcanti, D.P. Revisiting the *Trypanosoma cruzi* Metacyclogenesis: Morphological and Ultrastructural Analyses during Cell Differentiation. *Parasit. Vectors* **2018**, *11*, 83. [CrossRef] [PubMed]
91. Elias, M.C.; Marques-Porto, R.; Freymüller, E.; Schenkman, S. Transcription Rate Modulation through the *Trypanosoma cruzi* Life Cycle Occurs in Parallel with Changes in Nuclear Organisation. *Mol. Biochem. Parasitol.* **2001**, *112*, 79–90. [CrossRef]
92. Ferreira, L.R.P.; Dossin, F.d.M.; Ramos, T.C.; Freymüller, E.; Schenkman, S. Active Transcription and Ultrastructural Changes during *Trypanosoma cruzi* Metacyclogenesis. *An. Acad. Bras. Cienc.* **2008**, *80*, 157–166. [CrossRef] [PubMed]
93. Lima, A.R.J.; Silva, H.G.d.S.; Poubel, S.; Rosón, J.N.; de Lima, L.P.O.; Costa-Silva, H.M.; Gonçalves, C.S.; Galante, P.A.F.; Holetz, F.; Motta, M.C.M.M.; et al. Open Chromatin Analysis in *Trypanosoma cruzi* Life Forms Highlights Critical Differences in Genomic Compartments and Developmental Regulation at tDNA Loci. *Epigenetics Chromatin* **2022**, *15*, 22. [CrossRef]
94. Díaz-Viraqué, F.; Chiribao, M.L.; Libisch, M.G.; Robello, C. Genome-Wide Chromatin Interaction Map for *Trypanosoma cruzi*. *Nat. Microbiol.* **2023**, *8*, 2103. [CrossRef] [PubMed]
95. Matthews, K.R.; Tschudi, C.; Ullu, E. A Common Pyrimidine-Rich Motif Governs Trans-Splicing and Polyadenylation of Tubulin Polycistronic Pre-mRNA in Trypanosomes. *Genes Dev.* **1994**, *8*, 491–501. [CrossRef]
96. Chávez, S.; Urbaniak, M.D.; Benz, C.; Smircich, P.; Garat, B.; Sotelo-Silveira, J.R.; Duhagon, M.A. Extensive Translational Regulation through the Proliferative Transition of *Trypanosoma cruzi* Revealed by Multi-Omics. *mSphere* **2021**, *6*, e0036621. [CrossRef] [PubMed]
97. Araújo, P.R.; Teixeira, S.M. Regulatory Elements Involved in the Post-Transcriptional Control of Stage-Specific Gene Expression in *Trypanosoma cruzi*: A Review. *Mem. Inst. Oswaldo Cruz* **2011**, *106*, 257–266. [CrossRef] [PubMed]
98. Lima, A.R.J.; de Araujo, C.B.; Bispo, S.; Patané, J.; Silber, A.M.; Elias, M.C.; da Cunha, J.P.C. Nucleosome Landscape Reflects Phenotypic Differences in *Trypanosoma cruzi* Life Forms. *PLoS Pathog.* **2021**, *17*, e1009272. [CrossRef] [PubMed]

99. Rosón, J.N.; Vitarelli, M.d.O.; Costa-Silva, H.M.; Pereira, K.S.; Pires, D.d.S.; Lopes, L.d.S.; Cordeiro, B.; Kraus, A.J.; Cruz, K.N.T.; Calderano, S.G.; et al. H2B.V Demarcates Divergent Strand-Switch Regions, Some tDNA Loci, and Genome Compartments in *Trypanosoma cruzi* and Affects Parasite Differentiation and Host Cell Invasion. *PLoS Pathog.* **2022**, *18*, e1009694. [CrossRef]
100. Nunes, V.S.; Moretti, N.S.; da Silva, M.S.; Elias, M.C.; Janzen, C.J.; Schenkman, S. Trimethylation of Histone H3K76 by Dot1B Enhances Cell Cycle Progression after Mitosis in *Trypanosoma cruzi*. *Biochim. Biophys. Acta Mol. Cell Res.* **2020**, *1867*, 118694. [CrossRef]
101. Martínez-Calvillo, S.; Vizuet-de-Rueda, J.C.; Florencio-Martínez, L.E.; Manning-Cela, R.G.; Figueroa-Angulo, E.E. Gene Expression in Trypanosomatid Parasites. *J. Biomed. Biotechnol.* **2010**, *2010*, 525241. [CrossRef]
102. Obado, S.O.; Bot, C.; Nilsson, D.; Andersson, B.; Kelly, J.M. Repetitive DNA Is Associated with Centromeric Domains in *Trypanosoma brucei* but Not *Trypanosoma cruzi*. *Genome Biol.* **2007**, *8*, R37. [CrossRef] [PubMed]
103. Obado, S.O.; Taylor, M.C.; Wilkinson, S.R.; Bromley, E.V.; Kelly, J.M. Functional Mapping of a Trypanosome Centromere by Chromosome Fragmentation Identifies a 16-Kb GC-Rich Transcriptional "Strand-Switch" Domain as a Major Feature. *Genome Res.* **2005**, *15*, 36. [CrossRef] [PubMed]
104. Smircich, P.; El-Sayed, N.M.; Garat, B. Intrinsic DNA Curvature in Trypanosomes. *BMC Res. Notes* **2017**, *10*, 585. [CrossRef] [PubMed]
105. Günzl, A. The Pre-mRNA Splicing Machinery of Trypanosomes: Complex or Simplified? *Eukaryot. Cell* **2010**, *9*, 1159. [CrossRef] [PubMed]
106. De Gaudenzi, J.G.; Noé, G.; Campo, V.A.; Frasch, A.C.; Cassola, A. Gene Expression Regulation in Trypanosomatids. *Essays Biochem.* **2011**, *51*, 31–46. [CrossRef] [PubMed]
107. Rastrojo, A.; Carrasco-Ramiro, F.; Martín, D.; Crespillo, A.; Reguera, R.M.; Aguado, B.; Requena, J.M. The Transcriptome of Leishmania Major in the Axenic Promastigote Stage: Transcript Annotation and Relative Expression Levels by RNA-Seq. *BMC Genom.* **2013**, *14*, 223. [CrossRef] [PubMed]
108. Kolev, N.G.; Franklin, J.B.; Carmi, S.; Shi, H.; Michaeli, S.; Tschudi, C. The Transcriptome of the Human Pathogen Trypanosoma Brucei at Single-Nucleotide Resolution. *PLoS Pathog.* **2010**, *6*, e1001090. [CrossRef] [PubMed]
109. Pech-Canul, Á.d.l.C.; Monteón, V.; Solís-Oviedo, R.-L. A Brief View of the Surface Membrane Proteins from *Trypanosoma cruzi*. *J. Parasitol. Res.* **2017**, *2017*, 3751403. [CrossRef]
110. Santi-Rocca, J.; Fernandez-Cortes, F.; Chillón-Marinas, C.; González-Rubio, M.-L.; Martin, D.; Gironès, N.; Fresno, M. A Multi-Parametric Analysis of *Trypanosoma cruzi* Infection: Common Pathophysiologic Patterns beyond Extreme Heterogeneity of Host Responses. *Sci. Rep.* **2017**, *7*, 8893. [CrossRef] [PubMed]
111. Poveda, C.; Herreros-Cabello, A.; Callejas-Hernández, F.; Osuna-Pérez, J.; Maza, M.C.; Chillón-Marinas, C.; Calderón, J.; Stamatakis, K.; Fresno, M.; Gironès, N. Interaction of Signaling Lymphocytic Activation Molecule Family 1 (SLAMF1) Receptor with *Trypanosoma cruzi* Is Strain-Dependent and Affects NADPH Oxidase Expression and Activity. *PLoS Negl. Trop. Dis.* **2020**, *14*, e0008608. [CrossRef]
112. De Pablos, L.M.; Osuna, A. Multigene Families in *Trypanosoma cruzi* and Their Role in Infectivity. *Infect. Immun.* **2012**, *80*, 2258–2264. [CrossRef] [PubMed]
113. Chiurillo, M.A.; Cortez, D.R.; Lima, F.M.; Cortez, C.; Ramírez, J.L.; Martins, A.G.; Serrano, M.G.; Teixeira, M.M.G.; da Silveira, J.F. The Diversity and Expansion of the Trans-Sialidase Gene Family Is a Common Feature in *Trypanosoma cruzi* Clade Members. *Infect. Genet. Evol. J. Mol. Epidemiol. Evol. Genet. Infect. Dis.* **2016**, *37*, 266–274. [CrossRef]
114. Lantos, A.B.; Carlevaro, G.; Araoz, B.; Ruiz Diaz, P.; Camara, M.d.l.M.; Buscaglia, C.A.; Bossi, M.; Yu, H.; Chen, X.; Bertozzi, C.R.; et al. Sialic Acid Glycobiology Unveils *Trypanosoma cruzi* Trypomastigote Membrane Physiology. *PLoS Pathog.* **2016**, *12*, e1005559. [CrossRef] [PubMed]
115. Ribeiro, K.S.; Vasconcellos, C.I.; Soares, R.P.; Mendes, M.T.; Ellis, C.C.; Aguilera-Flores, M.; de Almeida, I.C.; Schenkman, S.; Iwai, L.K.; Torrecilhas, A.C. Proteomic Analysis Reveals Different Composition of Extracellular Vesicles Released by Two *Trypanosoma cruzi* Strains Associated with Their Distinct Interaction with Host Cells. *J. Extracell. Vesicles* **2018**, *7*, 1463779. [CrossRef] [PubMed]
116. Schenkman, S.; Eichinger, D.; Pereira, M.E.; Nussenzweig, V. Structural and Functional Properties of Trypanosoma Trans-Sialidase. *Annu. Rev. Microbiol.* **1994**, *48*, 499–523. [CrossRef] [PubMed]
117. Frasch, A.C. Functional Diversity in the Trans-Sialidase and Mucin Families in *Trypanosoma cruzi*. *Parasitol. Today* **2000**, *16*, 282–286. [CrossRef]
118. Pereira, M.E.A.; Loures, M.A.; Villalta, F.; Andrade, A.F.B. Lectin Receptors as Markers for *Trypanosoma cruzi*. Developmental Stages and a Study of the Interaction of Wheat Germ Agglutinin with Sialic Acid Residues on Epimastigote Cells. *J. Exp. Med* **1980**, *152*, 1375–1392. [CrossRef]
119. Pereira-Chioccola, V.L.; Acosta-Serrano, A.; Correia de Almeida, I.; Ferguson, M.A.; Souto-Padron, T.; Rodrigues, M.M.; Travassos, L.R.; Schenkman, S. Mucin-like Molecules Form a Negatively Charged Coat That Protects *Trypanosoma cruzi* Trypomastigotes from Killing by Human Anti-Alpha-Galactosyl Antibodies. *J. Cell Sci.* **2000**, *113 Pt 7*, 1299–1307. [CrossRef] [PubMed]

120. Nardy, A.F.F.R.; Freire-de-Lima, C.G.; Pérez, A.R.; Morrot, A. Role of *Trypanosoma cruzi* Trans-Sialidase on the Escape from Host Immune Surveillance. *Front. Microbiol.* **2016**, *7*, 348. [CrossRef] [PubMed]
121. Pereira, M.E. A Developmentally Regulated Neuraminidase Activity in *Trypanosoma cruzi*. *Science* **1983**, *219*, 1444–1446. [CrossRef] [PubMed]
122. Freitas, L.M.; dos Santos, S.L.; Rodrigues-Luiz, G.F.; Mendes, T.A.O.; Rodrigues, T.S.; Gazzinelli, R.T.; Teixeira, S.M.R.; Fujiwara, R.T.; Bartholomeu, D.C. Genomic Analyses, Gene Expression and Antigenic Profile of the Trans-Sialidase Superfamily of *Trypanosoma cruzi* Reveal an Undetected Level of Complexity. *PLoS ONE* **2011**, *6*, e25914. [CrossRef] [PubMed]
123. Oliveira, I.A.; Freire-de-Lima, L.; Penha, L.L.; Dias, W.B.; Todeschini, A.R. *Trypanosoma cruzi* Trans-Sialidase: Structural Features and Biological Implications. *Subcell. Biochem.* **2014**, *74*, 181–201. [CrossRef] [PubMed]
124. Bartholomeu, D.C.; Cerqueira, G.C.; Leão, A.C.A.; daRocha, W.D.; Pais, F.S.; Macedo, C.; Djikeng, A.; Teixeira, S.M.R.; El-Sayed, N.M. Genomic Organization and Expression Profile of the Mucin-Associated Surface Protein (Masp) Family of the Human Pathogen *Trypanosoma cruzi*. *Nucleic Acids Res.* **2009**, *37*, 3407. [CrossRef] [PubMed]
125. Bayer-Santos, E.; Aguilar-Bonavides, C.; Rodrigues, S.P.; Cordero, E.M.; Marques, A.F.; Varela-Ramirez, A.; Choi, H.; Yoshida, N.; da Silveira, J.F.; Almeida, I.C. Proteomic Analysis of *Trypanosoma cruzi* Secretome: Characterization of Two Populations of Extracellular Vesicles and Soluble Proteins. *J. Proteome Res.* **2013**, *12*, 883–897. [CrossRef] [PubMed]
126. Seco-Hidalgo, V.; De Pablos, L.M.; Osuna, A. Transcriptional and Phenotypical Heterogeneity of *Trypanosoma cruzi* Cell Populations. *Open Biol.* **2015**, *5*, 150190. [CrossRef]
127. dos Santos, S.L.; Freitas, L.M.; Lobo, F.P.; Rodrigues-Luiz, G.F.; Mendes, T.A.d.O.; Oliveira, A.C.S.; Andrade, L.O.; Chiari, E.; Gazzinelli, R.T.; Teixeira, S.M.R.; et al. The MASP Family of *Trypanosoma cruzi*: Changes in Gene Expression and Antigenic Profile during the Acute Phase of Experimental Infection. *PLoS Negl. Trop. Dis.* **2012**, *6*, e1779. [CrossRef]
128. Chuenkova, M.V.; PereiraPerrin, M. Chagas' Disease Parasite Promotes Neuron Survival and Differentiation through TrkA Nerve Growth Factor Receptor. *J. Neurochem.* **2004**, *91*, 385–394. [CrossRef]
129. Espinoza, B.; Martínez, I.; Martínez-Velasco, M.L.; Rodríguez-Sosa, M.; González-Canto, A.; Vázquez-Mendoza, A.; Terrazas, L.I. Role of a 49 kDa *Trypanosoma cruzi* Mucin-Associated Surface Protein (MASP49) during the Infection Process and Identification of a Mammalian Cell Surface Receptor. *Pathogens* **2023**, *12*, 105. [CrossRef] [PubMed]
130. Leão, A.C.; Viana, L.A.; Fortes de Araujo, F.; de Lourdes Almeida, R.; Freitas, L.M.; Coqueiro-Dos-Santos, A.; da Silveira-Lemos, D.; Cardoso, M.S.; Reis-Cunha, J.L.; Teixeira-Carvalho, A.; et al. Antigenic Diversity of MASP Gene Family of *Trypanosoma cruzi*. *Microbes Infect.* **2022**, *24*, 104982. [CrossRef] [PubMed]
131. Díaz Lozano, I.M.; De Pablos, L.M.; Longhi, S.A.; Zago, M.P.; Schijman, A.G.; Osuna, A. Immune Complexes in Chronic Chagas Disease Patients Are Formed by Exovesicles from *Trypanosoma cruzi* Carrying the Conserved MASP N-Terminal Region. *Sci. Rep.* **2017**, *7*, 44451. [CrossRef] [PubMed]
132. De Pablos, L.M.; Díaz Lozano, I.M.; Jercic, M.I.; Quinzada, M.; Giménez, M.J.; Calabuig, E.; Espino, A.M.; Schijman, A.G.; Zulantay, I.; Apt, W.; et al. The C-Terminal Region of *Trypanosoma cruzi* MASPs Is Antigenic and Secreted via Exovesicles. *Sci. Rep.* **2016**, *6*, 27293. [CrossRef]
133. Acosta-Serrano, A.; Almeida, I.C.; Freitas-Junior, L.H.; Yoshida, N.; Schenkman, S. The Mucin-like Glycoprotein Super-Family of *Trypanosoma cruzi*: Structure and Biological Roles. *Mol. Biochem. Parasitol.* **2001**, *114*, 143–150. [CrossRef] [PubMed]
134. Schenkman, S.; Ferguson, M.A.; Heise, N.; de Almeida, M.L.; Mortara, R.A.; Yoshida, N. Mucin-like Glycoproteins Linked to the Membrane by Glycosylphosphatidylinositol Anchor Are the Major Acceptors of Sialic Acid in a Reaction Catalyzed by Trans-Sialidase in Metacyclic Forms of *Trypanosoma cruzi*. *Mol. Biochem. Parasitol.* **1993**, *59*, 293–303. [CrossRef]
135. Cánepa, G.E.; Mesías, A.C.; Yu, H.; Chen, X.; Buscaglia, C.A. Structural Features Affecting Trafficking, Processing, and Secretion of *Trypanosoma cruzi* Mucins. *J. Biol. Chem.* **2012**, *287*, 26365–26376. [CrossRef]
136. Bunkofske, M.E.; Perumal, N.; White, B.; Strauch, E.-M.; Tarleton, R. Epitopes in the GPI Attachment Signal Peptide of *Trypanosoma cruzi* Mucin Proteins Generate Robust but Delayed and Nonprotective $CD8^+$ T Cell Responses. *J. Immunol.* **2023**, *210*, 420. [CrossRef] [PubMed]
137. Buscaglia, C.A.; Campo, V.A.; Frasch, A.C.C.; Di Noia, J.M. *Trypanosoma cruzi* Surface Mucins: Host-Dependent Coat Diversity. *Nat. Rev. Microbiol.* **2006**, *4*, 229–236. [CrossRef] [PubMed]
138. Mucci, J.; Lantos, A.B.; Buscaglia, C.A.; Leguizamón, M.S.; Campetella, O. The *Trypanosoma cruzi* Surface, a Nanoscale Patchwork Quilt. *Trends Parasitol.* **2016**, *33*, 102. [CrossRef]
139. Campo, V.; Di Noia, J.M.; Buscaglia, C.A.; Agüero, F.; Sánchez, D.O.; Frasch, A.C.C. Differential Accumulation of Mutations Localized in Particular Domains of the Mucin Genes Expressed in the Vertebrate Host Stage of *Trypanosoma cruzi*. *Mol. Biochem. Parasitol.* **2004**, *133*, 81–91. [CrossRef] [PubMed]
140. Soares, R.P.; Torrecilhas, A.C.; Assis, R.R.; Rocha, M.N.; Moura e Castro, F.A.; Freitas, G.F.; Murta, S.M.; Santos, S.L.; Marques, A.F.; Almeida, I.C.; et al. Intraspecies Variation in *Trypanosoma cruzi* GPI-Mucins: Biological Activities and Differential Expression of α-Galactosyl Residues. *Am. J. Trop. Med. Hyg.* **2012**, *87*, 87. [CrossRef]

141. Cánepa, G.E.; Degese, M.S.; Budu, A.; Garcia, C.R.S.; Buscaglia, C.A. Involvement of TSSA (Trypomastigote Small Surface Antigen) in *Trypanosoma cruzi* Invasion of Mammalian Cells. *Biochem. J.* **2012**, *444*, 211–218. [CrossRef]
142. Noia, J.M.D.; Buscaglia, C.A.; Marchi, C.R.D.; Almeida, I.C.; Frasch, A.C. A *Trypanosoma cruzi* Small Surface Molecule Provides the First Immunological Evidence That Chagas' Disease Is Due to a Single Parasite Lineage. *J. Exp. Med.* **2002**, *195*, 401. [CrossRef] [PubMed]
143. Cámara, M.d.l.M.; Cánepa, G.E.; Lantos, A.B.; Balouz, V.; Yu, H.; Chen, X.; Campetella, O.; Mucci, J.; Buscaglia, C.A. The Trypomastigote Small Surface Antigen (TSSA) Regulates *Trypanosoma cruzi* Infectivity and Differentiation. *PLoS Negl. Trop. Dis.* **2017**, *11*, e0005856. [CrossRef] [PubMed]
144. Cámara, M.d.l.M.; Balouz, V.; Centeno Cameán, C.; Cori, C.R.; Kashiwagi, G.A.; Gil, S.A.; Macchiaverna, N.P.; Cardinal, M.V.; Guaimas, F.; Lobo, M.M.; et al. *Trypanosoma cruzi* Surface Mucins Are Involved in the Attachment to the Triatoma Infestans Rectal Ampoule. *PLoS Negl. Trop. Dis.* **2019**, *13*, e0007418. [CrossRef] [PubMed]
145. Yoshida, N. Molecular Basis of Mammalian Cell Invasion by *Trypanosoma cruzi*. *An. Acad. Bras. Cienc.* **2006**, *78*, 87–111. [CrossRef]
146. Onofre, T.S.; Loch, L.; Rodrigues, J.P.F.; Macedo, S.; Yoshida, N. Gp35/50 Mucin Molecules of *Trypanosoma cruzi* Metacyclic Forms That Mediate Host Cell Invasion Interact with Annexin A2. *PLoS Negl. Trop. Dis.* **2022**, *16*, e0010788. [CrossRef] [PubMed]
147. Mortara, R.A.; da Silva, S.; Araguth, M.F.; Blanco, S.A.; Yoshida, N. Polymorphism of the 35- and 50-Kilodalton Surface Glycoconjugates of *Trypanosoma cruzi* Metacyclic Trypomastigotes. *Infect. Immun.* **1992**, *60*, 4673–4678. [CrossRef]
148. Urban, I.; Santurio, L.B.; Chidichimo, A.; Yu, H.; Chen, X.; Mucci, J.; Agüero, F.; Buscaglia, C.A. Molecular Diversity of the *Trypanosoma cruzi* TcSMUG Family of Mucin Genes and Proteins. *Biochem. J.* **2011**, *438*, 303–313. [CrossRef] [PubMed]
149. Nogueira, N.F.S.; Gonzalez, M.S.; Gomes, J.E.; de Souza, W.; Garcia, E.S.; Azambuja, P.; Nohara, L.L.; Almeida, I.C.; Zingales, B.; Colli, W. *Trypanosoma cruzi*: Involvement of Glycoinositolphospholipids in the Attachment to the Luminal Midgut Surface of Rhodnius Prolixus. *Exp. Parasitol.* **2007**, *116*, 120–128. [CrossRef]
150. Chanda, I.; Pan, A.; Saha, S.K.; Dutta, C. Comparative Codon and Amino Acid Composition Analysis of Tritryps-Conspicuous Features of Leishmania Major. *FEBS Lett.* **2007**, *581*, 5751–5758. [CrossRef]
151. Lander, N.; Bernal, C.; Diez, N.; Añez, N.; Docampo, R.; Ramírez, J.L. Localization and Developmental Regulation of a Dispersed Gene Family 1 Protein in *Trypanosoma cruzi*. *Infect. Immun.* **2010**, *78*, 231–240. [CrossRef]
152. Kawashita, S.Y.; da Silva, C.V.; Mortara, R.A.; Burleigh, B.A.; Briones, M.R.S. Homology, Paralogy and Function of DGF-1, a Highly Dispersed *Trypanosoma cruzi* Specific Gene Family and Its Implications for Information Entropy of Its Encoded Proteins. *Mol. Biochem. Parasitol.* **2009**, *165*, 19–31. [CrossRef] [PubMed]
153. Munjal, G.; Hanmandlu, M.; Srivastava, S. Phylogenetics Algorithms and Applications. *Ambient Commun. Comput. Syst.* **2018**, *904*, 187–194. [CrossRef]
154. Ramírez, J.L. The Elusive *Trypanosoma cruzi* Disperse Gene Protein Family (DGF-1). *Pathogens* **2023**, *12*, 292. [CrossRef] [PubMed]
155. Bao, Y.; Weiss, L.M.; Braunstein, V.L.; Huang, H. Role of Protein Kinase A in *Trypanosoma cruzi*. *Infect. Immun.* **2008**, *76*, 4757–4763. [CrossRef] [PubMed]
156. Atwood, J.A.; Minning, T.; Ludolf, F.; Nuccio, A.; Weatherly, D.B.; Alvarez-Manilla, G.; Tarleton, R.; Orlando, R. Glycoproteomics of *Trypanosoma cruzi* Trypomastigotes Using Subcellular Fractionation, Lectin Affinity, and Stable Isotope Labeling. *J. Proteome Res.* **2006**, *5*, 3376–3384. [CrossRef] [PubMed]

pathogens

MDPI

Article

In Vivo Imaging of Cardiac Attachment of TcI and TcII Variants of *Trypanosoma cruzi* in a Zebrafish Model

Victoria E. Rodriguez-Castellanos [1,†], Cristhian David Perdomo-Gómez [2,3], Juan Carlos Santos-Barbosa [1], Manu Forero-Shelton [3], Verónica Akle [2] and John M. González [1,*]

1 Biomedical Sciences Laboratory (CBMU), School of Medicine, Universidad de Los Andes, Bogotá D.C 111711, Colombia; ve.rodriguezc@uniandes.edu.co (V.E.R.-C.); js.barbosa@uniandes.edu.co (J.C.S.-B.)
2 Laboratory of Neurosciences and Circadian Rhythms, School of Medicine, Universidad de Los Andes, Bogotá D.C 111711, Colombia; cd.perdomo10@uniandes.edu.co (C.D.P.-G.); v.akle@uniandes.edu.co (V.A.)
3 Biophysics Group, Department of Physics, Universidad de Los Andes, Bogotá D.C 111711, Colombia; anforero@uniandes.edu.co
* Correspondence: johgonza@uniandes.edu.co
† Current affiliation: Center for Systems Biology, Massachusetts General Hospital Research Institute, 185 Cambridge Street, Suite 5.210, Boston, MA 02114, USA.

Abstract: *Trypanosoma cruzi*, the etiological agent of Chagas disease, is a parasite known for its diverse genotypic variants, or Discrete Typing Units (DTUs), which have been associated with varying degrees of tissue involvement. However, aspects such as parasite attachment remain unclear. It has been suggested that the TcI genotype is associated with cardiac infection, the most common involved site in chronic human infection, while TcII is associated with digestive tract involvement. Traditional models for *T. cruzi* infection provide limited in vivo observation, making it challenging to observe the dynamics of parasite-host interactions. This study evaluates the cardiac attachment of trypomastigotes from TcI and TcII DTUs in zebrafish larvae. Labeled trypomastigotes were injected in the duct of Cuvier of zebrafish larvae and tracked by stereomicroscopy and light-sheet fluorescence microscopy (LSFM). Remarkably, it was possible to observe TcI parasites adhered to the atrium, atrioventricular valve, and circulatory system, while TcII trypomastigotes demonstrated adhesion to the atrium, atrioventricular valve, and yolk sac extension. When TcI and TcII were simultaneously injected, they both attached to the heart; however, more of the TcII trypomastigotes were observed attached to this organ. Although TcII DTU has previously been associated with digestive tissue infection, both parasite variants showed cardiac tissue attachment in this in vivo model.

Keywords: Chagas disease; parasitic disease; intravital microscopy; zebrafish; tissue attachment

1. Introduction

One of the most intriguing aspects of parasitic infectious diseases is the attachment and tropism of the parasites, which can occur at different levels, such as in tissues, cells, and even subcellular compartments, as in the case of an intracellular parasite [1]. The molecular mechanisms underlying pathogen preference for specific tissues remain poorly understood but are probably multifactorial, including many pathways and interaction sites between host and parasitic molecules [2]. Tissue attachment is contingent on the genetic variability of the parasite, host genetic factors, and immunocompetence; furthermore, tropism can change throughout the course of a particular infection [3]. In Chagas disease, caused by

Trypanosoma cruzi, the hemoflagellate parasite mainly resides in the cardiac tissue during chronic infection, but it can also be found in the central nervous system of chronically infected immunocompromised individuals [4]. Tissue attachment in Chagas disease is not well understood, and genetic variability has been described as one of the factors that influence *T. cruzi* infection and tissue tropism [2,3].

The infective form of the *T. cruzi* parasite is the metacyclic trypomastigote, which is found in the feces of the triatomine vector. This form arises from the differentiation of epimastigotes, another extracellular and flagellated stage of the parasite within the vector. The vector defecates near the biting site, and the parasite enters the host via the damaged skin or intact mucous membranes. Subsequently, the parasite attaches to and infects cells, primarily macrophages and dendritic cells, and potentially various other mammalian nucleated cells. Within these cells, the parasite loses its external flagella and replicates asexually through binary fission. The resulting amastigotes eventually transform into trypomastigotes, which burst from the infected cells and spread through the bloodstream to infect other tissues. In addition to vector transmission, it can also be transmitted orally by consuming food contaminated with the vector's feces; also, infection can be via blood transfusions, transplants, vertical transmission, and laboratory accidents [1]. For successful infection, *T. cruzi* must adhere to the host cell membrane to pursue cellular invasion [2].

T. cruzi genotypes are classified as Discrete Typing Units (DTUs), and currently, seven DTUs have been described (TcI to TcVI and TcBat). Among them, TcI and TcII are considered the two major parasite lineages, whereas the subsequent TcIII to TcVI variants are derived from the TcII lineage, constituting independent subgroups [5]. TcBat was initially described in Chiroptera with subsequent reports of infection in humans. TcI is mainly found in Central America and northern South America, and it has been associated with cardiac disease, causing cardiomyopathy in 90% of the chronic symptomatic cases, being the most common cause of infectious myocarditis. In contrast, TcII has a more limited geographic distribution in the central and southern regions of South America and has been associated with gastrointestinal diseases, causing megaesophagus and megacolon in less than 10% of chronic symptomatic cases [6].

To date, *T. cruzi* has been studied in multiple animal models, including mice, dogs, rabbits, and zebrafish [7]. Animal studies have different purposes, either to study cell attachment, motility [8], disease development, and immunity or response to treatment [9]. In mammals, the parasite can enter through the skin and mucous membranes, and in humans, there is an additional oral infection route where the parasite invades through the gastric mucosa. In both infection routes, the binding of *T. cruzi* to host cells relies on interactions between parasite ligands and cellular receptors [2]. Several of these ligands, mainly mucins and glycoproteins, are involved in binding to carbohydrate residues on mammalian cells, with gp85/trans-sialidase being one of the primary molecules involved [10]. Inside the mammalian hosts, the parasite travels through the vasculature to its preferred infection site [11]. Few studies have been able to study the behavior of *Trypanosomas* spp. inside living organisms [12]. A recent study in zebrafish showed that *T. carassii*, a natural host of zebrafish, moves by passive dragging with the blood flow in small capillaries or by active swimming against the current [13]. We have previously established zebrafish larvae as a model to study the motility of *T. cruzi* and found adhesion of the TcI trypomastigote to the heart, although *T. cruzi* is not a natural parasite of zebrafish [8].

The aim of this study is to assess the attachment of the *T. cruzi* parasite to various zebrafish tissues, with a TcII isolate, in relation to a TcI isolate. We focused on the tissue adhesion of different genotypes, using an established zebrafish model for *T. cruzi* motility from our lab. To achieve this, we used microinjection of fluorescently labeled parasites and monitored their dynamics inside zebrafish using stereomicroscopy and light sheet fluorescence microscopy (LSFM).

2. Materials and Methods

2.1. Type of Study

This experimental study was approved by the Institutional Animal Care and Use Committee (CICUAL) through the application form with code C.FUA_14-017 at the Universidad de Los Andes in Bogotá DC, Colombia.

2.2. Fish Breeding and Maintenance

Adult fish were placed in breeding tanks, and three females per two males were used to increase the number of eggs. The next day, eggs were collected using a strainer and then transferred to Petri dishes filled with egg water (a solution with a concentration of 0.6 g/L in deionized water with 0.01 mg/L of methylene blue with a pH between 7.2 and 7.9). Approximately 50 fertilized eggs were placed in each Petri dish, dechorionated using surgical forceps, and then placed in an incubator at 33 °C. Although the optimal temperature for the fish was 28 °C, the temperature was increased to 33 °C to maintain the parasites in an environment closer to the mammalian temperature. Before injection, the larvae were immobilized on 1% agarose (Amresco Inc., Solon, OH, USA).

2.3. Parasite Culture and Labeling

Isolated *T. cruzi* DA (MHOM/CO/01/DA) TcI and Y (MHOM/BR/00/Y) TcII were maintained in culture of a human astrocytoma cell line CRL-1718 (American Tissue Typing Culture, Manassas, VA, USA), in RPMI 1640 (Sigma, St. Louis, MI, USA) with 10% fetal bovine serum (FBS) (Eurobio, Les Ulis, France) incubated at 37 °C and 5% CO_2. For labeling, trypomastigotes were collected from the culture supernatants, washed, and centrifuged for counting in a Neubauer chamber to obtain approximately 5×10^5 to 1×10^6 parasites per tube. The parasites were distributed in three different conditions: one for control (unlabeled) and the other for labeling with either CellTraceTM carboxyfluorescein succinimidyl ester (CTCFSE, Thermo Fisher Scientific, Cleveland, OH, USA) or CellTraceTM FarRed (CTFR, Thermo Fisher Scientific, Cleveland, OH, USA). The parasites were placed in 4 mL tubes, washed with 0.01 M PBS (1 × PBS) pH 7.4, and centrifuged at 1300 g for 5 min. The pellet was labeled with 1 mL of 1 × PBS containing 1 µL of CTCFSE or CTFR, incubated in the dark at room temperature for 10 min, washed with 1 × PBS, and centrifuged as described above. Parasite pellets were resuspended in 1 × PBS at a concentration of 100–200 parasites per 1 nL. For each condition, a sample was obtained for flow cytometry (FacsCanto II, BD Bioscience, San Jose, CA, USA) with FacsDiva software version 6.1. Fluorescence readings were performed at 510 nm for CTCFSE and 670 nm for CTFR PMTs. At least 10,000 parasites were examined under each condition. The gating and analysis strategies are illustrated in Figure 1. CTCFSE parasite labeling was also described previously [8].

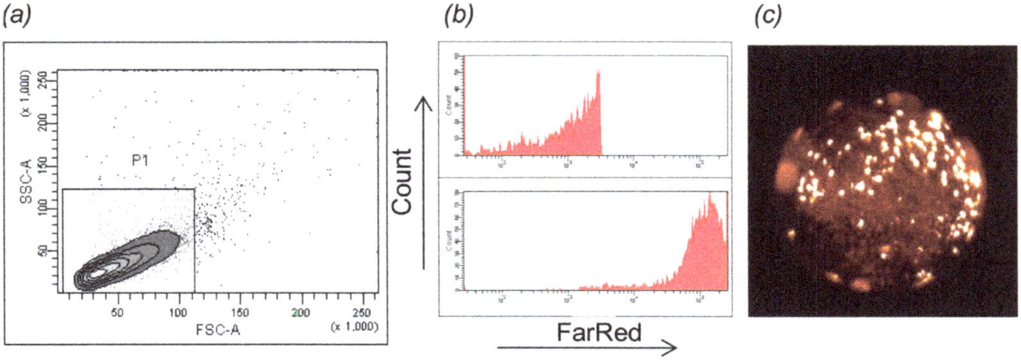

Figure 1. Detection of fluorescently labeled *T. cruzi* trypomastigotes. (**a**) Flow cytometry dot plot of forward scatter (FSC) versus side scatter (SSC) of TcI trypomastigotes. (**b**) Upper panel histogram of unlabeled TcI trypomastigotes as the control group, At 670 nm detection. Lower panel histogram of TcI trypomastigotes labeled with CTFR at 670 nm detections. (**c**) Standardized droplet of 160 μm diameter for microinjection under stereomicroscopy to estimate live parasites by their movement.

2.4. Zebrafish Microinjection and Microscopic Follow-Up

Needles were pulled at 60 °C with borosilicate glass microcapillaries in a micropipette puller PC-10 (Narishige, Tokyo, Japan) with two light weights and one heavy weight. Each needle tip was observed and opened under a stereomicroscope and cut to 5–10 uM diameter with forceps. The needles were filled with 10 μL of the parasite solution at a concentration of about 100–200 trypomastigotes per nL and placed in a micromanipulator. For the microinjection, each larva was placed on an agarose mold containing egg water. The larvae were anesthetized with 0.002% tricaine solution (Fluka, Buchs, Switzerland) diluted in E3 water. Larval microinjection was performed in the duct of Cuvier, observed under stereomicroscopy, and the micromanipulator was placed at ~45° and parallel to the agarose mold. Once injected with 3–5 nL, each larva was placed separately in 6-well culture plates containing egg water. About 30 min after injection, zebrafish larvae were screened using a fluorescent stereomicroscope. An ET-Far Red fluorescence cube (filters: ET630/20x (EX), T647lpxr (BS), ET667/30m (EM); Chroma, VT, USA) was used to follow up CFTR-labeled trypomastigotes, while an ET-EGFP fluorescence cube (filters: ET470/40x (EX), T495lpxr (BS), ET525/50m (EM); Chroma, VT, USA) was used for CT CAFE-labeled trypomastigotes. Images and videos of each specimen were acquired using NIS-Elements software (Nikon, Tokyo, Japan). Then, larvae containing parasites adhered to any tissue of interest were transferred for follow-up using light-sheet fluorescence microscopy (LSFM). This occurred from 2 to 4 h after injection.

2.5. Detection by Light Sheet Fluorescence Microscopy (LSFM)

We used a custom-made LSFM configuration as detailed [8]. Briefly, for the illumination arm, 488 nm and 633 nm diode lasers (06-MLD, Cobolt AB, Solna, Sweden) were used for CTCFSE and CTFR, respectively. A Galilean telescope was used to expand the laser beam, composed of f = 30 mm and f = 500 mm achromat lenses (Edmund Optics, Barrington, NJ, USA), and the expanded beam was then focused with a f = 75 cylindrical lens (Thorlabs, Newton, NJ, USA) into the back focal plane of an N Plan 10× NA0.25 air objective (Leica, Wetzlar, Germany), which was used to create the light sheet. For the detection arm, a water immersion CFI Plan Fluorite 40× NA0.8 objective (Nikon, Tokyo, Japan) and HQ525/50M and ET700/75 emission filters (Chroma, Bellows Falls, VT, USA) were used to capture the signals of CTCFSE and CTFR, respectively. Images were projected onto an sCMOS camera

(Neo 4.5, Andor, Belfast, Ireland) using an f = 200 achromat tube lens (Nikon, Tokio, Japan). The injected larvae were mounted inside agarose-filled borosilicate glass capillaries and placed inside a chamber filled with egg water maintained at 33 °C. The intensity of light that reached each larva was set to 2 mW, videos were acquired at 50 ms exposure, and the sensor was cooled to −40 °C. The microscope was controlled with a custom LabView application (National Instruments, Austin, TX, USA), and ImageJ 1.53 software [14] was used to process and analyze the data.

2.6. Injected Zebrafish Larvae Enzymatic Digestion for Flow Cytometry

The inoculated larvae were processed by enzymatic digestion for flow cytometry analysis. The procedure was conducted following Vargas-Patron 2019. In brief, anesthetized larvae were transferred to a 1.5 mL vial with calcium-free Ringer's solution (NaCl 116 mM, KCl 2.9 mM, HEPES 5 mM) for 15 min. They were then transferred to a Petri dish, where 2 mL of 0.5% trypsin (Gibco, Waltham, MA, USA) solution was added. Then Petri dishes were incubated at 37 °C until larval cell dissociation was observed under a microscope. Trypsin was neutralized with 4 mL of a 5% FBS solution in 1 × PBS [15]. The solution was transferred to a 4 mL tube and centrifuged at 1300× g for 5 min. The pellet was resuspended in 250 µL of FACSFlow (BD Biosciences, Milpitas, CA, USA) and analyzed using a cytometer to detect the presence of labeled parasites.

2.7. Data Analysis and Presentation

This was an experimental descriptive study. All videos and image stacks were captured using the aforementioned systems and stored in the raw format as they were acquired. The images selected for publication were converted to .tiff format to preserve quality, with no manipulation applied to any of the presented images. This study provides a qualitative description of the tissues to which the parasites attach, depending on their genotype. Consequently, inferential statistics were not used. The success of the injection was categorized as unsuccessful if no parasite was observed under stereomicroscopy, successful if any parasite was observed, and migration to an organ if a successful injection resulted in parasite migration and attachment to a specific organ. The four potential categories of tissue migration were classified as follows: eye or brain (due to proximity), heart or pericardium, GI tract, and circulation.

3. Results

Samples of labeled parasites in suspension were analyzed by flow cytometry to confirm their detection and later the specific fluorescence (Figure 1a). The flow cytometry histogram shows the absence of fluorescence in the control tube (Figure 1b, upper panel) and the fluorescence intensity at 670 nm when labeled with CTFR (Figure 1b, lower panel). A droplet of labeled parasite approximately 160 µm in diameter was standardized (Figure 1c) with injection pressures between 6 and 15 psi, depending on the needle used, to achieve this droplet diameter. Then, 178 larvae (48–72 hpf) were microinjected with either *T. cruzi* TcI (DA) (81 larvae), TcII (Y) isolates (64 larvae), or both simultaneously (33 larvae) into the duct of Cuvier, which is a part of the larval circulatory system. For the control, 1× PBS (the vehicle for the parasites) was microinjected. After 30 to 60 min, under fluorescence stereoscopy using a 40× lens (40× magnification), the migration and attachment of the parasite to different tissues and organs of the zebrafish were observed, from the duct of Cuvier onto the cardiac circulation, and from there to the cardinal vein (Video S1). The success rate of TcI injections was 91% (74 larvae), with organ migration observed in 28% (23 larvae) of the successful cases. For TcII, the success rate was 81% (52 larvae), with tissue migration occurring in 28% (18 larvae) of injections. Simultaneous injection of both TcI and

TcII resulted in a 100% success rate (33 larvae), with the detection of either TcI, TcII, or both genotypes, and migration was observed in 30% (10 larvae) of the injections (Table 1).

Under stereomicroscopy at 40× magnification, parasites of the TcI genotype attached to the heart were observed to move with the heartbeat (Video S2). Using LSFM and at higher magnification, the parasites in the heart were found to be attached mainly to the heart valves and remained adhered to the atrioventricular valve of zebrafish with each heartbeat. Out of all the larvae injected with TcI, 35% were found attached to cardiac tissue (Table 2). In addition, TcII parasites were also found adhered to the cardiac tissue (Figure 2a). They were mainly found in the atrium, as observed under stereomicroscopy, which was later confirmed with LSFM (Video S3). Among all larvae in which TcII parasites were observed, cardiac parasite attachment was observed in 39% (Table 2). In the case of TcI trypomastigotes, up to two parasites were found attached simultaneously in different cardiac locations, whereas for TcII, it was possible to detect up to four attached parasites.

Table 1. Parasite injection outcome in zebrafish larvae by DTU TcI and TcII.

Parasite DTU	TcI	TcII	Both **
No. Injected larvae	81	64	33
Injections			
Unsuccessful	7 (9%)	12 (19%)	0 (0%)
Successful	74 (91%) *	52 (81%)	33 (70%)
Migration to organ	23 (28%)	18 (28%)	10 (30%)

* Values represent the total and percentage of injections where the parasite was detected. Success includes migration. ** Simultaneous injection of TcI and TcII trypomastigotes.

Table 2. Detection of parasite in organs according to DTU TcI and TcII.

Parasite DTU	TcI	TcII	Both **
No. Injected larvae *	74	52	33
Injections			
Eye/Brain	4 (17%)	3 (17%)	3 (30%)
Heart/Pericardium	8 (35%)	7 (39%)	4 (40%)
GI tract	5 (22%)	5 (28%)	1 (10%)
Circulation	6 (26%)	3 (17%)	2 (20%)

* Successful injections: parasite found inside the larvae. ** Simultaneous injection of TcI and TcII trypomastigotes.

Using stereomicroscopy, TcII parasites migrating from the duct of Cuvier to the yolk sac extension, a structure that originates in the digestive tract of zebrafish, were observed (Figure 2b upper panel). These larvae were subsequently followed by LSFM, and detectable parasite migration towards the yolk sac extension was found (Figure 2a, lower panel). Remarkably, TcI adhesion to the yolk sac extension was detected in 22% of the larvae that were positive for parasite migration, while TcII adhesion was observed in 28% (Table 2). In two larvae with TcI parasites adhered to their hearts, trypomastigotes were also observed in the peripheral vasculature. This location was determined by transmitted-light stereomicroscopy, as shown in Figure 2b, where a parasite was observed attached to the posterior cardinal vein. Additionally, in another larva, a trypomastigote was detected in the endothelium of the dorsal aorta (Figure 2c). Attachment to the zebrafish peripheral vasculature was observed in 26% and 17% of cases for TcI and TcII, respectively (Table 2).

Parasite attachment of both DTUs was assessed by microinjecting them simultaneously into the same larvae (33 larvae) with different dyes in order to distinguish them. Trypomastigote labeling with each dye was confirmed by flow cytometry, and three populations can be distinguished: the population in region P1 without labeling, used as a control (Figure 3a); the population in region P2 consists of TcII parasites labeled with CTCFSE (CFSE-green channel); and the population in region P3, TcI labeled with CTFR (Figure 3b).

After larval microinjection, fluorescence was checked under stereo-microscopy and then at high resolution by LSFM. Both TcI and TcII parasites were simultaneously detected attached to cardiac tissue in different structures. For instance, in one larva, two TcI trypomastigotes were attached to the atrioventricular valves, while a TcII trypomastigote adhered to the pericardium of the atrium (Video S4). Figure 4 provides a summary of the anatomic location of the trypomastigotes TcI and TcII detected in the heart structure of zebrafish larvae in all the experiments.

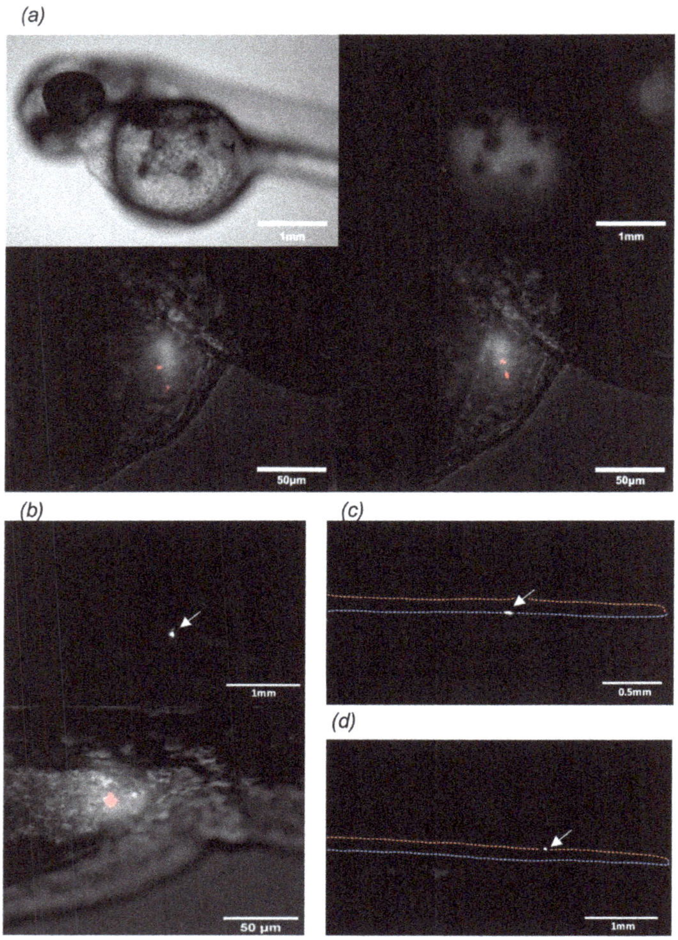

Figure 2. (**a**) Zebrafish larvae were injected with *T. cruzi* TcII trypomastigotes found attached to the heart valves of the larva under stereomicroscopy in the upper panel. Two trypomastigotes can be observed under LSFM 4 h after microinjection in the lower panel. (**b**) The upper panel shows the migration of the parasite towards the yolk sac extension under stereomicroscopy. The lower panel shows the same larva 4 h after microinjection under LSFM with a subtle movement towards the end of the yolk sac extension. (**c**) Zebrafish larvae injected with TcI trypomastigotes, one of which was attached to the cardinal vein (dotted blue line), observed under stereomicroscopy. (**d**) TcI trypomastigote attached to the dorsal aorta (dotted red line), observed under stereomicroscopy. The images shown came from different larvae.

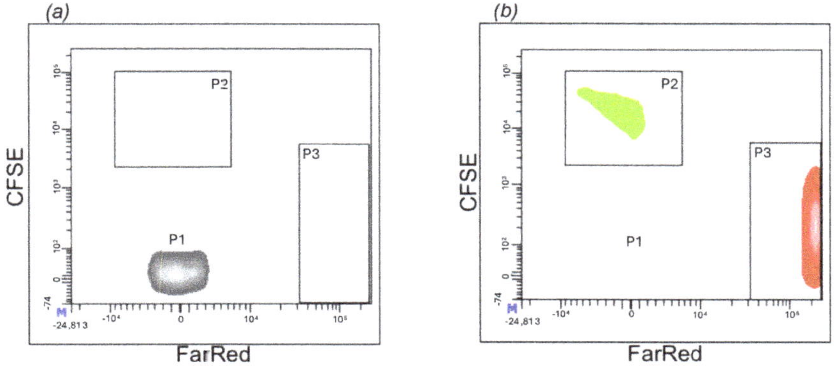

Figure 3. Dot plot of flow Cytometry assay with *T. cruzi* trypomastigotes. (**a**) The unlabeled population (P1) served as the control. (**b**) Region P2 represents TcII trypomastigotes labeled with CTCFSE, while region P3 represents TcI trypomastigotes labeled with CTFR.

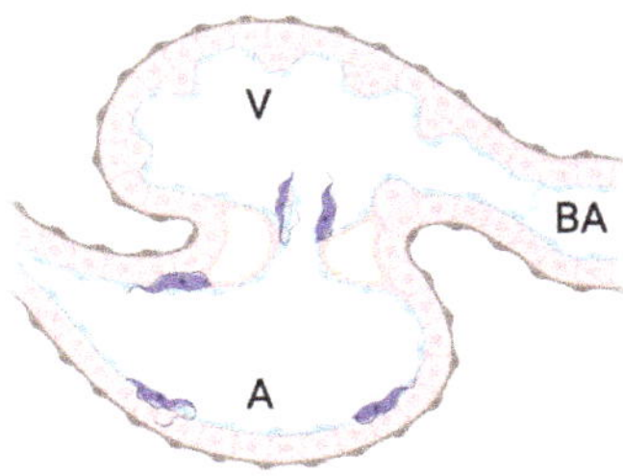

Figure 4. Schematic representation of zebrafish larvae heart showing the locations where trypomastigotes were detected using LSFM. Parasites (purple figures) were observed to be attached mainly to the heart valves and atrium, independent of DTU. V: Ventricle, A: Atrium, BA: Bulbus arteriosus. Figure was made using BioRender.com.

Finally, to establish the average number of parasites recovered after larvae injection, 20 fish larvae were injected: 18 with TcI parasites and 2 with 1× PBS as a control. After enzymatic digestion of the larvae, 5 out of 18 fish contained parasites, with an average of 2.6 parasites per larva. The highest parasite number found was nine parasites, whereas microinjection was calculated to have an input of 100 parasites per nL. The low number of detected parasites may be attributed to the large amount of injected solution, immediately ejected from the larvae after microinjection.

4. Discussion

The zebrafish model used here is versatile, cost-effective, space-saving, and relatively easy to handle. One of the advantages of this model is the easy visualization given the translucent properties of zebrafish, which makes it suitable for in vivo visualization of larvae through optical microscopy, using techniques such as LSFM. This advantage makes it an ideal in vivo model to study pathogen-host interactions, as shown in a previous study by our group, which provided visual evidence of the attachment of *T. cruzi* TcI trypomastigotes to heart tissue in zebrafish after microinjection [8]. The present study expands on this prior observation by documenting in vivo attachment of TcI and TcII parasites in zebrafish cardiac tissue, as well as the yolk sac and circulatory system.

TcI and TcII are the main *T. cruzi* lineages, hence the importance of understanding their migration and adhesion patterns in vivo [5]. In our study, microinjected larvae survived long enough to observe the *T. cruzi* migration from the duct of Cuvier, which is the initial injection site, to the target tissue. Studies in humans have suggested a possible association between TcI infection, primarily in the cardiac tissue, and TcII infection in the gastrointestinal tract [6]. It was found that the TcI genotype circulates and then attaches to the cardiac tissue, which is consistent with the published literature in which chagasic heart disease is associated with this genotype [16]. Indeed, cytokine profile and heart disease are more associated with TcI chronic infection [17]. Here, TcII trypomastigotes attached to the cardiac structures of zebrafish larvae were found. As it is shown in the videos, the trypomastigote can be seen moving in synchrony with each heartbeat, indicating that the parasite adheres to the cardiac structures. To our knowledge, this is the first in vivo demonstration of TcII adhesion to heart tissue, and interestingly, TcII infection has been associated mainly with chronic mega-visceral Chagas disease [18]. Of note, TcII is easier to visualize than TcI (DA), because of the relatively larger size of the specific (isolate Y) parasite used. Remarkably, a high percentage of TcII parasites were found in the yolk sac, consistent with its association with digestive system infection. Gastrointestinal tissue attachment has previously been imaged in vivo in mice using bioluminescent *T. cruzi* [12].

The coexistence of both TcI and TcII in patients is poorly understood; however, in this in vivo model we document that both genotypes can be found attached to cardiac structures. This observation was confirmed by LSFM microscopy, where the parasite showed strong adherence to the heart, tolerating high blood flow driven by heartbeats. Given that, LSFM microscopy was executed more than 2 h after injection, parasites were attached for over 2 h, from their initial adhesion observed 30 to 60 min after injection.

When a single injection was performed with both genotypes individually labeled and mixed in a single tube, only the TcI genotype parasites migrated along the zebrafish circulation. Multiple individual parasites were detected at the posterior ends of the larvae. Additionally, both the TcI and TcII genotypes were identified in cardiac tissue. In mixed parasitic infections, there is a likelihood of misidentification of minor parasite genotypes because populations that grow at a faster rate are more likely to be detected given the available typification approaches [19]. This could explain why the correlation between heart disease and infection with TcI is typically more prevalent in some regions, as it is often regarded as more infectious and pathogenic than TcII [6]. Interestingly, TcII was the most detected DTU in isolates from Brazilian patients with either cardiac, indeterminate, or cardio-digestive forms [20].

The reasons behind the parasite's preference for a specific tissue are not fully understood but are likely to be multifactorial. Tissue attachment depends not only on the genetic variability of the parasite but also on host genetic factors, host immune system status [4], and parasite replication sites [21], which can even change during infection [1]. In a mouse model, *T. cruzi* showed identical tissue attachment in mice with the same H-2 haplotype, H-2^d (BALB/c and DBA-2 mice), but differed in H-2^b (C57BL/6 mice) using two different DTUs: Col 1.7G2 (TcI) and JG (TcII). In this case, it was suggested that major histocompatibility complex variability influenced the differential tissue distribution of each parasite variant [3].

Host factors, including lectins, particularly human galectin-3, a β-galactosyl-binding lectin, have been identified to play a role in the attachment and invasion of *T. cruzi* into dendritic and endothelial cells [22]. Cardiac attachment in *T. cruzi* seems to be influenced by the peptide motif FLY, which is conserved in all gp85/trans-sialidases and interacts with the vascular endothelium. Interestingly, this motif has greater avidity for cardiac vasculature than for vessels in other organs [10].

Cell surface glycosaminoglycans (GAGs), primarily heparan sulfate, play a crucial role in *T. cruzi* cell attachment and invasion [23]. In an in vitro model, heparan sulfate proteoglycans mediate the attachment of *T. cruzi* to cardiomyocytes [24]. Heparan sulfate is one of the most conserved molecules across species [25], and similar interactions with parasite proteins could occur during attachment to human and zebrafish hearts, although specific *T. cruzi* proteins involved have yet to be identified. Genomic comparisons between the human and zebrafish genomes have shown approximately 70% homology [26], and zebrafish have been used as a model for studying genetic cardiac diseases [27].

Although zebrafish are not commonly used as an infection model for certain pathogens, they are valuable for studying parasite attachment. In addition to *T. cruzi*, zebrafish models have been established for *Toxoplasma gondii* infection, demonstrating its interaction with macrophages and successful replication [28]. Lastly, it is essential to acknowledge that while the tissue niche of trypanosomiasis is relatively well known, a comprehensive understanding of the pathways governing its attachment is still lacking. Similarly, the invasion and pathogenesis of *Trypanosoma brucei* in the central nervous system remain largely unknown. A study with two-photon microscopy provided insights into specific tissue locations and the involvement of immune cells of this parasite during infection [29]. Moreover, despite advancements in understanding the cellular pathways facilitating parasite adherence to host cells, it is crucial to clarify that adherence does not necessarily imply invasion by the parasite [30].

These findings underscore the importance of further investigating the molecular mechanisms behind tissue specificity and attachment in *T. cruzi* infections. The zebrafish model, with its unique advantages, provides a promising platform for such studies, potentially leading to new insights into the pathogenesis and progression of Chagas disease. Future research could focus on identifying the specific host-parasite interactions that drive the differential tissue tropism observed in TcI and TcII, which may contribute to the development of targeted therapies aimed at preventing tissue-specific damage in infected individuals.

5. Conclusions

Our study has provided evidence of the differential tissue migration and attachment of *T. cruzi* TcI and TcII genotypes within the zebrafish model, highlighting its utility for in vivo investigations of pathogen-host interactions. We document parasite adherence to multiple locations, not only within the heart but also the yolk sac and circulatory system, and the observation of cardiac involvement by both TcI and TcII DTUs after simultaneous injection. The simultaneous detection of TcI and TcII attachment to cardiac structures in zebrafish larvae offers new insights into the potential for co-infection. This model thus serves as a valuable tool for further dissecting the molecular mechanisms underlying tissue migration and attachment during *T. cruzi* infections and could aid in developing targeted therapeutic interventions.

Supplementary Materials: The following supporting information can be downloaded at: https://www.mdpi.com/article/10.3390/pathogens14010025/s1, Video S1: Fluorescence Imaging of TcI Trypomastigote Migration in Zebrafish Larva, Video S2: TcI Trypomastigote Attachment to Zebrafish Cardiac Tissue Observed by Stereomicroscopy and LSFM, Video S3: TcII Trypomastigote Cardiac Attachment in Zebrafish Larva: Stereomicroscopy and LSFM Analysis, Video S4: Dual Imaging of TcI and TcII Trypomastigote Attachment in Zebrafish Cardiac Tissue using LSFM.

Author Contributions: Conceptualization, J.M.G., M.F.-S., V.A. and V.E.R.-C.; methodology, C.D.P.-G., J.C.S.-B., J.M.G., M.F.-S. and V.E.R.-C.; validation, C.D.P.-G., J.C.S.-B., J.M.G., M.F.-S., V.A. and V.E.R.-C.; formal analysis, C.D.P.-G., J.C.S.-B., J.M.G., M.F.-S., V.A. and V.E.R.-C.; investigation, C.D.P.-G., J.C.S.-B., J.M.G., M.F.-S., V.A. and V.E.R.-C.; writing—original draft preparation, J.M.G. and V.E.R.-C.; writing—review and editing, C.D.P.-G., J.C.S.-B., J.M.G., M.F.-S., V.A. and V.E.R.-C.; visualization J.M.G., C.D.P.-G. and V.E.R.-C.; supervision, J.M.G., M.F.-S. and V.A.; project administration, J.M.G.,

M.F.-S. and V.A.; funding acquisition, J.M.G., M.F.-S. and V.A. All authors have read and agreed to the published version of the manuscript.

Funding: M.F.-S. and C.D.P.-G. received support through a grant from the Faculty of Sciences at Universidad de Los Andes, with additional funding allocated for microscopic components also covered by the same project. Financial support was extended from the School of Medicine to both the Biomedical Sciences Laboratory (CBMU) and the Laboratory of Neuroscience and Circadian Rhythms. The funding for this project does not come from large, specific grants but rather from internal financial resources provided by the departments and faculties involved. These internal contributions reflect institutional support and underscore the collaborative nature of the initiative.

Institutional Review Board Statement: This experimental study was approved by the Institutional Animal Care and Use Committee (CICUAL) through a general application form for the use of animals (FUA) with code C.FUA_14-017 at the Universidad de Los Andes in Bogotá DC, Colombia.

Informed Consent Statement: Not applicable.

Data Availability Statement: The data are presented in this paper.

Acknowledgments: The authors would like to thank Yeferzon Ardila B. Sc from the Laboratory of Neurosciences and Circadian Rhythms, School of Medicine, Universidad de Los Andes for his technical support.

Conflicts of Interest: The authors declare no conflicts of interest.

References

1. Pereira, S.S.; Trindade, S.; De Niz, M.; Figueiredo, L.M. Tissue Tropism in Parasitic Diseases. *Open Biol.* **2019**, *9*, 190036. [CrossRef]
2. de Souza, W.; de Carvalho, T.M.U.; Barrias, E.S. Review on *Trypanosoma cruzi*: Host Cell Interaction. *Int. J. Cell Biol.* **2010**, *2010*, 295394. [CrossRef]
3. Andrade, L.O.; Machado, C.R.S.; Chiari, E.; Pena, S.D.J.; Macedo, A.M. *Trypanosoma cruzi*: Role of Host Genetic Background in the Differential Tissue Distribution of Parasite Clonal Populations. *Exp. Parasitol.* **2002**, *100*, 269–275. [CrossRef]
4. Shelton, W.J.; Gonzalez, J.M. Outcomes of patients in Chagas disease of the central nervous system: A systematic review. *Parasitology* **2024**, *151*, 15–23. [CrossRef] [PubMed]
5. Tibayrenc, M.; Ayala, F.J. Microevolution and subspecific taxonomy of *Trypanosoma cruzi*. *Infect. Genet. Evol.* **2022**, *103*, 105344. [CrossRef]
6. Zingales, B.; Miles, M.A.; Campbell, D.A.; Tibayrenc, M.; Macedo, A.M.; Teixeira, M.M.G.; Schijman, A.G.; Llewellyn, M.S.; Lages-Silva, E.; Machado, C.R.; et al. The Revised *Trypanosoma cruzi* Subspecific Nomenclature: Rationale, Epidemiological Relevance and Research Applications. *Infect. Genet. Evol.* **2012**, *12*, 240–253. [CrossRef]
7. Mateus, J.; Guerrero, P.; Lasso, P.; Cuervo, C.; González, J.M.; Puerta, C.J.; Cuéllar, A. An Animal Model of Acute and Chronic Chagas Disease with the Reticulotropic Y Strain of *Trypanosoma cruzi* That Depicts the Multifunctionality and Dysfunctionality of T Cells. *Front. Immunol.* **2019**, *10*, 918. [CrossRef] [PubMed]
8. Akle, V.; Agudelo-Dueñas, N.; Molina-Rodriguez, M.A.; Kartchner, L.B.; Ruth, A.M.; González, J.M.; Forero-Shelton, M. Establishment of Larval Zebrafish as an Animal Model to Investigate *Trypanosoma cruzi* Motility in vivo. *J. Vis. Exp.* **2017**, *2017*, e56238. [CrossRef]
9. Chatelain, E.; Konar, N. Translational Challenges of Animal Models in Chagas Disease Drug Development: A Review. *Drug Des. Dev. Ther.* **2015**, *9*, 4807–4823. [CrossRef] [PubMed]
10. Tonelli, R.R.; Giordano, R.J.; Barbu, E.M.; Torrecilhas, A.C.; Kobayashi, G.S.; Langley, R.R.; Arap, W.; Pasqualini, R.; Colli, W.; Alves, M.J.M. Role of the Gp85/Trans-Sialidases in *Trypanosoma cruzi* Tissue Tropism: Preferential Binding of a Conserved Peptide Motif to the Vasculature in Vivo. *PLoS Negl. Trop. Dis.* **2010**, *4*, e864. [CrossRef]
11. Pérez-Mazliah, D.; Ward, A.I.; Lewis, M.D. Host-Parasite Dynamics in Chagas Disease from Systemic to Hyper-Local Scales. *Parasite Immunol.* **2021**, *43*, e12786. [CrossRef]
12. Silberstein, E.; Serna, C.; Fragoso, S.P.; Nagarkatti, R.; Debrabant, A. A Novel Nanoluciferase-Based System to Monitor *Trypanosoma cruzi* Infection in Mice by Bioluminescence Imaging. *PLoS ONE* **2018**, *13*, e0195879. [CrossRef] [PubMed]
13. Dóró, É.; Jacobs, S.H.; Hammond, F.R.; Schipper, H.; Pieters, R.P.; Carrington, M.; Wiegertjes, G.F.; Forlenza, M. Visualizing Trypanosomes in a Vertebrate Reveals Novel Swimming Behaviours, Adaptations and Attachment Mechanisms. *eLife* **2019**, *24*, e48388. [CrossRef] [PubMed]

14. Schneider, C.A.; Rasband, W.S.; Eliceiri, K.W. NIH Image to ImageJ: 25 Years of Image Analysis. *Nat. Methods* **2012**, *9*, 671–675. [CrossRef] [PubMed]
15. Vargas-Patron, L.A.; Agudelo-Dueñas, N.; Madrid-Wolff, J.; Venegas, J.A.; González, J.M.; Forero-Shelton, M.; Akle, V. Xenotransplantation of Human Glioblastoma in Zebrafish Larvae: In Vivo Imaging and Proliferation Assessment. *Biol. Open* **2019**, *8*, bio043257. [CrossRef] [PubMed]
16. Messenger, L.A.; Miles, M.A.; Bern, C. Between a Bug and a Hard Place: *Trypanosoma cruzi* Genetic Diversity and the Clinical Outcomes of Chagas Disease. *Expert. Rev. Anti Infect. Ther.* **2015**, *13*, 995–1029. [CrossRef]
17. Poveda, C.; Fresno, M.; Gironès, N.; Martins-Filho, O.A.; Ramírez, J.D.; Santi-Rocca, J.; Marin-Neto, J.A.; Morillo, C.A.; Rosas, F.; Guhl, F. Cytokine profiling in Chagas disease: Towards understanding the association with infecting *Trypanosoma cruzi* discrete typing units (a BENEFIT TRIAL sub-study). *PLoS ONE* **2014**, *9*, e91154. [CrossRef]
18. Cura, C.I.; Lucero, R.H.; Bisio, M.; Oshiro, E.; Formichelli, L.B.; Burgos, J.M.; Lejona, S.; Brusés, B.L.; Hernández, D.O.; Severini, G.V.; et al. *Trypanosoma cruzi* Discrete Typing Units in Chagas Disease Patients from Endemic and Non-Endemic Regions of Argentina. *Parasitology* **2012**, *139*, 516–521. [CrossRef] [PubMed]
19. Ledezma, A.P.; Blandon, R.; Schijman, A.G.; Benatar, A.; Saldaña, A.; Osuna, A. Mixed Infections by Different *Trypanosoma cruzi* Discrete Typing Units among Chagas Disease Patients in an Endemic Community in Panama. *PLoS ONE* **2020**, *15*, e0241921. [CrossRef] [PubMed]
20. Nielebock, M.A.P.; Moreira, O.C.; das Chagas Xavier, S.C.; de Freitas Campos Miranda, L.; de Lima, A.C.B.; de Jesus Sales Pereira, T.O.; Hasslocher-Moreno, A.M.; Britto, C.; Sangenis, L.H.C.; Saraiva, R.M. Association between *Trypanosoma cruzi* DTU TcII and Chronic Chagas Disease Clinical Presentation and Outcome in an Urban Cohort in Brazil. *PLoS ONE* **2020**, *15*, e0243008. [CrossRef]
21. McCall, L.I.; Siqueira-Neto, J.L.; McKerrow, J.H. Location, Location, Location: Five Facts about Tissue Tropism and Pathogenesis. *PLoS Pathog.* **2016**, *12*, 15–18. [CrossRef] [PubMed]
22. Vray, B.; Camby, I.; Vercruysse, V; Mijatovic, T.; Bovin, N.V.; Ricciardi-Castagnoli, P.; Kaltner, H.; Salmon, I.; Gabius, H.J.; Kiss, R. Up-Regulation of Galectin-3 and Its Ligands by *Trypanosoma cruzi* Infection with Modulation of Adhesion and Migration of Murine Dendritic Cells. *Glycobiology* **2004**, *14*, 647–657. [CrossRef] [PubMed]
23. de Oliveira, F.O.R.; Alves, C.R.; Calvet, C.M.; Toma, L.; Bouças, R.I.; Nader, H.B.; de Castro Côrtes, L.M.; Krieger, M.A.; Meirelles, M.d.N.S.L.; de Souza Pereira, M.C. *Trypanosoma cruzi* Heparin-Binding Proteins and the Nature of the Host Cell Heparan Sulfate-Binding Domain. *Microb Pathog.* **2008**, *44*, 329–338. [CrossRef]
24. Calvet, C.M.; Toma, L.; De Souza, F.R.; De Meirelles, M.D.N.S.L.; Pereira, M.C.S. Heparan Sulfate Proteoglycans Mediate the Invasion of Cardiomyocytes by *Trypanosoma cruzi*. *J. Eukaryot. Microbiol.* **2003**, *50*, 97–103. [CrossRef] [PubMed]
25. Esko, J.D.; Lindahl, U. Molecular Diversity of Heparan Sulfate. *J. Clin. Investig.* **2001**, *108*, 169–173. [CrossRef]
26. Howe, K.; Clark, M.D.; Torroja, C.F.; Torrance, J.; Berthelot, C.; Muffato, M.; Collins, J.E.; Humphray, S.; McLaren, K.; Matthews, L.; et al. The Zebrafish Reference Genome Sequence and Its Relationship to the Human Genome. *Nature* **2013**, *496*, 498–503. [CrossRef] [PubMed]
27. Tu, S.; Chi, N.C. Zebrafish Models in Cardiac Development and Congenital Heart Birth Defects. *Differentiation* **2012**, *84*, 4–16. [CrossRef]
28. Yoshida, N.; Domart, M.C.; Peddie, C.J.; Yakimovich, A.; Mazon-Moya, M.J.; Hawkins, T.A.; Collinson, L.; Mercer, J.; Frickel, E.M.; Mostowy, S. The Zebrafish as a Novel Model for the in Vivo Study of *Toxoplasma gondii* Replication and Interaction with Macrophages. *DMM Dis. Models Mech.* **2020**, *13*, dmm043091. [CrossRef]
29. Coles, J.A.; Myburgh, E.; Ritchie, R.; Hamilton, A.; Rodgers, J.; Mottram, J.C.; Barrett, M.P.; Brewer, J.M. Intravital Imaging of a Massive Lymphocyte Response in the Cortical Dura of Mice after Peripheral Infection by Trypanosomes. *PLoS Negl. Trop. Dis.* **2015**, *9*, e0003714. [CrossRef]
30. Salassa, B.N.; Romano, P.S. Autophagy: A Necessary Process during the *Trypanosoma cruzi* Life-Cycle. *Virulence* **2019**, *10*, 460–469. [CrossRef]

pathogens

MDPI

Review

Fighting Strategies Against Chagas' Disease: A Review

Andrea Hernández-Flores [1,†], Debora Elías-Díaz [2,†], Bernadeth Cubillo-Cervantes [2], Carlos N. Ibarra-Cerdeña [3], David Morán [4], Audrey Arnal [1,5,6,‡] and Andrea Chaves [7,8,*,‡]

1 Departamento de Etología, Fauna Silvestre y Animales de Laboratorio, Facultad de Medicina Veterinaria y Zootecnia, Universidad Nacional Autónoma de México (UNAM), Ciudad Universitaria, Av. Universidad #3000, Mexico City 04510, Mexico; andreahf.mvz@gmail.com (A.H.-F.); audrey.arnal@ird.fr (A.A.)
2 Sistema de Estudios de Posgrado Posgrado en Biología, Universidad de Costa Rica, San José 11501-206, Costa Rica; debaelias@gmail.com (D.E.-D.); bernadethcubillo21@gmail.com (B.C.-C.)
3 Departamento de Ecología Humana, Centro de Investigación y de Estudios Avanzados (Cinvestav), Unidad Mérida, Merida 97205, Mexico; ibarra.cerdena@gmail.com
4 Unidad de Ecología y Epidemiología, Programa Arbovirus y Zoonoses, Centro para Estudios de Salud, Universidad del Valle de Guatemala, Guatemala City 01015, Guatemala; dmoran@uvg.edu.gt
5 MIVEGEC, IRD, CNRS, Université de Montpellier, 34394 Montpellier, France
6 International Joint Laboratory IRD/UNAM ELDORADO, Merida 97205, Mexico
7 Centro Nacional de Innovaciones Biotecnológicas (CENIBiot), CeNAT, Conare, San José 1174-1200, Costa Rica
8 Escuela de Biología, Universidad de Costa Rica, San José 11501-206, Costa Rica
* Correspondence: andreachaves.biol@gmail.com
† These authors contributed equally to this work.
‡ These authors contributed equally to this work.

Abstract: Chagas disease, caused by *Trypanosoma cruzi*, remains a significant public health challenge, particularly in Latin America, where it is one of the most neglected diseases and is primarily transmitted by triatomine insects. The disease exhibits complexity due to its diverse transmission routes, including vectorial and non-vectorial mechanisms such as blood transfusions and congenital transmission. Effective monitoring and control strategies are critical to mitigating its impact. This review focuses on current monitoring and control efforts, emphasizing the importance of enhanced surveillance systems, improved risk assessments, and integrated vector control programs. Surveillance plays a pivotal role in early detection and timely intervention, particularly in endemic regions, while vector control remains central to reducing transmission. Moreover, the development of novel diagnostic tools, treatments, and vaccines is a crucial step in advancing control efforts. This review also highlights the involvement of local governments, international organizations, and civil society in executing these strategies, stressing the need for sustained political commitment to ensure the success of public health programs. By addressing key challenges in monitoring, control, and prevention, this review aims to provide insights and recommendations to further global efforts in reducing the burden of Chagas disease.

Keywords: *Trypanosoma cruzi*; vector control; surveillance systems; public health strategies; neglected diseases

1. Introduction

Chagas disease, also known as "American trypanosomiasis", is an important zoonotic disease transmitted by vectors to humans, caused by the hemoflagellate protists *Trypanosoma cruzi*. It is part of a group of neglected diseases that primarily affect vulnerable populations and are increasingly expanding their distribution as a consequence of urbanization and migration [1,2]. Its cycle was fully described in 1909 by the Brazilian physician Carlos Ribeiro Justiniano Chagas [3]. Currently, this disease is endemic in 21 countries in

Latin America (Argentina, Belize, Bolivia, Brazil, Chile, Colombia, Costa Rica, Ecuador, El Salvador, French Guiana, Guatemala, Guyana, Honduras, Mexico, Nicaragua, Panama, Paraguay, Peru, Suriname, Venezuela and Uruguay), as well as in the United States, particularly in the southern states, where—though rare—estimates suggest it could affect up to 10,000 individuals [4–6].

Chagas disease evolves in phases, with the acute phase usually presenting few or no particular symptoms; some patients experience fever, lymphadenopathy, splenomegaly and/or edema, but rarely with severe disease [7]. After the acute phase, individuals infected with *T. cruzi* evolve to a clinically silent phase [7,8]. However, over a period of 10 to 30 years, 20% to 35% of patients develop chronic symptoms of Chagas disease, which is mainly characterized by affecting the cardiovascular, gastrointestinal and/or central nervous systems [8,9]. The remaining infected individuals may persist asymptomatic throughout their lifetime.

According to estimates by the World Health Organization (WHO), which classifies Chagas disease as one of the "neglected tropical diseases", it is a major public health problem, primarily in Latin America. An estimated 6 to 7 million people could be infected with Chagas disease [5,10], causing 22,000 to 50,000 deaths per year [11,12].

In the last 20 years, many factors have contributed to geographic variation in the burden of Chagas disease [13–16]. For example, due to increased migration and international travel, Chagas disease is spreading to other geographic areas, mainly in Europe [17], Canada [18,19], New Zealand, and Australia [20]. Estimates suggest that 80,000 to 120,000 immigrants infected with *T. cruzi* live in Europe, and 300,000 live in the United States [21].

Chagas disease is integrated into an eco-epidemiological network, making it one of the most complex human vector-borne diseases [22]. Its causative agent, Trypanosoma cruzi, is a genetically diverse protist capable of infecting a broad spectrum of vertebrates, including humans and more than 100 other mammalian species [23–26].

This protist is primarily transmitted to its host community through the autoinoculation of contaminated feces from a diverse array of hematophagous insect species within the subfamily Triatominae (Hemiptera: Reduviidae) [27,28]. However, it can even spread by non-vectorial mechanisms, such as blood transfusion [8,29], congenital transmission [8,16], organ/tissue/cell transplantation [8,30], oral transmission because of food contaminated by infected triatomine feces [8,31,32], laboratory accidents [8], vertical or placental transmission, and through the birth canal at birth [33].

To address the complexity of the eco-epidemiology of Chagas disease, it has been essential to implement efficient control strategies that address the factors driving disease risk across spatial and temporal ranges. The establishment of novel control strategies is even more important knowing that there is a growing consensus that the disease is currently a global public health problem and will remain so for the foreseeable future. For example, the development of an effective vaccine or other improved and novel therapies is the next crucial step in the fight against Chagas disease [34–36]. It has also been essential to develop sustainable vector control strategies based on improved risk assessment of infection and disease [37,38] or to improve surveillance programs in all endemic regions [14,39]. Finally, another concern of relevance in the effort to combat Chagas disease is to maintain a sufficiently high level of political priority to promote and fund surveillance as well as research to avoid failures in the success of public health programs and a possible evolution of the disease [14,40,41]. The future success of the fight against Chagas disease depends on efficient management of newly emerging infectious foci, maintenance of high levels of public awareness and governmental interest in controlling the disease, and continuous improvements in diagnostic, therapeutic and surveillance tools [42]. The objective of this review is to describe the current status of strategies to control Chagas disease and highlight new research topics to reduce the burden of this public health problem.

Specifically, we address selected aspects that include the biology of *T. cruzi* to better understand the dynamics of its transmission, the identification of methods focused on the prevention and control of the disease developed by local governments, international organizations, public health and civil society with both scientific and traditional or home-made approaches, as well as the recognition of the main social and ecological factors that have historically played a role in favoring or reducing the presence of the disease vector in the Latin American region. Altogether, this will serve as a basis for proposing recommendations directed under an integral vision.

2. Review

2.1. *Trypanosoma cruzi Taxonomy*

The taxonomy of *Trypanosoma cruzi* remains a subject of ongoing debate, with increasing genomic and molecular evidence challenging traditional classifications. Currently, *T. cruzi* is divided into six discrete typing units (DTUs), TcI to TcVI, based on multilocus sequence typing and phylogenetic analyses. However, recent studies suggest a more complex population structure with high genetic diversity and potential hybridization events [43]. In this context, the development of a comprehensive phylogenetic framework for the Trypanosomatidae family, as proposed by Kostygov et al. [44], is crucial for exploring trait evolution and refining taxonomic classifications. Advances in whole-genome sequencing have significantly expanded our understanding of *T. cruzi's* genetic variability, particularly through large-scale genomic projects that aim to elucidate the mechanisms underlying its high genetic plasticity and host adaptation [45]. Comparative genomics studies have revealed extensive gene copy number variations, repetitive sequences, and large segmental duplications, which contribute to the parasite's adaptability and differential virulence among DTUs [46]. Moreover, transcriptomic and epigenomic approaches have provided insights into gene regulation and host–parasite interactions, highlighting the role of gene expression variability in pathogenesis and transmission cycles [47]. With recent technological advances, including long-read sequencing and single-cell genomics, ongoing projects continue to refine our understanding of *T. cruzi* diversity and its epidemiological implications [48].

A clearer understanding of *T. cruzi* taxonomy has direct implications for parasite transmission dynamics, diagnostics, and treatment strategies. The DTUs differ in their geographic distribution, host range, and clinical outcomes. For example, TcI is the most widely distributed DTU in the Americas and is associated primarily with synanthropic transmission cycles and chronic cardiomyopathy in humans [49]. In contrast, TcII and TcV are predominant in the Southern Cone and are linked to severe cardiac manifestations of Chagas disease [50]. Moreover, TcIV, which circulates in both North and South America, has been implicated in atypical Chagas disease cases, including oral transmission outbreaks in the Amazon region [51]. DTU distribution is also closely linked to ecological differences in host use and habitat preferences; for instance, TcI is frequently found in marsupials and armadillos in sylvatic cycles, whereas TcII, TcV, and TcVI are more commonly associated with a great variety of terrestrial mammals, including synanthropic rodents and domestic animals, facilitating their transmission in human environments [23]. These ecological associations not only shape the parasite's geographical distribution, but also influence the epidemiological risk for humans by determining the likelihood of spillover from wildlife to domestic transmission cycles. Understanding these host–parasite relationships is critical for designing effective control strategies as reservoir host diversity directly affects transmission dynamics and human infection risk.

DTU-specific differences also impact diagnosis and treatment. Variability in antigenic profiles between DTUs affects the sensitivity of serological tests, sometimes leading to false-negative results, particularly in cases involving TcI or hybrid DTUs such as TcV and TcVI [52]. From a therapeutic perspective, susceptibility to benznidazole and nifurtimox is known to vary among DTUs, with TcI strains often exhibiting higher resistance compared to TcII and TcVI [53]. Such variations highlight the need for DTU-targeted treatment approaches and the development of new chemotherapeutic options. Additionally, the application of omics tools, including cytogenomics and genomics, to the study of Chagas disease vectors has enhanced our understanding of vector biology and taxonomy, which is essential for developing effective control measures [54]. These findings emphasize the necessity for a more dynamic and integrative classification framework that accounts for the parasite's genetic complexity and potential cryptic diversity.

2.2. Vector Transmission

The transmission of *T. cruzi* in animals and humans occurs through hematophagous insect vectors whose contaminated feces come into contact with bite sites or mucous membranes. Currently, 144 species of triatomines have been described, with firm evidence that more than a half of them can carry *T. cruzi* [11]. Triatomines belong to the Reduviidae family with the most epidemiologically important genera being *Rhodnius*, *Triatoma* and *Panstrongylus*, also known as "kissing bugs" [32,55]. Although the transmission of the disease is mostly associated with triatomines in domestic habitats, some wild triatomine species have a relevant role in the transmission of *T. cruzi* to humans, especially in rural and peri-urban areas [41]. However, only a limited number of species are associated with environments close to humans and are considered of medical importance [56].

Triatoma infestans and *Rhodnius prolixus* are the most efficient and relevant vectors in the transmission of *T. cruzi* due to their tolerance to humans and their high adaptability to households. These two species vary in their habitats, between sylvatic, peri-urban and domestic environments. *T. infestans* and *R. prolixus* are predominantly domestic, although both species maintain sylvatic populations. In contrast, other species, such as *T. dimidiata* and *P. megistus*, are primarily sylvatic, with some populations exhibiting human tolerance or a high degree of domestication [14]. Triatomines feed mainly on the blood of humans, birds or mammals, although there are some reports that mention that some species can even feed on the blood of reptiles or amphibians. They have an incomplete metamorphosis since development includes the egg stage, five nymphal stages and adults, thus being considered hemimetabolous insects [57]. The nymphal stages include five developmental phases before reaching adulthood, characterized by a progressive increase in size and blood-feeding capacity. Nymphs emerge from the egg during the first stage, being small and vulnerable, feeding only once before molting. In subsequent stages, they grow larger and more mobile (second and third stages), acquiring a greater ability to feed on blood. During the fourth stage, nymphs reach a size close to that of adults, but lack sexual maturity. Finally, in the fifth stage, known as the pre-adult stage, they prepare for the final molt, during which they develop functional wings and fully formed sexual organs. These stages have variable durations depending on the species, temperature, and food availability [58,59].

Triatomines require at least one blood meal during the five nymphal stages in order to continue their development. The female may deposit hundreds of eggs during her lifetime. These eggs hatch approximately 10 to 30 days later, depending on the temperature and humidity of the environment [14]. Generally, triatomines feed during the night, although, under conditions of food stress, they may come out during the day in search of food. In laboratory-reared colonies, triatomines feed equally during the day and at night [55].

After feeding, they defecate near the wound, and if the host scratches or presents an open lesion, the parasite present in the insect's feces or urine can enter the body through the skin or mucous membranes, causing infection [60]. The main Latin American vectors, such as *T. infestans*, *R. prolixus*, *T. dimidiata*, *P. megistus*, *T. brasiliensis* and *T. sordida* and *R. pallescens*, defecate during or immediately after ingesting blood. It has been reported that most sylvatic triatomine species can colonize the nests of their food sources, establishing close associations with rodent and marsupial species, which are recognized as potential wild hosts. These hosts frequently move between human dwellings and forested areas, acting as reservoirs that facilitate the movement of *T. cruzi* between the sylvatic and domestic environments [61].

Population growth, migration and the high concentration of humans have seriously affected the natural mechanisms for vector control, related to environmental transformations, including factors such as climate change, land use change, uncontrolled urbanization and deforestation [62]. These have favored the redistribution and adaptation of Chagas vectors to new areas and environments and consequently led to an increase in the risk of parasite transmission to humans [63]. Therefore, vector control has become a challenge because its presence depends on social, biological and environmental interactions [60,62,64]. Both environmental impact and socioeconomic inequity drive parasite infection rates, and may determine its distribution and the mortality it causes [61].

In line with the above, the presence of the principal vectors of Chagas disease pathogens in people has historically been linked to dwellings established in rural areas. This type of domicile has been the main target for the detection of vector infestation processes. However, based on the multiple impacts caused at the ecosystemic level, which have consequently affected the natural dynamics of vectors, more and more urban and peri-urban infestations are being reported [65].

Therefore, traditional vector control and prevention programs have focused on the eradication of these organisms in rural dwellings due to their physical vulnerability. However, vector control programs now also exist at the urban level. In both rural and urban areas, home improvements at the infrastructure level are promoted, as well as health education, manual collections and mainly the use of chemical insecticides [66].

Poverty and infrastructure conditions promote the permanence of the vector because the homes, generally infested, are usually built with precarious materials (adobe, wood, thatched roofs) that have cracks or holes in which triatomines can hide and reproduce, as well as dirt floors [67]. These conditions provide them with viable shelters for nesting, which is why control and prevention methods suggest focusing on improving housing conditions, which can be classified mainly into physical and chemical methods, without leaving aside other methods such as biological control and genetic control [66].

2.2.1. Traditional or Home-Based Vector Control and Prevention Methods

Traditional or home-based methods have long been used by society, particularly in rural and indigenous communities, to prevent infestations of disease vectors such as insects and small mammals. These practices, rooted in local knowledge, include both physical and chemical approaches applied inside or around dwellings based on household needs. While their effectiveness may vary and often lacks comprehensive scientific validation, they remain valuable preventive measures, especially in areas with limited access to formal control programs [68–71].

One widely used approach in rural localities involves the application of insecticide sprays, which serve as a practical and accessible solution. Additionally, preventive measures emphasize household cleanliness, such as plastering floors, removing accumulated materials outside the house, and maintaining clean yards and corrals to minimize breeding sites for insects. These measures reflect a cultural association between vector presence and household hygiene [68,70].

Likewise, in Latin American rural communities, the role of women in vector control and insect prevention in the home is highlighted as they are responsible for carrying out these measures. For example, in the Guaraní indigenous communities of Gutiérrez, Bolivia, women play an active role in the prevention and control of the disease, highlighting that they are primarily responsible for carrying out the cleaning of homes, mainly the yard and corrals, in places identified as appropriate for vectors to inhabit [66]. In turn, women play a central role in household insect control measures in the villages of Bokoba, Teya, and Sudzal, located in the central part of Yucatan, Mexico, driven primarily by their motivation to protect their children. Their proactive efforts, including spraying insecticides, sealing windows and doors, and employing repellents, reflect their pivotal contributions to family health and well-being [70]. These traditional methods, while sometimes viewed as supplementary to formal control strategies, highlight the ingenuity and resourcefulness of communities in safeguarding their health and environments.

2.2.2. Physical Methods

Home improvement is one of the most effective ways to reduce the vector population, especially in rural areas. Examples include plastering walls and sealing crowded spaces as the vector hides in cracks and gaps in walls and ceilings; so, sealing walls and ceilings helps reduce its habitat. The type of roof and wall material is important, not because of the material itself, but because of the way it is used; adobe, cane and mud or stick and pique can enable the appearance of cracks that serve as protection for the vector. The same occurs with thatched or reed roofs [72].

Kissing bugs, due to their naturally nocturnal habits, tend to prefer dark environments, and thus lighting in homes with warm or blue lights, by keeping some lights on outdoors at night, while not eliminating the problem, may help deter these [73]. In contrast, another physical method of control is homemade UV or CO_2-emitting light traps. These are traps that are made from simple materials (such as plastic bottles) and emit CO_2 by mixing yeast and sugar, which attract bed bugs, as does UV light. Triatomines are attracted to certain wavelengths of light, especially in the ultraviolet (UV) spectrum. Light traps emit light in this range, which causes insects to approach and become trapped. These traps are especially useful in dark areas or at night, when insect activity is highest. Light traps are popular in endemic areas because they allow vectors to be captured economically and without pesticides [73]. Although these traps do not eradicate the infestation, they can help monitor the presence of these insects and reduce their proliferation.

Thermal control is also an important element in vector control since the triatomine bug is sensitive to extreme temperatures, and thus applying intense heat to places where they live can eliminate larvae and adults. *T. infestans* is able to tolerate a minimum temperature close to 0 °C, while *R. prolixus* does not tolerate temperatures below 5 °C. However, the difference is smaller at the maximum temperature, as both tolerate up to approximately 50 °C [73,74]. However, this is less common in mass control campaigns due to the difficulty of applying it effectively [73].

2.2.3. Chemical Methods

Although household insecticides (such as common insect sprays) are not as effective as residual insecticides, they appear to help if applied regularly to specific locations where triatomine bugs may hide [66,75]. There are sources of plant origin that seem to be useful for the control of kissing bugs, such as essential oils with repellent function, and there is knowledge of the effective toxicity of eucalyptus and neem (*Azadirachta indica*) oils on triatomine bugs [66]. Another example is cinnamon oil, which inhibits molting, and citronella oil, which causes insects not to feed, leading to their premature death. This is because plants produce secondary metabolites, many of which have an insecticidal or repellent effect. For example, the monoterpenes eucalyptol and geraniol have lethal effects on some species of kissing bugs. Other compounds, such as carvacrol, carveol, citronellol, eugenol, linalool, menthol, a-terpineol, thymol and verbenol, have repellent effects against triatomine bug nymphs [66,76].

Due to the volatile property of essential oils, they may have poor environmental stability, leading to a short period of effective control and subsequently increasing the cost of triatomine control [76]. Cleaning with vinegar and water solutions appears to be another effective home control method as vinegar acts as a repellent due to its odor and acidity. Acidic components, such as acetic acid, generate a strong odor that is unpleasant to many insects, disrupting their ability to detect olfactory cues, such as carbon dioxide or sweat, which often attract them to humans or animals [77].

The use of insecticides that have a prolonged effect, applied on walls, ceilings and cracks in and around dwellings, is common. Residual-action insecticides act for weeks or even months after application, eliminating bedbugs that come into contact with treated surfaces [78]. Some of the most common compounds are pyrethroids, such as deltamethrin and permethrin. They are effective and have low toxicity to humans and domestic animals, being the most widely used insecticides in vector control programs [75]. Organophosphates such as malathion, although their use has decreased due to their higher toxicity and lower residual effectiveness compared to pyrethroids [75], have historically also been used. Complete fumigation of dwellings and adjacent structures is especially useful for treating large areas in a short time but requires that people and domestic animals be temporarily absent from their homes to avoid exposure [78].

Although the use of baits is the least common method in eradication campaigns, some bedbug traps use insecticides in the baits. When the vectors are attracted and come in contact with the baits, they die from poisoning. This method is more common in research or as a control measure in small areas [79]. Outdoor fogging is recommended only in situations of high risk for disease transmission, where the intensity of the infestation is very high or uncontrollable by traditional methods and the vulnerability of the community justifies the use of such an intensive technique [70].

In recent years, there has been increasing research on the use of entomopathogenic fungi, nematodes, viruses, endosymbionts, and parasitoids as biological control of vectors [75,80]. Biological control of a vector refers to the use of living organisms, such as predators, parasites, pathogens, or competitors, to reduce the population or limit the ability of a vector to transmit disease [41]. The use of two fungal agents that have exhibited mortality in the nymphal stages of the triatomine bug has been proposed. Metarhizium anisopliae and *Ixora javanica* have been widely used as effective biological control agents with chinch bugs in agricultural environments [80]. Both entomopathogenic fungi are an effective alternative to control adults and immature stages of one of the main vectors of Chagas disease [80].

Research on the effects of entomopathogenic nematodes against triatomines is limited. Entomopathogenic nematodes have an almost unique combination of desirable attributes that make them successful bioregulators [81]. Two nematode genera, *Heterorhabditis* and *Steinernema*, in a nematode–bacterium complex (*Photorhabdus luminescens*), have demonstrated the ability to kill *T. infestans* and *R. prolixus* under laboratory conditions [75,81]. Heterorhabditis indica could be a viable biological control for adults of *T. dimidiata*, especially in areas with resistance to chemical insecticides. However, further studies are needed to evaluate its effectiveness and specificity under field conditions and its impact on non-target insects [82]. *Triatoma* virus (TrV) is a non-enveloped virus of the family Dicistroviridae [83,84]. Although there is no evidence to suggest that TrV suppresses *T. cruzi* in triatomines, it can induce high mortality in *T. infestans* and subsequently reduce populations, which is promising for use as a biological control for triatomines and Chagas disease [75].

Endosymbionts as genetic control is another way to achieve vector population control or make populations refractory to the transmission of disease-causing agents [75]. Some entomopathogenic endosymbionts have shown potential, mainly nematode-associated bacteria, such as *Photorhabdus luminescens* and *Xenorhabdus* spp., which live in symbiosis with nematodes of the families Heterorhabditidae and Steinernematidae. These endosymbionts release toxins into the vector's digestive system, as in the case of *T. infestans*, producing high mortality [85]. The field release of genetically modified bacteria is risky. An additional barrier to the implementation of this intervention is the possibility of undesirable environmental impacts. Paratransgenesis can threaten the diversity of the microbiome and influence horizontal gene transfer [75]. Parasitoids are organisms, which spend a significant part of their life cycle in or on the body of a host, eventually causing its death, and unlike parasites, do not cause the death of the host [73,86]. Multiple parasitoid species of triatomines have been reported, such as *Telenomus fariai Costa Lima* (Hymenoptera: Scelionidae), *T. costalimai*, *Megaselia scalaris* and *Aprostocetus asthenogmus* [75].

In this context, the fight against the vector of Chagas disease, such as *T. infestans* and other triatomine species, is generally led by international agencies such as the World Health Organization (WHO), Pan American Health Organization (PAHO), Drugs for Neglected Diseases Initiative (DNDi), national governments, public health organizations and community initiatives that conduct specific programs for the control and surveillance of Chagas disease [87–93]. An example of this is the formalization of epidemiological and entomological surveillance activities for Chagas disease at the national level through a program governed by the Ministry of Public Health in Ecuador [15]. These programs are crucial for the development of surveillance activities and provide insight into the current situation of countries in the fight against neglected diseases such as Chagas disease.

However, vector-borne transmission of Chagas disease has not been effectively controlled in many regions of Latin America, and the application of strategies to suppress vector populations has not been fully effective. A case study developed in the Gran Chaco ecoregion, Argentina, through community spraying and intensive monitoring for about 3 years shows that selective insecticide spraying, previously recognized as an effective method of control, did not perform efficiently in the detected vector foci. Therefore, the authors highlight the need to improve not only management and surveillance actions at the household level, but also to reduce the operational, economic, and political constraints that contribute to increase habitat suitability for the vector [94].

Current control programs in Latin American countries have focused on this method together with health education for the interruption of domestic and peridomestic cycles of transmission involving vectors [95]. A success story is the control program implemented in Brazil since 1975, which has led to a significant reduction in the *T. infestans* vector, with only 485 insects nationwide in 1998 [95]. However, although much of the Southern Cone countries have focused their efforts on the surveillance phase and focal spraying, when necessary, several differential factors converge to maintain the infestation of dwellings and consequently the transmission of the disease [93,95], such as re-infestation of dwellings by triatomines [91], resistance to pyrethroid insecticides in *T. infestans* [93,96], and non-vectorial transmission [8,16,29,31].

2.3. Non-Vector Transmission

Chagas disease can also spread by non-vectorial mechanisms, such as blood transfusion [8,28], congenital transmission [8,16], organ/tissue/cell transplantation [8,30], food contaminated by infected triatomine feces (oral transmission) [8,31], laboratory accidents [8], vertical or placental transmission, and through the birth canal at birth [33]. Blood transfusion and the transplacental or breast milk route are the main transmission mechanisms in non-endemic countries [57]. The risk of blood transfusion has decreased due to screening in all blood banks in Latin American countries [97].

Maternal–fetal transmission of *T. cruzi*, also known as vertical transmission, occurs when live parasites persist and multiply in the fetus or after birth, leading to congenital infection, and can take place prenatally (in utero), perinatally (during delivery), or postnatally (after birth) [98]. A possible mode of maternal–fetal transmission could be through parasites released in the amniotic fluid. The latter could contaminate fetuses via oral or pulmonary routes, or eventually by cutaneous penetration (fetuses bathe in amniotic fluid and continuously absorb it) [98]. Transmission through contact with equipment or handling of materials infected with viable parasites in research or clinical laboratories may represent a non-vectorial source of risk for human infection with *T. cruzi*. Because they can occur in many circumstances and can often go undetected or undiagnosed, laboratory-acquired cases of *T. cruzi* infection are likely to be underreported [98].

Finally, oral transmission of *Trypanosoma cruzi* has emerged as a significant route of Chagas disease transmission, particularly in regions where vector control measures have reduced traditional vector-borne transmission. The primary causes of oral transmission include the ingestion of contaminated food or beverages containing *T. cruzi*-infected triatomine feces, crushed triatomine insects, or contaminated secretions from infected mammalian reservoirs [31,99]. This mode of transmission is most frequently reported in the Amazon and other forested areas of South America, where the parasite is maintained in complex sylvatic cycles involving a diverse range of wildlife reservoirs and vector species [100]. Freshly prepared fruit juices, particularly açaí pulp, sugarcane juice, and other unprocessed food items, have been implicated in multiple outbreaks, highlighting the role of food hygiene and processing in disease prevention [101,102].

Poor sanitary conditions, inadequate food handling practices, and limited awareness contribute to the risk of contamination, particularly in endemic rural and peri-urban areas [103]. Additionally, oral transmission can occur through the consumption of raw or undercooked meat from infected wild mammals, including marsupials and rodents, which serve as natural reservoirs for *T. cruzi* [104].

A comprehensive review identified 32 outbreaks of orally transmitted Chagas disease between 1965 and 2010, with a notable increase in cases reported between 2006 and 2010 [105]. These outbreaks, primarily occurring in Brazil, Colombia, and Venezuela, underscore the growing importance of oral transmission in the epidemiology of Chagas disease. The rising frequency of oral transmission cases emphasizes the need for enhanced surveillance, food safety regulations, and public health campaigns to mitigate this overlooked yet growing route of Chagas disease transmission.

Methods of Control and Prevention of Non-Vector Transmission

The prevention of non-vectorial transmission of Chagas disease requires strategies to minimize the risk of infection through alternative routes, such as blood transfusions and organ transplants. Mandatory screening of blood and organ donations using specific tests to detect *T. cruzi* is essential in endemic regions and countries receiving migrants from affected areas, as it ensures the exclusion of positive donors, significantly reduces the risk of transmission, and requires careful monitoring of both donors and recipients in organ transplants [106]. Recent studies indicate that the implementation of comprehensive screening programs in healthcare systems has nearly eliminated the transmission of *T. cruzi* via transplantation in structured settings [106,107]. These preventive measures, combined with community education and awareness, play a vital role in controlling the spread of Chagas disease [5,98].

In the case of congenital transmission, early diagnosis in pregnant women, especially in high-prevalence areas, is essential. Timely treatment of infected neonates is highly effective and prevents long-term complications [98]. Women of childbearing age from endemic regions should undergo screening, and antiparasitic treatment should ideally be completed before pregnancy to reduce the risk of fetal infection [5]. Neonatal screening is equally important for early detection and treatment in infants born to infected mothers. Examining neonates from infected mothers using PCR or microhematocrit tests ensures the early identification and treatment of infections, which is crucial for preventing disease progression [107,108]. Moreover, follow-up care for children of HIV-positive mothers ensures the identification and management of vertically acquired infections [96]. Both benznidazole and nifurtimox are highly effective when administered during the first year of life for treating congenital Chagas disease [32]. The effectiveness of these interventions is supported by evidence indicating that treatment prior to pregnancy reduces congenital transmission nearly to 100%, while neonatal screening and prompt treatment in infected infants achieve similar rates of cure [5,106].

Consumption of food or beverages contaminated with infected material, although usually less frequent, represents another important avenue of prevention. Preventive measures include ensuring proper food hygiene, the use of safe drinking water, along with vector control in food production areas, and the safe processing of products such as cane or fruit juices, particularly in rural areas of Latin America. Maintaining strict food and beverage hygiene, coupled with public awareness campaigns, has proven effective in reducing cases. [109].

Finally, in laboratory settings, the implementation of strict biosafety standards and specialized training for handling potentially infectious samples significantly reduces the risk of accidents that can result in occupational transmission. These measures include the use of appropriate personal protective equipment, standardized procedures for handling and discarding contaminated material, and ongoing training of personnel in safe practices [110]. In addition, strengthening health systems through community education and access to modern diagnostic resources is essential to address non-vectorial transmission comprehensively and effectively [109]. These strategies not only decrease the incidence of new infections, but also contribute to a broader control of Chagas disease in diverse epidemiological contexts.

2.4. Host

As part of the cycle, in addition to the vector, the host plays a fundamental role in the maintenance of the disease. The relationship between *T. cruzi*, the causative agent of the disease, and its hosts is fundamental to understanding the epidemiology and transmission of the disease.

The hosts of Chagas disease are diverse and include both wild and domestic animals, as well as humans [3]. In the domestic cycle of the disease, the main hosts include dogs, cats, farm animals (cattle, sheep, goats, pigs) and poultry [67]. Dogs and cats are easily infected and can harbor the parasite in the bloodstream, allowing triatomines to become infected by feeding on them [110]. Dogs are considered the most important peridomestic reservoirs as they develop symptoms similar to humans and can increase the risk of transmission if they are in close contact with bed bugs inhabiting dwellings or nearby areas.

Infection in cats has also been identified, although it is less studied [111]. Farm animals can be infected by *T. cruzi*, especially in rural areas where these animals coexist with triatomines. While their role in direct transmission to humans is minor, they contribute to the transmission cycle by enabling the feeding of infected and uninfected triatomines [41]. Although birds are not direct hosts of *T. cruzi*, because the protists cannot survive in them, they may play an indirect role. Triatomines often feed on birds, which allows them to survive and thrive in rural and urban areas. In this way, birds can help maintain triatomine populations near homes and in contact with other animals that can be infected [41].

To date, 37 species of wild mammals belonging to the orders Marsupiales, Cingulata, Edentata, Chiroptera, Carnivora, Arthiodactyla, Rodentia and Primates have been confirmed positive for the disease. These include animals inhabiting sylvatic or peridomestic environments, including rodents, raccoons (*Procyon lotor*), and opossums (*Didelphis virginiana*), as well as Ocelots (*Leopardus pardalis*), coatis (*Nasua narica*) [4], and bats [112,113]. These animals act as key reservoirs, as the parasite can persist in their organisms without causing them great effects, thus facilitating transmission through hematophagous vectors that infect them and then transmit the parasite to other hosts, including humans [41].

Since rodents are at the base of many food chains and inhabit a wide range of environments, they are critical links between the sylvatic and domestic transmission cycle of *T. cruzi*. Their importance is due to biological, ecological and behavioral factors, which make them efficient reservoirs in different environments, making them relevant species for understanding the transmission dynamics of the disease and foci of attention in control and surveillance programs [114]. Due to the limited availability of studies on infection in wildlife, the role of *T. cruzi* may be underestimated. Although wildlife species play important roles as sylvatic reservoirs, investigations of *T. cruzi* pathology in wildlife are limited to a few studies documenting histological lesions in opossums and raccoons [115]. It has been found that infected animals such as coyotes (*Canis latrans*) can show cardiac inflammation; however, it has not been attributable to Chagas disease, which may lead to severe underestimation due to the existence of unknown mammalian hosts [94].

2.5. Experimental Drugs and Plant-Based Drugs for Treatment and Prevention

Experimental drugs are pharmaceutical compounds that have been authorized by the United States Food and Drug Administration (FDA) for human testing in specific conditions but have not yet received approval for commercial distribution or widespread use. These drugs undergo rigorous evaluation through clinical trials, which are structured research studies involving human participants to assess their safety, efficacy, and potential therapeutic benefits. If the outcomes of these trials confirm their safety and effectiveness, regulatory agencies may approve them for market release [116,117]. Similarly, experimental vaccines are immunological preparations that are still under investigation to determine their capacity to induce protective immunity against specific pathogens. These vaccines progress through preclinical and clinical phases before they can be approved for general use. On the other hand, plant-based drugs are therapeutic agents derived from natural plant sources, utilizing bioactive compounds identified through traditional medicine and scientific research. These compounds have demonstrated potential for treating various diseases, including neglected tropical diseases such as Chagas disease, by targeting key biological pathways involved in pathogen survival and host–pathogen interactions [118].

2.5.1. Dewormers (Human and Animal)

In some regions, domestic animals, such as dogs, have been found to play an important role as a reservoir of the parasite and its transmission to humans [36,119]. Therefore, infection control in these animals has been proposed as an important complementary strategy to vector control, for example, by using collars impregnated with insecticides such as deltamethrin (DTDC) [120], and topical application of the combination of fipronil and permethrin to vector-exposed dogs [121]. The topical formulation of imidacloprid was effective against pyrethroid-resistant *T. infestans* populations when applied to pigeons [122] at the laboratory level. Likewise, xenointoxication has been suggested for triatomine control [123], with products such as ivermectin applied to dogs [75,124], cypermethrin in poultry chickens [125,126] and the use of fluralaner for vector control in both chickens and dogs [126–129].

In the treatment of Chagas disease at the human level, two drugs are currently used: Nifurtimox and benznidazole, effective for the acute phase [130] and congenital cases [131]. Benznidazole (which acts by damaging the parasite's mitochondrial DNA) and Nifurtimox (which induces oxidative stress, inhibiting parasite growth) are the only drugs with proven efficacy [58]. However, in recent years, it has been demonstrated that benznidazole, in the treatment of patients with chagasic cardiomyopathy, can work very well in curing patients during the early stages of the disease, especially newly infected children and young adults [132].

2.5.2. Vaccination Projects

The first attempts to develop a Chagas disease vaccine started by inoculating mice with strains of *T. cruzi* in 1913, just four years after the discovery and report of the parasite causing the disease [133]. In these assays, the mice received only partial immunity, with a reduction in mortality rate but persistent parasitemia. Since then, the development of a vaccine for Chagas disease has faced several challenges, and some of the main issues to be addressed (but not limited to) include the complex parasite life cycle (different stages and multiple forms of the parasite), a limited understanding of the protective immunity mechanisms (precise immune response required to protect against the disease is not fully understood), a lack of predictive animal models (mice do not reflect the human disease, dogs are better models, but may not fully recapitulate the dynamics of the disease in humans) and limited commercial interest (primarily affects marginalized and economically disadvantaged populations in Latin America) [36,134–136].

Vaccines have been developed to date, including trials of attenuated vaccines, DNA vaccines, viral vectors, recombinant protein or peptide vaccines, glycoconjugates, multivalent vaccines, heterologous vaccines and mRNA vaccines [137–140]. An extensive and very comprehensive comparison amongst the known vaccine trials, phase of the disease analyzed, time of vaccine initiation in the case of therapy, the animals used, the strain of *T. cruzi* and doses used, and the level of significance observed for each outcome was reviewed first by Quijano Hernández and Dumonteil [36], and updated by Farany and collaborators [135].

Given the need to improve the treatment of the disease and provide timely care of the infection, the search for new drugs and, more recently, the development of vaccines that are effective for its control at all levels of infection caused by *T. cruzi* has been crucial. It has been estimated that the advantages of having a vaccine that prevents Chagas disease are significant not only in terms of public health, but also in terms of economic and social development [98]. In this sense, the development of proposals for the creation of vaccines has been a significant advance but so far limited to a few emerging studies. Nevertheless, these findings have allowed for the recognition of certain characteristics and chemical compounds of importance, which provide a positive response for the prophylactic treatment of Chagas disease.

Among these advances, the development of vaccines could help control the infection and pathology of the disease [141]. For example, it has been recognized that anti-*T.cruzi* vaccines should have certain characteristics at the molecular level that make them suitable candidates for incorporation. These characteristics include being expressed in all stages of *T. cruzi*, being conserved and expressed in its different strains, and being crucial for the development and survival of the pathogen [139,142]. In this scenario, the cruzipain enzyme has been considered a target for developing prophylactic treatment because it is found within the three main developmental stages of *T. cruzi* in all strains that have been tested, thus being crucial to achieve internalization into host cells [143]. Likewise, the development of an effective vaccine against *T. cruzi* must consider the high levels of genetic variability in these protists due to the restrictions in protective efficacy reported in some strains [144,145] that may disrupt vaccine efficacy.

Studies at the molecular level have led to an important basis in the recognition of other key molecules such as the Tc52 molecule, a glutathione S-transferase for the development of prophylactic vaccines, since this molecule is vital for the survival of *T. cruzi* as it is expressed during all phases of development, although mainly in the replicative stages: epimastigote and amastigote. Immunization based on this molecule enabled the production of anti-Tc52 antibodies, thus showing defense against infection [139]. In addition, it has been evaluated that the enzyme prolyl oligopeptidase Tc80 could be considered another candidate to be used as a prophylactic since it is expressed in the blood and amastigote trypomastigote, allowing for invasion into host cells as it participates in the degradation of the extracellular matrix, collagen and fibronectin [137]. The use of this molecule elicited a strong immune response both humoral and cellular since anti-Tc80 antibodies caused enzymatic inhibition and lysis of trypomastigotes by activating the complement system [132].

Who Is the Vaccination Aimed at? Examples of Animal Experiments

In line with the above, the development of preventive and therapeutic vaccines against *T. cruzi* evaluated in mice has led to the recognition and testing of various formulations, mainly based on the application of whole organisms (live attenuated *T. cruzi* and live *T. rangeli*), targeting reservoir species as a means to alleviate the burden of disease [146,147], recombinant proteins that induce immunity [143,148–154], viral vectors [155–157], heterologous primary drivers [158,159], and DNA [148,160–163]. These findings have shown diverse results in the immune response and parasitemia of the disease; in most cases, survival was increased and, in a few cases, it was variable according to the challenge route [36].

Certain therapeutic vaccines have also been evaluated very recently in rhesus macaques [161] and historically in mice [164]. Particularly, disease development in dogs presents a much more similar evolution to that in humans [36,134,165]. Therefore, dogs have been considered candidates to evaluate these vaccines [36]. The main formulations evaluated include recombinant proteins and DNA [162,166–168]. However, a single DNA-based formulation with the TSA-1 antigen, Tc24 and Tc52, has resulted in a decrease in parasitemia and parasite load in the heart, with an increase in survival [162].

The immunization of mice with heterologous Tc24 mRNA protein is a new approach based on a new generation of RNA-based vaccines that have demonstrated the ability to induce protective immunity, inducing strong antigen-specific CD8+ T cell responses and effective CD4+ T cell responses [146]. In this study, it was possible to verify that heterologous mRNA protein vaccination with Tc24 mRNA to prime and Tc24-C4 protein (a genetically modified cysteine-free polypeptide construct) to stimulate promotes a cellular immune response against *T. cruzi*, characterized mainly by an increase in the level of polyfunctional CD8+ T cells [169].

Another prevention strategy is the development of a canine vaccine to control the transmission and incidence of Chagas disease in endemic countries [170–176]. Given that dogs have been identified as a key host for the parasite and peridomestic reservoir of

Chagas disease, they play an important role in pathogen transmission to humans in several countries and epidemiological settings [110,169,177–179].

In Argentina, the application of a vaccine with attenuated parasites to domestic dogs and two boosters every two and fourteen months, demonstrated, by parasitological evaluation by xenodiagnoses, that vaccination reduced natural infection by *T. cruzi* from 26.7% to 12.3% after one year [170]. Also, in a trial with DNA vaccines, the authors found that the prophylactic effect of DNA vaccines induced moderate protection in immunized dogs by preventing cardiac chamber enlargement, poor contractility and heart failure [180]. Despite attempts in the development of a vaccine against Chagas disease, the results, which mainly focus on the immune response and cardiac damage, have been varied; however, none have prevented infection and the degree of protection has been mild to moderate [168,172,174,181].

Vaccines Linked to Chemotherapy

The current elective treatment of Chagas disease is oral benznidazole (BNZ), which is most effective in the acute phase of disease. However, BZN has dermatologic and neurologic side effects, and BZN can also produce liver toxicity due to the formation of reactive metabolites, which can damage liver cells, leading to liver injury [135]. A new approach to treat Chagas disease is the use of vaccine-linked chemotherapy, combining the antiparasitic drug with the vaccine antigen. This enables a reduction in the parasite load and an increase in the immune responses in the patient [182,183]. The studies focused on vaccine-linked chemotherapy evaluated the recombinant protein vaccine Tc24-C4 + TLR4 adjuvant in mice and showed a reduction in parasite burden and reduction in cardiac inflammation and fibrosis [182–184]. This strategy has been tested on acute and chronic infection models using the H1 strain of *T. cruzi*. This strategy allowed for a reduction in the dose of BZN, which will benefit the patients by minimizing the side effects of the treatment. The test of this regimen in a chronic model of infection showed improved cardiac structure and function and cardiac metabolism [153]. Similarly, the recombinant protein vaccine antigen TS + ISPA was evaluated in conjunction with the curative dose of BZN and showed a reduction in parasite burdens and reduced cardiac arrhythmias. The studies showed that combining vaccine-linked chemotherapy enhanced efficacy in BNZ treatment and offered a dose-sparing approach that may improve overall cardiac, liver health and clinical outcomes.

2.6. Environmental Factors Associated with Human Transmission

Chagas disease is endemic in vectors and wild reservoirs throughout the Americas, having spread from the southern half of the United States to Argentina. To date, three main types of transmission cycles for *T. cruzi* have been described, including the sylvatic, domestic, and peridomestic cycle. The domestic cycle is characterized as being maintained or established between interactions by humans, domestic animals such as rats, dogs, and their vectors, mainly triatomine bugs that have adapted to human dwellings [185]. This cycle is considered the most important for maintaining the circulation of the pathogen in human populations because it has concentrated the disease burden in human communities [186].

2.6.1. Sylvatic vs. Domestic Cycle and Its Relationship with the Environment

The sylvatic cycle involves wild mammals such as rodents and marsupials that are infected by individuals from wild triatomine populations [187]. This sylvatic transmission cycle includes raccoons, opossums, skunks, and wild and captive neotropical nonhuman primates from the New and Old World. Considering that this cycle has not been evaluated as important in the generation of human infections, current environmental changes (deforestation, agricultural and livestock expansion) are altering the dynamics of *T. cruzi* transmissions.

There is a type of cycle that has been considered a bridge between the other two cycles of sylvatic and domestic transmission. This cycle arises as a result of the origin of sylvatic transmission of the infection and is maintained by domestic animals in areas close to human dwellings through the action of peridomestic triatomines (for example, dogs and cats that hunt wild animals or wild animals that enter areas close to human dwellings) [188].

Recent studies have shown that there is a direct relationship between the impact of deforestation on ecosystems and the incidence of the disease [1], the favoring of wild vector populations [189], disease prevalence, and its association with vector abundance [190,191]. On the other hand, the impact on natural ecosystems generated by changes in land use has also led to effects on host infection rates, mainly rodents [192]. In particular, a predictive study developed in Mexico on the potential changes in the range of disease transmission cycles as a function of changes in land use, land cover and climate change shows a significant expansion of the range of the disease in urban environments, especially because the transmission cycles of *T. cruzi* are adapting to human dwellings and domestic species that may be their main food sources under this scenario [193].

These studies highlight a pronounced need for disease prevention, better monitoring of the management of environmental variables such as landscape disturbance, allowing for close monitoring of the impacts of its transformation on human health [1]. Prioritization of strategies for surveillance, control and involvement of the environmental sector to strengthen and carry out joint efforts are needed [194]. Changes need to be driven in socioeconomic and development policies around sustainable objectives to enact disease prevention and avoid an increase in *T. cruzi* transmission cycles in suitable habitats [193].

An increasing environmental challenge is the proximity of human settlements, both urban and rural, to natural habitats. This has led to multiple difficulties in the long-term control of wild populations of the vector, which frequently infest dwellings near fragmented habitats [167]. As a consequence, these conditions have especially favored the potential exchange between wild triatomines and mammals carrying *T. cruzi* [195], giving rise to sylvatic–domestic interfaces of the disease. This scenario not only highlights the relevance of ecosystem conservation, but also the need for timely host monitoring, vector control and prevention measures against possible interactions between sylvatic and domestic cycles, such as those that have been previously documented in a future context of changes in land use and climate change [193].

2.6.2. Agricultural and Livestock Expansion

In particular, Chagas disease has been favored by agricultural and livestock expansion by fragmenting and transforming natural habitats. These events have allowed for the movement of vectors and their wild reservoirs, such as rodents and marsupials, into inhabited areas, increasing their contact with humans and domestic animals, and facilitating the transmission of *T. cruzi* [41]. This land use change also creates sylvatic–domestic interfaces, where triatomines can move between domestic and wild animals, enhancing the risk of infection. As in fragmented landscapes, where plants that are suitable refuges for disease vectors can be found, agricultural infrastructures, such as corrals and barns, play an important role as shelter and food to triatomines, promoting their establishment in inhabited areas and facilitating the colonization of dwellings by vectors and domestic transmission [61].

2.6.3. Climate Change

Climate change has the potential to alter or expand the natural ranges of vectors, hosts, and pathogens and to convert some previously uninhabitable regions into suitable habitats [196–199]. Although triatomines prefer dry environments, moisture in certain locations can create favorable conditions for their refugia. Moist locations, such as areas near sources of standing water, can support the growth of plants and structures that provide shelter for these insects. Social and ecological factors interact together to influence the presence of the vector and the spread of Chagas disease. The integration of social and educational interventions with environmental control strategies is key to effectively addressing the disease.

2.6.4. Factors Favoring the Presence of the Vector or the Disease

Social vulnerability influences the presence of the disease, showing a higher prevalence even in the low abundance of infected vectors [142]. Lack of basic services such as drinking water, sanitation and electricity can contribute to unsanitary conditions that favor the presence of triatomines. However, nowadays, with the invasion of wildlife habitats caused by accelerated urban growth, it is no longer considered a rural disease.

It is estimated that if urban growth continues, in the future, without adequate control of the disease, it may become a more acute public health problem in densely populated cities. Domestic mammals (dogs, cats) and species such as rats could play an important role as blood sources for triatomine bugs in the absence of wild mammals. The low level of awareness may lead to people not being aware of the necessary preventive measures or the importance of hygiene and housekeeping.

Cultural practices that favor close coexistence with domestic or wild animals, which may serve as a reservoir for triatomines, may increase the risk of infestation. Likewise, lack of cooperation with vector control programs due to cultural beliefs or distrust of health authorities can reduce the effectiveness of control measures. Migration and displacement of communities due to work, education, conflicts or natural disasters lead to the spread of the disease to areas where vectors were not present, thus facilitating the spread of the disease.

2.6.5. Factors That Reduce the Presence of the Vector or the Disease

The intervention of public health programs involves access to education campaigns, where educational programs are provided to inform the population about Chagas disease, its vectors and preventive measures, increasing participation in control efforts. Housing improvement programs, such as rehabilitation and construction with appropriate materials, help to reduce the available shelters for the vector. Availability of access to health services for early diagnosis and treatment is critical. Improved sanitation and waste management, such as garbage disposal and reduction in shelters for triatomines, is one way to control the spread of the disease.

Fumigation and vector control programs, implemented regularly and effectively, can significantly reduce the triatomine population. These programs often require the active participation of communities to be effective. Government policies that support vector control, disease education and housing improvement can be very effective. Implementation and enforcement of these policies help create a more favorable environment for disease control. Proper surveillance and monitoring of vector presence and disease, in conjunction with research into new methods of control and prevention, are critical to disease and/or vector control.

2.7. Integral Recommendations for Chagas Disease Prevention

2.7.1. Challenges of Wild and Domestic Triatomines in the Control of Chagas Disease

The control and eventual eradication of Chagas disease face significant challenges due to the complex interactions between wild and domestic triatomines. Wild triatomines remain a persistent obstacle as they exhibit high adaptability, enabling them to reinfest areas where domestic transmission has been previously reduced [195]. For instance, studies in the Gran Chaco region have shown that wild triatomines recolonize areas even after extensive vector control efforts, driven by their ability to exploit ecological disruptions such as deforestation and agricultural expansion [199]. This adaptability complicates control measures and underscores the need for continuous monitoring and innovative strategies. Their ecological behavior and wide distribution across diverse habitats—from forested ecosystems to peri-urban environments—pose a continuous risk for *T. cruzi* transmission.

Similarly, domestic triatomines present challenges linked to their close association with human dwellings and domestic animals, particularly in rural and peri-urban regions. Poor housing conditions, such as cracks in walls, thatched roofs, and inadequate sanitation, create ideal environments for their colonization and reproduction [67]. For instance, sealing cracks and crevices with cement, replacing thatched roofs with durable materials like metal sheets, and installing fine-mesh screens on windows and doors have been proven effective in minimizing vector entry and habitation. Additionally, ensuring proper sanitation and waste management in and around homes further disrupts triatomine breeding grounds. Additionally, the coexistence of domestic animals (e.g., dogs, chickens, and pigs) serves as a reservoir for *T. cruzi*, perpetuating transmission cycles within households [200].

2.7.2. Challenges Demanding an Integrated and Differentiated Approach to Chagas Control

Wild triatomines: Specific control strategies must focus on detailed studies of the biology, ecology, and population dynamics of wild triatomines to understand dispersal patterns and reinfestation risks. The use of geospatial tools, such as GIS-based mapping, can help identify areas of greatest risk and design targeted interventions. Strategies may include environmental management, monitoring sylvatic populations, and mitigating land-use changes that increase human–vector contact.

Domestic triatomines: Control measures should prioritize housing improvement programs, including sealing entry points, replacing thatched roofs with solid materials, and applying insecticidal treatments where necessary. Community-based surveillance systems can play a key role in detecting and addressing infestations early. Moreover, efforts to manage domestic animal reservoirs through health checks, vaccination development, and exclusion from sleeping areas are essential.

2.7.3. Effective Results and Responsible Organizations

Successful control of vector transmission has been demonstrated through sustained, region-specific interventions. For example, in Brazil's northeastern region, the implementation of housing improvement programs combined with community surveillance has significantly reduced triatomine infestations [197]. These efforts, led by partnerships between local health departments and international organizations, provide a model of effective collaboration and highlight the importance of tailored strategies in endemic areas. For instance, community engagement programs in endemic areas have effectively reduced vector densities by over 70% in targeted regions. Similarly, housing improvement initiatives, such as those supported by international organizations like the PAHO and local health ministries, have significantly reduced domestic infestations. These efforts highlight the critical role of partnerships among government health agencies, academic institutions, and non-governmental organizations (NGOs) in achieving measurable outcomes. Strengthening collaboration and resource allocation among these stakeholders remains crucial to sustaining these achievements [201].

2.7.4. Importance of Joint Community and Professional Preventive Strategies

Community-based preventive strategies are fundamental for the sustainable control of Chagas disease; however, they must be complemented by professional interventions to ensure long-term effectiveness. Effective integration can be achieved through initiatives that combine local knowledge with scientific expertise [202]. For example, while community members can be trained to identify and report triatomines, entomologists and health professionals can support these efforts by providing advanced diagnostic tools and conducting targeted insecticide applications. Additionally, joint workshops that bring together community leaders, health workers, and researchers can foster mutual understanding and the co-creation of tailored strategies that address both local practices and scientific insights. By aligning grassroots actions with professional interventions, these integrated approaches can enhance the sustainability and efficacy of Chagas prevention efforts. Home-based methods, such as the use of natural repellents and house cleaning, while valuable, cannot substitute systematic and scientifically based approaches. These include regular monitoring for the presence of triatomines, and the strengthening of entomological surveillance systems. To achieve comprehensive prevention, it is essential to implement joint strategies that bridge professional expertise with community empowerment. These strategies must consider the following key dimensions:

Cultural aspects: Tailor health education programs to local customs, beliefs, and languages to ensure that preventive measures resonate with the affected populations. For example, involving community leaders and respecting cultural practices around housing construction and maintenance will improve acceptance and adoption of interventions.

Scientific and technical solutions: Ensure that professional interventions, such as targeted insecticide applications and environmental management, are based on scientific evidence and adapted to local ecological conditions. Combining these with innovations such as geospatial tools for vector mapping and monitoring will enhance the precision and impact of control efforts.

Social empowerment: Promote active community participation through educational programs that emphasize the role of families and individuals in preventing infestations. Empowering local populations to identify and report triatomines, maintain clean and vector-proof homes, and manage domestic animals effectively creates a sense of ownership over disease control.

Political and governmental leadership: Advocate for strong political will and governmental commitment to support Chagas control programs. This includes securing consistent funding, establishing policies that prioritize housing improvements, and integrating Chagas disease prevention into national and regional health agendas. Additionally, fostering intersectoral collaboration among health, education, housing, and agricultural ministries will ensure coordinated and sustained interventions.

Economic considerations: Invest in infrastructure improvements, such as housing rehabilitation programs to eliminate cracks and replace vulnerable materials (e.g., thatched roofs), which serve as habitats for domestic triatomines. Allocating resources for community surveillance systems and local health workers can ensure regular monitoring and early detection of vector presence.

3. Conclusions

The control of Chagas disease remains a significant challenge due to the complex interplay of factors involving wild and domestic triatomines, socio-economic conditions, and ecological dynamics. For example, socio-economic conditions such as poverty, inadequate housing, and limited access to healthcare contribute to persistent domestic vector presence, while ecological factors like deforestation and urban expansion disrupt natural habitats, increasing human–vector contact. While substantial progress has been made in reducing transmission in many regions, the persistence of wild vectors, poor housing conditions, and inadequate health infrastructure continue to hinder eradication efforts. To effectively address these challenges, integrated approaches combining vector control, housing improvements, community education, and scientific innovation are essential. Sustained collaboration among governments, academic institutions, non-governmental organizations, and communities will be critical to achieving the long-term goal of Chagas disease elimination. By fostering intersectoral partnerships and promoting the development of novel tools and strategies, the global health community can ensure more effective and sustainable control efforts.

Author Contributions: Conceptualization, A.A. and A.C.; Writing—Original Draft Preparation, A.H.-F., D.E.-D. and B.C.-C.; Writing—Review and Editing, A.C., A.A., C.N.I.-C. and D.M. All authors have read and agreed to the published version of the manuscript.

Funding: This research received no external funding.

Institutional Review Board Statement: Not applicable.

Informed Consent Statement: Not applicable.

Data Availability Statement: Not applicable.

Acknowledgments: We would like to express our deepest appreciation of the members of the research team for their invaluable contributions and collaborative spirit.

Conflicts of Interest: The authors declare no conflicts of interest.

References

1. De Souza, R.D.; Gorla, D.E.; Chame, M.; Jaramillo, N.; Monroy, C.; Diotaiuti, L. Chagas disease in the context of the 2030 agenda: Global warming and vectors. *Mem. Inst. Oswaldo Cruz* **2022**, *117*, e200479. [CrossRef] [PubMed]
2. Cucunubá, Z.M.; Gutiérrez-Romero, S.A.; Ramírez, J.D.; Velásquez-Ortiz, N.; Ceccarelli, S.; Parra-Henao, G.; Henao-Martínez, A.F.; Rabinovich, J.; Basáñez, M.G.; Nouvellet, P.; et al. The epidemiology of Chagas disease in the Americas. *Lancet Reg. Health–Am.* **2024**, *37*, 100881. [CrossRef] [PubMed]
3. Chagas, C. Nova trypanozomiaze humana: Estudos sobre a morfolojia e o ciclo evolutivo do *Schizotrypanum cruzi* n. gen, n. spec, ajente etiolojico de nova entidade morbida de hörnern. *Mem. Inst. Oswaldo Cruz* **1909**, *1*, 159–218. [CrossRef]
4. Bern, C.; Kjos, S.; Yabsley, M.J.; Montgomery, S.P. *Trypanosoma cruzi* and Chagas' disease in the United States. *Clin. Microbiol. Rev.* **2011**, *24*, 655–681. [CrossRef]
5. Pan American Health Organization. *Guidelines for the Diagnosis and Treatment of Chagas Disease*; PAHO/WHO: Washington, DC, USA, 2019. Available online: https://www.paho.org/en/topics/chagas-disease (accessed on 21 October 2024).
6. Forsyth, C.; Agudelo Higuita, N.I.; Hamer, S.A.; Ibarra-Cerdeña, C.N.; Valdez-Tah, A.; Stigler Granados, P.; Hamer, G.L.; Vingiello, M.; Beatty, N.L. Climate change and *Trypanosoma cruzi* transmission in North and central America. *Lancet Microbe* **2024**, *5*, 100946. [CrossRef]
7. Prata, A. Clinical and epidemiological aspects of Chagas disease. *Lancet Infect. Dis.* **2001**, *1*, 92–100. [CrossRef]
8. Rassi, A., Jr.; Rassi, A.; Marcondes de Rezende, J. American trypanosomiasis (Chagas disease). *Infect. Dis. Clin. N. Am.* **2012**, *26*, 275–291. [CrossRef]
9. Castillo, M.D.L.A.; Cando, W.; Cando, J.; Palma Mera, F. Factores que Influyen en la Persistencia de la Enfermedad de Chagas en la Costa Ecuatoriana. *Reciamuc* **2022**, *6*, 687–695. [CrossRef]
10. World Health Organization. *Guidelines for Drinking-Water Quality: Fourth Edition Incorporating the First Addendum*; WHO: Geneva, Switzerland, 2017; pp. 1–240.
11. De Fuentes-Vicente, J.; Gutiérrez-Cabrera, A.; Flores-Villegas, A.; Lowenberger, C.; Benelli, G.; Salazar-Schettino, P.; Córdoba-Aguilar, A. What Makes an Effective Chagas Disease Vector? Factors Underlying *Trypanosoma cruzi*-Triatomine Interactions. *Acta Trop.* **2018**, *183*, 23–31. [CrossRef]
12. Lidani, K.C.F.; Andrade, F.A.; Bavia, L.; Damasceno, F.S.; Beltrame, M.H.; Messias-Reason, I.J.; Sandri, T.L. Chagas disease: From discovery to a worldwide health problem. *Front. Public Health* **2019**, *7*, 166. [CrossRef]
13. Stanaway, J.D.; Roth, G. The burden of Chagas disease: Estimates and challenges. *Glob. Heart* **2015**, *10*, 139–144. [CrossRef] [PubMed]
14. Moncayo, A.; Silveira, A. Current epidemiological trends for Chagas disease in Latin America and future challenges in epidemiology, surveillance and health policy. *Mem. Inst. Oswaldo Cruz* **2009**, *104* (Suppl. S1), 17–30. [CrossRef] [PubMed]
15. Quinde-Calderón, L.; Rios-Quituizaca, P.; Solorzano, L.; Dumonteil, E. Ten years (2004–2014) of Chagas disease surveillance and vector control in Ecuador: Successes and challenges. *Trop. Med. Int. Health* **2016**, *21*, 84–92. [CrossRef] [PubMed]

16. Carlier, Y.; Sosa-Estani, S.; Luquetti, A.; Buekens, P. Congenital Chagas disease: An update. *Mem. Inst. Oswaldo Cruz* **2015**, *110*, 363–368. [CrossRef]
17. Brenière, S.F.; Villacís, A.; Aznar, C. Vector transmission: How it works, what transmits, where it occurs. In *American Trypanosomiasis Chagas Disease*; Elsevier: Amsterdam, The Netherlands, 2017; pp. 497–515.
18. Kirchhoff, L.; Pearson, R. The emergence of Chagas disease in the United States and Canada. *Curr. Infect. Dis. Rep.* **2009**, *9*, 347–350. [CrossRef]
19. Schipper, H.; McClarty, B.M.; McRuer, K.E.; Nash, R.A.; Penney, C.J. Tropical diseases encountered in Canada: 1. Chagas' disease. *Can. Med. Assoc. J.* **1980**, *122*, 165.
20. Jackson, Y.; Pinto, A.; Pett, S. Chagas disease in Australia and New Zealand: Risks and needs for public health interventions. *Trop. Med. Int. Health* **2014**, *19*, 212–218. [CrossRef]
21. Soriano-Arandes, A.; Angheben, A.; Serre-Delcor, N.; Treviño-Maruri, B.; Gomez i Prat, J.; Jackson, Y. Control and management of congenital chagas disease in Europe and other non-endemic countries: Current policies and practices. *Trop. Med. Int. Health* **2016**, *21*, 590–596. [CrossRef]
22. Flores-Ferrer, A.; Waleckx, E.; Rascalou, G.; Dumonteil, E.; Gourbière, S. *Trypanosoma cruzi* Transmission Dynamics in a Synanthropic and Domesticated Host Community. *PLOS Neglected Trop. Dis.* **2019**, *13*, e0007902. [CrossRef]
23. Izeta-Alberdi, A.; Ibarra-Cerdeña, C.N.; Moo-Llanes, D.A.; Ramsey, J.M. Geographical, landscape and host associations of *Trypanosoma cruzi* DTUs and lineages. *Parasites Vectors* **2016**, *9*, 631. [CrossRef]
24. Jansen, A.M.; Roque, A.L.R. Domestic and wild mammalian reservoirs. In *American Trypanosomiasis*; Elsevier: Amsterdam, The Netherlands, 2010; pp. 249–276. [CrossRef]
25. World Health Organization. *Control of Chagas Disease: Second Report of the WHO Expert Committee*; World Health Organization: Geneva, Switzerland, 2002; Volume 2, pp. 249–276.
26. Ibarra-Cerdena, C.N.; Valiente-Banuet, L.; Sanchez-Cordero, V.; Stephens, C.R.; Ramsey, J.M. *Trypanosoma cruzi* reservoir—Triatomine vector co-occurrence networks reveal meta-community effects by synanthropic mammals on geographic dispersal. *PeerJ* **2017**, *5*, e3152. [CrossRef] [PubMed]
27. Gourbière, S.; Dorn, P.; Tripet, F.; Dumonteil, E. Genetics and evolution of triatomines: From phylogeny to vector control. *Heredity* **2012**, *108*, 190–202. [CrossRef] [PubMed]
28. Justi, S.; Galvão, C.; Schrago, C. Geological Changes of the Americas and their Influence on the Diversification of the Neotropical Kissing Bugs (Hemiptera: Reduviidae: Triatominae). *PLOS Neglected Trop. Dis.* **2016**, *10*, e0004527. [CrossRef]
29. Wendel, S. Transfusion transmitted Chagas disease: Is it really under control? *Acta Trop.* **2010**, *115*, 28–34. [CrossRef]
30. Chin-Hong, P.V.; Schwartz, B.S.; Bern, C.; Montgomery, S.P.; Kontak, S.; Kubak, B.; Ison, M.G. Screening and treatment of chagas disease in organ transplant recipients in the United States: Recommendations from the chagas in transplant working group. *Am. J. Transplant.* **2011**, *11*, 672–680. [CrossRef]
31. Shikanai-Yasuda, M.A.; Carvalho, N.B. Oral transmission of Chagas disease. *Clin. Infect. Dis.* **2012**, *54*, 845–852. [CrossRef]
32. De Sousa, A.; Vermeij, D.; Ramos, A.; Luquetti, A. Chagas Disease. *Lancet* **2024**, *403*, 203–218. [CrossRef]
33. Coura, J.R. The Main Sceneries of Chagas Disease Transmission: The Vectors, Blood, and Oral Transmissions—A Comprehensive Review. *Mem. Inst. Oswaldo Cruz* **2014**, *110*, 277–282. [CrossRef]
34. Le Loup, G.; Pialoux, G.; Lescure, F.X. Update in treatment of Chagas disease. *Curr. Opin. Infect. Dis.* **2011**, *24*, 428–434. [CrossRef]
35. Lee, B.Y.; Bacon, K.M.; Wateska, A.R.; Bottazzi, M.E.; Dumonteil, E.; Hotez, P.J. Modeling the economic value of a Chagas' disease therapeutic vaccine. *Hum. Vaccines Immunother.* **2012**, *8*, 1293–1301. [CrossRef]
36. Quijano-Hernandez, I.; Dumonteil, E. Advances and challenges towards a vaccine against Chagas disease. *Hum. Vaccines* **2011**, *7*, 1184–1191. [CrossRef] [PubMed]
37. Frade, A.F.; Pissetti, C.W.; Ianni, B.M.; Saba, B.; Lin-Wang, H.T.; Nogueira, L.G.; Chevillard, C. Genetic susceptibility to Chagas disease cardiomyopathy: Involvement of several genes of the innate immunity and chemokine-dependent migration pathways. *BMC Infect. Dis.* **2013**, *13*, 587. [CrossRef] [PubMed]
38. Nouvellet, P.; Dumonteil, E.; Gourbière, S. The improbable transmission of *Trypanosoma cruzi* to human: The missing link in the dynamics and control of Chagas disease. *PLoS Neglected Trop. Dis.* **2013**, *7*, e2505. [CrossRef] [PubMed]
39. Coura, J.R. Chagas disease: What is known and what is needed-A background article. *Mem. Inst. Oswaldo Cruz* **2007**, *102*, 113–122. [CrossRef] [PubMed]
40. Coura, J.R.; Borges-Pereira, J. Chagas disease: 100 years after its discovery. A systemic review. *Acta Trop.* **2010**, *115*, 5–13. [CrossRef]
41. Gürtler, R.; Yadon, Z. Eco-bio-social research on community-based approaches for Chagas disease vector control in Latin America. *Trans. R. Soc. Trop. Med. Hyg.* **2015**, *109*, 91–98. [CrossRef]
42. Bonney, K.M. Chagas disease in the 21st century: A public health success or an emerging threat? *Parasite* **2014**, *21*, 11. [CrossRef]
43. Zingales, B.; Pereira, M.E.; Almeida, I.C.; Ferguson, M.A.; Brener, Z. *Trypanosoma cruzi* taxonomy revisited: Current perspectives and new challenges. *Trends Parasitol.* **2023**, *40*, 3–14.

44. Kostygov, A.Y.; Albanaz, A.T.S.; Butenko, A.; Gerasimov, E.S.; Lukeš, J.; Yurchenko, V. Phylogenetic framework to explore trait evolution in Trypanosomatidae. *Trends Parasitol.* **2024**, *40*, 96–99. [CrossRef]
45. Talavera-López, C.; Messenger, L.A.; Lewis, M.D.; Schwede, A. Whole genome analysis of *Trypanosoma cruzi* reveals independent hybrids and diverse mitochondrial DNA haplotypes. *Am. J. Trop. Med. Hyg.* **2021**, *104*, 247–258.
46. Berná, L.; Rodríguez, M.; Chiribao, M.L.; Parodi-Talice, A.; Pita, S.; Rijo, G.; Alvarez-Valin, F. Genomic organization and evolution of the *Trypanosoma cruzi* nuclear genome. *Infect. Genet. Evol.* **2022**, *102*, 105344.
47. Rojas Andrade, R.; Carvalho-Pereira, T.S.; Lima, M.M.; Magalhães, L.K.; Abad-Franch, F. Molecular epidemiology and transmission dynamics of *Trypanosoma cruzi* in Latin America: Insights from high-throughput sequencing. *Life* **2023**, *13*, 2339.
48. Crispim, B.A.; de Azevedo, L.G.; Ribeiro, V.; Sangenito, L.S.; Menna-Barreto, R.F.S.; Santos, V.S.; Batista, M.M. Advances in the development of new drugs for Chagas disease. *Pathogens* **2024**, *14*, 72.
49. Zingales, B.; Miles, M.A.; Campbell, D.A.; Tibayrenc, M.; Macedo, A.M.; Teixeira, M.M.; Guhl, F. The revised *Trypanosoma cruzi* DTU nomenclature: Rationale, epidemiological relevance and research applications. *Infect. Genet. Evol.* **2012**, *12*, 240–253. [CrossRef]
50. Burgos, J.M.; Diez, C.; Vigliano, C.; Bisio, M.; Risso, M.G.; Duffy, T.; Schijman, A.G. Molecular identification of *Trypanosoma cruzi* discrete typing units in end-stage chronic Chagas heart disease and reactivation after heart transplantation. *Clin. Infect. Dis.* **2017**, *64*, 1247–1253.
51. Ramírez, J.D.; Hernández, C.; Montilla, M.; Zambrano, P.; Flórez, C.; Parra, E.; González, M. Molecular epidemiology of *Trypanosoma cruzi* in Colombia and its association with oral Chagas disease outbreaks. *Infect. Genet. Evol.* **2021**, *92*, 104986.
52. Guhl, F.; Ramírez, J.D. The evolutionary history of *Trypanosoma cruzi*: A proposal for a new nomenclature of discrete typing units. *Trends Parasitol.* **2011**, *27*, 538–543.
53. Moraes, C.B.; Franco, C.H.; Figueira, L.P.; Chatelain, E.; Freitas-Junior, L.H.; Soeiro, M.N.C. Drug resistance in *Trypanosoma cruzi*: The problem and prospects for new approaches. *Int. J. Parasitol. Drugs Drug Resist.* **2014**, *4*, 17–22.
54. Borsatto, K.C.; Coronado, M.A.; Arni, R.K.; Cesar, C.A.K. Omics tools applied to the study of Chagas disease vectors: Cytogenomics and genomics. *Am. J. Trop. Med. Hyg.* **2021**, *104*, 1973. [CrossRef]
55. Campos, A.; Rubio, M.; Martínez, T.; Hernández Osorio, L.A.; Martínez Calvillo, S.; Manning Cela, R.G. Enfermedad de Chagas: Vectores. *Ciencia* **2017**, *68*, 30–33.
56. Rojas de Arias, A.; Monroy, C.; Guhl, F.; Sosa-Estani, S.; Souza, S.; Abad-Franch, F. Chagas disease control-sur-veillance in the Americas: The multinational initiatives and the practical impossibility of interrupting vector-borne *Trypanosoma cruzi* transmission. *Mem. Inst. Oswaldo Cruz* **2022**, *117*, e210130. [CrossRef] [PubMed]
57. Peña-Callejas, G.; Instituto de Investigaciones Jurídicas, Universidad Nacional Autónoma de México. Entre el Mar y las Montañas: Política, Comercio y Contrabando en la Costa Oaxaqueña, Siglos XV-XVIII. Bachelor's Thesis, Universidad Nacional Autónoma de México, Mexico City, Mexico, 2022. Available online: http://132.248.9.195/ptd2022/agosto/0829260/Index.html (accessed on 10 October 2024).
58. Peña-Callejas, G.; González, J.; Jiménez-Cortés, J.; Fuentes-Vicente, G.; Salazar-Schettino, J.; Bucio-Torres, P.; Cabrera-Bravo, M.; Flores-Villegas, A. Enfermedad de Chagas: Biología y transmisión de *Trypanosoma cruzi*. TIP. *Rev. Espec. Cienc. Quim.-Biol.* **2022**, *25*, e449. [CrossRef]
59. World Health Organization (WHO). *Chagas Disease (American Trypanosomiasis)*; World Health Organization: Geneva, Switzerland, 2023. Available online: https://www.who.int/es/news-room/fact-sheets/detail/chagas-disease-(american-trypanosomiasis) (accessed on 25 September 2024).
60. Pereyra, N.; Lobbia, P.; Mougabure-Cueto, G. Effects of the infection with *Trypanosoma cruzi* on the feeding and excretion/defecation patterns of *Triatoma infestans*. *Bull. Entomol. Res.* **2020**, *110*, 169–176. [CrossRef]
61. Briceño-León, R. La Enfermedad de Chagas en las Américas: Una Perspectiva de Ecosalud. *Cad. Saúde Pública* **2009**, *25*, 71–82. [CrossRef] [PubMed]
62. Córdoba-Aguilar, A.; Ibarra-Cerdeña, C.N.; Castro-Arellano, I.; Suzan, G. Tackling zoonoses in a crowded world: Lessons to be learned from the COVID-19 pandemic. *Acta Trop.* **2021**, *214*, 105780. [CrossRef] [PubMed]
63. Ruiz-Guevara, R.; Alarcón de Noya, B.; Mendoza, I.; Rojas, C.; Machado, I.; Díaz-Bello, Z.; Muñoz-Calderón, A.; Castro, J.; Noya, O. Ten years follow-up of the largest oral Chagas disease outbreak: Cardiological prospective cohort study. *PLOS Neglected Trop. Dis.* **2023**, *17*, e0011643. [CrossRef]
64. Dorn, P.; Monroy, M.; Stevens, L. Sustainable, Integrated Control of Native Vectors: The Case of Chagas Disease in Central America. *Front. Trop. Dis.* **2022**, *3*, 971000. [CrossRef]
65. Montes de Oca-Aguilar, A.C.; González-Martínez, A.; Chan-González, R.; Ibarra-López, P.; Smith-Ávila, S.; Córdoba-Aguilar, A.; Ibarra-Cerdeña, C.N. Signs of urban evolution? Morpho-functional traits co-variation along a nature-urban gradient in a Chagas disease vector. *Front. Ecol. Evol.* **2022**, *10*, 805040. [CrossRef]
66. Vargas-Abasolo, R.; Gutiérrez-Cabrera, A.; Cruz-López, L.; Alavez-Rosas, D.; Benelli, G.; Córdoba-Aguilar, A. Chagas disease vector control strategies: Where we are and where we should go from here. *Entomol. Gen.* **2023**, *43*, 771–788. [CrossRef]

67. Ramírez Solano, A.; Chamizo García, H. Determinantes sociales de la enfermedad de Chagas en Costa Rica. *Poblac. Y Salud Mesoam.* **2023**, *21*, 80–98. [CrossRef]
68. Ávila Montes, G.; Martínez Hernández, M.; Ponce, C.; Ponce, E.; Soto Hernández, R. La Enfermedad de Chagas en la Zona Central de Honduras: Conocimientos, Creencias y Prácticas. *Rev. Panam. Salud Pública* **1998**, *3*, 158–163. [CrossRef] [PubMed]
69. Verdú, J.; Ruiz, M. Control del Chagas en comunidades guaraníes: Conocimiento y hábitos higiénicos dentro del Proyecto de Mejoramiento de Viviendas en Bolivia. *Gac. Sanit.* **2003**, *17*, 166–168. [CrossRef] [PubMed]
70. Rosecrans, K.; Cruz-Martin, G.; King, A.; Dumonteil, E. Opportunities for improved Chagas disease vector control based on knowledge, attitudes and practices of communities in the Yucatan Peninsula, Mexico. *PLoS Neglected Trop. Dis.* **2014**, *8*, e2763. [CrossRef] [PubMed]
71. Gaspe, M.; Provecho, Y.; Fernández, M.; Vassena, C.; Orihuela, P.; Gürtler, R. Beating the odds: Sustained Chagas disease vector control in remote indigenous communities of the Argentine Chaco over a seven-year period. *PLOS Neglected Trop. Dis.* **2018**, *12*, e0006804. [CrossRef]
72. Verhasselt, Y.; Curto de Casas, S.I.; Carcavallo, R.U.; Boffi, R. Geografía de la Salud: Algunos Factores Ambientales de Riesgo para Enfermedades Transmitidas por Mosquitos (Salvador Mazza, Salta, Argentina). *An. Soc. Argent. Estud. Geogr.* **1997**, *20*, 288–297.
73. Clavijo-Baquet, S.; Cavieres, G.; González, A.; Cattan, P.; Bozinovic, F. Thermal performance of the Chagas disease vector, *Triatoma infestans*, under thermal variability. *PLoS Neglected Trop. Dis.* **2021**, *15*, e0009148. [CrossRef]
74. SA. Vichuncas Aguantadoras. Nexciencia. 2015. Available online: https://nexciencia.exactas.uba.ar/vinchucas-chagas-tolerancia-termica-pablo-schilman-gerardo-de-la-vega (accessed on 10 October 2024).
75. Tian, Y.; Durden, C.; Hamer, G. A scoping review of triatomine control for Chagas disease prevention: Current and developing tools in Latin America and the United States. *J. Med. Entomol.* **2024**, *61*, 1290–1308. [CrossRef]
76. Moretti, A.; Zerba, E.; Alzogaray, R. Lethal and sublethal effects of eucalyptol on *Triatoma infestans* and *Rhodnius prolixus*, vectors of Chagas disease. *Entomol. Exp. Appl.* **2015**, *154*, 62–70. [CrossRef]
77. Oramahi, H.; Yoshimura, T. Antifungal and antitermitic activities of wood vinegar from *Vitex pubescens* Vahl. *J. Wood Sci.* **2013**, *59*, 344–350. [CrossRef]
78. Ávila Montes, G.; Ponce, C.; Ponce, E.; Martínez Hernández, M.; Flores, M. Insecticidal Paint and Fumigant Canisters for Chagas' Disease Control: Community Acceptance in Honduras. *Rev. Panam. Salud Pública* **1999**, *6*, 311–320. [CrossRef]
79. Angulo, V.; Esteban, L. Nueva Trampa para la Captura de Triatominos en Hábitats Silvestres y Peridomésticos. *Biomédica* **2011**, *31*, 264–268. [CrossRef] [PubMed]
80. Toriello, C.; Brunner-Mendoza, C.; Navarro-Barranco, M. *Metarhizium* and *Isaria* as biological control agents against *Meccus* vector of Chagas disease. *Int. J. Trop. Insect Sci.* **2022**, *42*, 3713–3719. [CrossRef]
81. Eliceche, D.; Achinelly, M.; Silvestre, C.; Micieli, M.; Martí, G. Nematodos Entomopatógenos (Heterorhabditidae y Steinernematidae), para el Control de Poblaciones de *Triatoma infestans* (Hemiptera: Reduviidae), Vector de la Enfermedad de Chagas. *Biol. Control* **2021**, *163*, 104742. [CrossRef]
82. Smith-Avila, S.; Ibarra-Cerdena, C.; Barranco-Florido, J.; Vidal-Martinez, V. *Heterorhabditis indica* (Nematoda: Rhabditida) a possible new biological control agent against the vector of Chagas disease. *Acta Trop.* **2024**, *256*, 107262. [CrossRef]
83. Ceccarelli, S.; Balsalobre, A.; Susevich, M.L.; Echeverria, M.G.; Gorla, D.E.; Marti, G.A. Modelling the Potential Geographic Distribution of Triatomines Infected by Triatoma Virus in the Southern Cone of South America. *Parasites Vectors* **2015**, *8*, 1–9. [CrossRef]
84. Susevich, M.L.; Metz, G.E.; Marti, G.A.; Echeverria, M.G. Dicistroviridae: A New Viral Purification Technique. *Rev. Argent. Microbiol.* **2017**, *49*, 311–314. [CrossRef]
85. Flores-Villegas, A.; Salazar-Schettino, P.; Cordoba-Aguilar, A.; Gutierrez-Cabrera, A.; Rojas-Wastavino, G.; Bucio-Torres, M.; Cabrera-Bravo, M. Immune Defence Mechanisms of Triatomines against Bacteria, Viruses, Fungi and Parasites. *Bull. Entomol. Res.* **2015**, *105*, 523–532. [CrossRef]
86. Godfray, H.C.J. *Parasitoids: Behavioral and Evolutionary Ecology*; Princeton University Press: Princeton, NJ, USA, 1994; Volume 5, p. e1207.
87. Chuit, R.; Paulone, I.; Wisnivesky-Colli, C.; Bo, R.; Perez, A.; Sosa-Stani, S.; Segura, E. Result of a First Step Toward Community-Based Surveillance of Transmission of Chagas' Disease with Appropriate Technology in Rural Areas. *Am. J. Trop. Med. Hyg.* **1992**, *46*, 444–450. [CrossRef]
88. Guillen, G.; Diaz, R.; Jemio, A.; Cassab, J.; Pinto, C.; Schofield, C. Chagas disease vector control in Tupiza, southern Bolivia. *Mem Inst. Oswaldo Cruz* **1997**, *92*, 1–8. [CrossRef]
89. Yamagata, Y.; Nakagawa, J. Control of Chagas disease. *Adv. Parasitol.* **2006**, *61*, 129–165. [CrossRef] [PubMed]
90. Gürtler, R.; Kitron, U.; Cecere, M.; Segura, E.; Cohen, J. Sustainable vector control and management of Chagas disease in the Gran Chaco, Argentina. *Proc. Natl. Acad. Sci. USA* **2007**, *104*, 16194–16199. [CrossRef]
91. Abad-Franch, F.; Vega, M.; Rolón, M.; Santos, W.; Arias, A.D. Community Participation in Chagas Disease Vector Surveillance: Systematic Review. *PLOS Neglected Trop. Dis.* **2011**, *5*, e1207. [CrossRef] [PubMed]

92. Hashimoto, K.; Yoshioka, K. Review: Surveillance of Chagas disease. In *Advances in Parasitology*; Rollinson, E.D., Hay, S.I., Eds.; Academic Press: Cambridge, MA, USA, 2012; pp. 375–428. [CrossRef]
93. Gurevitz, J.; Ceballos, L.; Gaspe, M.; Alvarado-Otegui, J.; Enríquez, G.; Kitron, U.; Gürtler, R. E Factors affecting infestation by *Triatoma infestans* in a rural area of the humid Chaco in Argentina: A multi-model inference approach. *PLOS Neglected Trop. Dis.* **2011**, *5*, e1349. [CrossRef] [PubMed]
94. Gurevitz, J.; Gaspe, M.; Enriquez, G.; Provecho, Y.; Kitron, U.; Gürtler, R. Intensified surveillance and insecticide-based control of the Chagas disease vector *Triatoma infestans* in the Argentinean Chaco. *PLOS Neglected Trop. Dis.* **2011**, *7*, e2158. [CrossRef]
95. Moncayo, Á.; Silveira, A. Current epidemiological trends of Chagas disease in Latin America and future challenges: Epidemiology, surveillance, and health policies. In *American Trypanosomiasis Chagas Disease*, 2nd ed.; Telleria, E.J., Tibayrenc, M., Eds.; Elsevier: Amsterdam, The Netherlands, 2017; pp. 59–88. [CrossRef]
96. Picolo, M.I.; Vassena, C.; Orihuela, P.S.; Barrios, S.; Zaidemberg, M.; Zerba, E. High resistance to pyrethroid insecticides associated with ineffective field treatments in *Triatoma infestans* (Hemiptera: Reduviidae) from Northern Argentina. *J. Med. Entomol.* **2005**, *42*, 637–642. [CrossRef]
97. Rojo-Medina, J.; González-Roldán, J.F.; Salazar-Schettino, P.M.; Ruiz-Matus, C. Enfermedad de Chagas en México. *Gac. Medica Mex.* **2018**, *154*, 605–612. [CrossRef]
98. Telleria, J.; Tibayrenc, M. *American Trypanosomiasis Chagas Disease: One Hundred Years of Research*; Elsevier: Amsterdam, The Netherlands, 2017. Available online: https://www.sciencedirect.com/book/9780128010297/american-trypanosomiasis-chagas-disease (accessed on 20 September 2024).
99. Pereira, K.S.; Schmidt, F.L.; Guaraldo, A.M.; Franco, R.M.B.; Dias, V.L.; Passos, L.A. Chagas' disease as a foodborne illness. *J. Food Prot.* **2009**, *72*, 441–446. [CrossRef]
100. Coura, J.R. The main sceneries of *Trypanosoma cruzi* transmission. The vectors, blood and oral transmissions—A comprehensive review. *Mem. Inst. Oswaldo Cruz* **2015**, *110*, 277–282. [CrossRef]
101. Nóbrega, A.A.; Garcia, M.H.; Tatto, E.; Obara, M.T.; Costa, E.; Sobel, J.; Araujo, W.N. Oral transmission of Chagas disease by consumption of açaí palm fruit, Brazil. *Emerg. Infect. Dis.* **2009**, *15*, 653–655.
102. Santos, V.S.; Farias, J.P.; Batista, M.M.; Macedo, G.C. The role of food contamination in the oral transmission of *Trypanosoma cruzi*: A systematic review. *Pathogens* **2023**, *12*, 582.
103. Xavier, S.C.C.; Roque, A.L.R.; Lima, V.D.S.; Monteiro, K.J.L.; Otaviano, J.C.; Jansen, A.M. Lower richness of small wild mammal species and Trypanosoma cruzi infection in an anthropized rural area of the Brazilian Pantanal region. *BioMed Res. Int.* **2014**, 696512.
104. Jansen, A.M.; Xavier, S.C.C.; Roque, A.L.R. *Trypanosoma cruzi* transmission in the wild and its most important reservoir hosts in Brazil. *Parasites Vectors* **2018**, *11*, 502. [CrossRef] [PubMed]
105. López-García, A.; Gilabert, J.A. Oral transmission of Chagas disease from a One Health approach: A systematic review. *Trop. Med. Int. Health* **2023**, *28*, 689–698. [CrossRef] [PubMed]
106. Carlier, Y.; Torrico, F.; Sosa-Estani, S.; Russomando, G.; Luquetti, A.; Freilij, H.; Albajar-Viñas, P. Congenital Chagas Disease: Recommendations for Diagnosis, Treatment and Control in Infants and Newborns. *J. Parasitol.* **2011**, *97*, 524–529. [CrossRef]
107. Buekens, P.; Cafferata, M.L.; Alger, J.; Althabe, F.; Belizán, J.M. Congenital Transmission of *Trypanosoma Cruzi* and Prospects for Treatment. *Clin. Infect. Dis.* **2018**, *67*, 569–576. [CrossRef]
108. Carlier, Y.; Truyens, C. Maternal–Fetal Transmission of Trypanosoma cruzi. In *American Trypanosomiasis*; Telleria, E.J., Tibayrenc, M., Eds.; Elsevier: Amsterdam, The Netherlands, 2010; pp. 539–581. [CrossRef]
109. Brito, R.; Gorla, D.; Diotaiuti, L.; Gomes, A.; Souza, R.C.; Abad-Franch, F. Drivers of House Invasion by Sylvatic Chagas Disease Vectors in the Amazon-Cerrado Transition: A Multi-Year, State-Wide Assessment of Municipality-Aggregated Surveillance Data. *PLOS Neglected Trop. Dis.* **2017**, *11*, e0006035. [CrossRef]
110. Esch, K.; Petersen, C. Transmission and Epidemiology of Zoonotic Protozoal Diseases of Companion Animals. *Clin. Microbiol. Rev.* **2013**, *26*, 58–85. [CrossRef]
111. Márquez, M. Chagas Disease in Dogs. *J. Vet. Healthc.* **2018**, *1*, 26–30. [CrossRef]
112. Marcili, A.; Lima, L.; Jr, M.C.; Junqueira, A.; Veludo, H.; Silva, F.; Campaner, M.; Paiva, F.; Nunes, V.; Teixeira, M. A new genotype of Trypanosoma cruzi associated with bats evidenced by phylogenetic analyses using SSU rDNA, cytochrome b and Histone H2B genes and genotyping based on ITS1 rDNA. *Parasitology* **2009**, *136*, 641–655. [CrossRef]
113. Pinto, C.; Kalko, E.; Cottontail, I.; Wellinghausen, N.; Cottontail, V. TcBat a bat-exclusive lineage of Trypanosoma cruzi in the Panama Canal Zone, with comments on its classification and the use of the 18S rRNA gene for lineage identification. *Infect. Genet. Evol.* **2012**, *12*, 1328–1332. [CrossRef]
114. Noireau, F.; Diosque, P.; Jansen, A. *Trypanosoma cruzi*: Adaptation to its vectors and its hosts. *Vet. Res.* **2009**, *40*, 26. [CrossRef]
115. Hodo, C.; Bañuelos, R.; Edwards, E.; Wozniak, E.; Hamer, S. Pathology and discrete typing unit associations of *Trypanosoma cruzi* infection in coyotes (*Canis latrans*) and raccoons (*Procyon lotor*) of Texas, USA. *J. Wildl. Dis.* **2020**, *56*, 134–144. [CrossRef] [PubMed]

116. Junior, P.A.S.; Molina, I.; Murta, S.M.F.; Sánchez-Montalvá, A.; Salvador, F.; Corrêa-Oliveira, R.; Carneiro, C.M. Experimental and clinical treatment of Chagas disease: A review. *Am. J. Trop. Med. Hyg.* **2017**, *97*, 1289. [CrossRef]
117. Clinicalinfo HIV. *Glossary of HIV/AIDS-Related Terms*, 9th ed.; U.S. Department of Health & Human Services: Washington, DC, USA, 2021. Available online: https://clinicalinfo.hiv.gov/sites/default/files/glossary/Glossary-English_HIVinfo.pdf (accessed on 10 October 2024).
118. Chaachouay, N.; Zidane, L. Plant-derived natural products: A source for drug discovery and development. *Drugs Drug Candidates* **2024**, *3*, 184–207. [CrossRef]
119. Gürtler, R.; Cardinal, M. Dogs and their role in the eco-epidemiology of Chagas disease. In *Dog Parasites Endangering Human Health*; Strube, E.C., Mehlhorn, H., Eds.; Springer International Publishing: Cham, Switzerland, 2021; pp. 73–106. [CrossRef]
120. Reithinger, R.; Ceballos, L.; Stariolo, R.; Davies, C.; Gürtler, R. Extinction of experimental Triatoma infestans populations following continuous exposure to dogs wearing deltamethrin-treated collars. *Am. J. Trop. Med. Hyg.* **2006**, *74*, 766–771. [CrossRef] [PubMed]
121. Ucan-Mézquita, A.; Jimenez-Coello, M.; Guzmán-Marín, E.; Gutierrez-Blanco, E.; Chan-Pérez, J.; Travi, B.; Hernandez-Cortazar, I.; Ortega-Pacheco, A. Efficacy of a topical combination of fipronil-permethrin against *Rhodnius prolixus* on dogs. *Vet. Parasitol.* **2019**, *276*, 108978. [CrossRef]
122. Carvajal, G.; Picollo, M.; Toloza, A. Is Imidacloprid an Effective Alternative for Controlling Pyrethroid-Resistant Populations of Triatoma infestans (Hemiptera: Reduviidae) in the Gran Chaco Ecoregion? *Mem. Inst. Oswaldo Cruz* **2014**, *109*, 761–766. [CrossRef]
123. Rokhsar, J.; Raynor, B.; Sheen, J.; Goldstein, N.; Levy, M.; Castillo-Neyra, R. Modeling the impact of xenointoxication in dogs to halt *Trypanosoma cruzi* transmission. *PLoS Comput. Biol.* **2023**, *19*, e1011115. [CrossRef]
124. Dias, J.; Schofield, C.; Machado, E.; Fernandes, A. Ticks, ivermectin, and experimental Chagas disease. *Mem. Inst. Oswaldo Cruz* **2005**, *100*, 829–832. [CrossRef]
125. Amelotti, I.; Catalá, S.; Gorla, D. The Effects of Cypermethrin Pour-On and Piperonyl Butoxide on Triatoma infestans under Laboratory Conditions. *J. Med. Entomol.* **2010**, *47*, 1135–1140. [CrossRef]
126. Laiño, M.; Cardinal, M.; Gaspe, M.; Enriquez, G.; Alvedro, A.; Macchiaverna, N.; Gürtler, R. Control of pyrethroid-resistant populations of *Triatoma infestans*, the main vector of *Trypanosoma cruzi*, by treating dogs with fluralaner in the Argentine Chaco. *Med. Vet. Entomol.* **2022**, *36*, 149–158. [CrossRef]
127. Queiroga, T.; Gomez, L.; de Sena, E.; dos Santos, W.; Ferreira, H.; de Araújo-Neto, V.; Barbosa-Silva, A.; Brito, C.R.; Lima, R.; Fagundes-Neto, J.; et al. Insecticidal efficacy of fluralaner (Bravecto®) against Triatoma brasiliensis, a major vector of Trypanosoma cruzi in Brazil. *Parasites Vectors* **2021**, *14*, 456. [CrossRef]
128. Ortega-Pacheco, A.; Poot-Ramos, A.; Chan-Pérez, J.I.; Gutiérrez-Blanco, E.; Acevedo-Arcique, C.; Baak-Baak, C. Evaluation of the effectiveness of fluralaner against adult stages of Rhodnius prolixus in dogs. *Parasitol. Int.* **2022**, *87*, 102508. [CrossRef] [PubMed]
129. Durden, C.; Tian, Y.; Knape, K.; Klemashevich, C.; Norman, K.; Carey, J. Fluralaner systemic treatment of chickens results in mortality in Triatoma gerstaeckeri, vector of the agent of Chagas disease. *Parasites Vectors* **2023**, *16*, 178. [CrossRef] [PubMed]
130. Mendes, F.; de Souza, N.S.; Perez-Molina, J.A.; Angheben, A.; Meymandi, S.K.; Sosa-Estani, S.; Molina, I. Critical analysis of Chagas disease treatment in different countries. *Memórias Do Inst. Oswaldo Cruz* **2022**, *117*, e210034. [CrossRef] [PubMed]
131. Fabbro, D.; Danesi, E.; Olivera, V.; Codebó, M.; Denner, S.; Heredia, C. Trypanocide treatment of women infected with Trypanosoma cruzi and its effect on preventing congenital Chagas. *PLoS Neglected Trop. Dis.* **2014**, *8*, e3312. [CrossRef] [PubMed]
132. Pecoul, B.; Batista, C.; Stobbaerts, E.; Ribeiro, I.; Vilasanjuan, R.; Gascon, J.; Pinazo, M.; Moriana, S.; Gold, S.; Pereiro, A.; et al. The BENEFIT Trial: Where Do We Go from Here? *PLOS Neglected Trop. Dis.* **2016**, *10*, e0004343. [CrossRef] [PubMed]
133. Brumpt, E. Immunité partielle dans les infections à *Trypanosoma cruzi*, transmission de ce trypanosome par *Cimex rotundus*. Rôle regulateur des hôtes en internédiaires. *Bull. Soc. Pathol. Exot.* **1913**, *6*, 93–100.
134. Plessmann-Camargo, E.P.; Gazzinelli, R.T.; Morel, C.M.; Precioso, A.R. Why do we still have not a vaccine against Chagas disease? *Mem. Inst. Oswaldo Cruz* **2022**, *117*, e200314. [CrossRef] [PubMed] [PubMed Central]
135. Farani, P.S.G.; Jones, K.M.; Poveda, C. Treatments and the Perspectives of Developing a Vaccine for Chagas Disease. *Vaccines* **2024**, *12*, 870. [CrossRef]
136. Pinazo, M.J.; Malchiodi, E.; Ioset, J.R.; Bivona, A.; Gollob, K.J.; Dutra, W.O. Challenges and advancements in the development of vaccines and therapies against Chagas disease. *Lancet Microbe* **2024**, *5*, 100972. [CrossRef]
137. Bivona, A.E.; Alberti, A.S.; Cerny, N.; Trinitario, S.N.; Malchiodi, E.L. Chagas disease vaccine design: The search for an efficient Trypanosoma cruzi immune-mediated control. *Biochim. Biophys. Acta Mol. Basis Dis.* **2020**, *1866*, 165658. [CrossRef]
138. Da Costa, K.M.; Marques da Fonseca, L.; dos Reis, J.S.; Santos, M.A.; Previato, J.O.; Mendonca-Previato, L.; Freire-de-Lima, L. Trypanosoma cruzi trans-Sialidase as a Potential Vaccine Target Against Chagas Disease. *Front. Cell. Infect. Microbiol.* **2021**, *11*, 768450. [CrossRef]
139. Maldonado, E.; Morales-Pison, S.; Urbina, F.; Solari, A. Vaccine Design against Chagas Disease Focused on the Use of Nucleic Acids. *Vaccines* **2022**, *10*, 587. [CrossRef]

140. Masip, Y.E.; Caeiro, L.D.; Cosenza, M.; Postan, M.; Molina, G.; Taboga, O.; Molinari, M.P.; Tekiel, V. Vaccination with parasite-specific TcTASV proteins combined with recombinant baculovirus as a delivery platform protects against acute and chronic Trypanosoma cruzi infection. *Front. Cell. Infect. Microbiol.* **2024**, *14*, 1297321. [CrossRef]
141. Dumonteil, E.; Herrera, C. The Case for the Development of a Chagas Disease Vaccine: Why? How? When? *Trop. Med. Infect. Dis.* **2021**, *6*, 16. [CrossRef]
142. Fernández, M.D.P.; Gaspe, M.S.; Sartor, P.; Gürtler, R.E. Human *Trypanosoma cruzi* Infection Is Driven by Eco-Social Interactions in Rural Communities of the Argentine Chaco. *PLoS Neglected Trop. Dis.* **2019**, *13*, e0007430. [CrossRef] [PubMed]
143. De Alencar, B.; Araújo, A.; Penido, M.; Gazzinelli, R.; Rodrigues, M. Cross-Priming of Long-Lived Protective CD8+ T Cells against *Trypanosoma cruzi* Infection: Importance of a TLR9 Agonist and CD4+ T Cells. *Vaccine* **2007**, *25*, 6018–6027. [CrossRef] [PubMed]
144. Haolla, F.; Claser, C.; de Alencar, B.; Tzelepis, F.; de Vasconcelos, J.; de Oliveira, G.; Silvério, J.; Machado, A.; Lannes-Vieira, J.; Bruna-Romero, O.; et al. Strain-specific protective immunity following vaccination against experimental *Trypanosoma cruzi* infection. *Vaccine* **2009**, *27*, 5644–5653. [CrossRef] [PubMed]
145. Arnal, A.; Villanueva-Lizama, L.; Teh-Poot, C.; Herrera, C.; Dumonteil, E. Extent of Polymorphism and Selection Pressure on the *Trypanosoma cruzi* Vaccine Candidate Antigen Tc24. *Evol. Appl.* **2020**, *13*, 2663–2672. [CrossRef]
146. Basso, B.; Moretti, E.; Fretes, R. Vaccination with Epimastigotes of Different Strains of *Trypanosoma rangeli* Protects Mice Against *Trypanosoma cruzi* Infection. *Mem. Inst. Oswaldo Cruz* **2008**, *103*, 370–374. [CrossRef]
147. Collins, M.H.; Craft, J.M.; Bustamante, J.M.; Tarleton, R.L. Oral exposure to *Trypanosoma cruzi* elicits a systemic CD8+ T cell response and protection against heterotopic challenge. *Infect. Immun.* **2011**, *79*, 3397–3406. [CrossRef]
148. Araújo, A.F.; de Alencar, B.C.; Vasconcelos, J.R.C.; Hiyane, M.I.; Marinho, C.R.; Penido, M.L.; Rodrigues, M.M. CD8+-T-cell-dependent control of *Trypanosoma cruzi* infection in a highly susceptible mouse strain after immunization with recombinant proteins based on amastigote surface protein 2. *Infect. Immun.* **2005**, *73*, 6017–6025. [CrossRef]
149. Hoft, D.F.; Eickhoff, C.S.; Giddings, O.K.; Vasconcelos, J.R.; Rodrigues, M.M. Trans-sialidase recombinant protein mixed with CpG motif-containing oligodeoxynucleotide induces protective mucosal and systemic *Trypanosoma cruzi* immunity involving CD8+ CTL and B cell-mediated cross-priming. *J. Immunol.* **2007**, *179*, 6889–6900. [CrossRef]
150. Giddings, O.K.; Eickhoff, C.S.; Sullivan, N.L.; Hoft, D.F. Intranasal vaccinations with the trans-sialidase antigen plus CpG Adjuvant induce mucosal immunity protective against conjunctival *Trypanosoma cruzi* challenges. *Infect. Immun.* **2010**, *78*, 1333–1338. [CrossRef]
151. Cazorla-Perfetti, D.J.; Nieves-Blanco, E.E. Triatominos de Venezuela: Aspectos taxonómicos biológicos-distribución geográfica e importancia médica. *Av. Cardiol.* **2010**, *30*, 347–369.
152. Eickhoff, C.S.; Giddings, O.K.; Yoshida, N.; Hoft, D.F. Immune responses to gp82 provide protection against mucosal *Trypanosoma cruzi* infection. *Mem. Inst. Oswaldo Cruz* **2010**, *105*, 687–691. [CrossRef] [PubMed]
153. Dzul-Huchim, V.M.; Ramirez-Sierra, M.J.; Martinez-Vega, P.P.; Rosado-Vallado, M.E.; Arana-Argaez, V.E.; Ortega-Lopez, J.; Villanueva-Lizama, L.E. Vaccine-linked chemotherapy with a low dose of benznidazole plus a bivalent recombinant protein vaccine prevents the development of cardiac fibrosis caused by *Trypanosoma cruzi* in chronically-infected BALB/c mice. *PLoS Neglected Trop. Dis.* **2022**, *16*, e0010258. [CrossRef] [PubMed]
154. Bontempi, I.; Leal, K.; Prochetto, E.; Díaz, G.; Cabrera, G.; Bortolotti, A.; Morbidoni, H.R.; Borsuk, S.; Dellagostin, O.; Marcipar, I. Recombinant Mycobacterium bovis BCG Is a Promising Platform to Develop Vaccines Against *Trypanosoma cruzi* Infection. *Clin. Exp. Immunol.* **2020**, *201*, 306–316. [CrossRef] [PubMed]
155. Miyahira, Y.; Takashima, Y.; Kobayashi, S.; Matsumoto, Y.; Takeuchi, T.; Ohyanagi-Hara, M.; Yoshida, A.; Ohwada, A.; Akiba, H.; Yagita, H.; et al. Immune Responses against a Single CD8+-T-Cell Epitope Induced by Virus Vector Vaccination Can Successfully Control *Trypanosoma cruzi* Infection. *Infect. Immun.* **2005**, *73*, 7356–7365. [CrossRef]
156. Machado, A.; Cardoso, J.; Claser, C.; Rodrigues, M.; Gazzinelli, R.; Bruna-Romero, O. Long-Term Protective Immunity Induced Against *Trypanosoma cruzi* Infection After Vaccination with Recombinant Adenoviruses Encoding Amastigote Surface Protein-2 and Trans-Sialidase. *Hum. Gene Ther.* **2006**, *17*, 898–908. [CrossRef]
157. Duan, X.; Yonemitsu, Y.; Chou, B.; Yoshida, K.; Tanaka, S.; Hasegawa, M.; Tetsutani, K.; Ishida, H.; Himeno, K.; Hisaeda, H. Efficient Protective Immunity against *Trypanosoma cruzi* Infection after Nasal Vaccination with Recombinant Sendai Virus Vector Expressing Amastigote Surface Protein-2. *Vaccine* **2009**, *27*, 6154–6159. [CrossRef]
158. Gupta, S.; Garg, N. Prophylactic efficacy of TcVac2 against *Trypanosoma cruzi* in mice. *PLOS Neglected Trop. Dis.* **2010**, *4*, e797. [CrossRef]
159. Rigato, P.O.; de Alencar, B.C.; de Vasconcelos, J.R.C.; Dominguez, M.R.; Araújo, A.F.; Machado, A.V.; Gazzinelli, R.T.; Bruna-Romero, O.; Rodrigues, M. Heterologous Plasmid DNA Prime-Recombinant Human Adenovirus 5 Boost Vaccination Generates a Stable Pool of Protective Long-Lived CD8+ T Effector Memory Cells Specific for a Human Parasite, *Trypanosoma cruzi*. *Infect. Immun.* **2011**, *79*, 2120–2130. [CrossRef]
160. Garg, N.; Tarleton, R. Genetic immunization elicits antigen-specific protective immune responses and decreases disease severity in *Trypanosoma cruzi* infection. *Infect. Immun.* **2002**, *70*, 5547–5555. [CrossRef]

161. Chou, B.; Hiromatsu, K.; Hisaeda, H.; Duan, X.; Imai, T.; Murata, S.; Himeno, K. Genetic immunization based on the ubiquitin-fusion degradation pathway against *Trypanosoma cruzi*. *Biochem. Biophys. Res. Commun.* **2010**, *392*, 277–282. [CrossRef]
162. Limon-Flores, A.; Cervera-Cetina, R.; Tzec-Arjona, J.; Ek-Macias, L.; Sánchez-Burgos, G.; Ramirez-Sierra, M.; Cruz-Chan, J.; VanWynsberghe, N.; Dumonteil, E. Effect of a combination DNA vaccine for the prevention and therapy of *Trypanosoma cruzi* infection in mice: Role of CD4+ and CD8+ T cells. *Vaccine* **2010**, *28*, 7414–7419. [CrossRef] [PubMed]
163. Arce-Fonseca, M.; Ballinas-Verdugo, M.; Zenteno, E.; Suárez-Flores, D.; Carrillo-Sánchez, S.; Alejandre-Aguilar, R.; Rosales-Encina, J.; Reyes, P.; Rodríguez-Morales, O. Specific Humoral and Cellular Immunity Induced by *Trypanosoma cruzi* DNA Immunization in a Canine Model. *Parasites Vectors* **2013**, *44*, 15. [CrossRef] [PubMed]
164. Prochetto, E.; Bontempi, I.; Rodeles, L.; Cabrera, G.; Vicco, M.; Cacik, P.; Pacini, M.; Pérez Gianeselli, M.; Pérez, A.; Marcipar, I. Assessment of a combined treatment with a therapeutic vaccine and benznidazole for the *Trypanosoma cruzi* chronic infection. *Acta Trop.* **2022**, *229*, 106334. [CrossRef]
165. Barr, S.; Gossett, K.; Klei, T. Clinical, Clinicopathologic, and Parasitologic Observations of Trypanosomiasis in Dogs Infected with North American *Trypanosoma cruzi* Isolates. *Am. J. Vet. Res.* **1991**, *52*, 954. [CrossRef]
166. Dumonteil, E.; Escobedo-Ortegon, J.; Reyes-Rodriguez, N.; Arjona-Torres, A.; Ramirez-Sierra, M.J. Immunotherapy of *Trypanosoma cruzi* Infection with DNA Vaccines in Mice. *Infect. Immun.* **2004**, *72*, 46–53. [CrossRef]
167. Zapata-Estrella, H.; Hummel-Newell, C.; Sanchez-Burgos, G.; Escobedo-Ortegon, J.; Ramirez-Sierra, M.; Arjona-Torres, A.; Dumonteil, E. Control of *Trypanosoma cruzi* infection and changes in T-cell populations induced by a therapeutic DNA vaccine in mice. *Immunol. Lett.* **2006**, *103*, 186–191. [CrossRef]
168. Duthie, M.; Kahn, M.; Zakayan, A.; White, M.; Kahn, S. Parasite-Induced Chronic Inflammation Is Not Exacerbated by Immunotherapy before or during *Trypanosoma cruzi* Infection. *Clin. Vaccine Immunol.* **2007**, *14*, 1005–1012. [CrossRef]
169. Durães-Oliveira, J.; Palma-Marques, J.; Moreno, C.; Rodrigues, A.; Monteiro, M.; Alexandre-Pires, G.; da Fonseca, I.P.; Santos-Gomes, G. Chagas Disease: A Silent Threat for Dogs and Humans. *Int. J. Mol. Sci.* **2024**, *25*, 3840. [CrossRef]
170. Basombrio, M.; Segura, M.; Mora, M.; Gomez, L. Field Trial of Vaccination Against American Trypanosomiasis (Chagas' Disease) in Dogs. *Am. J. Trop. Med. Hyg.* **1993**, *49*, 143–151. [CrossRef]
171. Quijano-Hernández, I.; Castro-Barcena, A.; Vázquez-Chagoyán, J.; Bolio-González, M.; Ortega-López, J.; Dumonteil, E. Preventive and therapeutic DNA vaccination partially protect dogs against an infectious challenge with *Trypanosoma cruzi*. *Vaccine* **2013**, *31*, 2246–2252. [CrossRef]
172. Basso, B.; Marini, V.; Gauna, D.; Frias, M. Vaccination of Dogs with *Trypanosoma rangeli* Induces Antibodies Against *Trypanosoma cruzi* in a Rural Area of Córdoba, Argentina. *Mem. Inst. Oswaldo Cruz* **2016**, *111*, 271–274. [CrossRef] [PubMed]
173. Floridia-Yapur, N.; Rumi, M.; Ragone, P.; Lauthier, J.; Tomasini, N.; D'amato, A.; Diosque, P.; Cimino, R.; Marco, J.; Barroso, P.; et al. The TcTASV Proteins Are Novel Promising Antigens to Detect Active *Trypanosoma cruzi* Infection in Dogs. *Parasitology* **2016**, *143*, 1382–1389. [CrossRef] [PubMed]
174. Arce-Fonseca, M.; Carbajal-Hernández, A.; Lozano-Camacho, M.; Carrillo-Sánchez, S.D.C.; Roldán, F.; Aranda-Fraustro, A.; Rosales-Encina, J.; Rodríguez-Morales, O. DNA Vaccine Treatment in Dogs Experimentally Infected with *Trypanosoma cruzi*. *J. Immunol. Res.* **2020**, *2020*, 9794575. [CrossRef] [PubMed]
175. Fiatsonu, E.; Deka, A.; Mbah, M. Assessing the Impact of a Potential Canine Vaccine for the Control of Chagas Disease: A Mathematical Modeling Study. *Front. Trop. Dis.* **2024**, *5*, 1421669. [CrossRef]
176. Khan, A.A.; Taylor, M.C.; Fortes Francisco, A.; Jayawardhana, S.; Atherton, R.L.; Olmo, F.; Lewis, M.D.; Kelly, J.M. Animal models for exploring Chagas disease pathogenesis and supporting drug discovery. *Clin. Microbiol. Rev.* **2024**, *37*, e00155-23. [CrossRef]
177. Montenegro, V.; Jiménez, M.; Dias, J.P.; Zeledón, R. Chagas Disease in Dogs from Endemic Areas of Costa Rica. *Mem. Inst. Oswaldo Cruz* **2002**, *97*, 491–494. [CrossRef]
178. Tenney, T.; Curtis-Robles, R.; Snowden, K.; Hamer, S. Shelter Dogs as Sentinels for *Trypanosoma cruzi* Transmission across Texas. *Emerg. Infect. Dis.* **2014**, *20*, 1323–1326. [CrossRef]
179. Travi, B. Considering Dogs as Complementary Targets of Chagas Disease Control. *Vector-Borne Zoonotic Dis.* **2019**, *19*, 90–94. [CrossRef]
180. Rodríguez-Morales, O.; Roldán, F.; Vargas-Barrón, J.; Parra-Benítez, E.; Medina-García, M.; Vergara-Bello, E.; Arce-Fonseca, M. Echocardiographic Findings in Canine Model of Chagas Disease Immunized with DNA *Trypanosoma cruzi* Genes. *Animals* **2020**, *10*, 648. [CrossRef]
181. Rodríguez-Morales, O.; Pérez-Leyva, M.; Ballinas-Verdugo, M.; Carrillo-Sánchez, S.; Rosales-Encina, J.; Alejandre-Aguilar, R.; Reyes, P.; Arce-Fonseca, M. Plasmid DNA immunization with *Trypanosoma cruzi* genes induces cardiac and clinical protection against Chagas disease in the canine model. *Vet. Res.* **2012**, *43*, 79. [CrossRef]
182. Jones, K.; Versteeg, L.; Damania, A.; Keegan, B.; Kendricks, A.; Pollet, J.; Cruz-Chan, J.V.; Gusovsky, F.; Hotez, P.J.; Bottazzi, M.E. Vaccine-linked chemotherapy improves benznidazole efficacy for acute Chagas disease. *Infect. Immun.* **2018**, *86*, e00876-17. [CrossRef] [PubMed]

183. Nguyen, D.M.; Poveda, C.; Pollet, J.; Gusovsky, F.; Bottazzi, M.E.; Hotez, P.J.; Jones, K.M. The impact of vaccine-linked chemotherapy on liver health in a mouse model of chronic Trypanosoma cruzi infection. *PLoS Neglected Trop. Dis.* **2023**, *17*, e0011519. [CrossRef] [PubMed]
184. Cruz-Chan, J.V.; Villanueva-Lizama, L.E.; Versteeg, L.; Damania, A.; Villar, M.J.; González-López, C.; Keegan, B.; Pollet, J.; Gusovsky, F.; Hotez, P.J.; et al. Vaccine-linked chemotherapy induces IL-17 production and reduces cardiac pathology during acute *Trypanosoma cruzi* infection. *Sci. Rep.* **2021**, *11*, 3222. [CrossRef]
185. Goicochea-Del Rosal, G.; Nogueda-Torres, B.; Villagrán, M.E.; de Diego-Cabrera, J.A.; Montañez-Valdez, O.D.; Martínez-Ibarra, J.A. Chagas Disease: Importance of Rats as Reservoir Hosts of *Trypanosoma cruzi* (Chagas, 1909) in Western Mexico. *J. Infect. Public Health* **2018**, *11*, 230–233. [CrossRef]
186. Organización Panamericana de la Salud. *La Enfermedad de Chagas en las Américas: Una Revisión de la Situación Actual de la Salud Pública y Su Visión para el Futuro*; Organización Panamericana de la Salud: Washington, DC, USA, 2018. Available online: https://www.paho.org/es/temas/enfermedad-chagas (accessed on 21 October 2024).
187. López-Cancino, S.A.; Tun-Ku, E.; De la Cruz-Felix, H.K.; Ibarra-Cerdeña, C.N.; Izeta-Alberdi, A.; Pech-May, A.; Mazariegos-Hidalgo, C.J.; Valdez-Tah, A.; Ramsey, J.M. Ecología del Paisaje de *Trypanosoma cruzi* en el Sur de la Península de Yucatán. *Acta Trop.* **2015**, *151*, 58–72. [CrossRef]
188. Hernández-Cortazar, I.; Guardia, K.; Torres-Castro, M.; Acosta-Viana, K.; Guzmán-Marín, E.; Chan-Pérez, J.; Ortega-Pacheco, A.; Rodríguez-Vivas, R.; Medina-Pinto, R.; Jiménez-Coello, M. Frecuencia de infección por Trypanosoma cruzi en roedores sinantrópicos y salvajes capturados en una comunidad rural del sureste de México. *Vet. Med. Int.* **2018**. Available online: https://www.researchgate.net/publication/333077158 (accessed on 21 September 2024).
189. Santos, W.; Gurgel-Gonçalves, R.; Garcez, L.; Abad-Franch, F. Deforestation effects on *Attalea* palms and their resident *Rhodnius*, vectors of Chagas disease, in eastern Amazonia. *PLoS ONE* **2021**, *16*, e0252071. [CrossRef]
190. Ramsey, J.; Schofield, C. Control of Chagas Disease Vectors. *Salud Pública Mex.* **2012**, *45*, S123–S128. [CrossRef]
191. Gottdenker, N.L.; Calzada, J.E.; Saldaña, A.; Carroll, C.R. Association of anthropogenic land use change and increased abundance of the Chagas disease vector *Rhodnius pallescens* in a rural landscape of Panama. *Am. J. Trop. Med. Hyg.* **2011**, *84*, 70. [CrossRef]
192. Grijalva, M.; Terán, D.; Dangles, O. Dynamics of sylvatic Chagas disease vectors in coastal Ecuador is driven by changes in land cover. *PLOS Neglected Trop. Dis.* **2014**, *8*, e2960. [CrossRef]
193. González-Salazar, C.; Meneses-Mosquera, A.; Aguirre-Peña, A.; Fernández-Castel, K.; Stephens, C.; Mendoza-Ponce, A.; Velasco, J.; Calderón-Bustamante, O.; Estrada, F. Toward new epidemiological landscapes of *Trypanosoma cruzi* (Kinetoplastida, Trypanosomatidae) transmission under future human-modified land cover and climatic change in Mexico. *Trop. Med. Infect. Dis.* **2022**, *7*, 221. [CrossRef]
194. Guhl, F.; Ramírez, J.D. Poverty, migration, and Chagas disease. *Curr. Trop. Med. Rep.* **2021**, *8*, 52–58. [CrossRef]
195. Gómez-Bravo, A.; Cirignoli, S.; Wehrendt, D.; Schijman, A.; León, C.M.; Flores-Chaves, M.; Nieto, J.; Kieran, T.J.; Abril, M.; Guhl, F. Zoonotic cycle of American Trypanosomiasis in an endemic region of the Argentine Chaco, factors that influenced a paradigm shift. *Insects* **2024**, *15*, 471. [CrossRef] [PubMed]
196. Lafferty, K.D. The ecology of climate change and infectious diseases. *Ecology* **2008**, *90*, 888–900. [CrossRef] [PubMed]
197. Short, E.; Caminade, C.; Thomas, B. Climate Change Contribution to the Emergence or Re-Emergence of Parasitic Diseases. *Infect. Dis.* **2017**, *10*, 1178633617732296. [CrossRef]
198. Gürtler, R. How is global change affecting Chagas disease landscapes? *Mem. Inst. Oswaldo Cruz* **2022**, *117*, e210479chgsa. [CrossRef]
199. Catalá, S.S.; Noireau, F.; Dujardin, J.P. Biology of triatominae. In *American Trypanosomiasis Chagas Disease*; Elsevier: Amsterdam, The Netherlands, 2017; pp. 145–167.
200. Barreto, M.L.; Teixeira, M.G.; Bastos, F.I.; Ximenes, R.A.; Barata, R.B.; Rodrigues, L.C. Successes and failures in the control of infectious diseases in Brazil: Social and environmental context, policies, interventions, and research needs. *Lancet* **2011**, *377*, 1877–1889. [CrossRef]
201. Wilson, A.L.; Courtenay, O.; Kelly-Hope, L.A.; Scott, T.W.; Takken, W.; Torr, S.J.; Lindsay, S.W. The importance of vector control for the control and elimination of vector-borne diseases. *PLoS Neglected Trop. Dis.* **2020**, *14*, e0007831. [CrossRef]
202. Cecere, M.C.; Rodríguez-Planes, L.I.; Vazquez-Prokopec, G.M.; Kitron, U.; Gürtler, R.E. Community-based surveillance and control of chagas disease vectors in remote rural areas of the Argentine Chaco: A five-year follow-up. *Acta Trop.* **2019**, *191*, 108–115. [CrossRef]

pathogens

Perspective

Epidemiology of Chagas Disease in the United States of America: A Short Review and Some Comments

Stephen A. Klotz

Division of Infectious Diseases, University of Arizona, 5701 North Campbell Ave., Tucson, AZ 85719,USA; sklotz@u.arizona.edu

Abstract: The epidemiology of Chagas disease in humans has markedly changed within the past several decades in the United States of America. This report discusses autochthonous cases of Chagas disease as well as disease in immigrants from Latin American countries. Suggestions for epidemiology research and medical care are discussed given the evolving epidemiology of the disease in the United States of America.

Keywords: Chagas disease; *Trypanosoma cruzi*; kissing bugs; epidemiology

The epidemiology of Chagas Disease (CD) in the United States of America (USA) has never been more important than at present. Accurate data is needed to carefully plan for and obtain the medical resources required to identify and treat people with CD in the USA. The influx of millions of immigrants within the past several decades, some of whom are infected with *Trypanosoma cruzi* (Chagas, 1909), has changed the nature of CD in this country. In addition, epidemiologic studies will also provide a direction for future research of CD in the USA. In this report, the discussion will be confined to the epidemiology of CD in humans, only referring to the enzootic relationships among mammals and kissing bugs as needed to knit together the complex life cycle of the pathogen.

Trypanosoma cruzi is found within three major reservoirs (hosts) in the USA. The first reservoir is the many species of small-to-medium-sized burrowing and nesting wild mammals. These animals, in turn, are fed upon by 11 species of hematophagous insects known as "kissing bugs" in the USA (family Reduviidae, subfamily Triatominae). Kissing bugs are scattered over the lower 2/3 tier of states and constitute the second reservoir. The third reservoir of *T. cruzi* is humans. Some hosts among mammals and humans are susceptible to disease once infected. At present, the human reservoir in the USA is preponderantly composed of infected immigrants from Latin American countries endemic for CD (these individuals may number in the thousands, perhaps hundreds of thousands). In addition, there is a smaller number of autochthonous cases of CD, numbering from fifty to a hundred [1]. Kissing bugs are the link or vector between the mammals and humans, with infection occurring via fecal contamination either orally or percutaneously. In addition, humans, through blood transfusions, transplants, and childbirth, can be a source of infection for other humans.

1. Chagas Disease in the USA

The first description of *T. cruzi* in the USA was provided by Kofoid and McCulloch, who reported the pathogen in the blood of wood rats (*Neotoma* spp.) in California in 1916 [2]. Unlike Chagas' discovery in Brazil a decade earlier, where he described the pathogen, insect vector, and human disease, all within several years [3–5], the first autochthonous human case of CD in the USA was not reported until 1955 [6]. This lapse of 39 years between the first description of disease in a wood rat and disease in a human in the USA provides an interesting contrast between the discovery of the pathogen in the savannah of Brazil and finding it in the xeric landscape of California. Geography is a contributing factor to differences in the incidence of CD between Latin America and the southern border of the USA. However, the US–Mexico border is a barrier in name only between the two countries of different cultures, different housing stock, different kissing bug species, and differing human genetic pools. The border allows for the almost free movement back and forth of mammals, insects, and humans, as well as their parasites. Although Mexico and the USA share three competent vectors of CD at the border, including *T. protracta* (Uhler), *T. recurva* (Stahl, 1868), and *T. rubida* (Uhler, 1894) [7–9], the autochthonous incidence of human CD is vastly different between the two countries. It has recently been estimated that 1,182,286 cases of CD are present in Mexico (0.96% of the population) [10]. Another species, *T. longipes* (Barber, 1937), was reported by Wood to occur in Arizona [11]; however, this species is considered synonymous to *T. recurva* [12].

The natural history of kissing bugs and their interaction with humans has been studied for these three species over the past 15 years in Arizona. We concluded that because there is little or no domiciliation of the bugs in the USA and, in addition, taking into account the costive defecatory patterns of Arizona bugs, infecting humans with CD in the USA seems unlikely [13].

2. Latin American Immigrants to the USA and Chagas Disease

The arrival of Chagas disease into the USA through infected Central American immigrants was reported in 1987 [14]. The authors studied 205 Salvadorian and Nicaraguan immigrants living in the Washington DC area and found that 10 (4.9%) had positive serologies for *T. cruzi*. Three individuals of six were also positive by xenodiagnoses. In concluding the paper, the authors stated "The general quality of housing and the absence of reduviid species that readily become domiciliary are the most likely explanations for the rarity of *T. cruzi* transmission to humans in the United States. The presence of additional numbers of *T. cruzi*-infected immigrants will not change this situation [15].". The first sentence explains a great deal about the rarity of autochthonous human cases in the USA, viz. that housing stock amongst the poor in Latin American countries allowed reduviids to domiciliate, thus giving them unrestricted access to human residents. This does not occur in the USA [13]. However, the concluding sentence (quoted above) is arguable, given the magnitude of immigration into the USA. Immigrants have brought along many new isolates ("near clades") [14] of *T. cruzi* that have the potential to change the behavior of the kissing bugs and parasites in the USA when acquired by indigenous kissing bugs [16–19].

Schmunis in 2007 attempted to quantitate the numbers of infected immigrants, estimating there were 56,028 to 357,205 infected individuals among 7.2 million legal immigrants to the USA from 1981 to 2005 [20]. Shortly thereafter, Bern and Montgomery published a figure of 300,167 cases of CD among immigrants to the USA [21]. This is the most cited report regarding infected immigrants. Lacking real numbers concerning cases of CD, these reports and others use aggregate prevalence data for each Latin American country from which the immigrants originated to determine (estimate) the number of cases of CD. In addition, the number of infections that are projected to progress to either cardiovascular or gastrointestinal syndromes are an estimate as well. These estimates usually range from 20 to 40% of those infected. The difficulties (inaccuracies) involving the use of estimates were noted and appreciated in the original 2009 publication. Unfortunately, in the absence of actual evidence (i.e., proven cases), not the fault of the authors, these estimates are often treated as fact by readers.

A problem plaguing all the epidemiological studies of CD is not only the use of estimations but that the data are derived mainly from asymptomatic people who have indeterminate or chronic infection, and there is no possibility of accurately determining when or where infection occurred. This fact makes determining whether positive serologies reported in the USA represent autochthonous infections or imported infections very difficult to determine. For example, the original case descriptions in the USA, without exception, described individuals with an acute illness accompanied by a peripheral blood smear demonstrating *T. cruzi* in the blood. Interestingly, of the first seven documented CD infections in the USA, four were acute CD in children or infants [22]. An example of the difficulty is a recent paper reporting 18 autochthonous cases of CD in Texas [23]. However, the border area between Mexico and Texas was intensively studied in the past and no cases of CD were described from 1909 participants living in "appalling" conditions in 1949 [24]. The possibility exists that these "new cases" may reflect the changing ethnic composition of the state rather than increased numbers of autochthonous cases of CD. However, as stated before, there is no way to accurately determine the date and location of the acquisition of CD by the use of serology unless other data are provided.

3. Given the Present Status of CD in the USA and What We Know Now, What Should Be Done?

Screening immigrants is a priority, but in addition, screening citizens identified as "at risk" by local authorities will be important as well. Hence, it is urgent that we identify one test or a battery of tests to serve as the "Gold Standard" for the serological diagnosis of CD. Choosing a gold standard that would be adopted countrywide would facilitate the identification of infected individuals to begin the process of screening [25]. Since testing at the Centers for Disease Control (CDC) is currently regarded as the "Gold Standard" in the USA, perhaps scaling up testing at the CDC would be the most efficacious approach. (Their requirements for a positive serology require that at least two of three different tests be positive.) It would be prudent if female immigrants from CD-endemic countries under 40 years of age and immigrant males under 20 years of age undergo serological testing for CD. The risk for delivering an infected newborn would be identified and treatment offered when needed. It is possible that other measures may be required in the future as the Chagas epidemic unfolds in the USA among permissive kissing bug species and wild mammals.

Since *T. cruzi* is mostly a clonal pathogen [26], it is important to determine what effect the introduction of many new clones of *T. cruzi* into the USA will have upon native kissing bugs as well as *T. cruzi* itself. Furthermore, what effect, if any, new clones of *T. cruzi* will have upon the clinical presentation of CD if acquired autochthonously should be closely watched.

A list of autochthonous USA cases should be compiled with pertinent clinical and historical findings and patients revisited over time to determine the outcome of the infection. For example, do the percentages of cardiac and gastrointestinal complications known to occur in Latin America predict what will occur in autochthonous cases in the USA? Autochthonous cases (treated and untreated) should be revisited regularly to determine how often serologies turn negative. If and when serologies turn negative, does this mean it is cured [27]?

Investigators should adopt stricter criteria for autochthonous cases in the USA. Individuals should not be counted as autochthonous if they spent two weeks or less in an endemic country, since an individual could, and undoubtedly has, visited an endemic area in Latin America and been infected on day one of their visit. Relying on some arbitrary timeline is tenuous at best and individuals with a travel history to endemic countries should be placed in a separate category. This also applies to identifying cases as autochthonous based upon the possession of European surnames, as was done in one report [28]. Problems of categorization will continue to arise because Chagas' infection when acquired is overwhelmingly asymptomatic. Unfortunately, a positive serology cannot tell the investigator when infection occurred, and this is a critical piece of information when deciding whether an infection is autochthonous or not.

Although eradicating kissing bugs in Latin America has been effective in reducing CD, there is probably little to be gained by trying to interrupt vector-borne disease in the USA, as domiciliation by the bugs is uncommon and home entry sporadic [29]. However, educating the public about the risk of transmission of *T. cruzi* to humans by kissing bugs as well as by mammals infected with *T. cruzi* is important. Home hygiene is a particularly important feature in southwestern USA, where home intrusions by kissing bugs occur yearly in great numbers.

Funding: This research received no external funding.

Institutional Review Board Statement: Not applicable.

Informed Consent Statement: Not applicable.

Data Availability Statement: Not applicable.

Conflicts of Interest: The author declares no conflict of interest.

References

1. Irish, A.; Whitman, J.D.; Clark, E.H.; Marcus, R.; Bern, C. Updated Estimates and Mapping for Prevalence of Chagas Disease among Adults, United States. *Emerg. Infect. Dis.* **2022**, *28*, 1313–1320. [CrossRef] [PubMed] [PubMed Central]
2. Kofoid, C.A.; Mcculloch, I. On Trypanosoma triatomae, a new flagellate from a hemipteran bug from the nests of the wood rat *Neotoma fuscipes*. *Univ. Calif. Pub. Zool.* **1916**, *16*, 113–126.
3. Chagas, C. Neue Trypanosomen: Vorläufige mitteilung. *Arch. Schiffs Trop.-Hyg.* **1909**, *13*, 120–122.
4. Chagas, C. Nova especie morbida do homem, produzida por um Trypanozoma (*Trypanozoma cruzi*): Nota previa. *Braz.-Med.* **1909**, *23*, 161.
5. Chagas, C. Ueber Eine Neue Trypanosomiasis des Menschen. *Mem. Inst. Oswaldo Cruz* **1909**, *1*, 159. [CrossRef]
6. Woody, N.C.; Woody, H.B. American trypanosomiasis (Chagas' disease); first indigenous case in the United States. *J. Am. Med. Assoc.* **1955**, *159*, 676–677. [CrossRef] [PubMed]
7. Forsyth, C.; Agudelo Higuita, N.I.; Hamer, S.A.; Ibarra-Cerdeña, C.N.; Valdez-Tah, A.; Stigler Granados, P.; Hamer, G.L.; Vingiello, M.; Beatty, N.L. Climate change and *Trypanosoma cruzi* transmission in North and central America. *Lancet Microbe* **2024**, *5*, 100946. [CrossRef] [PubMed]
8. Martínez-Ibarra, J.A.; Paredes-González, E.; Licón-Trillo, Á.; Montañez-Valdez, O.D.; Rocha-Chávez, G.; Nogueda-Torres, B. The biology of three Mexican-American species of Triatominae (*Hemiptera: Reduviidae*): *Triatoma recurva*, *Triatoma protracta* and *Triatoma rubida*. *Mem. Inst. Oswaldo Cruz.* **2012**, *107*, 659–663. [CrossRef] [PubMed]
9. Hill, J.; Teal, E.; Cross, C.L.; Sanchez, Z.; Webber, M.M.; Oxborough, R.M.; Messenger, L.A. Using iNaturalist presence data to produce suitability maps for *Triatoma protracta*, *T. rubida* and *T. recurva* in the American Southwest, Texas and northern Mexico, to identify potential transmission zones of Chagas disease. *Sci. Rep.* **2024**, *14*, 26879. [CrossRef] [PubMed]
10. Cucunubá, Z.M.; Gutiérrez-Romero, S.A.; Ramírez, J.D.; Velásquez-Ortiz, N.; Ceccarelli, S.; Parra-Henao, G.; Henao-Martínez, A.F.; Rabinovich, J.; Basáñez, M.G.; Nouvellet, P.; et al. The epidemiology of Chagas disease in the Americas. *Lancet Reg. Health Am.* **2024**, *37*, 100881. [CrossRef] [PubMed] [PubMed Central]
11. Wood, F.D.; Wood, S.F. Present knowledge of the distribution of *Trypanosoma cruzi* in reservoir animals and vectors. *Am. J. Trop. Med.* **1941**, *21*, 335. [CrossRef]
12. de Paiva, V.F.; Belintani, T.; de Oliveira, J.; Galvão, C.; da Rosa, J.A. A review of the taxonomy and biology of Triatominae subspecies (*Hemiptera: Reduviidae*). *Parasitol. Res.* **2022**, *121*, 499–512. [CrossRef] [PubMed]
13. Behrens-Bradley, N.; Smith, S.; Beatty, N.L.; Love, M.; Ahmad, N.; Dorn, P.L.; Schmidt, J.O.; Klotz, S.A. Kissing Bugs Harboring *Trypanosoma cruzi*, Frequently Bite Residents of the US Southwest But Do Not Cause Chagas Disease. *Am. J. Med.* **2020**, *133*, 108–114.e13. [CrossRef] [PubMed]
14. Kirchhoff, L.V.; Gam, A.A.; Gilliam, F.C. American trypanosomiasis (Chagas' disease) in Central American immigrants. *Am. J. Med.* **1987**, *82*, 915–920. [CrossRef] [PubMed]

15. Tibayrenc, M.; Ayala, F.J. Reproductive clonality of pathogens: A perspective on pathogenic viruses, bacteria, fungi, and parasitic protozoa. *Proc. Natl. Acad. Sci. USA* **2012**, *109*, E3305–E3313. [CrossRef] [PubMed] [PubMed Central]
16. Dorn, P.; Buekens, P.; Hanford, E. Whac-a-mole: Future trends in Chagas transmission and the importance of a global perspective on disease control. *Future Microbiol.* **2007**, *2*, 365–367. [CrossRef]
17. Guarneri, A.A.; Lorenzo, M.G. Triatomine physiology in the context of trypanosome infection. *J. Insect Physiol.* **2017**, *97*, 66–76. [CrossRef] [PubMed]
18. De Fuentes-Vicente, J.A.; Cabrera-Bravo, M.; Enríquez-Vara, J.N.; Bucio-Torres, M.I.; Gutiérrez-Cabrera, A.E.; Vidal-López, D.G.; Martínez-Ibarra, J.A.; Salazar-Schettino, P.M.; Córdoba-Aguilar, A. Relationships between altitude, triatomine (*Triatoma dimidiata*) immune response and virulence of *Trypanosoma cruzi*, the causal agent of Chagas' disease. *Med. Vet. Entomol.* **2017**, *31*, 63–71. [CrossRef] [PubMed]
19. Guzman-Marin, E.; Jimenez-Coello, M.; Puerto-Solis, M.; Ortega-Pacheco, A.; Acosta-Viana, K.Y. Influence of *Triatoma dimidiata* in Modulating the Virulence of *Trypanosoma cruzi* Mexican Strains. *Interdiscip. Perspect. Infect. Dis.* **2012**, *2012*, 328091. [CrossRef] [PubMed] [PubMed Central]
20. Schmunis, G.A. Epidemiology of Chagas disease in non-endemic countries: The role of international migration. *Mem. Inst. Oswaldo Cruz* **2007**, *102* (Suppl. S1), 75–85, Erratum in *Mem. Inst. Oswaldo Cruz* **2007**, *102*, 1009. [CrossRef]
21. Bern, C.; Montgomery, S.P. An estimate of the burden of Chagas disease in the United States. *Clin. Infect. Dis.* **2009**, *49*, e52–e54. [CrossRef] [PubMed]
22. Klotz, S.A.; Dorn, P.L.; Mosbacher, M.; Schmidt, J.O. Kissing bugs in the United States: Risk for vector-borne disease in humans. *Environ. Health Insights* **2014**, *8* (Suppl. S2), 49–59. [CrossRef] [PubMed] [PubMed Central]
23. Lynn, M.K.; Dye-Braumuller, K.C.; Beatty, N.L.; Dorn, P.L.; Klotz, S.A.; Stramer, S.L.; Townsend, R.L.; Kamel, H.; Vannoy, J.M.; Sadler, P.; et al. Evidence of likely autochthonous Chagas disease in the southwestern United States: A case series of *Trypanosoma cruzi* seropositive blood donors. *Transfusion* **2022**, *62*, 1808–1817. [CrossRef] [PubMed] [PubMed Central]
24. Faust, E.C. The etiologic agent of Chagas' disease in the United States. *Bull. Pan Am. Sanit. Bureau* **1949**, *28*, 455–461.
25. Agudelo Higuita, N.I.; Beatty, N.L.; Forsyth, C.; Henao-Martínez, A.F.; Manne-Goehler, J.; US Chagas Research Consortium. Chagas disease in the United States: A call for increased investment and collaborative research. *Lancet Reg. Health Am.* **2024**, *34*, 100768. [CrossRef] [PubMed] [PubMed Central]
26. Tibayrenc, M.; Ward, P.; Moya, A.; Ayala, F.J. Natural populations of *Trypanosoma cruzi*, the agent of Chagas' disease, have a complex multiclonal structure. *Proc. Natl. Acad. Sci. USA* **1986**, *83*, 115–119. [CrossRef]
27. Zeledon, R.; Dias, J.C.P.; Brilla-Salazar, A.; Marcondes de Rezende, J.; Vargas, L.G.; Urbina, A. Does a spontaneous cure for Chagas" exist? *Rev. Soc. Bras. Med. Trop.* **1988**, *21*, 15–20. [CrossRef]
28. Cantey, P.T.; Stramer, S.L.; Townsend, R.L.; Kamel, H.; Ofafa, K.; Todd, C.W.; Currier, M.; Hand, S.; Varnado, W.; Dotson, E.; et al. The United States *Trypanosoma cruzi* Infection Study: Evidence for vector-borne transmission of the parasite that causes Chagas disease among United States blood donors. *Transfusion* **2012**, *52*, 1922–1930. [CrossRef] [PubMed]
29. de Arias, A.R.; Monroy, C.; Guhl, F.; Sosa-Estani, S.; Santos, W.S.; Abad-Franch, F. Chagas disease control-surveillance in the Americas: The multinational initiatives and the practical impossibility of interrupting vector-borne *Trypanosoma cruzi* transmission. *Mem. Inst. Oswaldo Cruz* **2022**, *117*, e210130. [CrossRef] [PubMed] [PubMed Central]

MDPI AG
Grosspeteranlage 5
4052 Basel
Switzerland
Tel.: +41 61 683 77 34

Pathogens Editorial Office
E-mail: pathogens@mdpi.com
www.mdpi.com/journal/pathogens

www.ingramcontent.com/pod-product-compliance
Lightning Source LLC
Chambersburg PA
CBHW061611220926
51850CB00005B/481
* 9 7 8 3 7 2 5 8 5 8 8 3 5 *